Molecular Biology of the Gene

3RD EDITION

Molecular Biology of the Gene

3RD EDITION

James D. Watson

HARVARD UNIVERSITY AND
COLD SPRING HARBOR LABORATORY

With illustrations by
Keith Roberts

W. A. BENJAMIN, INC.
Menlo Park, California
Reading, Massachusetts
London · Amsterdam
Don Mills, Ontario · Sydney

Various editions of this work appear in the following translations:

French
Inter European Editions
De Lairessestraat 90
Amsterdam 1007, The Netherlands

Hungarian
Medicina Publishers
c/o Artisjus
H-1364 Post Box 67
Budapest V. Hungary

Italian
Nicola Zanichelli
via Irnerio 34
Bologna, Italy

Japanese
Kagaku Dojin
Yanagino-Banba
Oike-Sagarie
Nakakyo-ku, Kyoto-shi
Japan

German
Inter European Editions
De Lairessestraat 90
Amsterdam 1007, The Netherlands

Polish
Panstwowe Wydawnictwo Naukowe
Miodowa 10
Warsaw 5, Poland

Romanian
Editura Stiintifica
Sectorul 4, Bd. Republicii 17
Bucharest, Romania

Serbo Croatian
Naucna Knjiga
c/o Jugoslovenska Autorska Agencija
Majke Jevrosime 38
Beograd, Yugoslavia

Spanish
Fondo Educativo Interamericano,
FEI, C.A.
Apartado del Este 62361
Caracas, Venezuela

Turkish
Hacettepe University
Hacettepe, Ankara, Turkey

Second printing, March 1977

Cover photograph: Electron micrograph of SV40 specific chromatin circles isolated from monkey cells infected with SV40. [Kindly supplied by O. Croissant, C. Cremisi, P. Pignatti, and M. Yaniv, Institut Pasteur, Paris.]

ISBN 0-8053-9609-8
BCDEFGHIJ-DO-798

To Liz

Preface

Perhaps the most striking aspect of molecular biology today is that it is not slowing down. Though all of us may occasionally wonder whether we are capable of new ways of thinking, we still feverishly rush to the latest issues of the major journals. And, more often than not, we come across new facts that we must digest if we are to remain practicing biologists.

As a result, we remain far from the day when the teaching of introductory biology will be a semistatic subject. Its year-to-year gyrations do not merely reflect the need for new formats to overcome the boredom that necessarily accompanies the use of last year's lecture notes. I thus suspected, almost from the moment the Second Edition came out, that a third edition of this book might soon be necessary. And I feared it might have to grow much in length as the past distinctions between molecular biology and cell biology were fading fast. I would necessarily have to broaden its scope in the direction of what in the past might be called the province of the pure biologist. I thus wondered whether I should condense, if not eliminate, much of the material of the earlier introductory chapters to keep the book to a comfortably readable length. But in the end I concluded that it still makes sense to offer a text in which the student not previously familiar with the essentials of genetics and biochemistry could start with their basic principles and so not be dependent upon other texts.

Most of what is here presented was given in lecture form to Harvard and Radcliffe undergraduates enrolled in our introductory course in biochemistry and molecular biology. It thus aims to bring together all those key facts which beginning university students should know before they move on to more specialized courses in biology and biochemistry.

Again I have tried to incorporate the most recent of key observations and so I stand the risk of reporting as hard facts observations that may not stand the test of time. I hope, however, that such instances will be rare and that what I have here written will be found to be a balanced account of the biology that now can be analyzed at the molecular level.

During the writing I have shown portions of this book to many friends for their comments and in particular I wish to thank Robert Goldman, Elias Lazarides, Robert Pollack, and Keith Roberts. I am also indebted to David Dressler for invaluable help both with the manuscript and in helping to eradicate misconceptions that persisted into the galleys. Much intelligent assistance in the early drafts was also given by Jan Connery of Radcliffe College. Two Harvard University students, Jay Baer and Robert Schick, played invaluable roles during the correction of the proofs and in the preparation of the glossary and index. As a result I now hope that the most serious errors both of language and fact are gone. But I often opted for my personal uses of style and fact, and I remain responsible for all the errors that may remain.

The reading of my handwriting is not a straightforward task and only with the cheerful and persistent competence of my secretary, Maria Hedges, could this book have emerged. The illustrations again are largely the work of Dr. Keith Roberts, now at the John Innis Institute of Norwich, England. Very few individuals are highly talented in both science and art and I was most fortunate in again obtaining his most unique assistance.

And I must mention the invaluable encouragement provided by my wife and two young sons who on all too many days accepted the fact that I must revise still again another paragraph or drawing.

J. D. Watson

Seven Gates Farm
Martha's Vineyard, Massachusetts
October 1975

Contents

Contents

Contents

6   THE CONCEPT OF TEMPLATE SURFACES   129

7   THE ARRANGEMENT OF GENES
    ON CHROMOSOMES   149

Contents

Contents

Contents

15 THE REPLICATION OF VIRUSES   411

16 THE ESSENCE OF BEING EUCARYOTIC 455

Contents

18   THE CONTROL OF CELL PROLIFERATION   547

Contents

19 THE PROBLEM OF ANTIBODY SYNTHESIS 591

20 THE VIRAL ORIGINS OF CANCER 641

Contents

The Mendelian View of the World

It is easy to consider man unique among living organisms. He alone has developed complicated languages that allow meaningful and complex interplay of ideas and emotions. Great civilizations have developed and changed our world's environment in ways inconceivable for any other form of life. Thus there has always been a tendency to think that something special differentiates man from everything else. This belief has found expression in man's religions, by which he tries to find an origin for his existence and, in so doing, to provide workable rules for conducting his life. It seemed natural to think that, just as every human life begins and ends at a fixed time, man had not always existed but was created at a fixed moment, perhaps the same moment for man and for all other forms of life.

This belief was first seriously questioned just over 100 years ago when Darwin and Wallace proposed their theories of evolution, based upon selection of the most fit. They stated that the various forms of life are not constant, but are continually giving rise to slightly different animals and plants, some of which are adapted to survive and to multiply more effectively. At the time of this theory, they did not know the origin of this continuous variation, but they did correctly realize that these new characteristics must persist in the progeny if such variations were to form the basis of evolution.

1

At first, there was a great deal of furor against Darwin, most of it coming from people who did not like to believe that man and the rather obscene-looking apes could have a common ancestor, even if this ancestor had occurred some 50 to 100 million years in the past. There was also initial opposition from many biologists, who failed to find Darwin's evidence convincing. Among these was the famous Swiss-born naturalist Agassiz, then at Harvard, who spent many years writing against Darwin and Darwin's champion, T. H. Huxley, the most successful of the popularizers of evolution. But by the end of the nineteenth century, the scientific argument was almost complete; both the current geographic distribution of plants and animals and their selective occurrence in the fossil records of the geologic past were explicable only by postulating that continuously evolving groups of organisms had descended from a common ancestor. Today, the theory of evolution is an accepted fact for everyone but a fundamentalist minority, whose objections are based not on reasoning but on doctrinaire adherence to religious principles.

An immediate consequence of the acceptance of Darwinian theory is the realization that life first existed on our Earth some 1 to 2 billion years ago in a simple form, possibly resembling the bacteria—the simplest variety of life now existing. Of course, the very existence of such small bacteria tells us that the essence of the living state is found in very small organisms. Nonetheless, evolutionary theory further affects our thinking by suggesting that the same basic principles of life exist in all living forms.

THE CELL THEORY

The same conclusion was independently reached by the second great principle of nineteenth century biology, the *cell theory*. This theory, first put forward convincingly

Figure 1–1
Electron micrograph of a thin section from a cell of the African violet. The thin primary cellulose cell wall and the nucleus, containing a prominent nucleolus, are clearly visible. The cytoplasmic ground substance is heavily laden with spherical particles, the ribosomes, visible as small black dots. The profiles of a network of hollow membranes, the endoplasmic reticulum, can be seen scattered throughout the cell (courtesy of Drs. K. R. Porter and M. C. Ledbetter, Biological Laboratories, Harvard University). ►

in 1839 by the German microscopists Schleiden and Schwann, proposes that all the plants and animals are constructed from small fundamental units called cells. All cells are surrounded by a membrane, and usually contain an inner body, the nucleus, which is also surrounded by a membrane, called the nuclear membrane (Figures 1–1 and 1–2). Most important, cells arise only from other cells by the process of cell division. Most cells are capable of growing and of splitting roughly equally to give two daughter cells. At the same time, the nucleus divides so that each daughter cell can receive a nucleus.

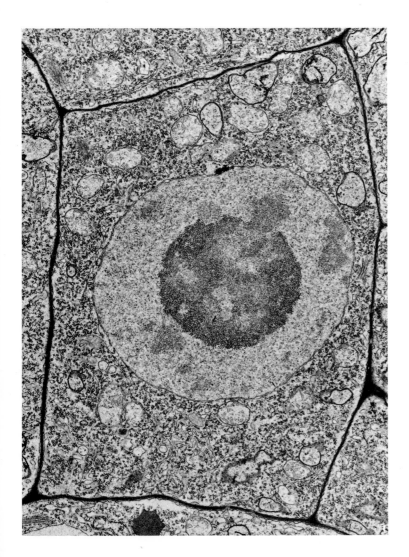

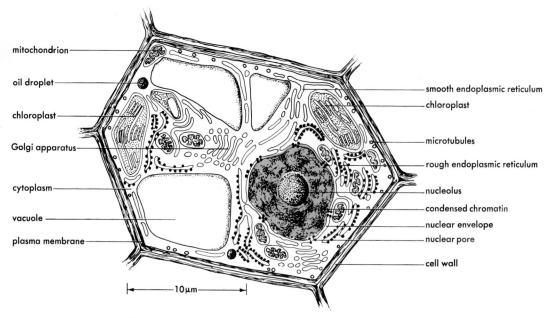

mitochondrion

oil droplet

chloroplast

Golgi apparatus

cytoplasm

vacuole

plasma membrane

smooth endoplasmic reticulum

chloroplast

microtubules

rough endoplasmic reticulum

nucleolus

condensed chromatin

nuclear envelope

nuclear pore

cell wall

⊢———10μm———⊣

Figure 1–2

A schematic view of the plant cell shown in Figure 1–1. The various components are not always drawn to scale.

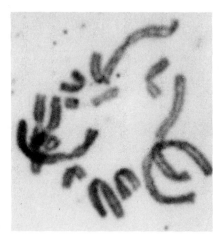

Figure 1–3

The haploid complement of chromosomes from the leopard frog (*Rana pipiens*), magnified 2125 times. This photograph was taken with a light microscope by T. E. Powell, of the Biological Laboratories, Harvard University. It shows the chromosomes when they have duplicated to form two chromatids held together by a single centromere.

MITOSIS MAINTAINS THE PARENTAL CHROMOSOME NUMBER

Each nucleus encloses a fixed number of linear bodies, called chromosomes (Figure 1–3). Before cell division, each chromosome divides to form two chromosomes identical to the parental body. This process, first accurately observed by Flemming in 1879, doubles the number of nuclear chromosomes. During nuclear division, one of each pair of daughter chromosomes moves into each daughter nucleus (Figure 1–4). As a result of these events (now collectively termed *mitosis*), the chromosomal complement of daughter cells is usually identical to that of the parental cells.

During most of a cell's life, its chromosomes exist in a highly extended linear form. Prior to cell division, however, they condense into much more compact bodies. The duplication of chromosomes occurs chiefly when they are in the extended state characteristic of interphase (the various stages of cell division are defined in Figure 1–4). One part of the chromosome, however, always duplicates during the contracted metaphase state; this is the *centromere*, a body that controls the movement of the chromosome during cell divisions. The centromere always has a

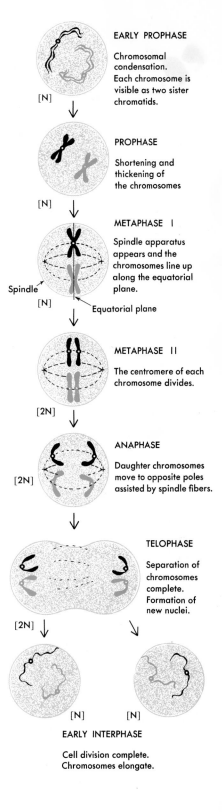

Figure 1–4
Diagram of mitosis in the nucleus of a haploid cell containing two nonhomologous chromosomes.

EARLY PROPHASE

Chromosomal condensation. Each chromosome is visible as two sister chromatids.

PROPHASE

Shortening and thickening of the chromosomes

METAPHASE I

Spindle apparatus appears and the chromosomes line up along the equatorial plane.

Spindle

Equatorial plane

METAPHASE II

The centromere of each chromosome divides.

ANAPHASE

Daughter chromosomes move to opposite poles assisted by spindle fibers.

TELOPHASE

Separation of chromosomes complete. Formation of new nuclei.

EARLY INTERPHASE

Cell division complete. Chromosomes elongate.

fixed location specific to a given chromosome; in some it is near one end, and in others it occupies an intermediate region.

When a chromosome is completely duplicated except for the centromere, it is said to consist of two *chromatids*. A chromatid is transformed into a chromosome as soon as its centromere has divided and is no longer shared with another chromatid. As soon as one centromere becomes two, the two daughter chromosomes begin to move away from each other.

The regular lining up of chromosomes during the metaphase stage is accompanied by the appearance of the *spindle*. This is a cellular region, shaped like a spindle, through which the chromosomes of higher organisms move apart during the anaphase stage. Much of the spindle region is filled with long, thin fibers, called microtubules. These fibers are largely responsible for chromosome movements on the spindle. Microtubules attach to the chromosomes at the centromeres and as the daughter chromosomes move toward the spindle poles during anaphase, the centromeres are in the lead.

Objects called the *nucleoli* are also present in the nucleus of practically every plant and animal cell. There is often one nucleolus per haploid set of chromosomes, and in some cells the nucleolus is connected to a specific chromosome. Until recently, the functional role of the nucleolus was completely obscure, though some biologists originally thought that it might be related to the formation of the spindle. Now, however, there are some strong hints that the nucleolus is involved in the synthesis of ribosomes, small particles within the cell upon which all proteins are synthesized.

MEIOSIS REDUCES THE PARENTAL CHROMOSOME NUMBER

One important exception was found to the mitotic process. After the conclusion of the two cell divisions (*meiosis*) that form the sex cells, the sperm and the egg, the number of chromosomes is reduced to one-half of its previous

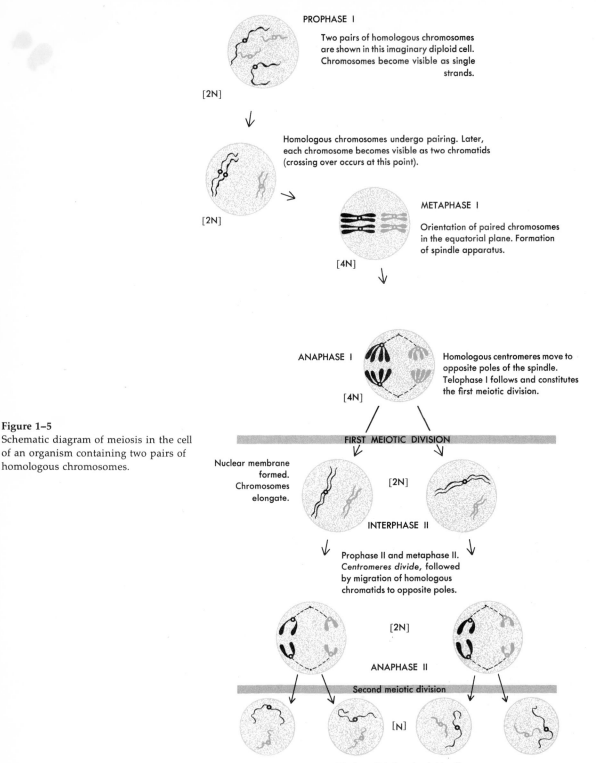

PROPHASE I

Two pairs of homologous chromosomes are shown in this imaginary diploid cell. Chromosomes become visible as single strands.

[2N]

Homologous chromosomes undergo pairing. Later, each chromosome becomes visible as two chromatids (crossing over occurs at this point).

[2N]

METAPHASE I

Orientation of paired chromosomes in the equatorial plane. Formation of spindle apparatus.

[4N]

ANAPHASE I

Homologous centromeres move to opposite poles of the spindle. Telophase I follows and constitutes the first meiotic division.

[4N]

FIRST MEIOTIC DIVISION

Nuclear membrane formed. Chromosomes elongate.

[2N]

INTERPHASE II

Prophase II and metaphase II. *Centromeres divide,* followed by migration of homologous chromatids to opposite poles.

[2N]

ANAPHASE II

Second meiotic division

[N]

Final result is four haploid cells.

Figure 1–5
Schematic diagram of meiosis in the cell of an organism containing two pairs of homologous chromosomes.

number (Figure 1–5). In higher plants and animals each specific type of chromosome is normally present in two copies: the homologous chromosomes (the *diploid* state). In sex-cell formation the resulting sperm and egg each usually encloses only one of each type (the *haploid* state). Union of sperm and egg during fertilization results in a fertilized egg (*zygote*) containing one homologous chromosome from the male parent and another from the female parent. Thus the normal diploid chromosome constitution is restored.

Although most cells are diploid in higher plants and animals, the haploid state is the most frequent condition in lower plants and bacteria, the diploid number existing only briefly following sex-cell fusion. Usually meiosis occurs almost immediately after fertilization to produce haploid cells (Figure 1–6).

The cell theory thus tells us that all cells come from preexisting cells. All the cells in adult plants and animals are derived from the division and growth of a fertilized egg, itself formed by the union of two other cells, the sperm and the egg. All growing cells contain chromosomes, usually two of each type, and here again, new chromosomes always arise through division of previously existing bodies.

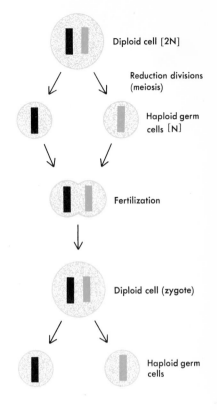

Figure 1–6
Diagram of the alternation of haploid and diploid states, which comprise the sexual cycle. The chromosome set derived from one parent is shown in black, that from the other parent in color.

THE CELL THEORY IS UNIVERSALLY APPLICABLE

Although the cell theory developed from observations about higher organisms, it holds with equal force for the more simple forms of life, such as protozoa and bacteria. Each bacterium or protozoan is a single cell, whose division ordinarily produces a new cell identical to its parent, from which it soon separates. In the higher organisms, on the other hand, the daughter cells not only often remain together, but also frequently differentiate into radically different cell types (such as nerve or muscle cells), while maintaining the chromosome complement of the zygote. Here, new organisms arise from the highly differentiated sperm and egg, whose union initiates a new cycle of division and differentiation.

Thus, although a complicated organism like man contains a large number of cells (up to 5×10^{12}), all these cells arise initially from a single cell. The fertilized egg contains all the information necessary for the growth and development of an adult plant or animal. Again the living state *per se* does not demand the complicated interactions that occur in complex organisms; its essential properties can be found in single growing cells.

MENDELIAN LAWS

The most striking attribute of a living cell is its ability to transmit hereditary properties from one cell generation to another. The existence of heredity must have been noticed by early man as he witnessed the passing of characteristics, like eye or hair color, from parents to their offspring. Its physical basis, however, was not understood until the first years of the twentieth century, when, during a remarkable period of creative activity, the chromosomal theory of heredity was established.

Hereditary transmission through the sperm and egg became known by 1860, and in 1868 Haeckel, noting that sperm consisted largely of nuclear material, postulated that the nucleus was responsible for heredity. Almost 20 years passed before the chromosomes were singled out as the active factors, because the details of mitosis, meiosis, and fertilization had to be worked out first.

When this was accomplished, it could be seen that, unlike other cell constituents, the chromosomes were equally divided between daughter cells. Moreover, the complicated chromosomal changes which reduce the sperm and egg chromosome number to the haploid number during meiosis became understandable as necessary for keeping the chromosome number constant. These facts, however, merely suggested that chromosomes carry heredity.

Proof came at the turn of the century with the discovery of the basic rules of heredity. These rules, named after their original discoverer, Mendel, had in fact been first proposed in 1865, but the climate of scientific opinion had not been ripe for their acceptance. They were completely ignored until 1900, despite some early efforts on Mendel's part to interest the prominent biologists of his time. Then de Vries, Correns, and Tschermak, all working independently, realized the great importance of Mendel's forgotten work. All three were plant breeders, doing experiments related to Mendel's, and each reached similar conclusions before they knew of Mendel's work.

PRINCIPLE OF INDEPENDENT SEGREGATION

Mendel's experiments traced the results of breeding experiments (genetic crosses) between strains of peas differing in well defined characteristics, like seed shape (round or wrinkled), seed color (yellow or green), pod shape (inflated or wrinkled), and stem length (long or short). His concentration on well defined differences was of great importance; many breeders had previously tried to follow the inheritance of more gross qualities, like body weight, and were unable to discover any simple rules about their transmission from parents to offspring. After ascertaining that each type of parental strain bred true (that is, produced progeny with particular qualities identical to those of the parents), Mendel made a number of crosses between parents (P) differing in single characteristics (such as seed shape *or* seed color). All the progeny (F_1 = first filial generation) had the appearance of *one* parent. For example, in a cross between peas having yellow seeds and peas having green seeds, all the progeny had yellow seeds. The trait that appears in the progeny is called *dominant*, whereas that not appearing in F_1 is called *recessive*.

The meaning of these results became clear when Mendel made genetic crosses between F_1 offspring. These crosses gave the most important result that the recessive trait reappeared in approximately 25 percent of the progeny, whereas the dominant trait appeared in 75 percent of them. For each of the seven traits he followed, the ratio in F_2 of dominant to recessive traits was always approximately 3:1. When these experiments were carried to a third (F_3) progeny generation, all the F_2 peas with recessive traits bred true (produced progeny with the recessive traits). Those with dominant traits fell into two groups: one-third bred true (produced only progeny with the dominant trait); the remaining two-thirds again produced mixed progeny in a 3:1 ratio of dominant to recessive.

Mendel correctly interpreted his results as follows (Figure 1-7): The various traits are controlled by pairs of factors (which we now call *genes*), one factor derived from the male parent, the other from the female. For example, pure-breeding strains of round peas contain two genes for roundness (RR), whereas pure-breeding wrinkled strains have two genes for wrinkledness (rr). The round-strain gametes each have one gene for roundness (R); the wrinkled-strain gametes each have one gene for wrinkledness (r). In a cross between RR and rr, fertilization produces an F_1

9

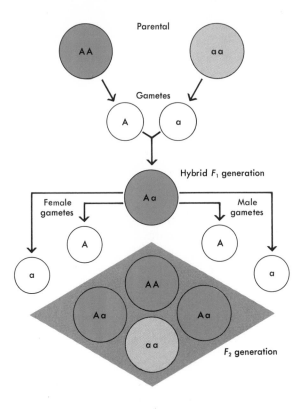

Figure 1–7
Representation of how Mendel's first law (independent segregation) explains the 3:1 ratio of dominant to recessive phenotypes among the F_2 progeny. (A) represents the dominant gene and (a) the recessive gene. The shaded circles represent dominance, the gray circles the recessive phenotype.

plant with both genes (Rr). The seeds look round because R is dominant over r. We refer to the appearance (physical structure) of an individual as its *phenotype*, and to its genetic composition as its *genotype*. Individuals with identical phenotypes may possess different genotypes; thus, to determine the genotype of an organism, it is frequently necessary to perform genetic crosses for several generations. The term *homozygous* refers to a gene pair in which both the maternal and paternal genes are identical (e.g., RR or rr). In contrast, those gene pairs in which paternal and maternal genes are different (e.g., Rr) are called *heterozygous*.

It is important to notice that a given gamete contains only one of the two genes present in the organism it comes from (for example, either the R or the r, but never both) and that the two types of gamete are produced in equal numbers. Thus there is a 50:50 chance that a given gamete from an F_1 pea will contain a particular gene (R or r). This choice is purely random. We do not expect to find *exact* 3:1 ratios when we examine a limited number of F_2 progeny. The ratio will sometimes be slightly higher and other times slightly lower. But as we look at increasingly larger samples, we expect that the ratio of peas with the dominant trait to peas with the recessive trait will approximate the 3:1 ratio more and more closely.

The reappearance of the recessive character in the F_2 generation indicates that recessive genes are neither modified nor lost in the *hybrid* (Rr) generation, but that the dominant and recessive genes are independently transmitted, and so are able to segregate independently during the formation of sex cells. *This principle of independent segregation is frequently referred to as Mendel's first law.*

SOME GENES ARE NEITHER DOMINANT NOR RECESSIVE

In the crosses reported by Mendel, one of each gene pair was clearly dominant, and the other recessive. Such behavior, however, is not universal. Sometimes the heterozygous phenotype is intermediate between the two homozygous phenotypes. For example, the cross between a pure-breeding red snapdragon (*Antirrhinum*) and a pure-breeding white variety gives F_1 progeny of the intermediate pink color. If these F_1 progeny are crossed among themselves, the resulting F_2 progeny contain red, pink, and white flowers in the proportion of 1:2:1 (Figure 1–8). Thus it is possible here to distinguish heterozygotes from homozygotes by their phenotype. We also see that Mendel's laws do not depend for their applicability on whether one *gene* of a gene pair is dominant over the other.

PRINCIPLE OF INDEPENDENT ASSORTMENT

Mendel extended his breeding experiments to peas differing by more than one character. As before, he started with two strains of peas, each of which bred pure when mated with itself. One of the strains had round yellow seeds, the other, wrinkled green seeds. Since round and

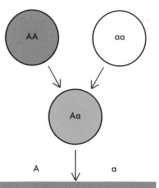

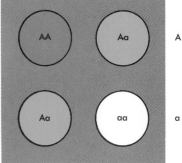

Figure 1–8

The inheritance of flower color in the snapdragon. One parent is homozygous for red flowers (AA) and the other homozygous for white flowers (aa). No dominance is present, and the heterozygous flowers are pink. The 1:2:1 ratio of red:pink:white flowers is shown by appropriate coloring.

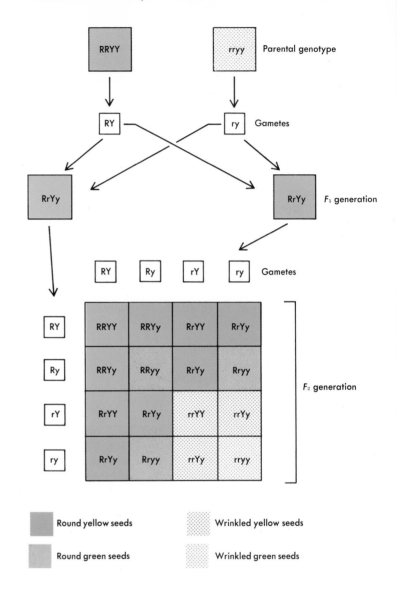

yellow are dominant over wrinkled and green, the entire F_1 generation produced round, yellow seeds. The F_1 generation was then crossed within itself to produce a number of F_2 progeny, which were examined for seed appearance (phenotype). In addition to the two original phenotypes (round yellow; wrinkled green), two new types (*recombinants*) emerged: wrinkled yellow and round green.

Again Mendel found he could interpret the results by the postulate of genes, if he assumed that, during sex-cell

◄ **Figure 1–9**
Schematic drawing of how Mendel's second law (independent assortment) operates. In this example, the inheritance of yellow (Y) and green
(y) seed color is followed together with the inheritance of round (R) and
wrinkled (r) seed shapes. The (R) and (Y) alleles are dominant over (r)
and (y). The genotypes of the various parents and progeny are indicated
by letter combinations, and four different phenotypes distinguished by
appropriate shading.

formation, each gene pair was independently transmitted
to the sex cell (gamete). This interpretation is shown in
Figure 1–9. Any one gamete contains only one type of
inherited factor from each gene pair. Thus the gametes
produced by an F_1 (RrYy) will have the composition RY,
Ry, rY, or ry, but never Rr, Yy, YY, or RR. Furthermore, in
this example, all four possible gametes are produced with
equal frequency. There is no tendency of the genes arising
from one parent to stay together. As a result, the F_2
progeny phenotypes appear in the ratio: 9 round yellow, 3
round green, 3 wrinkled yellow, and 1 wrinkled green. *This
phenomenon of independent assortment is frequently called
Mendel's second law.*

CHROMOSOMAL THEORY OF HEREDITY

A principal reason for the original failure to appreciate
Mendel's discovery was the absence of firm facts about the
behavior of chromosomes during meiosis and mitosis.
This knowledge was available, however, when Mendel's
laws were reannounced in 1900, and was seized upon in
1903 by the American Sutton. In his classic paper, *The
Chromosomes in Heredity,* he emphasized the importance of
the fact that the diploid chromosome group consists of two
morphologically similar sets and that, during meiosis,
every gamete receives only one chromosome of each homologous pair. He then used this fact to explain Mendel's
results by the assumption that genes are parts of the chromosome. He postulated that the yellow- and green-seed
genes are carried on a certain pair of chromosomes, and
that the round- and wrinkled-seed genes are carried on
a different pair. This hypothesis immediately explains
the experimentally observed 9:3:3:1 segregation ratios.
Though Sutton's paper did not prove the chromosomal
theory of heredity, it was immensely important; it brought
together for the first time the independent disciplines of

13

genetics (the study of breeding experiments) and cytology (the study of cell structure).

CHROMOSOMAL DETERMINATION OF SEX

There exists one important exception to the rule that all chromosomes of diploid organisms are present in two copies. It was observed as early as 1890 that one chromosome (then called an accessory chromosome and now the X chromosome) does not always possess a morphologically identical mate. The biological significance of this ob-

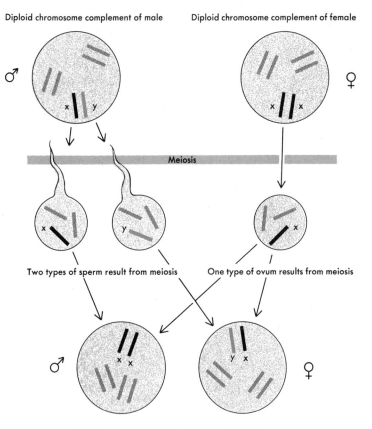

Figure 1–10

Schematic representation of how sex chromosomes operate. Here is shown a case in which males contain one X and one Y chromosome, and females, two X chromosomes. This is the situation in both humans and *Drosophila*. In some other species there is no Y chromosome, so that diploid male cells contain one less chromosome than diploid female cells.

servation was clarified by the American cytologist Wilson and his student Stevens, in 1905. They showed that, although the female contains a pair of X chromosomes, the male contains only one. In addition, in some species (including man), the male cells contain a unique chromosome, not found in females, called the Y chromosome. They pointed out how this situation provides a simple method of sex determination; whereas every egg will contain one X chromosome, only half the sperms will carry one. Fertilization of an ovum by an X-bearing sperm leads to an XX zygote, which becomes a female; fertilization by a sperm cell lacking an X chromosome gives rise to male offspring (Figure 1–10). These observations provided the first clear linking of a definite chromosome to a hereditary property. In addition they elegantly explained how male and female zygotes are created in equal numbers.

THE IMPORTANCE OF *DROSOPHILA*

Initially, all breeding experiments used genetic differences already existing in nature. For example, Mendel used seeds obtained from seed dealers who must have obtained them from farmers. The existence of alternative forms of the same gene (alleles) raises the question of how they arose. One obvious hypothesis states that genes can change (mutate) to give rise to new genes (mutant genes). This hypothesis was first seriously tested, beginning in 1908, by the great American biologist Morgan and his young collaborators, the geneticists Bridges, Muller, and Sturtevant. They worked with the tiny fly *Drosophila*. This fly, which normally lives on fruit, was found to be easily maintained under laboratory conditions, where a new generation can be produced every 14 days. Thus by using *Drosophila* instead of more slowly multiplying organisms like peas, it was possible to work at least 25 times faster, and also much more economically. The first mutant found was a male with white eyes instead of the normal red eyes. It spontaneously appeared in a culture bottle of red-eyed flies. Because essentially all *Drosophila* found in nature have red eyes, the gene leading to red eyes was referred to as the *wild type* gene; the gene leading to white eyes was called a *mutant gene* (allele).

The white-eye mutant gene was immediately used in breeding experiments (Figure 1–11a and b), with the striking result that the behavior of the allele completely paralleled the distribution of an X chromosome (i.e., was sex-linked). This immediately suggested that this gene

Figure 1–11
The inheritance of a sex-linked gene in *Drosophila*. Genes located on sex chromosomes can express themselves differentially in male and female progeny because, if there is only one X chromosome present, recessive genes present on this chromosome are always expressed. Here are shown two crosses, both involving a recessive gene (w, for white eye) located on the X chromosome. In (a) the male parent is a white-eyed (wY) fly, and the female, homozygous for red eye (WW). In (b) the male has red eyes (WY) and the female, white eyes (ww). The letter (Y) stands, here, not for an allele, but for the Y chromosome, present in male *Drosophila* in place of a homologous X chromosome. There is no gene on the Y chromosome corresponding to the (w) or (W) gene on the X chromosome.

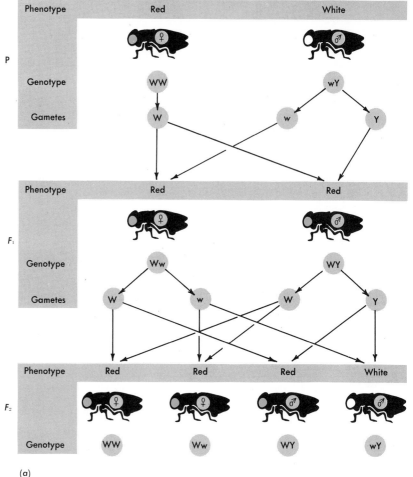

(a)

might be located on the X chromosome, together with those genes controlling sex. This hypothesis was quickly confirmed by additional genetic crosses using newly isolated mutant genes. Many of these additional mutant genes also were sex-linked.

GENE LINKAGE AND CROSSING-OVER

Mendel's principle of independent assortment is based on the fact that genes located on different chromosomes behave independently during meiosis. Often, however, two genes do not assort independently, because they are located on the same chromosome (*linked genes*). Numerous examples of nonrandom assortment were found as soon as

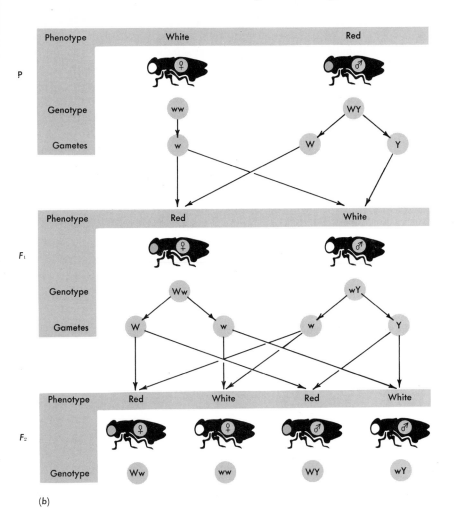

(b)

a large number of mutant genes became available for breeding analysis. In every well-studied case, the number of linked groups was identical with the haploid chromosome number. For example, there are four groups of linked genes in *Drosophila* and four morphologically distinct chromosomes in a haploid cell.

Linkage, however, is in effect never complete. The probability that two genes on the same chromosome will remain together during meiosis ranges from just less than 100% to about 50%.

This means that a mechanism must exist for exchanging genes on homologous chromosomes. This mechanism is called *crossing over*. Its cytological basis was first described by the Belgian cytologist Janssens. At the

Synapsis of duplicated chromosomes
to form tetrads

↓

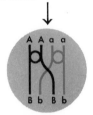

Two chromatids bend across one another

↓

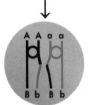

Each chromatid breaks at point of contact
and fuses with a portion of the other

Figure 1–12
Janssens' theory of crossing over.

start of meiosis, the homologous chromosomes form pairs (*synapse*) with their long axes parallel. At this stage, each chromosome has duplicated to form two chromatids. Thus synapsis brings together four chromatids (a tetrad), which coil about each other. Janssens postulated that, possibly because of tension resulting from this coiling, two of the chromatids might sometimes break at a corresponding place on each. This could create four broken ends, which might rejoin crossways, so that a section of each of the two chromatids would be joined to a section of the other (Figure 1–12). Thus recombinant chromatids might be produced that contain a segment derived from each of the original homologous chromosomes.

Morgan and his students were quick to exploit the implication of Janssens' still unproved theory: that genes located close to each other on a chromosome would assort with each other much more regularly (close linkage) than genes located far apart on a chromosome. This immediately suggested a way to locate (map) the relative positions of genes on the various chromosomes (see Chapter 7 for details). By 1915, more than 85 mutant genes in *Drosophila* had been assigned locations, each a distinct spot on one of the four linkage groups or chromosomes (Table 1–1). The definitive volume which Morgan then published, *The Mechanism of Mendelian Heredity*, showed the general validity of the chromosomal basis of heredity, a concept ranking with the theories of evolution and the cell as one of the main achievements of the biologist's attempt to understand the nature of the living world.

MANY GENES CONTROL THE RED EYE

Mere inspection of the list of mutant genes in Table 1–1 reveals an important fact; many different genes act to influence a single character. For example, 13 of the genes discovered by 1915 affect eye color. When a fly is homozygous for a mutant form of any of these genes, the eye color is not red, but a different color, distinct for the mutant gene (e.g., carnation, vermilion). Thus there is no one-to-one correspondence between genes and complex characters like eye color or wing shape. Instead, the development of each character is controlled by a series of events, each of which is controlled by a gene. We might make a useful analogy with the functioning of a complex machine like the automobile: There are clearly a number of separate parts, like the motor, the brakes, the radiator, and the fuel tank, all of which are essential for proper opera-

Table 1–1 The Eighty-Five Mutant Genes Reported in
Drosophila Melanogaster in 1915*

Name	Region Affected	Name	Region Affected
Group I			
Abnormal	Abdomen	Lethal, 13	Body, death
Bar	Eye	Miniature	Wing
Bifid	Venation	Notch	Venation
Bow	Wing	Reduplicated	Eye color
Cherry	Eye color	Ruby	Leg
Chrome	Body color	Rudimentary	Wing
Cleft	Venation	Sable	Body color
Club	Wing	Shifted	Venation
Depressed	Wing	Short	Wing
Dotted	Thorax	Skee	Wing
Eosin	Eye color	Spoon	Wing
Facet	Ommatidia	Spot	Body color
Forked	Spine	Tan	Antenna
Furrowed	Eye	Truncate	Wing
Fused	Venation	Vermilion	Eye color
Green	Body color	White	Eye color
Jaunty	Wing	Yellow	Body color
Lemon	Body color		
Group II			
Antlered	Wing	Jaunty	Wing
Apterous	Wing	Limited	Abdominal band
Arc	Wing	Little crossover	II chromosome
Balloon	Venation	Morula	Ommatidia
Black	Body color	Olive	Body color
Blistered	Wing	Plexus	Venation
Comma	Thorax mark	Purple	Eye color
Confluent	Venation	Speck	Thorax mark
Cream II	Eye color	Strap	Wing
Curved	Wing	Streak	Pattern
Dachs	Leg	Trefoil	Pattern
Extra vein	Venation	Truncate	Wing
Fringed	Wing	Vestigial	Wing
Group III			
Band	Pattern	Pink	Eye color
Beaded	Wing	Rough	Eye
Cream III	Eye color	Safranin	Eye color
Deformed	Eye	Sepia	Eye color
Dwarf	Size of body	Sooty	Body color
Ebony	Body color	Spineless	Spine
Giant	Size of body	Spread	Wing
Kidney	Eye	Trident	Pattern
Low crossing over	III chromosome	Truncate intensf.	Wing
Maroon	Eye color	Whitehead	Pattern
Peach	Eye color	White ocelli	Simple eye
Group IV			
Bent	Wing	Eyeless	Eye

* The mutations fall into four linkage groups. Since four chromosomes were cytologically observed, this indicated that the genes are situated on the chromosomes. Notice that mutations in various genes can act to alter a single character, such as body color, in different ways.

tion of the vehicle. Although a fault in any one part may cause the car to stop functioning properly, there is no reason to believe that the presence of that component alone is sufficient for proper functioning.

ORIGIN OF GENETIC VARIABILITY THROUGH MUTATIONS

It now became possible to understand the hereditary variation that is found throughout the biological world and that forms the basis of the theory of evolution. Genes are normally copied exactly during chromosome duplication. Rarely, however, changes (*mutations*) occur in genes to give rise to altered forms most, *but not all,* of which function less well than the wild-type alleles. This process is necessarily rare; otherwise many genes would be changed during every cell cycle, and offspring would not ordinarily resemble their parents. There is instead a strong advantage in there being a small but finite mutation rate; it provides a constant source of new variability, necessary to allow plants and animals to adapt to a constantly changing physical and biological environment.

Surprisingly, however, the results of the Mendelian geneticists were not avidly seized upon by the classical biologists, then the authorities on the evolutionary relations between the various forms of life. Doubts were raised about whether genetic changes of the type studied by Morgan and his students were sufficient to permit the evolution of radically new structures, like wings or eyes. Instead, they believed that there must also exist more powerful "macromutations," and that it was these which allowed great evolutionary advances.

Gradually, however, doubts vanished, largely as a result of the efforts of the mathematical geneticists Wright, Fisher, and Haldane. Considering the great age of the earth, they showed that the relatively low mutation rates found for *Drosophila* genes, together with only mild selective advantages, would be sufficient to allow the gradual accumulation of new favorable attributes. By the 1930's, biologists themselves began to reevaluate their knowledge on the origin of species, and to understand the work of the mathematical geneticists. Among these new Darwinians were the biologist Julian Huxley (a grandson of Darwin's original publicist T. H. Huxley), the geneticist Dobzhansky, the paleontologist Simpson, and the ornithologist Mayr. In the 1940's, all four wrote major works, each showing from his special viewpoint how Mendelianism and Darwinism were indeed compatible.

EARLY SPECULATIONS ABOUT WHAT GENES ARE AND HOW THEY ACT

Almost immediately after the rediscovery of Mendel's laws, geneticists began to speculate about both the chemical structure of the gene and how it acts. No real progress could be made, however, since the chemical identity of the genetic material remained unknown. Even the realization that both nucleic acids and proteins are present in chromosomes did not really help, since the structure of neither was at all understood. The most fruitful speculations focused attention on the fact that genes must be, in some sense, self-duplicating: Their structure must be exactly copied every time one chromosome becomes two. This fact immediately raised the profound chemical question of how a complicated molecule could be precisely copied to yield exact replicas.

Some physicists also became intrigued with the gene, and when quantum mechanics burst on the world in the late 1920's the possibility arose that, to understand the gene, it would be necessary to master the subtleties of the most advanced theoretical physics. Such thoughts, however, never really took root, since it was obvious that even the best physicists or theoretical chemists could not worry about a substance whose structure still awaited elucidation. There was only one fact which they might ponder: Muller's 1927 discovery that x-rays induce mutation. Since there is a greater probability that an x-ray will hit a larger gene than a smaller one, the frequency of mutations induced in a given gene by a given x-ray dose yields an estimate of the size of this gene. But even here so many special assumptions had to be made that virtually no one, not even the estimators themselves, took the estimates very seriously.

PRELIMINARY ATTEMPTS TO FIND A GENE–PROTEIN RELATIONSHIP

The most fruitful endeavors to find a relationship between genes and proteins examined the ways in which gene changes affect what proteins are present in the cell. At first this study was difficult, since no one really knew anything about the proteins that were present in structures such as the eye or the wing. It soon became obvious that genes with simple metabolic functions would be easier to study than genes affecting gross structures. One of the first useful examples came from a study of an hereditary disease affecting amino acid metabolism. Spontaneous mutations occur in humans, affecting the ability to metabolize the amino acid phenylalanine. When individuals homozygous for the mutant trait eat food containing phenylalanine, their inability to convert it to tyrosine causes a toxic level of phenylpyruvic acid to build up in the bloodstream. The existence of such diseases, an example of the so-called "inborn errors of metabolism," suggested as early as 1909 to the physician Garrod that the wild-type gene is responsible for the presence of a particular enzyme and that, in a homozygous mutant, the enzyme is congenitally absent.

Garrod's general hypothesis of a gene–enzyme relationship was extended in the 1930's by work on flower pigments and the pigments of insect eyes. In both cases evidence was obtained that a particular gene affected a particular step in the formation of the pigment. However, the absence of fundamental knowledge about the structures of the relevant proteins ruled out deeper examination of the gene–protein relationship, and no assurance could be given either that most genes control the synthesis of proteins (by then it was suspected that all enzymes were proteins), or that all proteins are under gene control.

As early as 1935, it became obvious to the Mendelian geneticists that future experiments of the sort successful in elucidating the basic features of Mendelian genetics were unlikely to yield productive evidence about how genes act. Instead it would be necessary to find biological objects more suitable for chemical analysis. They were aware, however, that the contemporary state of nucleic acid and protein chemistry was completely inadequate for a fundamental chemical attack on even the most suitable biological systems. Fortunately, however, the limitations in chemistry did not deter them from learning how to do genetic experiments with chemically simple molds, bacteria, and viruses. As we shall see, the necessary chemical facts became available almost as soon as the geneticists were ready to use them.

SUMMARY

The study of living organisms at the biological level has led to three great generalizations: (1) Darwin's and Wallace's theory of evolution by natural selection, which tells us that today's complex plants and animals are derived by a continuous evolutionary progression from the first primitive organisms; (2) the cell theory, the realization that all organisms are built up of cells; (3) the chromosomal theory of heredity, the under-

standing that the function of chromosomes is the control of heredity.

All cells contain chromosomes, normally duplicated prior to a cell-division process (mitosis) *which produces two daughter cells, each with a chromosomal complement identical to that of the parental cell. In haploid cells there is just one copy of each type of chromosome; in diploid cells there are usually two copies (pairs of homologous chromosomes). A diploid cell arises by fusion of a male and a female haploid cell (fertilization), whereas haploid cells are formed from a diploid cell by a distinctive form of cell division* (meiosis), *which reduces the chromosome number to one-half of its previous number.*

Chromosomes control heredity because they are the cellular locations of genes. Genes were first discovered by Mendel in 1865, but their importance was not realized until the start of the twentieth century. Each gene can exist in a variety of different forms called alleles. *Mendel proposed that a gene for each hereditary trait is given by each parent to each of its offspring. The physical basis for this behavior is in the distribution of homologous chromosomes during meiosis: One (randomly chosen) of each pair of homologous chromosomes is distributed to each haploid cell. When two genes are on the same chromosome, they tend to be inherited together (linked genes). Genes affecting different characters are sometimes inherited independently of each other: this is because they are located on different chromosomes. In any case, linkage is seldom complete, because homologous chromosomes attach to each other during meiosis and often break at identical spots and rejoin crossways (crossing over). This attaches genes initially found on a paternally derived chromosome to gene groups originating from the maternal parent.*

Different alleles of the same gene arise by inheritable changes (mutations) in the gene itself. Normally genes are extremely stable and are exactly copied during chromosome duplication; mutation normally occurs only rarely and usually has harmful consequences. It does, however, play a positive role, since the accumulation of the rare favorable mutations provides the basis for the genetic variability that the theory of evolution presupposes.

For many years the structure of the genes and the chemical way in which they control cellular characteristics were a mystery. As soon as large numbers of spontaneous mutations had been described, it became obvious that a one gene–one character relationship does not exist, but that all complex characters are under the control of many genes. The most sensible idea, postulated clearly by Garrod as early as 1909, was that genes affect the synthesis of enzymes. However, in general, the tools of the Mendelian geneticists, organisms such as the corn plant, the mouse, and even the fruit fly, Drosophila, *were not suitable for chemical investigations of gene–protein relations. For this type of analysis, work with much simpler microorganisms became indispensable.*

REFERENCES

SWANSON, C. P., *The Cell,* 3rd ed., Prentice-Hall, Englewood Cliffs, N.J., 1969. An introductory survey of the cell theory.

MOORE, J., *Heredity and Development,* 2nd ed., Oxford University Press, 1972. An elegant introduction to genetics and embryology with emphasis on the historical approach.

STURTEVANT, A. H., AND G. W. BEADLE, *An Introduction to Genetics,* Dover, New York, 1962. Now available in paperback form, this book, originally published in 1939, remains a classic statement of the results of *Drosophila* genetics.

SRB, A., R. OWEN, AND R. EDGAR, *General Genetics,* 2nd ed., Freeman, San Francisco, 1965. Though badly out of date, this text remains a nice introduction to the world of genetics.

PETERS, J. A., *Classic Papers in Genetics,* Prentice-Hall, Englewood Cliffs, N.J., 1959. A collection of reprints of many of the most significant papers in the history of genetics, up to Benzer's fine-structure analysis of the gene.

MOORE, J., *Readings in Heredity and Development,* Oxford University Press, 1972. Many of the key papers in the development of the chromosomal theory of heredity are reproduced here.

STURTEVANT, A. H., *A History of Genetics,* Harper and Row, 1965. History as seen by one of the major contributors.

CARLSON, E. J., *The Gene Theory: A Critical History,* W. B. Saunders, Philadelphia, 1966. Surveys the development of genetical ideas from Mendel to the present.

MAYR, E., *Animal Species and Evolution,* Harvard University Press, Cambridge, 1963. The most complete statement of the facts supporting the theory of evolution.

LEWONTIN, R. C., *The Genetic Basis of Evolutionary Change,* Columbia University Press, 1974. The most recent astute look at the evolutionary process.

Cells Obey the Laws of Chemistry

In Darwin's time chemists were already asking whether living cells worked by the same chemical rules as non-living systems. By then, cells had been found to contain no atoms peculiar to living material. Also recognized early was the predominant role of carbon, a major constituent of almost all types of biological molecules. A reflection of the initial tendency to distinguish between carbon compounds like those in living matter and all other molecules is retained in the division of modern chemistry into organic chemistry (the study of most compounds containing carbon atoms) and inorganic chemistry. Now we know that this distinction is artificial and has no biological basis. There is no purely chemical way to decide whether a compound has been synthesized in a cell or in a chemist's laboratory.

Nonetheless, through the first quarter of this century, a strong feeling existed in many biological and chemical laboratories that some vital force outside the laws of chemistry differentiated between the animate and the inanimate. Part of the reason for the persistence of this "vitalism" was that the success of the biologically oriented chemists (now usually called biochemists) was limited. Although the techniques of the organic chemists were sufficient to work out the structures of relatively small molecules like glucose (Table 2–1), there was increasing awareness that many of the most important molecules in the cell

25

Table 2–1 Some Important Classes of Small Biological Molecules

Class	Characteristics	Example
Aliphatic hydrocarbon	Linear or branched molecules containing only carbon and hydrogen	Ethane
Aromatic hydrocarbon	Ring-shaped hydrocarbons containing alternating single and double bonds	Benzene
Pyrimidine	An aromatic compound of the formula $C_4H_4N_2$ (or a derivative thereof)	Uracil
Purine	An aromatic compound of the formula $C_5H_4N_4$ (or a derivative)	Guanine
Alcohol	A hydrocarbon skeleton substituted with one to several OH groups	Ethanol
Phosphate ester	Molecule formed from alcohols and phosphoric acids with the elimination of H_2O	Glucose-1-Ⓟ
Nucleoside	Contains a pentose sugar linked to either a purine or pyrimidine base, through a C—N bond	Adenosine

Class	Characteristics		Example
Nucleotide	Phosphate ester of a nucleoside	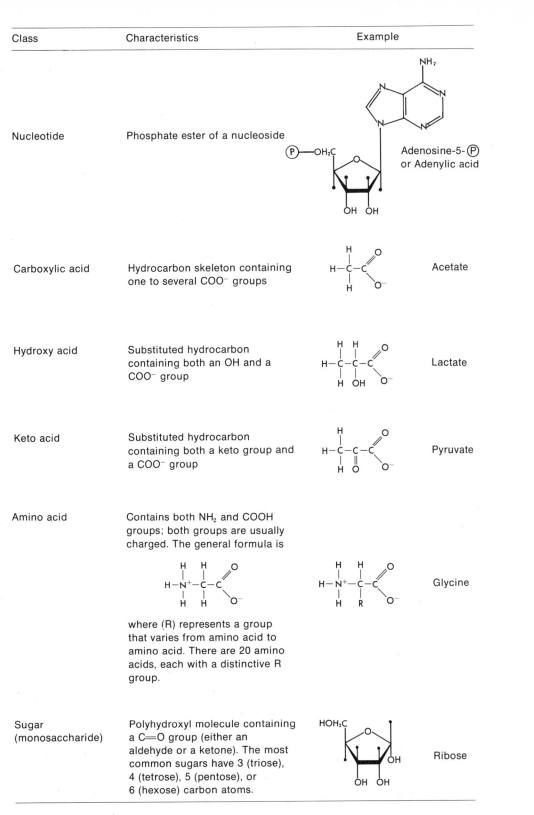	Adenosine-5-P or Adenylic acid
Carboxylic acid	Hydrocarbon skeleton containing one to several COO^- groups		Acetate
Hydroxy acid	Substituted hydrocarbon containing both an OH and a COO^- group		Lactate
Keto acid	Substituted hydrocarbon containing both a keto group and a COO^- group		Pyruvate
Amino acid	Contains both NH_2 and COOH groups; both groups are usually charged. The general formula is where (R) represents a group that varies from amino acid to amino acid. There are 20 amino acids, each with a distinctive R group.		Glycine
Sugar (monosaccharide)	Polyhydroxyl molecule containing a $C=O$ group (either an aldehyde or a ketone). The most common sugars have 3 (triose), 4 (tetrose), 5 (pentose), or 6 (hexose) carbon atoms.		Ribose

were very large—the so-called macromolecules—too large to be pursued by even the best of organic chemists.

The most important group of macromolecules was for many years believed to be the proteins, because of the growing evidence that all enzymes are proteins. Initially, there was controversy as to whether enzymes were small molecules or macromolecules. It was not until 1926 that the enzymatic nature of a crystalline protein was demonstrated by the American biochemist Sumner; the controversy was then practically settled. But even this important discovery did not dispel the general aura of mystery about proteins. Then, the complex structures of proteins were undecipherable by available chemical tools, so it was still possible, as late as 1940, for some scientists to believe that these molecules would eventually be shown to have features unique to living systems.

The general belief also existed that the genes, like the enzymes, might be proteins. There was no direct evidence, but the high degree of specificity of genes suggested to most people who speculated on their nature that they could only be proteins, by then known to occur in the chromosomes. Another class of molecules, the nucleic acids, were also found to be a common chromosomal component, but these were thought to be relatively small and incapable of carrying sufficient genetic information.

Besides general ignorance of the structures of the large molecules, the feeling was often expressed that something unique about the three-dimensional organization of the cell gave it its living feature. This argument was sometimes phrased in terms of the impossibility of ever understanding all the exact chemical interactions of the cell. More frequently, however, it took the form of the prediction that some new natural laws, as important as the cell theory or the theory of evolution, would have to be discovered before the essence of life could be understood. But these almost mystical ideas never led to meaningful experiments and, in their vague form, could never be tested. Progress was made instead only by biologically oriented chemists and physicists patiently attempting to devise new ways of solving more and more complex biological structures. But for many years, there were no triumphs to shout. The chemists and biologists usually moved in different and sometimes hostile worlds, the biologist often denying that the chemist would ever provide the real answers to the important riddles of biology. Always not too far back in some biologists' minds was the feeling, if not the hope, that something more basic than mere complexity and size separated biology from the bleak, inanimate world of a chemical laboratory.

THE CONCEPT OF INTERMEDIARY METABOLISM

As soon as the organic chemists began to identify some of the various cellular molecules, it became clear that food molecules are extensively transformed after they enter an organism. In no case does a food source contain all the different molecules present in a cell. On the contrary, in some cases practically all the organic molecules within an organism are synthesized inside it. This point is easily seen by observing cellular growth on well defined food sources: for example, the growth of yeast cells using the simple sugar glucose as the sole source of carbon. Here, soon after its cellular entry, glucose is chemically transformed into a large variety of molecules necessary for the building of new structural components. Usually these chemical transformations do not occur in one step; instead intermediate compounds are produced. These intermediate compounds often have no cellular function besides forming part of a pathway leading to the synthesis of a necessary structural component like an amino acid.

The sum total of all the various chemical reactions occurring in a cell is frequently referred to as the *metabolism* of the cell. Correspondingly, the various molecules involved in these transformations are often called *metabolites*. *Intermediary metabolism* is the term used to describe the various chemical reactions involved in the transformation of food molecules into essential cellular building blocks.

ENERGY GENERATION
BY OXIDATION–REDUCTION REACTIONS

By the middle of the nineteenth century, it was known that the food (initially of plant origin) eaten by animals and bacteria is only in part transformed into new cellular building blocks, some of it being burned by combustion with oxygen to yield CO_2 and H_2O and energy. At the same time it was becoming clear that the reverse process also operates in green plants.

Respiration:

$C_6H_{12}O_6$ (glucose) $+ 6O_2 \rightarrow 6CO_2 + 6H_2O +$ energy (in form of heat)

(occurs both in plants and animals) (2–1)

Photosynthesis:

$6CO_2 + 6H_2O +$ energy (from the sun) $\rightarrow C_6H_{12}O_6$ (glucose) $+ 6O_2$

(occurs only in plants) (2–2)

Table 2–2 Important Functional Groups in Biological Molecules

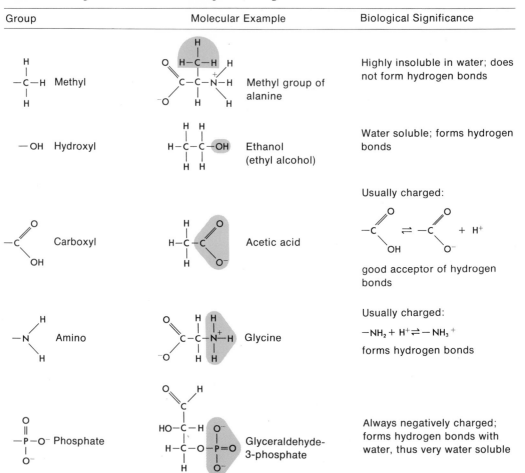

Group	Molecular Example	Biological Significance
—C—H Methyl	Methyl group of alanine	Highly insoluble in water; does not form hydrogen bonds
—OH Hydroxyl	Ethanol (ethyl alcohol)	Water soluble; forms hydrogen bonds
—C(=O)(OH) Carboxyl	Acetic acid	Usually charged: $-C{<}^{O}_{OH} \rightleftharpoons -C{<}^{O}_{O^-} + H^+$ good acceptor of hydrogen bonds
—N(H)(H) Amino	Glycine	Usually charged: $-NH_2 + H^+ \rightleftharpoons -NH_3^+$ forms hydrogen bonds
—P(=O)(O⁻)—O⁻ Phosphate	Glyceraldehyde-3-phosphate	Always negatively charged; forms hydrogen bonds with water, thus very water soluble

Both these equations can be thought of as the sum total of a lengthy series of oxidation–reduction reactions.

In respiration, organic molecules such as glucose are oxidized by molecular oxygen to form C=O bonds (Table 2–2), which contain *less* usable energy (energy which can do work) than the starting C—H, C—OH, and C—C bonds. Energy is given off in respiration, just as it is when glucose burns at high temperatures outside the cell to produce CO_2, H_2O, and energy in the form of heat. In contrast, during photosynthesis, the energy from the light quanta of the sun is used to reduce CO_2 to molecules which contain *more* usable energy.

When these relationships were first worked out, no one knew how the energy obtained during respiration was put to advantage. It was clear that somehow a useful form of energy had to be available to enable living organisms to

Table 2–2 Continued

Group	Molecular Example	Biological Significance

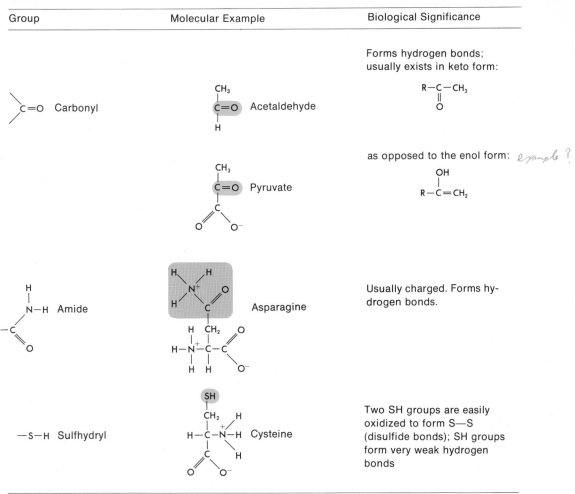

Forms hydrogen bonds; usually exists in keto form:

as opposed to the enol form: *example ?*

Usually charged. Forms hydrogen bonds.

Two SH groups are easily oxidized to form S—S (disulfide bonds); SH groups form very weak hydrogen bonds

carry out a variety of forms of work, such as muscular contraction and selective transport of molecules across cell membranes. Even then it seemed unlikely that the energy obtained from food was first released as heat, since, at the temperature at which life exists, heat energy cannot be effectively used to synthesize new chemical bonds. Thus since the awakening of an interest in the chemistry of life, a prime challenge has been to understand the generation of energy in a useful form.

MOST BIOLOGICAL OXIDATIONS OCCUR WITHOUT DIRECT PARTICIPATION OF OXYGEN

Because oxygen is so completely necessary for the functioning of animals, it was natural to guess that oxygen would participate directly in all oxidations of carbon com-

31

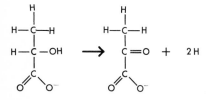

Figure 2–1
Oxidation of an organic molecule by removal of a pair of hydrogen atoms. This figure shows the oxidation of lactate to pyruvate.

Figure 2–2
The participation of NAD in the oxidation of lactate to pyruvate. The oxidizing agent here is NAD, and the hydrogen donor is lactate.

pounds. Actually, most biological oxidations occur in the *absence* of oxygen. This is possible because, as first proposed around 1912 by the German biochemist Wieland, most biological oxidations are actually dehydrogenations. A compound is oxidized when we remove a pair of hydrogen atoms from it (Figure 2–1). It is not possible, however, merely to remove the hydrogen atoms. They must be transferred to another molecule, which is then said to be reduced (Figure 2–2). In these reactions, as in all other oxidation–reduction reactions, every time one molecule is oxidized, another must be reduced. There are several different molecules whose role is to receive hydrogen atoms. All are medium-sized (MW ~ 500) organic molecules that associate with specific proteins to form active enzymes. The protein components alone have no enzymatic activity. Only when the small partner is present will activity be present. Hence these small molecules are named *coenzymes* (we should note that not all coenzymes participate in oxidation–reduction reactions; some coenzymes function in other types of metabolic reactions).

Although the involvement of coenzymes in oxidative reactions was hinted at by 1910, it was not until the early 1930's that their cardinal significance was appreciated. Then the work of the great German biochemist, Otto Warburg, and the Swedish chemists, von Euler and Theorell, established the structure and action of several of the most

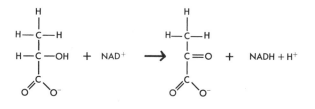

important coenzymes: nicotinamide adenine dinucleotide (NAD, earlier called diphosphopyridine nucleotide or DPN, Figure 2–3), flavin mononucleotide (FMN), and flavin adenine dinucleotide (FAD).

Coenzymes, like enzymes, function many different times and are not used up in the course of functioning. This is because the hydrogen atoms transferred to them do not remain permanently attached, but are transferred by a second oxidation–reduction reaction, usually to another coenzyme, or to oxygen itself (Figure 2–4). Coenzymes are thus being continually oxidized and reduced. Furthermore, we see that, although oxygen is not directly necessary for a

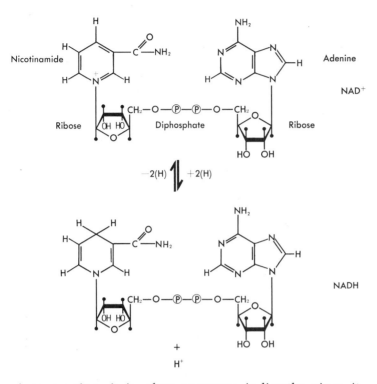

Figure 2–3
The oxidation and reduction of a coenzyme. Shown here are both the oxidized and reduced forms of the very important coenzyme nicotinamide adenine dinucleotide, NAD (the oxidized form), an acceptor of hydrogen atoms, and NADH₂ (the reduced form), a donor of hydrogen atoms. The release of hydrogen atoms decreases the free energy of a molecule; the acceptance of them increases its free energy.

given reaction, it is often necessary indirectly, since it must be available to oxidize the coenzyme molecules to make them available for accepting additional pairs of hydrogen atoms (or electrons).

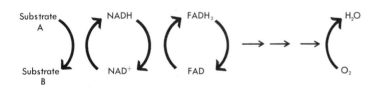

Figure 2–4
The transfer of a pair of hydrogen atoms from one coenzyme to another. In this series of reactions, the final hydrogen acceptor is oxygen (O_2). FAD is flavin adenine dinucleotide, which exists not free, but usually in combination with a specific protein, to form a flavoprotein.

THE BREAKDOWN OF GLUCOSE

Much of the early work in intermediary metabolism dealt with the transformation of glucose into other molecules. Glucose was emphasized, not only because it played a central role in the economy of cells, but also for a practical reason: The alcohol (ethanol) produced when wine is made from grapes is derived from the breakdown of glucose. As early as 1810, the chemist Gay-Lussac demonstrated the production of ethyl alcohol by this process, and by 1837 the essential role of yeast was established. The production of the alcohol in wine is not a spontaneous process but normally requires the presence of living yeast cells.

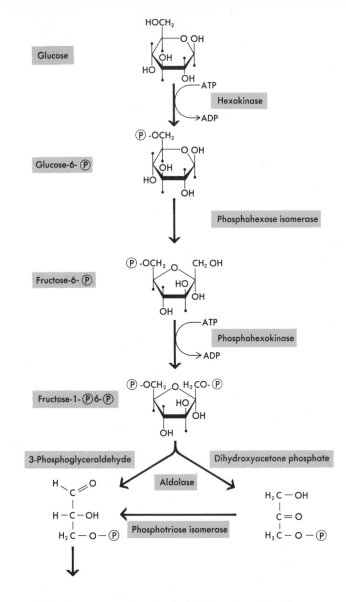

Figure 2–5
The stepwise degradation of glucose to pyruvic acid. This collection of consecutive reactions is often called the Embden–Meyerhof pathway.

Also important in the initial work with glucose was the French microbiologist Pasteur, who discovered that the process does not require air; to distinguish it from reactions requiring oxygen, he used the term *fermentation*. He also showed that ethanol is not the only product of glucose fermentation, but that there are other products, such as lactic acid and glycerol.

The next great advance came with Buchner's discovery in 1897 that the living cell *per se* is not necessary for fermentation, but that a cell-free extract from yeast can, by itself, transform glucose into ethanol. This step was not only conceptually important, but also provided a much

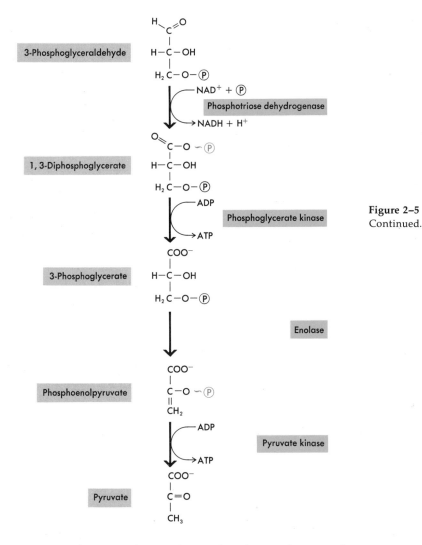

3-Phosphoglyceraldehyde

$NAD^+ + P$

Phosphotriose dehydrogenase

$NADH + H^+$

1, 3-Diphosphoglycerate

ADP

Phosphoglycerate kinase

ATP

3-Phosphoglycerate

Enolase

Phosphoenolpyruvate

ADP

Pyruvate kinase

ATP

Pyruvate

Figure 2–5
Continued.

more practical system for studying the chemical steps of fermentation. When working with cell-free systems, it is relatively easy to add or subtract components thought to be involved in the reaction; when living cells are being used, it is often very difficult, and sometimes impossible, to transfer specific compounds in an unmodified form across the cell membrane.

Over the next 40 years, cell-free extracts were used by a large number of distinguished biochemists, including the Englishmen Harden and Young, and the Germans Embden and Meyerhof, to work out the exact chemical pathways of glucose degradation (Figure 2–5). During this period, the important generalization emerged that the reactions involved (collectively called the Embden–Meyerhof pathway) were not peculiar to alcoholic fermentation in yeast, but occurred in many other cases of glucose utilization as well. Perhaps the most significant discovery was made by Meyerhof. He showed that, when muscles contract in the absence of oxygen, the carbohydrate food reserve of glycogen is broken down, via glucose, to lactic acid (anaerobic *glycolysis*). Thus it became clear that, not only can microorganisms obtain their energy and carbon via the Embden–Meyerhof pathway, but energy involved in the contraction of muscles is also generated by the same pathway.

A feeling often expressed as "the unity of biochemistry" began to develop. By this we mean the realization that the basic biochemical reactions upon which cell growth and division depend are the same, or very similar, in all cells, those of microorganisms as well as of higher plants and animals. This unity was not surprising to the more astute biologists, many of whom were largely preoccupied with the consequences of evolutionary theory: Given that a man and a fish are descended from a common ancestor, it should not be surprising that many of their cell constituents are similar.

INVOLVEMENT OF PHOSPHORUS AND THE GENERATION OF ATP

As early as 1905 the phosphorus atom was implicated in a vital role in metabolism. Then Harden and Young found that alcoholic fermentation occurs only when inorganic phosphate (PO_4^{3-}) is present. This discovery was followed by the eventual isolation of a large number of intermediary metabolites containing PO_4^{3-} () groups attached to carbon atoms by phosphate ester linkages,

The significance of phosphorylated intermediates was unclear for 25 years. Then in the 1930's, Meyerhof and Lipmann had the crucial insight that phosphate esters enable cells to trap much of the energy of the chemical bonds present in their food molecules. During fermentation several intermediates (Figure 2–6) are created (e.g., D-1,3-diphosphoglyceric acid), which contain what are popularly known as high-energy phosphate bonds (see Chapter

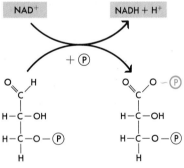

Figure 2–6
The formation of an energy-rich phosphate ester bond, coupled with the oxidation of 3-phosphoglyceraldehyde by NAD.

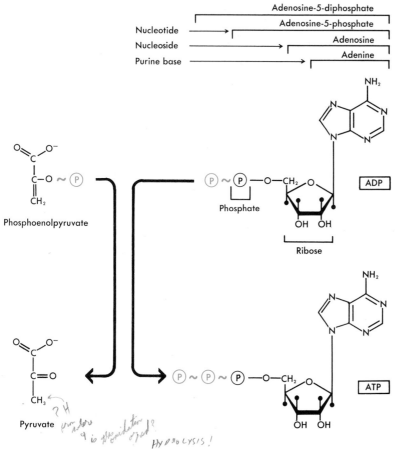

Figure 2–7

The formation of ATP (adenosine-5'-triphosphate) from ADP and an energy-rich phosphate bond. Here the donor of the high-energy bond is phosphoenolpyruvate. The symbol ~ signifies that the bond is of the high-energy variety.

5 for more details). These high-energy phosphate groups are usually transferred to acceptor molecules, where they can serve as sources of chemical energy for vital cellular processes, such as motion, generation of light, and (as we shall see in Chapters 5 and 6) the efficient biosynthesis of necessary cellular molecules. The most important of the acceptor molecules is adenosine diphosphate (ADP, Figure 2–7). Addition of a high-energy Ⓟ group to ADP forms *adenosine triphosphate* (ATP).

$$\text{ADP} + Ⓟ \rightleftarrows \text{ATP}$$

The discovery of the role of ADP as an acceptor molecule and that of ATP as a donor of high-energy phosphate groups was one of the most important discoveries of modern biology. Until the roles of these molecules were known, there was complete mystery about how cells obtained energy. There was constant speculation about how cellular existence was compatible with the second law of thermodynamics [in a closed system the amount of dis-

37

order (entropy) invariably increases]. What was conceivably a paradox ceased to exist as soon as it was seen how animal cells could trap and utilize the energy in food molecules. At that time the mechanism by which the sun's energy was trapped in photosynthesis was not known. Here again, the primary action of the sun's energy is now known to be the generation of ATP.

MOST SPECIFIC CELLULAR REACTIONS REQUIRE A SPECIFIC ENZYME

The idea that most specific metabolic steps require a specific enzyme was realized only when a number of specific reactions were unraveled. As the Embden–Meyerhof pathway was being worked out, it became clear that each step required a separate enzyme (see Figure 2–5). Each of the enzymes acts by combining with the molecules involved in the particular reaction (the *substrates* of the enzymes). For example, glucose and ATP are substrates for the enzyme hexokinase. When these molecules interact on the surface of hexokinase, the terminal P of ATP is transferred to a glucose molecule to form glucose-6-$\circledP$.

The essence of an enzyme is its ability to speed up (catalyze) a reaction involving the making or breaking of a specific covalent bond (a bond in which atoms are held together by the sharing of electrons). In the absence of enzymes, most of the covalent bonds of biological molecules are very stable, and decompose only under high nonphysiological temperatures; only at several hundred degrees centigrade is glucose, for example, appreciably oxidized by O_2 in the absence of enzymes. Enzymes must therefore act by somehow lowering the temperature at which a given bond is unstable. A physical chemist would say that an enzyme lowers the "activation energy." How this is done is just being understood at the molecular level. The 3-D structure of several enzymes is now known, and plausible chemical theories exist about how they work (see Figure 2–15). Thus, there is no reason to suspect that still undiscovered laws of chemistry underlie enzyme action. Numerous examples already exist in pure chemistry where well defined molecules speed up reactions between other molecules.

A very important characteristic of enzymes is that they are never consumed in the course of reaction; once a reaction is complete, they are free to adsorb new molecules and function again (Figure 2–8). On a biological time scale (seconds to years), enzymes can work very fast, some being

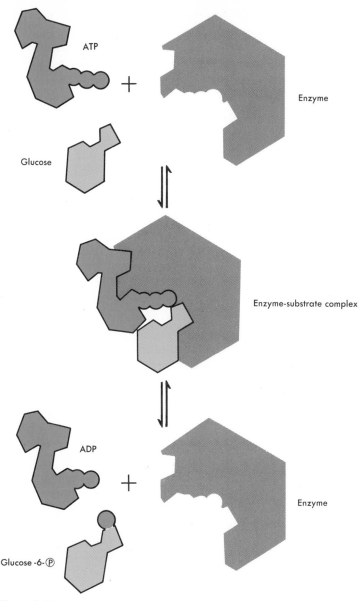

Figure 2—8
The formation of an enzyme–substrate complex, followed by catalysis.

able to catalyze as many as 10^6 reactions per minute; often no successful collision of substrates will occur in this time interval when enzymes are absent.

Not all enzymatic reactions, however, are specific. There exist, for example, various enzymes which break down a variety of different proteins to their component amino acids. They are specific only in the sense that they catalyze the breakdown of a specific type of covalent bond, the peptide bond, and will not, for example, degrade the phosphodiester linkages of the nucleic acids.

THE KEY ROLE OF PYRUVATE:*
ITS UTILIZATION VIA THE KREBS CYCLE

Attempts to understand the generation of ATP in the presence of oxygen occurred parallel with the study of fermentation and glycolysis. It was immediately obvious from consideration of the amounts of ATP generated by fermentation and glycolysis that these processes could account for only a small fraction of total ATP production in the presence of oxygen. This means that ATP production in the presence of oxygen does not cease once glucose has been degraded as far as pyruvic acid (pyruvate), but that pyruvic acid itself must be further transformed, via energy-yielding reactions requiring the presence of oxygen.

The first real breakthrough in understanding how this happens came with the discoveries made by the biochemists Szent-Györgyi, Martius, and Krebs. Their work revealed the existence of a cyclic series of reactions (now usually called the *Krebs cycle*) by which pyruvate is oxidatively broken down to yield carbon dioxide (CO_2) and a series of pairs of hydrogen atoms that attach to oxidized coenzyme molecules. Before pyruvate enters the Krebs cycle, it is transformed into a key molecule called acetyl-CoA (Figure 2–9), known before its chemical identification as "active acetate." This important intermediate, discovered in 1949 by Lipmann, working in Boston, then combines with oxaloacetate to yield citrate. A series of at least nine additional steps (see Figure 2–10) then occur to yield four pairs of H atoms and two molecules of CO_2. The pairs of hydrogen atoms never exist free, but are transferred to specific coenzyme molecules.

The Krebs cycle should be viewed as a mechanism for breaking down acetyl-CoA to two types of products: the completely oxidized CO_2 molecules, which cannot be used as energy sources, and the reduced coenzymes, whose further oxidation yields most of the energy used by organisms growing in the presence of oxygen.

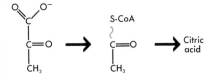

Figure 2–9

The transformation of pyruvate to acetyl-CoA. CoA refers to coenzyme A. The transformation, as written, is greatly simplified. Several steps are required in which the coenzymes thiamine pyrophosphate and lipoic acid are both involved. Acetyl-CoA is an extremely important intermediate, for it is formed not only from glucose via pyruvate, but also by the degradation of fatty acids.

* The terms pyruvic acid and pyruvate are used interchangeably. Technically, pyruvate refers to the negatively charged ion. Likewise, lactic acid is often called lactate; glutamic acid, glutamate; citric acid, citrate, etc.

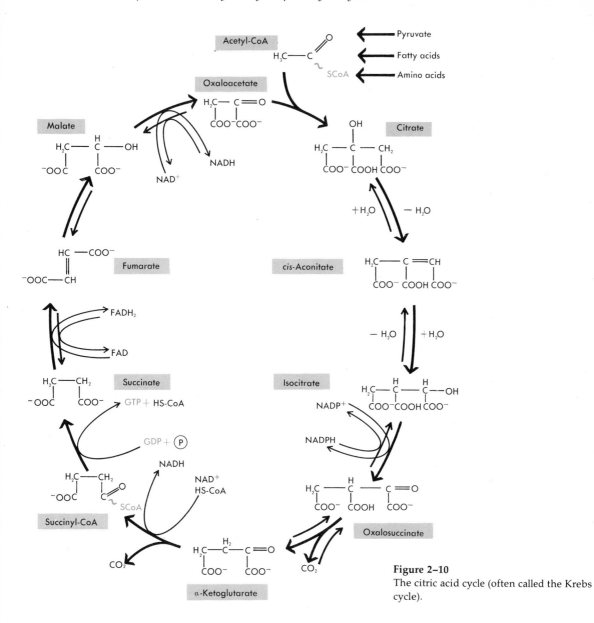

Figure 2–10
The citric acid cycle (often called the Krebs cycle).

OXIDATION OF REDUCED COENZYMES BY RESPIRATORY ENZYMES

During the functioning of the Krebs cycle there is no direct involvement of molecular oxygen. Oxygen is involved only after the hydrogen atoms (or electrons) have been transferred through an additional series of oxidation–reduction reactions that involve a series of closely linked enzymes, all of which contain iron atoms. These enzymes are often collectively called the respiratory enzymes.

41

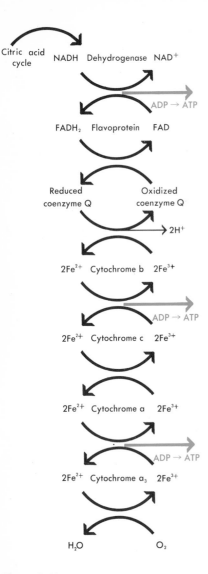

Figure 2–11

The respiratory chain. Here are shown schematically the successive oxidation–reduction reactions which release, in small packets, the energy present in NADH molecules. Whether all the cytochromes operate in a single cycle is not known. The exact sites of ATP formation have not been unambiguously determined.

Their existence was hinted at late in the nineteenth century, but it was not until the period of 1925 to 1940 that their significance was appreciated, largely as a result of the work of Warburg and the Polish-born David Keilin, who spent most of his scientific life in England. Even today there remains uncertainty about the exact number of enzymes involved. Nevertheless, the correctness of the general picture is not disputed. Figure 2–11 shows the general features of the respiratory chain.

The chain operates by a series of coupled oxidation–reduction reactions, during each of which energy is released. Thus the energy present in the reduced coenzymes is released not all at once but in a series of small packets. If NADH were instead directly oxidized by molecular oxygen, a great amount of energy would be released, which it would be impossible to couple efficiently with the formation of the high-energy bonds of ATP.

SYNTHESIS OF ATP IN THE PRESENCE OF OXYGEN (OXIDATIVE PHOSPHORYLATION)

During the period 1925 to 1940, most biochemists concentrated on following the path of hydrogen atoms (or electrons) through the linked, energy-yielding oxidation–reduction reactions. Until the end of this period, only slight attention was given to how the energy was released in a useful form. Then the Dane Kalckar and the Russian Belitzer observed ATP formation coupled with oxidation–reduction reactions in cell-free systems (1938–1940).

Further understanding did not come quickly, since most of the enzymes involved could not be obtained in pure soluble form. These troubles were not resolved until it was realized that the normal sites of *oxidative phosphorylation* in plant and animal cells are large, highly organized subcellular particles, the *mitochondria*. Using intact mitochondria, it is easy to observe the oxidative generation of ATP; this was first demonstrated in 1947 by the Americans Lehninger and Kennedy. Now there is evidence for the generation of three ATP molecules for each passage of a pair of hydrogens through the respiratory chain. There are believed to exist, however, at least six separate oxidation–reduction steps in the chain. Future work may reveal that one ATP molecule is generated during each distinct oxidation–reduction step.

Roughly 20 times more energy is released by the respiratory chain than by the initial breakdown of glucose to pyruvate. This explains why growth of cells under aerobic conditions is so much more efficient than growth without air: If oxygen is not present, pyruvate cannot accumulate, since its formation (see Figure 2–5) demands a supply of unreduced NAD. The amount of NAD within cells, like that of other coenzymes, is, however, very small. In the absence of oxygen, it is rapidly converted to NADH as glucose is oxidized by the Embden–Meyerhof pathway. Thus for the continued generation of ATP without oxygen (fermentation), a device must be used to oxidize the reduced NADH. This is often achieved by reducing pyruvate itself, using NADH as the hydrogen donor. This explains the appearance of lactic acid both during the anaerobic contraction of muscles and during the anaerobic growth of many bacteria (Figure 2–12).

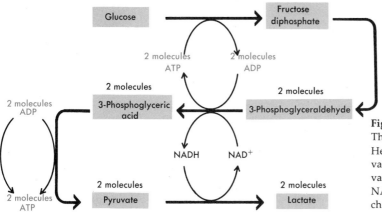

Figure 2–12
The fermentation of glucose to yield lactate. Here the NADH produced during pyruvate formation is oxidized to reduce pyruvate to lactate. When oxygen is present, the NADH is oxidized through the respiratory chain, and no lactate is produced.

GENERATION OF ATP DURING PHOTOSYNTHESIS

Today the ultimate source of the various food molecules used by microorganisms and animals is the photosynthetic plants. Thus the energy of the sun's light quanta must somehow be converted into the energy present in covalent bonds. How this happens chemically did not become clear until the basic energy relations within animals and bacteria were understood. Largely as a result of the work of the American biochemist Arnon, it was discovered in 1959 that the primary action of the sun's light quanta is to phosphorylate ADP to ATP.

This phosphorylation takes place in the chloroplasts, which are complicated, chlorophyll-containing cellular particles found in most cells capable of usefully trapping the energy of the sun. Thus we realize that the controlled release of energy in plants depends upon the same energy carrier important in bacterial and animal cells, ATP. The

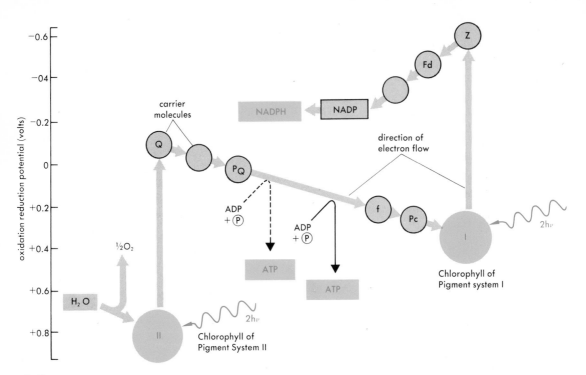

Figure 2–13

A simplified scheme to show the passage of a pair of electrons from water to NADP during photosynthesis, coupled with the formation of one or possibly two molecules of ATP. The NADPH and ATP formed are then used to incorporate CO_2 into carbohydrates. An electron, excited by the absorption of a photon by the pigment system II, is then passed via a series of carrier molecules, by oxidation–reduction reactions, to the chlorophyll of pigment system I. This replaces an electron, which is emitted following the absorption of another photon, and which is passed by further carriers to reduce NADP. ATP is generated somewhere between the two carriers PQ and f. The electron emitted from system II is replaced by one taken from water, coupled with the release of oxygen. (PQ = plastoquinone, f = cytochrome f, PC = plastocyanin, Fd = ferrodoxin.)

details of the process remain unclear. An initial step must be the capture of the light quantum by the green pigment molecule chlorophyll, to excite one of its electrons to a high-energy state.

Then, in a way we still do not yet fully understand, this energy is released gradually by a series of coupled oxidation–reduction reactions mediated by specific cytochrome molecules (Figure 2–13). Oxidative phosphorylation and photophosphorylation thus turn out to be surprisingly similar processes essentially differing only in the way energy is initially injected.

It is most important to realize the uniqueness of photosynthesis. It is the only significant cellular event that utilizes any energy source other than the covalent bond. All other important cellular reactions are accompanied by a decrease in the energy included in covalent bonds. Superficially, we might thus guess that the ability to photosynthesize must have been a primary feature of the first forms of life. It is, however, difficult to imagine that very early forms possessed the complicated chloroplast structures necessary for photosynthesis. Instead there is good reason to believe that, early in the history of the earth, a unique chemical environment allowed the creation of a large number of carbon-containing compounds using the energy of light quanta originating from the sun. These spontaneously formed organic molecules then served as the energy (food) supply for the first forms of life. As these original organic molecules were depleted and living matter increased, a strong selective advantage developed for those cells which evolved a photosynthetic structure to provide a means of increasing the amount of organic molecules. Today essentially all glucose molecules are formed using chemical energy originating in photosynthesis.

THE CHEMIOSMOTIC GENERATION OF ATP FROM ADP AND PHOSPHATE

At first it seemed obvious that the energy released by the cytochrome-mediated oxidation–reduction reactions would momentarily be stored in high-energy compounds capable of phosphorylating ADP to ATP. But some twenty years of search have failed to yield convincing evidence for even one such intermediate. Equally troublesome has been the observation that only intact mitochondria or chloroplasts make ATP. Unbroken surrounding membranes are required for the coupling of the oxidation–reduction reactions and ATP generation. Resolution of these unexpected complications came in the early 1960's from the work of the English biochemist Peter Mitchell. He proposed that the location of the cytochromes within membranes allows the generation of gradients in which more H^+ ions are present outside mitochondria (chloroplasts) than within. Such gradients form because the individual cytochromes' components are so placed that three H^+ pairs are displaced outwardly as each pair of electrons goes down the respiratory chain from NAD to O_2. Existence of these H^+ ion gradients is then used to provide the energy for the union of ADP and phosphate to form ATP (Figure 2–14). This explanation, known as *chemiosmotic theory*, was originally greeted with much skepticism by a biochemical world molded by a succession of new high-energy intermediates. Now, however, the Mitchell idea is recognized as one of the major conceptual advances in the development of modern biochemistry.

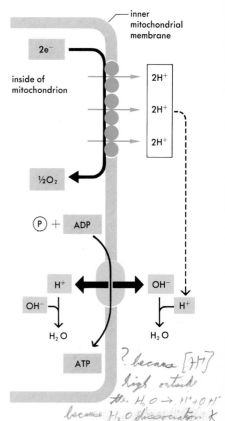

Figure 2–14 Simplified scheme showing the chemiosmotic theory of the generation of ATP from ADP and P. Two electrons pass down a chain of membrane-bound cytochromes and generate a hydrogen-ion gradient across the membrane. This gradient then drives the synthesis of ATP, coupled with the formation of water.

VITAMINS AND GROWTH FACTORS

Although some microorganisms, such as the bacteria *Escherichia coli,* can use glucose as their sole carbon and energy source, not all bacteria and none of the higher animals can use glucose to synthesize all the necessary metabolites. For example, rats are unable to synthesize 11 of the 20 amino acids present in their proteins; thus their food supply must contain substantial amounts of these molecules.

In addition to dietary requirements (necessary growth factors) for compounds with important structural roles, requirements often exist for very small amounts of certain specific organic molecules. These molecules, needed in just trace amounts, are called vitamins (vital molecules). For many years they seemed quite mysterious. Now we realize that the vitamins are closely related to the coenzymes. Some are precursors of coenzymes and some are coenzymes themselves. For example, the vitamin niacin is used in the synthesis of NAD. The fact that coenzymes, like enzymes, are able to function over and over explains why they are needed only in trace amounts.

Thus there is nothing unusual about the fact that a molecule is sometimes required as a growth factor or a vitamin; such requirements are fully explicable in chemical terms. It seems most likely that the genes necessary for the synthesis of certain molecules were lost during evolution, bringing specific growth-factor requirements into existence. There would be no selective advantage to an organism's retaining a specific gene if the corresponding metabolite were always available in its food supply.

THE LABILITY OF LARGE MOLECULES

In striking contrast with the splendid success of biochemists in understanding the behavior of small molecules like the amino acids and nucleotides, scientists interested in the large molecules had arrived at only partial answers before 1950. Their only real success involved a number of polysaccharide molecules (e.g., glycogen). These were relatively easy to understand, because they are built up by the regular polymerization of a smaller subunit (e.g., glucose). Their understanding, however, had no real biological impact because molecules like glycogen are, in most respects, structurally uninteresting. Their sole purpose is to serve as a reserve form of energy-yielding glucose residues. The molecules that people most wanted to unravel were the proteins, because many of them are enzymes, and the nucleic acids, because they were thought to be involved in the hereditary mechanism. Both these classes of molecules were, however, initially refractory to investigation.

One major reason why they were difficult to study was that they appeared to be much less stable (more labile) than most small molecules. Extremes of temperature and pH (acidity or alkalinity) cause them to lose their natural shapes (*denaturation*) and sometimes to precipitate irreversibly out of solution in an inactive form. Thus great care had to be taken in isolating them; sometimes it was necessary to perform the entire isolation process at temperatures near 0° C. At first it was thought that only proteins were subject to denaturation, but now it is clear that nucleic acid molecules also denature during isolation if proper precautions are not taken.

Until 1945 (after the Second World War), the commonly used techniques of organic chemistry were the main tools for studying most of the small molecules. Thus the success of a biochemist working on intermediary metabolism often depended on his ability as an organic chemist. In work with proteins and nucleic acids, however, most of the initial stages of research did not rely on the analytical techniques of the organic chemist. Instead the protein chemist, before he could even start worrying about the detailed structure of a protein, needed to work very hard to be sure his protein was both chemically pure and biologically active. He had to devise gentle techniques for isolation, which avoided the usual strong acids and alkalies of organic analysis. Then he needed techniques to reveal whether his product was homogeneous, and hopefully, also to provide data on molecular size. For this sort of answer, the help of physical chemists was indispensable, and there developed a well-recognized new line of research investigating the physical–chemical properties of macromolecules in solution. It concerned itself with topics like the osmotic properties of macromolecular solutions and the movement of macromolecules under electrical and centrifugal forces.

Perhaps the most striking contribution of physical chemistry to the study of biological macromolecules was the development, in the 1920's, of centrifuges that rotated at high speed (ultracentrifuges) and that could cause the rapid sedimentation of proteins and nucleic acids. The development of the ultracentrifuge was the work of the Swede Svedberg, after whom the unit of sedimentation (S = Svedberg) was named. Ultracentrifuges equipped with optical devices to observe exactly how fast the molecules sedimented were extremely valuable in obtaining data on the molecular weight of proteins and establishing the concept that proteins, like smaller biological molecules, are of discrete molecular weights and shapes. This work revealed that the sizes of proteins vary greatly, with a continuous range in weights between the extremes of approximately 10,000 and 1,000,000.

IMPLICATIONS OF CHROMATOGRAPHY

Early in the twentieth century, the work of the great German chemist Emil Fischer had established that protein molecules are largely composed of amino acids linked together by peptide bonds. The determination of the exact way in which the amino acids are linked together to form proteins remained, however, a great puzzle until 1951. This was partly because there are 20 different amino acids and their proportions vary from one type of protein to another. Until about 1942, the methodological problems involved in amino acid separation and identification were formidable, and most organic chemists chose to work with simpler molecules.

This state of affairs changed completely in 1942, when the Englishmen Martin and Synge developed separation methods that depended on the relative solubilities of the several amino acids in two different solvents (partition chromatography). Particularly useful were separation methods by which the amino acids were separated on strips of paper. With these new tricks, it became a routine matter to separate quantitatively the 20 amino acids found in proteins. These methods were quickly seized upon by the English biochemist Sanger, who used them to establish all the covalent linkages in the protein hormone insulin (see Chapter 6 for details of the insulin structure). Sanger's work was a milestone in the study of proteins, for it demonstrated that each type of protein contains a specific arrangement of amino acids.

THE 25-YEAR LONELINESS OF THE PROTEIN CRYSTALLOGRAPHERS

An equally significant step in understanding macromolecules was the effective extension of x-ray crystallographic techniques to their study. This approach utilizes the diffraction of x-rays by crystals to give precise data about the three-dimensional arrangement of the atoms in molecules. The first successful use of x-ray diffraction was in 1912, when the Englishman Bragg solved the NaCl structure. This success immediately initiated research on the structures of molecules of increasing complexity.

The technique used in the initial x-ray diffraction studies of small molecules consisted of guessing the structure, calculating the theoretical diffraction pattern predicted by this structure, and comparing the calculated with the observed pattern. This method was practical for studying relatively simple structures, but was not usually useful in the study of larger structures. It took much insight on the part of Bragg and the great American chemist Pauling, in the 1920's, to solve the structures of some complicated inorganic silicate molecules. Clearly, however, proteins were too complicated for even the best chemist to guess their 3-D structures. Thus the early protein crystallographers knew that, until new methods for structural determination were found, they would have no results to present to the impatient biochemists, who were increasingly anxious to know what proteins actually looked like.

The first serious x-ray diffraction studies on proteins began in the mid-1930's in Bernal's laboratory in Cambridge, England. Here it was found that although dry protein crystals gave very poor x-ray patterns, wet crystals often gave beautiful pictures. Unfortunately, however, there was no logical method available for their interpretation. Nonetheless, Bernal's student Perutz, an Austrian then in England, slowly increased the pace of his work (begun in 1937) with the oxygen-carrying blood protein, hemoglobin. He had chosen hemoglobin for several reasons: Not only is it one of the most important of all animal proteins, but it is also easy to obtain, and forms crystals that lend themselves well to crystallographic analysis. For many years, however, no very significant results emerged either from Perutz's work on hemoglobin structure or from that begun in 1947 by Kendrew, on the structure of the muscle protein myoglobin. This protein, which, like hemoglobin, combines with oxygen, had the added advantage of being four times smaller (MW = 17,000) than hemoglobin.

During the lonely period of no real results there was only one triumph. Pauling correctly guessed from stereochemical considerations that amino acids linked together by peptide bonds would sometimes tend to assume helical configurations, and proposed in 1951 that a helical configuration, which he called the alpha helix (see Chapter 4 for details), would be an important element in protein structure. Support for Pauling's α-helix theory came soon after its announcement when Perutz demonstrated that several synthetic polypeptide chains containing only one type of amino acid exist as α-helices.

It was not until 1959 that Perutz and Kendrew got their answers. An essential breakthrough, which occurred in 1953, showed how the attachment of heavy atoms to protein molecules could logically lead from the diffraction data to the correct structures. For the next several years, these heavy-atom methods were exploited at a pace undreamed of 20 years before, largely thanks to the availability of high-speed electronic computers. Then, to every-

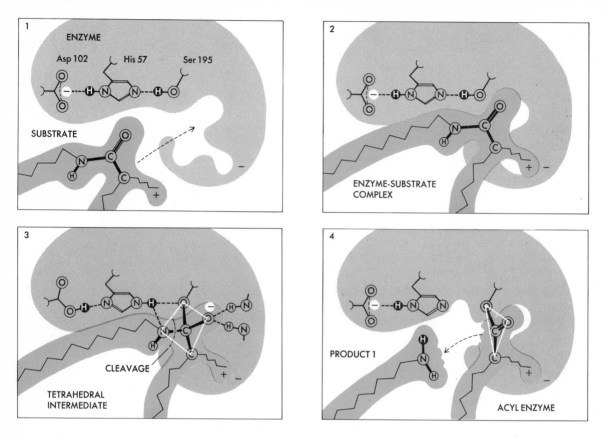

Figure 2–15

Involvement of specific amino acid residues in the catalytic breakdown of peptide bond by trypsin. The catalytic mechanism of trypsin accelerates the hydrolysis of peptide bonds by providing intermediate states for the reaction and by smoothing the transition from one intermediate to the next. Enzyme and substrate must first come together (1) to form a precisely oriented complex (2). The oxygen atom of serine 195 then bonds covalently to the substrate carbon, forming a tetrahedral intermediate (3); the proton, or hydrogen ion, from serine 195 is transferred to the substrate hydrogen. The proton transfer breaks the peptide bond, and the first product is liberated (4). The remaining complex is called

one's delight, the x-ray diffraction measurements could at last be translated into the arrangement of atoms in myoglobin and in hemoglobin. Both molecules were found to be enormously complicated, with their amino acid chains folded as α-helices in some regions, and very irregularly in others. Furthermore, their molecular configurations were found to obey in every respect the chemical laws that govern the shape of smaller molecules. Absolutely no new laws of nature are involved in the construction of proteins; this was no surprise to the biochemists.

VISUALIZATION OF THE "ACTIVE SITES" OF ENZYMES

Though for decades organic chemists had made speculative guesses on how specific amino acid side groups might speed up certain chemical reactions, convincing explanations had to await the existence of detailed 3-D structural analysis. This was first accomplished in London in 1967 when D. C. Phillips and his colleagues of the Royal Institution worked out the structure of the enzyme lysozyme. Possession of its 3-D conformation enabled them to propose a precise chemical mechanism for the way lysozyme

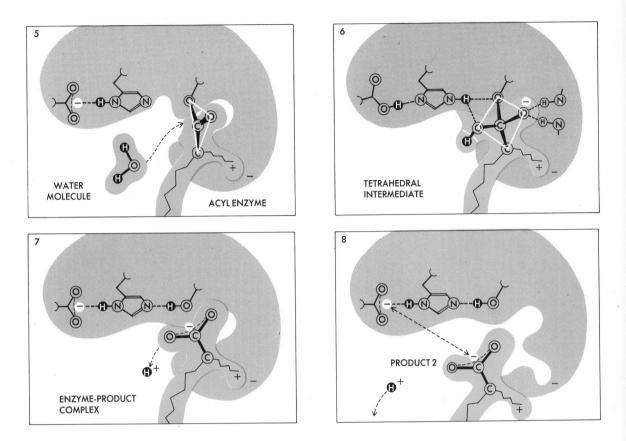

breaks down specific polysaccharide chains into their component sugar groups. This work, together with that of other scientists who have since established the conformation of over fifteen additional enzymes, indicates that the chemical mechanisms of enzyme-catalyzed reactions can be as well understood as the best-known chemical reactions of the pure organic chemist.

Particularly pleasing has been the unambiguous establishment of the way a number of proteolytic enzymes catalyze specific breaks in polypeptide chains. The recent working out of the 3-D structures of trypsin and chymotrypsin reveals very similar molecular shapes, with their respective active sites sharing many common amino acids. Three of these amino acids, serine, histidine, and aspartic acid, are directly involved in the catalytic process, with each residue playing identical roles in both enzymes (Figure 2–15). We now see that amino acid sequence data by itself seldom can reveal profound information on how proteins function. An increasing number of pure-protein chemists are thus becoming expert crystallographers rather than face loss of their profession.

an acyl enzyme; it breaks down to regenerate free enzyme in steps that are symmetrical with those of the first half of the process. A water molecule enters the reaction (5), its hydroxyl group forming with the substrate another tetrahedral intermediate (6). The hydrogen ion from the water molecule is transferred to serine 195, breaking the covalent bond between enzyme and substrate (7). The second product is now freed (8); its departure is hastened by repulsion between negatively charged carboxyl groups of product and aspartic acid 102. Histidine 57 and aspartic acid 102 participate in the proton transfers. (Redrawn from *A Family of Protein-cutting Proteins* by R. Stroud. Copyright © 1974 by *Scientific American*. All rights reserved.

AVERY'S BOMBSHELL: NUCLEIC ACIDS CAN CARRY GENETIC SPECIFICITY

Until 1944, the number of chemists working on the nucleic acids was but a tiny fraction of the number attempting to understand proteins. Two nucleic acids, DNA (*deoxyribonucleic acid*) and RNA (*ribonucleic acid*), were known to exist, but the general features of their chemical structures had not been elucidated. Although DNA was found only in nuclei (hence the name nucleic acids), there was general agreement that it probably was not a genetic substance, since chemists thought that its four types of nucleotide (see Chapter 3 for details of their structures) were present in equal amounts, giving DNA a repetitive structure like that of glycogen (the tetranucleotide hypothesis).

In the middle 1930's, the Swedish chemists Hammarsten and Caspersson found by physical–chemical techniques that DNA molecules, prepared by gentle procedures, have molecular weights even larger (>500,000) than most proteins. At the same time, chemical analysis of purified plant viruses by the American Stanley and the Englishmen Bawden and Pirie suggested the generalization that all viruses contain nucleic acid, hinting that nucleic acids might have a genetic role.

The first real proof, however, of the genetic role for nucleic acids came from the work of the noted American microbiologist Avery and his colleagues MacLeod and McCarty at the Rockefeller Institute in New York. They made the momentous discovery in 1944 that the hereditary properties of pneumonia bacteria can be specifically altered by the addition of carefully prepared DNA of high molecular weight.

Even though there was momentary hesitation in accepting its implications, their discovery provided great stimulation for a detailed chemical investigation of nucleic acids. Here also paper chromatography became immensely useful, and quickly allowed the biochemist Chargaff, then working in New York, to analyze the nucleotide composition of DNA molecules from a number of different organisms. In 1947, his experiments showed not only that the four nucleotides are not present in equal amounts, but also that exact ratios of the four nucleotides varied from one species to another. This finding meant that much more variation was possible among DNA molecules than the tetranucleotide hypothesis had allowed, and immediately opened up the possibility that the precise arrangement of nucleotides within a molecule is related to its genetic specificity.

It also became obvious from Chargaff's work in the next several years that the relative ratios of the four bases were not random. The amount of adenine in a DNA sample was always found to be equal to the amount of thymine, and the amount of guanine equal to the amount of cytosine. The fundamental significance of these relationships did not become clear, however, until serious attention was given to the 3-D structure of DNA.

THE DOUBLE HELIX

Parallel with work on the x-ray analysis of protein structure, a still smaller number of scientists concentrated on trying to solve the x-ray diffraction pattern of DNA. The first diffraction patterns were taken in 1938 by the Englishman Astbury and used DNA supplied by Hammarsten and Caspersson. It was not until after the war (1950–1952), however, that high-quality photographs were taken, by Wilkins and Franklin, working in London at King's College. Even then, however, the chemical bonds linking the various nucleotides were not unambiguously established. This was accomplished in 1952 by a group of organic chemists working in the Cambridge, England, laboratory of Alexander Todd.

Because of interest in Pauling's α-helix, in 1951 an elegant theory of the diffraction of helical molecules was developed. The existence of this theory made it easy to test possible DNA structures on a trial-and-error basis. The correct solution, a complementary double helix (see Chapter 9 for details), was found in 1953 by Crick and Watson, then working in England in the laboratory of Perutz and Kendrew. Their arrival at the correct answer was in large part dependent on finding the stereochemically most favorable configuration compatible with the x-ray diffraction data of the King's College group.

The establishment of the double helix immediately initiated a profound revolution in the way in which many geneticists analyzed their data. The gene was no longer a mysterious entity whose behavior could be investigated only by breeding experiments. Instead it quickly became a real molecular object about which chemists could think objectively in the same manner as smaller molecules, such as pyruvate or NAD. Most of the excitement, however, came not merely from the fact that the structure was solved, but also from the nature of the structure. Before the answer was known, there had always been the mild fear that it would turn out to be dull, and reveal nothing about how genes replicate and function. Fortunately, however, the answer was immensely exciting. The structure appeared to be two intertwined strands of complementary structures, suggesting that one strand serves as the specific surface (*template*) upon which the other strand is made. If this hypothesis were true (which it is now known to be!), then the fundamental problem of gene replication, about which the geneticists had puzzled for so many years, was, in fact, solved.

There were thus initiated over the past twenty-two years a variety of experiments designed to study, at a molecular level, how DNA molecules control what a cell is like. These studies have brought many discoveries, unforeseen in 1953, about how the genetic material functions. Because these answers are, for the first time, consistently at the molecular level, it is convenient to refer to the subject matter at this level as molecular genetics.

THE GOAL OF MOLECULAR BIOLOGY

Until recently, heredity has always seemed the most mysterious of life's characteristics. The current realization that the structure of DNA already allows us to understand practically all its fundamental features at the molecular level is thus most significant. We see not only that the laws of chemistry are sufficient for understanding protein structure, but also that they are consistent with all known hereditary phenomena. Complete certainty now exists among essentially all biochemists that the other characteristics of living organisms (for example, selective permeability across cell membranes, muscle contraction, nerve conduction, and the hearing and memory processes) will all be completely understood in terms of the coordinative interactions of small and large molecules. So we have complete confidence that further research, of the intensity recently given to genetics, will eventually provide man with the ability to describe with completeness the essential features that constitute life.

SUMMARY

The growth and division of cells are based upon the same laws of chemistry that control the behavior of molecules outside of cells. Cells contain no atoms unique to the living state; they can synthesize no molecules which the chemist, with inspired, hard work, cannot some day make. Thus there is no special chemistry of living cells. A biochemist is not someone who studies unique types of chemical laws, but a chemist interested in learning about the behavior of molecules found within cells (biological molecules).

The growth and division of cells depend upon the availability of a usable form of chemical energy. The energy now initially comes from the energy of the sun's light quanta, which is converted by photosynthetic plants into cellular molecules, some of which are then used as food sources by various microorganisms and animals. The most striking initial triumphs of the biochemists told us how food molecules are transformed into other cellular molecules and into useful forms of chemical energy. The energy within food molecules largely resides in the covalent bonds of reduced carbon compounds; it is released when these molecules are transformed by oxidation–reduction reactions to carbon compounds of a higher degree of oxidation. For most forms of life, the ultimate oxidizing agent is molecular O_2. The products of the complete oxidation of organic molecules like glucose are CO_2 and H_2O.

Most organic molecules, however, are not oxidized directly by oxygen. They are oxidized instead by diverse organic molecules, often coenzymes, such as the coenzyme NAD. The reduced coenzyme (for example, NADH) is itself oxidized by another molecule (such as FAD) to yield a new reduced coenzyme (FADH$_2$) and the original coenzyme, in the oxidized form (NAD). After several such cycles, molecular oxygen directly participates, to end the oxidation–reduction chain, giving off water (H_2O).

The energy released during the oxidation–reduction cycles is not released entirely as heat. Instead, more than half the energy is converted into new chemical bonds. Phosphorus atoms play a key role in this transformation. Phosphate esters are formed that have a higher usable energy content than most covalent bonds. These phosphate groups are transferred in a high-energy form to acceptor molecules. The most important acceptor of such groups is ADP; a phosphate group is added to ADP to yield ATP. Very recently, experiments revealed that the phosphorylation of ADP to ATP is a primary step in photosynthesis, where it is called photophosphorylation. *The ADP$\rightarrow$ ATP transformation is at the heart of energy relations in all cells.*

Until a few years ago, the chemists' understanding of the cell's very large molecules, the proteins and nucleic acids, was much less firm than it is now. Most of these molecules are "small molecules" studied by organic chemistry (molecules of protein and nucleic acid run from MW 10^4 to 10^9). Both proteins and nucleic acids are complex, and only recently have *physical and chemical techniques been developed to allow a concerted attack on their structure. Among the most important techniques have been partition chromatography, analytical ultracentrifugation, and x-ray crystallography as extended to the study of large molecules. Now we know practically all the important features of many large proteins like myoglobin, hemoglobin, and trypsin, as well as that of the primary genetic material DNA. In all cases the chemical laws applicable to small molecules also apply. So far the greatest impact on biological thought has come from the realization that DNA has a complementary double-helical structure. This structure immediately suggested a mechanism for the replication of the gene, and initiated a revolution in the way biologists think of heredity. These successes have created a firm belief that the current extension of our understanding of biological phenomena to the molecular level (molecular biology) will soon enable us to understand all the basic features of the living state.*

REFERENCES

LEHNINGER, A. L., *Bioenergetics,* 2nd ed., Benjamin, Menlo Park, California, 1971. A nice introduction to the chemical reactions by which cells trap, store, and utilize chemical energy. Emphasis is placed on generation of ATP in mitochondria and chloroplasts.

CONN, E. E., AND P. K. STUMPF, *Outlines of Biochemistry,* 3rd ed., Wiley, New York, 1971. An introduction to metabolism now in its third edition.

LOEWY, A., AND P. SIEKEVITZ, *Cell Structure and Function,* 2nd ed., Holt, New York, 1969. Essentials of the physiology of the cell, often with emphasis on the problems of multicellular organization.

EDELSTEIN, S. J., *Introductory Biochemistry,* Holden-Day, San Francisco, 1973. An introductory text that equally emphasizes the metabolism of small and large molecules.

BALDWIN, E., *Dynamic Aspects of Biochemistry,* 3rd ed., Cambridge, New York, 1959. This is one of the few texts in biochemistry deserving to be called a classic. Now it is best read for the way in which coenzymes participate in hydrogen transfers.

KALCKAR, H. M., *Biological Phosphorylations,* Prentice-Hall, Englewood Cliffs, N. J., 1969. Contains a collection of many of the classic papers which led to the understanding of the manyfold importance of high-energy bonds.

LIPMANN, F., *Wanderings of a Biochemist,* John Wiley, 1971. A look back at the author's key role in the discovery of high-energy bonds.

FRUTON, J. S., *Molecules and Life,* Wiley-Interscience, 1972. A detailed guide to the history of biochemistry.

WATSON, J. D., *The Double Helix,* Atheneum, New York, 1968. A personal account of how the double helix was discovered.

OLBY, R., *The Path to the Double Helix,* University of Washington Press, 1975. A superior retelling of DNA research by a professional historian of science.

A Chemist's Look at the Bacterial Cell

The most important aspect of living cells is their tendency to grow and divide. In this process, food molecules are absorbed from the external environment and transformed into cellular constituents. The rates of cell growth vary tremendously, but in general the smallest cells grow the fastest. Under optimal conditions some bacteria double their number every 20 minutes, whereas most larger mammalian cells can divide only once about every 24 hours. But, independent of the length of the time interval, growth and division necessarily demand that the number of cellular molecules double with each cell generation. One way, therefore, of asking the question, "What is life?" is to ask how a cell doubles its molecular content, that is, how biological molecules are replicated as a cell grows.

CHAPTER **3**

BACTERIA GROW UNDER SIMPLE, WELL-DEFINED CONDITIONS

Until recently most serious questions about cell growth and division were studied by using microorganisms, especially the bacteria. The tendency to concentrate on microorganisms did not arise from a belief that bacteria are fundamentally more important than higher organisms. The converse is obviously true to human beings, naturally curious to know about themselves and anxious to use information about their own chemical makeup to combat the various diseases threatening their existence. Nonetheless, upon even a superficial examination, the difficulties of thoroughly mastering the chemical events in a higher organism are staggering. There are about 5×10^{12} cells in a human being, each of whose existence is intimately related to the behavior of many other cells. It is therefore difficult to study the growth of a single cell within a multicellular organism without taking into consideration the influence of its surrounding cells.

Much effort has been devoted to learning how to grow the cells of multicellular organisms in an isolated system. In this work (often called *cell culture*) small groups of cells, or sometimes single cells, are removed from a plant or animal and placed under controlled laboratory conditions in a solution containing a variety of food molecules. In the first experiments, the isolated cells almost invariably died, but now, partly because of a better understanding of the nutritional requirements of cells, they often grow and divide to form large numbers of new cells. Such freely growing plant and animal cells have been of much value in showing how cells aggregate to form organized groups of similar cells (tissues), and even more striking, how tissues unite in the test tube to form bodies morphologically identical to small regions of organs, such as the liver or the kidney. On the other hand, higher cells growing in culture are not ordinarily easy objects to use in studying cell growth and division. Even though many cells of higher organisms will grow in isolation, it must be remembered that this is not their normal way of existence, and, unless precautions are taken, they tend to aggregate quickly into multicellular groups. Thus even today the isolated growth of cells from higher organisms can be difficult and time-consuming.

In contrast, the cells of many microorganisms normally grow free, as single cells, separating from each other as soon as cell division occurs. It is thus fairly easy to grow such single-celled organisms under well defined laboratory conditions, since the conditions of growth in the scientist's test tube are not radically different from the conditions under which they normally grow outside the laboratory. In contrast to mammalian cells, which require a large variety of growth supplements, many bacteria will grow on a simple, well defined diet or medium. For example, the bacteria *Escherichia coli* will grow on an aqueous solution containing just glucose and several inorganic ions (Table 3–1).

The growth of a specific bacterium is usually not dependent on the availability of a specific carbon source.

Table 3–1 A Simple Synthetic Growth Medium for *E. coli**

NH_4Cl	1.0 g
$MgSO_4$	0.13 g
KH_2PO_4	3.0 g
Na_2HPO_4	6.0 g
Glucose	4.0 g
Water	1000 ml

* Traces of other ions (e.g., Fe^{2+}) are also required for growth. Usually these are not added separately, since they are normally present as contaminants in either the added inorganic salts or the water itself.

Most bacteria are highly adaptable as to which organic molecules they can use as their carbon and energy sources. Glucose can be replaced by a variety of other organic molecules, and the greater the variety of food molecules supplied, the faster a cell generally grows. For example, if *E. coli* grows upon only glucose, about 60 minutes are required at 37°C to double the cell mass. But if glucose is supplemented by the various amino acids and purine and pyrimidine bases (the precursors of nucleic acids), then only 20 minutes are necessary for the doubling of cell mass. This effect is due to the direct incorporation of these components into proteins and nucleic acids, sparing the cell the task of carrying out the synthesis of the building blocks. There is a lower limit, however, to the time necessary to double the cell mass (often called the *generation time*): No matter how favorable the growth conditions, bacteria are unable to divide more than once every twenty minutes.

E. COLI IS THE BEST UNDERSTOOD ORGANISM AT THE MOLECULAR LEVEL!

Over the past thirty years, work with the bacterium *E. coli* and evolutionarily related organisms has dominated much of biochemistry. Because of its small size, normal lack of pathogenicity to any common organism, and ease of growth under laboratory conditions, *E. coli* is now the most intensively studied organism except for man. Many other bacteria besides *E. coli* possess the same favorable attributes, and the original reasons for choosing *E. coli* are essentially accidental. Once serious work had started on *E. coli,* however, it obviously made no sense to switch to another organism if *E. coli* could be used. Even now the tendency to concentrate on *E. coli* is increasing because, parallel with the chemical studies, extensive successful genetic analysis has also been carried out. Our knowledge of the genetics of *E. coli* is thus much more complete than our knowledge of that of any other bacterium or lower plant. As we shall see, the combined methods of genetics and biochemistry are so powerful that it is often just not sensible to use in biochemical studies an organism with which genetic analysis is not possible.

The average *E. coli* cell (Figure 3–1) is rod-shaped and about 2μ in length and 1μ in diameter. It grows by increasing in length, followed by a fission process that generates two cells of equal length. Growth occurs best at temperatures about 37°C, perhaps to suit it for existence in the intestines of higher mammals, where it is frequently

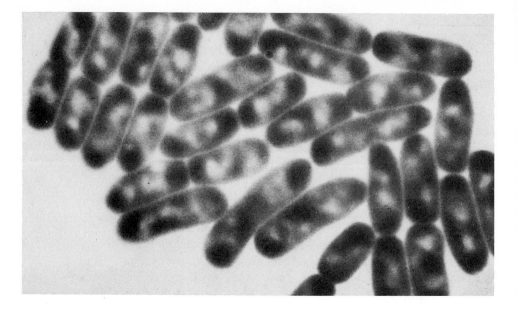

Figure 3–1

Electron micrograph of a group of intact *E. coli* cells. The light regions inside the bacteria represent areas where DNA is concentrated. Magnification is 12,000. This photograph was taken in the laboratory of E. Kellenberger, University of Geneva (reproduced with permission).

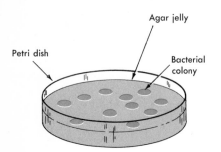

Figure 3–2

The multiplication of single bacterial cells to form colonies. *E. coli* cells are usually not motile. Thus when a cell has divided on a solid surface, the two daughter cells and all their descendants will tend to remain next to each other. After 24 hours at 37°C, each initial living cell has given rise to a solid mass of cells.

found as a harmless parasite. It will, however, regularly grow and divide at temperatures as low as 20°. Cell growth proceeds much more slowly at these low temperatures; the generation time under otherwise optimal conditions is about 120 minutes at 20°C.

Cell number and size are often measured by observation under the light microscope (and occasionally the electron microscope). Such observation, however, cannot reveal whether a visible cell is alive or dead. This can be determined only by seeing whether a given cell forms daughter cells. This determination is usually made by spreading a small number of cells on top of a solid agar surface (Figure 3–2), which has been supplemented with the nutrients necessary for cell growth. If a cell is alive, it will grow to form two daughter cells which in turn give rise to subsequent generations of daughter cells. The net result after 12 to 24 hours of incubation at 37°C is discrete masses (*colonies*) of bacterial cells. Provided that the colonies do not overlap, each must have arisen from the initial presence of a single bacterial cell.

The growth of bacteria may also be followed in liquid nutrient solutions. If a nutrient medium is inoculated with

Figure 3–3
Growth curve of *E. coli* cells at 37°C. The gray line shows the increase in ▶
cell number following the inoculation of a sterile, nutrient-rich solution
(glucose, salts, amino acids, purines, pyrimidines) with 10^5 cells from an
E. coli culture in an exponential phase of growth. If this growth curve
had been started, instead, from cells in a slow-multiplying, nearly satu-
rated culture, the growth would not have begun immediately, but rather
(colored curve) a lag period of approximately 1 hour would have preceded exponential growth.

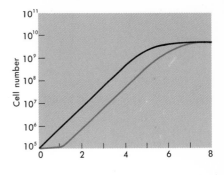

a small number of rapidly dividing bacteria from a similar
medium, the bacteria will continue dividing with a constant division time, doubling the number of bacteria each
generation time. Thus the number of bacteria increases in
an exponential (logarithmic) fashion (Figure 3–3). *Exponential growth* continues until the number of cells reaches such
a high level that the initial optimal nutritional conditions
no longer exist. One of the first factors that usually limits
growth is the supply of oxygen. When the number of cells
is low, the oxygen available by diffusion from the liquid
interface is sufficient, but as the number of cells rises additional oxygen is needed. It is often supplied either by
bubbling oxygen through the solution or by shaking the
solution rapidly. Even with violent aeration, the growth
rates begin to slow down after the cell density reaches
about 10^9 cells per milliliter, and a tendency develops for
the cells being produced to be shorter. Finally, at cell densities of about 5×10^9 cells per milliliter, cell growth is dis-
continued, for still unclear nutritional reasons. The term
growth curve is frequently used to describe the increase of
cell numbers as a function of time.

In most growing bacterial cultures, the exact division
time of the cells varies, so that even if a culture has started
from a single cell, after a few generations cells can be
found at various stages of the division cycle at any given
moment. Such growth is frequently called *unsynchronized
growth*. Over the past 10 years, tricks have been developed
to isolate bacterial cells at the same stage of the cell cycle.
These can be used to obtain several generations of *synchro-
nized cell growth* (Figure 3–4). Then, because of slightly
unequal divisions, the resulting growth curve again acquires an unsynchronized appearance.

During exponential growth each cell contains between two and four chromosomes. All these chromosomes

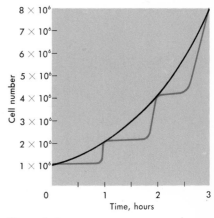

Figure 3–4
The growth curve (colored line) of a syn-
chronized *E. coli* culture growing upon glucose as the sole carbon source. The gray
line shows the increase in cell number of
an unsynchronized culture. Here we show
an example in which the degree of synchronization noticeably lessens in the second
and third cycles of growth.

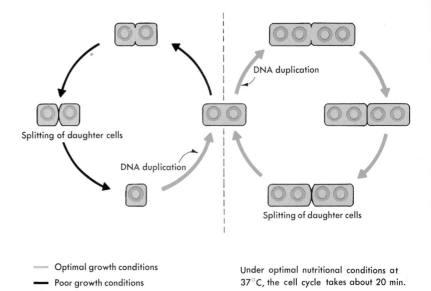

DNA duplication

Splitting of daughter cells

DNA duplication

Splitting of daughter cells

——— Optimal growth conditions

▬▬▬ Poor growth conditions

Under optimal nutritional conditions at 37°C, the cell cycle takes about 20 min.

have identical genetic compositions since they are all descendants of the same parental chromosome. Why each healthy cell contains several identical chromosomes is not known. It is not an intrinsic feature of the *E. coli* life cycle, since when the nutritional conditions are poor, chromosome duplication sometimes lags behind cell division (Figure 3–5), resulting in viable cells with just one chromosome apiece.

In *E. coli*, as in most if not all bacteria, the chromosome is not enclosed within a nuclear membrane. There is no structural distinction between the nucleus and the cytoplasm. Nonetheless, bacterial cytologists often refer to the region occupied by the chromosomes as the *nuclear region*. This term, however, is a misnomer. It is still very unclear what separates the chromosomes during cell division. There is no evidence of a spindle-like region even under the electron microscope, which hints that the spindle is a specialized structure developed at the point in evolution when the nuclear and cytoplasmic regions were differentiated. Thus the only essential parallel between mitotic divisions in higher organisms and in the bacteria is that in both the accurate duplication and partition of the chromosomes is essential.

For many years it was believed that bacteria lacked sexuality. In 1946, however, a cell fusion process was discovered in *E. coli* which produces diploid cells, and soon afterwards separate male and female cells were identified (see Chapter 7 for details). Diploid bacterial cells usually

◀ Figure 3–5

The life cycle of *E. coli* cells. Under optimal growth conditions the average cell contains from two to four chromosomes, depending upon its exact stage in the division cycle. All these chromosomes are descended from the same parental chromosome, and so are genetically identical.

do not exist for long intervals, segregating out haploid cells within one or two cell division cycles. Here, as in the case of bacterial mitosis, no evidence is available about the mechanics of chromosome separation, which underlies the reduction of chromosome number back to the haploid condition.

EVEN SMALL CELLS ARE COMPLEX

Even cells as small as those of *E. coli* present great difficulties when we study them at the molecular level. At first sight, the problem of soon, if ever, understanding the essential features of *E. coli* should seem insuperable to an honest chemist. He realizes immediately that, on a chemical scale, even the smallest cells are fantastically large. Although an *E. coli* cell is about 500 times smaller than an average cell in a higher plant or animal (which has a diameter of approximately 10μ), it nonetheless weighs approximately 2×10^{-12} gram (MW $\sim 10^{12}$ daltons; a dalton has a MW $= 1$). This number, which initially may seem very small, is immense on the chemist's scale, since it is 6×10^{10} times greater than the weight of a water molecule (MW $= 18$). Furthermore, this mass reflects the highly complex arrangement of a large number of different carbon-containing molecules.

There is also seemingly infinite variety in the chemical nature of these molecules. But fortunately it is possible to distribute most molecules in terms of mass into several well defined classes possessing common arrangements of some atoms. These classes are the carbohydrates, lipids, proteins, and nucleic acids (Table 3–2). Many molecules possess chemical groups common to several of these categories, so that the classification of such molecules is necessarily arbitrary. Also in the cell are many smaller molecules, such as amino acids, purine and pyrimidine nucleotides, various coenzymes, very small molecules (e.g., O_2 and CO_2), and numerous electrically charged inorganic ions (e.g., Na^+, K^+ and PO_4^{3-}). Finally, there is H_2O, the most common molecule in all cells, and a solvent for most biological molecules, through which diffusion from one cellular location to another can occur quickly.

Table 3–2   The Main Classes of Biological Molecules

	General description	Functions
Proteins	Molecules containing C, H, O, N, and sometimes S, which are built up from amino acids.	Most proteins are enzymes—some, usually present in very large numbers per cell, are used to build up essential structures, such as the cell wall, the cell membrane, the ribosomes, muscle fiber, nerves, etc.
Lipids	Molecules insoluble in water, that are sometimes built up by the combination of glycerol and three long-chain fatty acids (triglyceride). Sometimes one fatty acid is replaced by choline (lecithins). Sometimes glycerol is replaced by sphingosine. Phosphorus is present in a large number of the lipids (phospholipids).	Triglycerides are a main storehouse of energy-rich food. They degrade to give acetyl-CoA. Phospholipids are an essential component of all membranes. Their insolubility in water is related to their control of permeability.
Carbohydrates	Molecules containing C, H, and O, usually in ratios near $1:2:1$; polysaccharides are built up from simple sugars (monosaccharides) such as glucose and galactose. In some cases the sugars contain amino groups (e.g., glucosamine).	Some, such as cellulose and pectin, are used to construct strong, protective cell walls; others, such as glycogen, provide a form of storing glucose.
Nucleic acids	Long, linear molecules containing P, as well as C, H, O, and N, which are built up from pentose (5-carbon) nucleotides.	There are two main classes of nucleic acids in cells: DNA is the primary genetic component of all cells; RNA usually functions in the synthesis of proteins. In some viruses, RNA is the genetic material.

A glimpse into the cellular organization of *E. coli* is shown in an electron micrograph of a very thin section cut from a rapidly growing cell (Figure 3–6). On the outside is the rigid cell wall, a 100-Å thick mosaic structure built up of protein, polysaccharide, and lipid molecules. Just inside the cell wall is a flexible, 100-Å thick cell membrane, largely composed of protein and lipid. This membrane is semipermeable and controls which molecules enter and leave the cell. Of vital importance is its ability to maintain a concentration gradient, since most molecules, both small and large, are present at much higher levels inside than outside the cell membrane. This is true for both inorganic ions (e.g., K^+ and Mg^{2+}) and most important organic molecules. The membrane must actively prevent molecules

Table 3–2 (*continued*)

Building Blocks

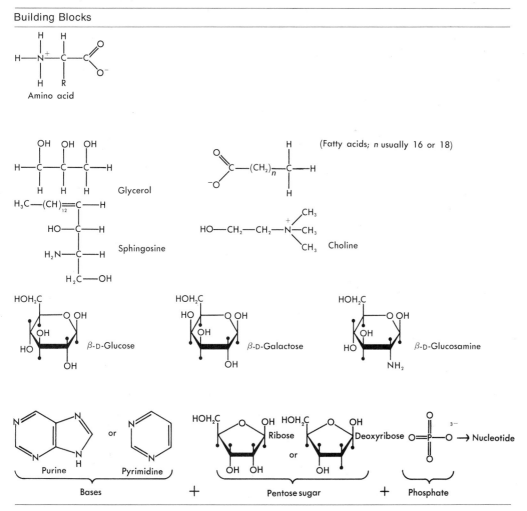

from diffusing into the outside area of very much lower concentrations.

A schematic view of a typical *E. coli* cell is shown in Figure 3–7. About one-fifth of the interior of the cell is occupied by deoxyribonucleic acid (DNA), the compound responsible for the transmission of genetic material from one cell to another. Immediately surrounding the DNA occur 20,000 to 30,000 spherical particles, 200 Å thick. These are the ribosomes, usually present in aggregates called polyribosomes. The cellular sites of protein synthesis, they contain approximately 40 percent protein and 60 percent ribonucleic acid (RNA). The remainder of the cell's interior is filled with water, water-soluble enzymes, and a large number of various small molecules.

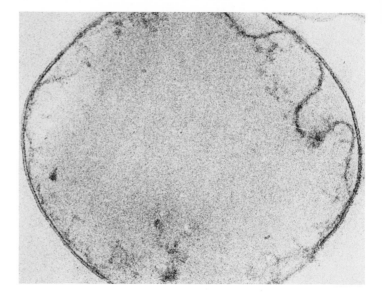

Figure 3–6
Electron micrograph of a very thin cross section of an *E. coli* cell. The magnification is ×105,000. The outer membrane is the protective cell wall; inside this is the cell membrane, which controls permeability. Normally the cell membrane lies tightly next to the cell wall; here, for illustrative purposes, we show an accidental situation in which the membrane has in a few places separated from the cell wall. (Supplied by E. Kellenberger; reproduced with permission.)

Figure 3–7
Schematic view of an *E. coli* cell containing two identical chromosomes.

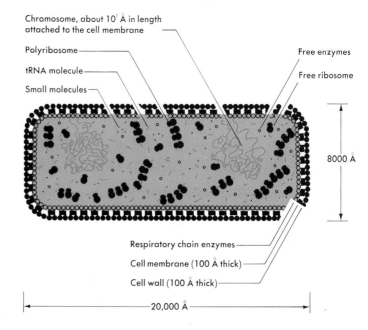

Chromosome, about 10^7 Å in length attached to the cell membrane
Polyribosome
tRNA molecule
Small molecules
Free enzymes
Free ribosome
8000 Å
Respiratory chain enzymes
Cell membrane (100 Å thick)
Cell wall (100 Å thick)
20,000 Å

At present we can make only an approximate guess of the number of chemically different molecules within a single *E. coli* cell. Each year many new molecules are discovered. The best guess is that between 3000 and 6000 different types of molecules are present (Table 3–3). Some of these, such as H_2O and CO_2, are chemically simple. Others, such as the common sugar glucose or the nitrogen-containing purine adenine, are more complex but none-

Table 3–3 Approximate Chemical Composition of a Rapidly Dividing *Escherichia coli* Cell*

Component	Percent of total cell weight	Average MW	Approximate number per cell	Number of different kinds
H_2O	70	18	4×10^{10}	1
Inorganic ions (Na^+, K^+, Mg^{2+}, Ca^{2+}, Fe^{2+}, Cl^-, PO_4^{4-}, SO_4^{2-}, etc.)	1	40	2.5×10^8	20
Carbohydrates and precursors	3	150	2×10^8	200
Amino acids and precursors	0.4	120	3×10^7	100
Nucleotides and precursors	0.4	300	1.2×10^7	200
Lipids and precursors	2	750	2.5×10^7	50
Other small molecules (heme, quinones, breakdown products of food molecules, etc.)	0.2	150	1.5×10^7	250
Proteins	15	40,000	10^6	2000 to 3000
Nucleic acids				
DNA	1	2.5×10^9	4	1
RNA	6			
16S rRNA		500,000	3×10^4	1
23S rRNA		1,000,000	3×10^4	1
tRNA		25,000	4×10^5	60
mRNA		1,000,000	10^3	1000

* Weight 10^{12} daltons.

theless rather easily studied by current chemical techniques. Still other cellular molecules, in particular the proteins and nucleic acids, are very large, and even today their chemical structures are immensely difficult to unravel. Most of these macromolecules are not being actively studied, since their overwhelming complexity has forced chemists to concentrate on relatively few of them. Thus we must immediately admit that the structure of a cell will never be understood in the same way as that of water or glucose molecules. Not only will the exact structures of most macromolecules remain unsolved, but their relative locations within cells can be only vaguely known.

It is thus not surprising that many chemists, after brief periods of enthusiasm for studying "life," silently return to the world of pure chemistry. Others, however, become more optimistic when they understand (1) that all macromolecules are polymeric molecules built up from smaller monomers, (2) that there exist well defined chains of successive chemical reactions in cells (metabolic pathways), and (3) that a limit is placed on the number of enzymes (and hence small molecules) that can exist in a cell by the fact that each contains a finite amount of DNA.

Table 3-4 Structural Organization of Several Important Biological Macromolecules

Macromolecule	Monomeric units	Number of different monomers
Glycogen (a polysaccharide)	Glucose	One
DNA (deoxyribonucleic acid)	Deoxynucleotides	Four: deoxyadenylate deoxyguanylate deoxythymidylate deoxycytidylate
RNA (ribonucleic acid)	Ribonucleotides	Four: adenylate guanylate uridylate cytidylate
Protein	L-Amino acids	Twenty: glycine, alanine, serine, etc.

MACROMOLECULES CONSTRUCTED BY LINEAR LINKING OF SMALL MOLECULES

Most of the mass of *E. coli* (excluding water), like that of all other cells, is composed of macromolecules. Most of these large molecules are proteins, about half of which function as enzymes. The remainder of the proteins are used to help construct the ribosomes, the cell wall, etc. Table 3–3 shows that the number of atoms in a macromolecule is 25 to 50 times the number in a small molecule. Thus one might initially guess that most biochemists concerned with synthesis would be directly concerned with the frightfully complicated task of understanding the atom-by-atom growth of big molecules. Furthermore, one might also guess that their relatively immense size would lead to a very slow pace of research. Fortunately, however, the existence of three simplifying structural generalizations reduces the problem to a difficult, but not impossible, task.

First, all macromolecules are polymeric molecules formed by the condensation of small molecules. Macro-

Table 3–4 (*continued*)

General mono-mer formula	Fixed or irregular chain length	Linkage between monomers
	Indefinite— may be > 1000	1–4-Glycosidic linkage
Purine-deoxy-ribose-P (or pyrimidine-deoxyribose-P)	Genetically fixed— may be > 10^7	3′–5′-Phosphodiester linkage
Purine-ribose-P (or pyrimidine-ribose-P)	Genetically fixed, often > 3000	3′–5′-Phosphodiester linkage
	Genetically fixed, usually varies between 100 and 1000	Peptide linkage

molecule biosynthesis thus occurs in two stages: (1) the formation of the smaller subunits, and (2) the systematic linking together of these subunits. An analogy can be made to the building of a house from preconstructed bricks.

Second, the building blocks for a given macromole-cule have common chemical groupings as illustrated in Table 3–4, which describes the main structural features of several important macromolecules. For example, proteins form by the condensation of nitrogen-containing organic molecules called amino acids. The chemical bond that links two amino acids together is the peptide bond. There are twenty important amino acids, each of which has part of its structure identical to that of comparable regions in the other amino acids. Attached to the regular region is a "side group," specific for each amino acid (Figure 3–8). Each amino acid thus has a *specific* region (the side group) and a *nonspecific* region. The nucleic acids, DNA and RNA, are also formed by the union of smaller molecules called nu-

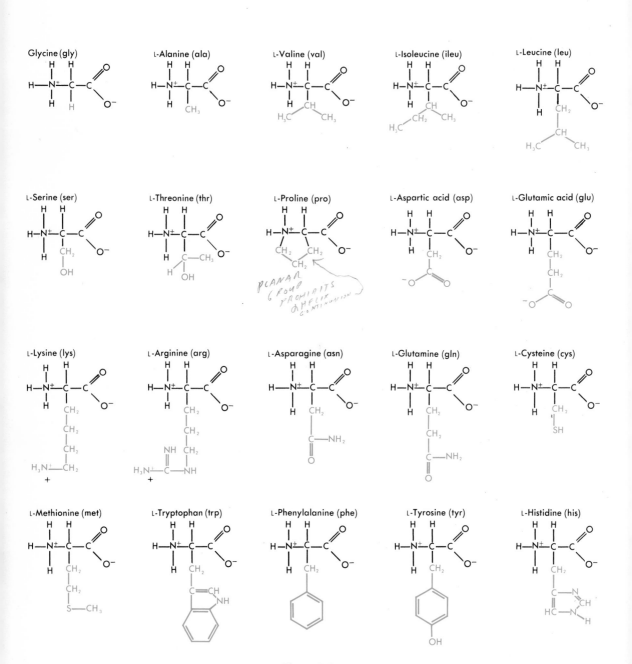

Figure 3–8
The twenty common amino acids found in proteins.

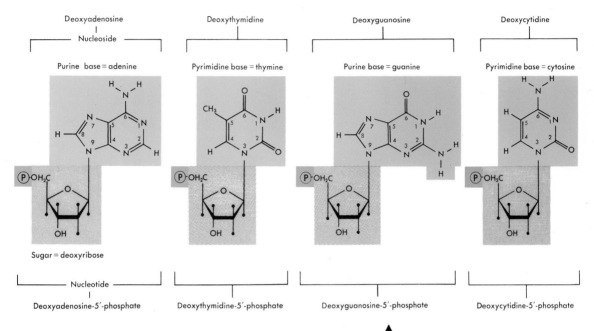

cleotides. The nucleotides of RNA, since they contain the sugar ribose, are called ribonucleotides, and those of DNA, which contain deoxyribose instead, are called deoxyribonucleotides (Figure 3–9; see Chapter 9 for details). The nucleotides are always linked together through a phosphate group and a hydroxyl group on the sugar component (Table 3–4). Hence these linkages are called phosphodiester bonds. Each nucleotide, like each amino acid, contains both a specific and a nonspecific region. The phosphate and sugar groups comprise the nonspecific portion of a nucleotide, while the purine and pyrimidine bases make up the specific portion. DNA and RNA each contain four main bases: two purines and two pyrimidines.

Third, most macromolecules (nucleic acids, proteins, and some polysaccharides) are *linear* aggregates in which the subunits are linked together by a chemical bond between atoms in the nonspecific regions. Their linear configuration follows from the fact that most subunits possess only two atoms that form bonds with other subunits. Thus a large fraction of most macromolecules consists of a repeating series of identical chemical groups (the *backbone*, Figure 3–10).

▲
Figure 3–9
The main nucleotide building blocks of DNA.

Figure 3–10
The structure of some biological macromolecules. The backbones are shown in color. (a) The structure of a portion of a glycogen chain composed of glucose subunits. (b) A portion of a polypeptide chain of a protein. The amino acid subunits are, from left to right: leucine, methionine, aspartic acid, tyrosine, and alanine. (c) A portion of the polynucleotide chain of a deoxyribonucleic acid.

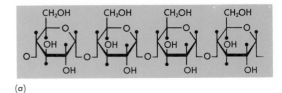

(a)

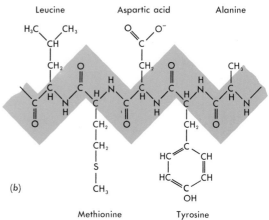

(b)

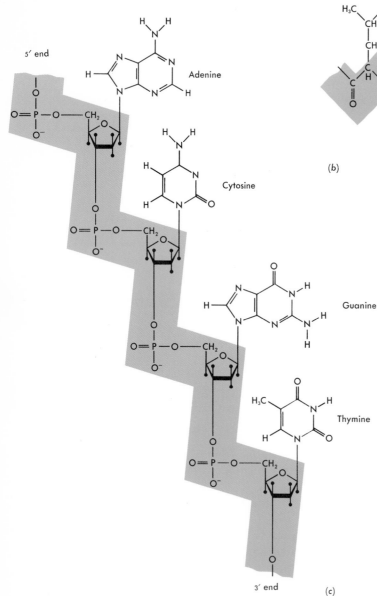

(c)

A feature common to all these biological polymers is that the individual subunits in the polymer chain contain two hydrogen atoms and one oxygen atom less than the simple monomers from which they are synthesized. Synthesis thus involves the release of water. When the polymers are degraded to yield smaller molecules, one H_2O molecule is incorporated by each peptide bond broken (Figure 3–11).

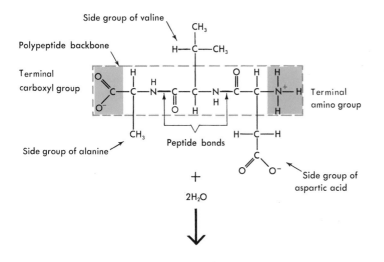

+

$2H_2O$

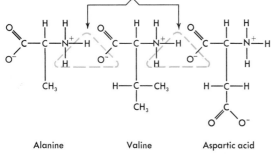

The breakdown of a peptide bond requires the addition of a water molecule. (It is thus an example of hydrolysis.)

Alanine   Valine   Aspartic acid

Figure 3–11
Hydrolysis of a polypeptide to form amino acids. Here we show a very simple polypeptide containing only three amino acids.

Degradative reactions in which H_2O uptake is required are known as hydrolytic reactions. There are many types of hydrolytic reactions, since numerous small molecules can be broken down to still smaller products by the addition of water. Under normal cell conditions, hydrolysis of any of the important polymers or small molecules is very rare. Hydrolysis is, however, speeded up by the presence of specific enzymes. For example, the enzymes pepsin and trypsin specifically catalyze the hydrolytic breakdown of proteins.

DISTINCTION BETWEEN REGULAR AND IRREGULAR POLYMERS

Table 3–4 also reveals an important difference between the polysaccharides, like glycogen, and the proteins and nucleic acids. Polysaccharides are usually constructed by *regular* (or semiregular) aggregation of one or two different kinds of monosaccharide building blocks. In contrast, proteins contain 20 different amino acids, and nucleic acids contain 4 different nucleotides. Moreover, in the nucleic acids and proteins, the order of subunits is highly *irregular* and varies greatly from one specific molecule to another. Polysaccharide synthesis from monosaccharides generally involves only the making of the same backbone bond; the synthesis of nucleic acids and proteins demands in addition a highly efficient mechanism for choosing and ordering the correct subunits.

METABOLIC PATHWAYS

We can see directly that all the molecules in a cell arise from cellular transformation of food molecules if we allow the bacterium *E. coli* to grow in a simple, well defined medium containing the sugar glucose (Table 3–1). Under these conditions glucose is the only organic source of carbon. In effect, all the carbon atoms of the *E. coli* molecules (a few are derived from CO_2) must result from chemical transformations by which glucose molecules are either broken down to smaller fragments or added to each other to form large molecules like the nucleotides or glycogen. The exact way in which all these transformations (collectively known as intermediary metabolism) occur is enormously complex, and most biochemists concern themselves with studying (or even knowing about!) only a small fraction of the total interactions.

Fortunately, some basic simplicity to the general pattern of metabolism is now beginning to emerge. Figure 3–12 shows some of the more important types of chemical events that occur after glucose is taken into an *E. coli* cell. Much of this information comes from experiments in which *E. coli* is fed molecules specifically labeled with radioactive isotopes. For example, if we expose *E. coli* for several seconds (a pulse) to C^{14}-labeled glucose, the radioactive atoms can be detected almost immediately in molecules chemically similar to glucose, such as glucose-6-phosphate. Only later do the labeled atoms find their way into the various amino acids and nucleotides. The amount of time before radioactivity appears in the various compounds corresponds roughly to the number of biochemical reactions separating glucose from the various metabolites.

Figure 3–12 shows that various key intermediates in glucose degradation often have several possible fates. They may be completely degraded via the Embden–Meyerhof pathway, the Krebs cycle, and the respiratory chain, to yield CO_2 and H_2O. During this process, ADP is converted to ATP. Alternatively, various intermediates may be used to initiate a series of successive chemical reactions that end with the synthesis of vital molecules, such as the amino acids or the nucleotides. For example, dihydroxyacetone-Ⓟ is used as a precursor for the lipid constituent glycerol, whereas 3-phosphoglyceric acid is the beginning metabolite in a series of reactions that leads to the amino acids serine, glycine, and cysteine.

Connected groups of biosynthetic (degradative) reactions are referred to as *metabolic pathways*. Once a molecule has started on a pathway it often has no choice but to undergo a series of successive transformations. Not all pathways, however, are necessarily linear. Some may be branched. Intermediates at branch points are transformed into one of two or more possible compounds. Those intermediates that are subject to several alternative fates are very important in cellular metabolism. Prominent among them are glucose-6-Ⓟ, pyruvate, α-ketoglutarate, and oxaloacetate; each serves as a starting point for several important pathways. Perhaps the most important such compound is acetyl-CoA, which not only is the main precursor of lipids, but also contains the acetate residue consumed by the citric acid cycle.

The metabolic fates of the majority of molecules are, however, much more limited. An average molecule can be either broken down to compound x or used as an intermediate in the biosynthesis of compound y; correspondingly, each such metabolite is able to combine specifically with only two different enzymes.

DEGRADATIVE PATHWAYS DISTINCT FROM BIOSYNTHETIC PATHWAYS

When *E. coli* is growing upon glucose as its sole carbon source, all its amino acids must be synthesized from metabolites derived from glucose. Thus there exists a distinct biosynthetic pathway for each of the 20 amino acids (Figure 3–12). *E. coli* can also grow, however, in the absence of a sugar, using any of the 20 amino acids as a *sole* carbon source. This means that there must also exist 20 pathways

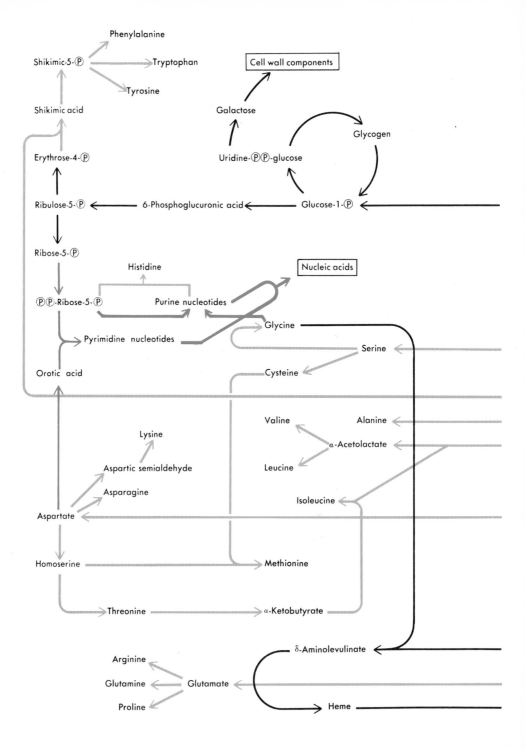

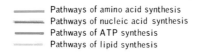

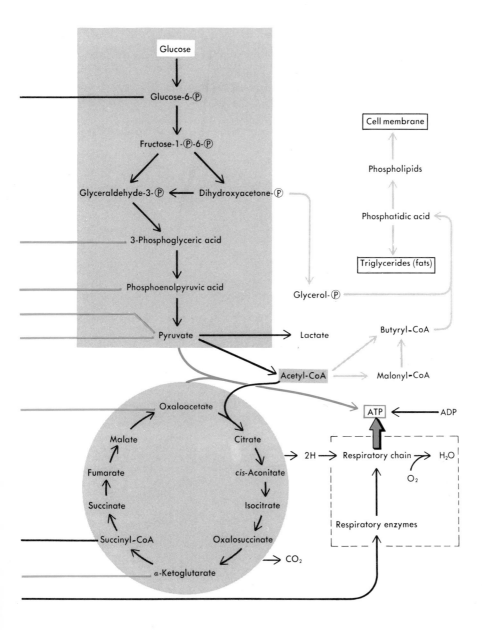

Figure 3–12
Schematic view of some of the main metabolic pathways in *E. coli.*

of amino acid degradation, by which the carbon and nitrogen atoms of the amino acids are usefully freed to form key metabolite compounds such as α-ketoglutarate and acetyl-CoA. These compounds can then be used in the synthesis of other amino acids. Degradative pathways also exist for various lipids, the purine and pyrimidine nucleotides, many pentose and hexose sugars, etc. Most of the degradative pathways are quite specific, and thus a very large number (perhaps 200 to 300) of different degradative intermediates may be found in *E. coli*. The number must be even larger in many other bacteria, particularly the Pseudomonads, since they degrade a larger, more varied collection of organic molecules than does *E. coli*.

The generalization is beginning to emerge that degradative pathways are usually quite different from pathways of biosynthesis. This observation is not surprising. As we shall see in Chapter 5, most biosynthetic reactions require energy, and often involve the breakdown of ATP, whereas degradative reactions, by their very function, must eventually generate ATP, in addition to supplying carbon and nitrogen skeletons.

THE SIGNIFICANCE OF A FINITE AMOUNT OF DNA

It is easy for the sophisticated pure chemist to look at Figure 3–12 with initial skepticism. Its neatness and clarity cannot obscure the fact that seemingly each week, a new enzyme, with its corresponding newly discovered metabolic reaction, is reported. The question arises whether Figure 3–12, by its simplification, completely misses the point of metabolism in *E. coli*. This would certainly be the case if there were not just one or two ways of degrading glucose, but 50 to 100 ways, and likewise if there were 20 different pathways leading to the biosynthesis of each of the nucleotides, amino acids, etc.

However, there exists a simple way to refute the heretical thought that only an insignificant fraction of the metabolic reactions that occur in *E. coli* have to date been described. The argument is based on the fact, which we shall prove in later chapters, that the sequence of nucleotides in DNA carries the genetic information that orders (codes) the sequence of amino acids in proteins. It is now clear that successive groups of three nucleotide pairs code for each amino acid. Thus the average-size protein, containing about 300 amino acids, requires a code of 900 nucleotide pairs. Since each nucleotide pair has a MW = 660, DNA units of MW = $(660)(900) \cong 6 \times 10^5$ are needed for each protein. Thus the number of different proteins within a cell can be no greater than the amount of haploid DNA/(6×10^5).

This figure can be used further to estimate the number of different types of small molecules a cell can possess. The majority of kinds of proteins in a cell are enzymes, each of which catalyzes a specific metabolic reaction. The approximate number of types of small molecules can be estimated if we know, on the average, how many specific enzymes are needed for the metabolism of the average small molecule. At present it seems a good guess that the number lies between one and two.

ONE-SIXTH TO ONE-THIRD OF THE CHEMICAL REACTIONS IN *E. COLI* ARE KNOWN

Our best estimate for the haploid MW of DNA in *E. coli* is $2.5 \times 10^9 \pm 0.5 \times 10^9$. This figure corresponds to 3000 to 4500 average-size protein molecules, and suggests that the number of different small molecules will be somewhat under 2000. Most pleasingly we find, by looking at Table 3–3, that our best guess of the number of different small metabolites involved in already known metabolic pathways is between 800 and 1000. This means that we already know at least 1/6, and maybe more than 1/3, of all the metabolic reactions that will ever be described in *E. coli*. The conclusion is most satisfying, for it strongly suggests that within the next 20 to 25 years we may approach a state in which it will be possible to describe essentially all the metabolic reactions involved in the life of an *E. coli* cell.

Therefore even a cautious chemist, when properly informed, need not look at a bacterial cell as a hopelessly complex object. Instead he might easily adopt an almost joyous enthusiasm, for it is clear that he, unlike his nineteenth-century equivalent, at last possesses the tools to describe completely the essential features of life.

SUMMARY

At the chemical level even the smallest cells are fantastically complicated. Most scientists interested in the essential chemical features of cell growth and division now concentrate on bacteria, since bacterial cells are about 500 times smaller than the average cell of a higher plant or animal. The most commonly employed bacterium, Escherichia coli, weighs about 2×10^{-12} gram (10^{12} daltons), of which about 70 percent is water. The number of different types of molecules within an E. coli cell *probably lies between 3000 and 6000. Approximately half are "small" molecules and the remainder, macromolecules. The large number of different macromolecules means that we shall not know in the near (or conceivably, even in the distant) future the exact 3-D structures of all the molecules in even the smallest cell.*

We do, however, know some rules about cell chemistry that make it possible to understand the growth of a cell without knowing the exact molecular structure of all its constituents. We know, for example, that all cellular macromolecules are polymeric molecules built up from much smaller monomers. Proteins are polymers containing amino acids as their monomers; the polymeric nucleic acids are built up by the linking of nucleotides. Further simplicity comes from the fact that most polymers, including all the proteins and nucleic acids, are essentially linear molecules.

Another simplifying rule concerns the complexity of intermediary metabolism. Generally compounds cannot be directly transformed into a large number of other compounds. Instead each compound comprises a step in a series of reactions (pathway) leading either to the degradation of a food molecule or to the biosynthesis of a necessary cellular molecule such as amino acid or a fatty acid. Cellular metabolism is the sum total of a large number of such pathways (metabolic maps) connected in such a way that products of degradative pathways can be used to initiate specific biosynthetic pathways.

The complexity of the metabolic map of an organism is related to the amount of genetic information (DNA) in the organism. The amount of DNA in a cell places an upper limit on the number of different enzymes the cell can produce. E. coli possess sufficient DNA to code for the amino acid sequence 3000 to 4500 different proteins. Some 800 different small molecules have now been detected in E. coli. This indicates that those metabolic reactions which we already know of in E. coli account for at least one-sixth of its total metabolism.

REFERENCES

SISTROM, W., *Microbial Life,* 2nd ed., Holt, New York, 1969. A brief paperback introduction to the biology and chemistry of microbes.

STANIER, R. Y., M. DOUDOROFF, AND E. A. ADELBERG, *The Microbial World,* 3rd ed., Prentice-Hall, Englewood Cliffs, N.J., 1970. A superb treatment of microbiology that can be read by any beginning college student.

DAVIS, B. D., R. DULBECCO, H. N. EISEN, H. S. GINSBERG, W. B. WOOD, JR., M. McCARTY, *Microbiology*, 2nd ed., Harper and Row, 1973. A massive, yet very good, introduction to microbiology, written for medical students, but valuable to a much broader audience.

The Importance of Weak Chemical Interactions

Until now we have focused our attention on the existence of discrete organic molecules and, following classical organic chemistry, have emphasized the covalent bonds which hold them together. It takes little insight, however, to realize that this type of analysis is inadequate for describing a cell, and that we must also concern ourselves with the exact shape of molecules and with the several factors which bind them together in an organized fashion. The distribution of molecules in cells is not random, and we must ask ourselves what chemical laws determine this distribution. Clearly, covalent bonding cannot be involved; by definition, atoms united by covalent bonds belong to the same molecule.

The arrangement of distinct molecules in cells is controlled instead by chemical bonds much weaker than covalent bonds. Atoms united by covalent bonds are capable of weak interactions with nearby atoms. These interactions, sometimes called "secondary bonds," occur not only between atoms in different molecules, but also between atoms in the same molecule. Weak bonds are important not just in deciding which molecules lie next to each other, but also in giving shape to flexible molecules such as the polypeptides and polynucleotides. It is, therefore, useful to have a feeling for the nature of weak chemical interactions and to understand how their "weak" character makes them indispensable to cellular existence. The most important include van der Waals bonds, hydrogen bonds, and ionic bonds.

DEFINITION AND SOME CHARACTERISTICS OF CHEMICAL BONDS

A chemical bond is an attractive force that holds atoms together. Aggregates of finite size are called molecules. Originally, it was thought that only covalent bonds hold atoms together in molecules, but now, as we shall show later in this chapter, weaker attractive forces are known to be important in holding together many macromolecules. For example, the four polypeptide chains of hemoglobin are held together by the combined action of several weak bonds. It is thus now customary also to call weak positive interactions chemical bonds, even though they are not strong enough, when present singly, to effectively bind two atoms together.

Chemical bonds are characterized in several ways. A most obvious characteristic of a bond is its strength. Strong bonds almost never fall apart at physiological temperatures. This is why atoms united by covalent bonds always belong to the same molecule. Weak bonds are easily broken, and when they exist singly, they exist fleetingly. Only when present in ordered groups do weak bonds exist for a long time. The strength of a bond is correlated with its length, so that two atoms connected by a strong bond are always closer together than the same two atoms held together by a weak bond. For example, two hydrogen atoms bound covalently to form a hydrogen molecule (H:H) are 0.74 Å apart, whereas the same two atoms, when held together by the van der Waals forces instead, are held 1.2 Å apart.

Another important bond characteristic is the maximum number of bonds that a given atom can make. The number of covalent bonds an atom forms is called its valence. Oxygen, for example, has a valence of two: It can never form more than two covalent bonds. There is more variability in the case of van der Waals bonds, where the limiting factor is purely steric: The number of possible bonds is limited only by the number of atoms that can simultaneously touch each other. The formation of hydrogen bonds is subject to more restrictions. A covalently bonded hydrogen atom usually participates in only one hydrogen bond, whereas an oxygen atom seldom participates in more than two hydrogen bonds.

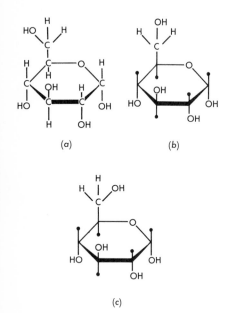

(a)   (b)

(c)

Figure 4–1
Rotation about the C_5—C_6 bond in glucose. This carbon–carbon bond is a single bond, and so any of the three configurations (a), (b), (c) may occur.

The angle between two bonds originating from a single atom is called the bond angle. The angle between two specific covalent bonds is always approximately the same. For example, when a carbon atom has four single bonds, they are directed tetrahedrally (bond angle = 109°). In contrast, the angles between weak bonds are much more variable.

Bonds differ also in the freedom of rotation they allow. Single covalent bonds permit free rotation of bound atoms (Figure 4–1), whereas double and triple bonds are quite rigid. For example, the carbonyl (C=O) and imino (N—H) groups bound together by the rigid peptide bond must lie in the same plane (Figure 4–2), because of the partial double-bond character of the peptide bond. Much weaker, ionic bonds show completely opposite behavior; they impose no restrictions on the relative orientations of bonded atoms.

CHEMICAL BONDS ARE EXPLAINABLE IN QUANTUM-MECHANICAL TERMS

The nature of the forces, strong as well as weak, that give rise to chemical bonds remained a mystery to chemists until the quantum theory of the atom (quantum mechanics) was developed in the 1920's. Then, for the first time, the various empirical laws about how chemical bonds are formed were put on a firm theoretical basis. It was realized that all chemical bonds, weak as well as strong, were based on electrostatic forces. Quantum-mechanical explanations were provided not only for covalent bonding by the sharing of electrons, but also for the formation of weaker bonds.

CHEMICAL-BOND FORMATION INVOLVES A CHANGE IN THE FORM OF ENERGY

The spontaneous formation of a bond between two atoms always involves the release of some of the internal energy of the unbonded atoms and its conversion to another energy form. The stronger the bond, the greater the amount of energy which is released upon its formation. The bonding reaction between two atoms A and B is thus described by

$$A + B \rightarrow AB + \text{energy} \qquad (4\text{--}1)$$

where AB represents the bonded aggregate. The rate of the reaction is proportional to the frequency of collision

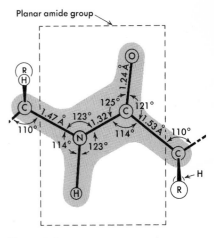

Figure 4–2
The planar shape of the peptide bond. Shown is a portion of an extended polypeptide chain. Almost no rotation is possible about the peptide bond because of its partial double-bond character:

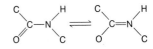

All the atoms in the grey must lie in the same plane. Rotation is possible, however, around the remaining two bonds, which make up the polypeptide configurations. (Redrawn from L. Pauling, *The Nature of the Chemical Bond*, 3rd ed., Cornell Univ. Press, Ithaca, N.Y., 1960, p. 498, with permission.)

between A and B. The unit most commonly used to measure energy is the calorie, the amount of energy required to raise the temperature of 1 gram of water from 14.5°C to 15.5°C. Since thousands of calories are usually involved in the breaking of a mole of chemical bonds, most chemical-energy changes in chemical reactions are expressed in kilocalories per mole.

Atoms joined by chemical bonds, however, do not forever remain together. There also exist forces which break chemical bonds. By far the most important of these forces arises from heat energy. Collisions with fast-moving molecules or atoms can break chemical bonds. During a collision, some of the kinetic energy of a moving molecule is given up as it pushes apart two bonded atoms. The faster a molecule is moving (the higher the temperature), the greater the probability that, upon collision, it will break a bond. Hence, as the temperature of a collection of molecules is increased, the stability of their bonds decreases. The breaking of a bond is thus always indicated by the formula

$$AB + \text{energy} \rightarrow A + B \qquad (4\text{--}2)$$

The amount of energy that must be added to break a bond is exactly equal to the amount which was released upon its formation. This equivalence follows from the first law of thermodynamics, which states that energy (except as it is interconvertible with mass) can be neither made nor destroyed.

EQUILIBRIUM BETWEEN BOND MAKING AND BREAKING

Every bond is thus a result of the combined actions of bond-making (arising from electrostatic interactions) and bond-breaking forces. When an equilibrium is reached in a closed system, the number of bonds forming per unit time will equal the number breaking. Then the proportion of bonded atoms is described by the following mass action formula:

$$K_{eq} = \frac{\text{conc}^{AB}}{\text{conc}^A \times \text{conc}^B} \qquad (4\text{--}3)$$

where K_{eq} is the equilibrium constant, and conc^A, conc^B, and conc^{AB} are the concentrations of A, B, and AB in moles per liter, respectively. Whether we start with only free A and B, with only the molecule AB, or with a combination of AB and free A and B, at equilibrium the proportions of A, B, and AB will reach the concentration given by K_{eq}.

THE CONCEPT OF FREE ENERGY

There is always a change in the form of energy as the proportion of bonded atoms moves toward the equilbrium concentration. Biologically, the most useful way to express this energy change is through the physical chemists' concept of *free energy, G.** Here we shall not give a rigorous description of free energy nor show how it differs from the other forms of energy. For this, the reader must refer to a chemistry text which discusses the second law of thermodynamics. We must suffice by saying that *free energy is energy that has the ability to do work.*

The second law of thermodynamics tells us that a decrease of free energy (ΔG is negative) always occurs in spontaneous reactions. When equilibrium is reached, there is no further change in the amount of free energy ($\Delta G = 0$). The equilibrium state for a closed collection of atoms is thus that state that contains the least amount of free energy.

The free energy lost as equilibrium is approached is either transformed into heat or used to increase the amount of entropy. Here, we shall not attempt to define entropy (again this task must be left to a chemistry text), except to say that the amount of entropy is a measure of the amount of disorder. The greater the disorder, the greater the amount of entropy. The existence of entropy means that many spontaneous chemical reactions do not proceed with an evolution of heat. For example, in the dissolving of NaCl in water, heat is absorbed. There is, nonetheless, a net decrease in free energy because of the increase of disorder of the Na^+ and Cl^- ions as they move from a solid to a liquid phase.

K_{eq} IS EXPONENTIALLY RELATED TO ΔG

It is obvious that the stronger the bond, and hence the greater the change in free energy (ΔG) which accompanies its formation, the greater the proportion of atoms that must exist in the bonded form. This common-sense idea is quantitatively expressed by the physical-chemical formula

$$\Delta G = - RT \ln K_{eq} \quad \text{or} \quad K_{eq} = e^{-\Delta G/RT} \qquad (4\text{--}4)$$

where R is the universal gas constant, T the absolute

---

* It was the custom in the United States until recently to refer to free energy by the symbol F. Now, however, most new texts have adopted the international symbol G, which honors the great nineteenth-century physicist Gibbs.

Table 4–1 The Numerical Relationship Between the Equilibrium Constant and ΔG at 25°C

K_{eq}	ΔG, kcal/mole
0.001	4.089
0.01	2.726
0.1	1.363
1.0	0
10.0	−1.363
100.0	−2.726
1000.0	−4.089

temperature, $e = 2.718$, ln the logarithm of K to the base e, and K_{eq} the equilibrium constant.

Insertion of the appropriate values of R ($= 1.987$ cal/deg·mole) and T ($= 298$ at 25°C) tells us (Table 4–1) that ΔG values as low as 2 kcal/mole can drive a bond-forming reaction to virtual completion if all reactants are present at molar concentrations.

COVALENT BONDS ARE VERY STRONG

The ΔG values accompanying the formation of covalent bonds from free atoms such as hydrogen or oxygen are very large and negative in sign, usually −50 to −110 kcal/mole. Application of Eq. (4–4) tells us that K_{eq} of the bonding reaction will be correspondingly large, and so the concentration of hydrogen or oxygen atoms existing unbound will be very small. For example, a ΔG value of −100 kcal/mole tells us that, if we start with 1 mole/liter of the reacting atoms, only one in 10^{40} atoms will remain unbound when equilibrium is reached.

WEAK BONDS HAVE ENERGIES BETWEEN 1 AND 7 KCAL/MOLE

The main types of weak bonds important in biological systems are the van der Waals bonds, hydrogen bonds, and ionic bonds. Sometimes, as we shall soon point out, the distinction between a hydrogen bond and an ionic bond is arbitrary. The weakest bonds are the van der Waals bonds. These have energies (1 to 2 kcal/mole) that are only slightly greater than the kinetic energy of heat motion. The energies of hydrogen and ionic bonds range between 3 and 7 kcal/mole.

In liquid solutions, almost all molecules are forming a number of weak bonds to nearby atoms. All molecules are able to form van der Waals bonds; hydrogen and ionic bonds can also form between molecules (ions) which have a net charge or in which the charge is unequally distributed. Some molecules thus have the capacity to form several types of weak bonds. Energetic considerations, however, tell us that molecules always have a greater tendency to form the stronger bond.

WEAK BONDS CONSTANTLY MADE AND BROKEN AT PHYSIOLOGICAL TEMPERATURES

The energy of the strongest weak bond is only about ten times larger than the average energy of kinetic motion (heat) at 25°C (0.6 kcal/mole). Since there is a significant spread in the energies of kinetic motion, many molecules with sufficient kinetic energy to break the strongest weak bond always exist at physiological temperatures.

ENZYMES NOT INVOLVED IN MAKING (BREAKING) OF WEAK BONDS

The average lifetime of a single weak bond is only a fraction of a second. Cells thus do not need a special mechanism to speed up the rate at which weak bonds are made and broken. Correspondingly, enzymes never participate in reactions of weak bonds.

DISTINCTION BETWEEN POLAR AND NONPOLAR MOLECULES

All forms of weak interactions are based upon attractions between electric charges. The separation of electric charges can be permanent or temporary, depending upon the atoms involved. For example, the oxygen molecule (O:O) has a symmetric distribution of electrons between its two oxygen atoms, and so each of its two atoms is uncharged. In contrast, there is a nonuniform distribution of charge in water (H:O:H), where the bond electrons are unevenly shared (Figure 4–3). They are held more strongly by the oxygen atom, which thus carries a considerable negative charge, whereas the two hydrogen atoms together have an equal amount of positive charge. The center of the positive charge is on one side of the center of the negative charge. A combination of separated positive and negative charges is called an electric dipole moment. Unequal electron sharing reflects dissimilar affinities of the bonding atoms for electrons. Atoms which have a tendency to gain electrons are called electronegative atoms. Electropositive atoms have a tendency to give up electrons.

Molecules such as H_2O, which have a dipole moment, are called *polar molecules. Nonpolar molecules* are those with no effective dipole moments. In CH_4 (methane), for example, the carbon and hydrogen atoms have similar affinities for their shared electron pairs, and so neither the carbon nor the hydrogen atom is noticeably charged.

The distribution of charge in a molecule can also be affected by the presence of nearby molecules, particularly if the affected molecule is polar. The effect may cause a nonpolar molecule to acquire a slight polar character. If the second molecule is not polar, its presence will still alter the nonpolar molecule, establishing a fluctuating charge distribution. Such induced effects, however, give rise to

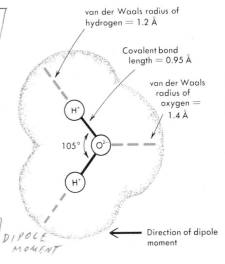

van der Waals radius of hydrogen = 1.2 Å

Covalent bond length = 0.95 Å

van der Waals radius of oxygen = 1.4 Å

Direction of dipole moment

Figure 4–3
The structure of a water molecule.

much smaller separation of charge than is found in polar molecules, thus resulting in smaller interaction energies and, correspondingly, weaker chemical bonds.

VAN DER WAALS FORCES

Van der Waals bonding arises from a nonspecific attractive force originating when two atoms come close to each other. It is based not upon the existence of permanent charge separations, but rather upon the induced fluctuating charges caused by the nearness of molecules. It therefore operates between all types of molecules, polar as well as nonpolar. It depends heavily upon the distance between the interacting groups, since the bond energy is inversely proportional to the sixth power of distance (Figure 4–4).

There also exists a more powerful van der Waals repulsive force, which comes into play at even shorter distances. This repulsion is caused by the overlapping of the outer electron shells of the atoms involved. The van der

Table 4–2 van der Waals Radii of the Atoms in Biological Molecules

Atom	van der Waals radius Å
H	1.2
N	1.5
O	1.4
P	1.9
S	1.85
CH$_3$ group	2.0
Half thickness of aromatic molecule	1.7

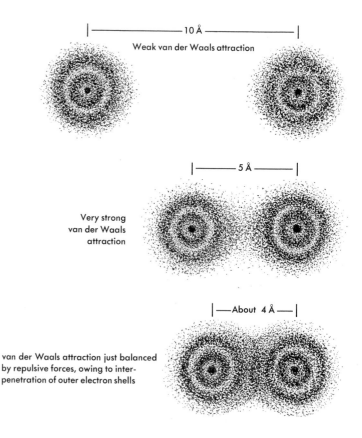

|——————— 10 Å ———————|

Weak van der Waals attraction

|——— 5 Å ———|

Very strong van der Waals attraction

|— About 4 Å —|

van der Waals attraction just balanced by repulsive forces, owing to interpenetration of outer electron shells

Figure 4–4
Diagram illustrating van der Waals attraction and repulsion forces in relation to electron distribution of monoatomic molecules on the inert rare gas argon. (Redrawn from *General Chemistry*, 2nd ed. by Linus Pauling. W. H. Freeman and Company, copyright © 1953, p. 322, with permission.)

Waals attractive and repulsive forces balance at a certain distance specific for each type of atom. This distance is the so-called van der Waals radius (Table 4–2 and Figure 4–5). The van der Waals bonding energy between two atoms separated by the sum of their van der Waals radii increases with the size of the respective atoms. For two average atoms it is only about 1 kcal/mole, which is just slightly more than the average thermal energy of molecules at room temperature (0.6 kcal/mole).

This means that van der Waals forces are an effective binding force at physiological temperatures only when several atoms in a given molecule are bound to several atoms in another molecule. Then the energy of interaction is much greater than the dissociating tendency resulting from random thermal movements. In order for several atoms to interact effectively, the molecular fit must be precise, since the distance separating any two interacting atoms must not be much greater than the sum of their van der Waals radii (Figure 4–6). The strength of interaction rapidly approaches zero when this distance is only slightly

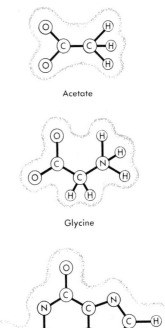

Acetate

Glycine

Guanine

Figure 4–5
Drawings of several molecules with the van der Waals radii of the atoms ▶ outlined.

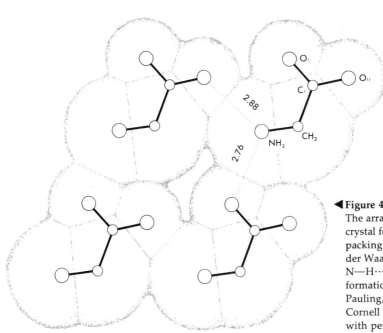

◀ **Figure 4–6**
The arrangement of molecules in a layer of a crystal formed by the amino acid glycine. The packing of the molecules is determined by the van der Waals radii of the groups, except for the N—H····O contacts, which are shortened by the formation of hydrogen bonds. (Redrawn from L. Pauling, *The Nature of the Chemical Bond*, 3rd ed., Cornell Univ. Press, Ithaca, N.Y., 1960, p. 262, with permission.)

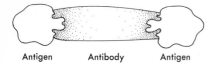

Antigen   Antibody   Antigen

Schematic drawing showing the complementary relation between the surface configurations of an antigen and an antibody, allowing the formation of secondary bonds between them.

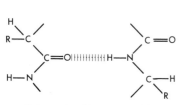

Hydrogen bond between peptide groups

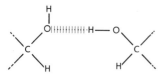

Hydrogen bond between two hydroxyl groups

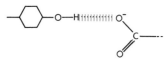

Hydrogen bond between a charged carboxyl group and the hydroxyl group of tyrosine

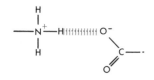

Hydrogen bond between a charged amino group and a charged carboxyl group

Figure 4–8
Examples of hydrogen bonds in biological molecules.

exceeded. Thus, the strongest type of van der Waals contact arises when a molecule contains a cavity exactly complementary in shape to a protruding group of another molecule (Figure 4–7). This is the type of situation thought to exist between an antigen and its specific antibody (see Chapter 19). In this instance, the binding energies sometimes can be as large as 10 kcal/mole, so that antigen–antibody complexes seldom fall apart. Many polar molecules are only seldom affected by van der Waals interactions, since such molecules can acquire a lower energy state (lose more free energy) by forming other types of bonds.

HYDROGEN BONDS

A hydrogen bond arises between a covalently bound hydrogen atom with some positive charge and a negatively charged, covalently bound acceptor atom (Figure 4–8). For example, the hydrogen atoms of the imino group (N—H) are attracted by the negatively charged keto oxygen atoms (C=O). Sometimes the hydrogen-bonded atoms belong to groups with a unit of charge (e.g., NH_3^+ or COO^-). In other cases, both the donor hydrogen atoms and the negative acceptor atoms have less than a unit of charge.

The biologically most important hydrogen bonds involve hydrogen atoms covalently bound to oxygen (O—H) or nitrogen atoms (N—H). Likewise, the negative acceptor atoms are usually nitrogen or oxygen. Table 4–3 lists some of the most important hydrogen bonds. Bond energies range between 3 and 7 kcal/mole, the stronger bonds involving the greater charge differences between donor and acceptor atoms. Hydrogen bonds are thus weaker than covalent bonds, yet considerably stronger than van der Waals bonds. A hydrogen bond, therefore, will hold two atoms closer together than the sum of their van der Waals radii, but not so close together as a covalent bond would hold them.

Hydrogen bonds, unlike van der Waals bonds, are highly directional. In optimally strong hydrogen bonds, the hydrogen atom points directly at the acceptor atom (Figure 4–9). If it points indirectly, the bond energy is much less. Hydrogen bonds are also much more specific

Figure 4–9
Directional properties of hydrogen bonds. In (a) the vector along the ▶
covalent O—H bond points directly at the acceptor oxygen, thereby
forming a strong bond. In (b) the vector points away from the oxygen
atom, resulting in a much weaker bond.

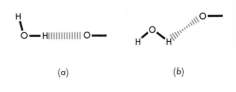

(a)   (b)

than van der Waals bonds, since they demand the existence of molecules with complementary donor hydrogen
and acceptor groups.

SOME IONIC BONDS ARE, IN EFFECT, HYDROGEN BONDS

Many organic molecules possess ionic groups that contain
one or more units of net positive or negative charge. The
negatively charged mononucleotides, for example, contain
phosphate groups (PO_3^{3-}) with three units of negative
charge, whereas each amino acid (except proline) has a
negative carboxyl group (COO^-) and a positive amino
group (NH_3^+), both of which carry a unit of charge. These
charged groups are usually neutralized by nearby, oppositely charged, groups. The electrostatic forces acting
between the oppositely charged groups are called ionic
bonds. Their average bond energy in an aqueous solution
is about 5 kcal/mole.

In many cases, either an inorganic cation like Na^+,
K^+, or Mg^{2+}, or an inorganic anion like Cl^- or SO_4^{2-}, neutralizes the charge of the ionized organic molecules. When
this happens in aqueous solution, the neutralizing cations
and anions do not occupy fixed positions, because inorganic ions are usually surrounded by shells of water
molecules and so do not directly bind to oppositely
charged groups. Thus it is now believed that, in water
solutions, electrostatic bonds to surrounding inorganic cations or anions are not of primary importance in determining the molecular shapes of organic molecules.

On the other hand, highly directional bonds result if
the oppositely charged groups can form hydrogen bonds
to each other. For example, both the COO^- and NH_3^+
groups are often held together by strong bonds. Since
these hydrogen bonds are stronger than those that involve
groups with less than a unit of charge, they are correspondingly shorter. A strong hydrogen bond can also form
between a group with a unit charge and a group having
less than a unit charge. For example, a hydrogen atom
belonging to an amino group ($-NH_2$) bonds strongly to
an oxygen atom of a carboxyl group (COO^-).

Table 4–3 Approximate Bond
Lengths of Biologically Important
Hydrogen Bonds

Bond	Approximate bond length, Å
O—H———O	$2.70 \pm .10$
O—H———O$^-$	$2.63 \pm .10$
O—H———N	$2.88 \pm .13$
N—H———O	$3.04 \pm .13$
N$^+$H ———O	$2.93 \pm .10$
N—H———N	$3.10 \pm .13$

WEAK INTERACTIONS DEMAND COMPLEMENTARY MOLECULAR SURFACES

Weak binding forces are effective only when the interacting surfaces are close. This proximity is possible only when the molecular surfaces have *complementary structures*, so that a protruding group (or positive charge) on one surface is matched by a cavity (or negative charge) on another; i.e., the interacting molecules must have a lock-and-key relationship. In cells this requirement often means that some molecules hardly ever bond to other molecules of the same kind, because such molecules do not have the properties of symmetry necessary for self-interaction. For example, some (polar) molecules contain donor hydrogen atoms and no suitable acceptor atoms, whereas others can accept hydrogen bonds but have no hydrogen atoms to donate. On the other hand, many molecules do exist with the symmetry to permit strong self-interaction in cells, water being the most important example.

H₂O MOLECULES FORM HYDROGEN BONDS

Under physiological conditions, water molecules rarely ionize to form H^+ and OH^- ions. Instead, they exist as polar H—O—H molecules. Both the hydrogen and oxygen atoms form strong hydrogen bonds. In each H_2O molecule, the oxygen atom can bind to two external hydrogen atoms, whereas each hydrogen atom can bind to one adjacent oxygen atom. These bonds are directed tetrahedrally (Figure 4–10), so that, in its solid and liquid forms, each water molecule tends to have four nearest neighbors, one in each of the four directions of a tetrahedron. In ice the bonds to these neighbors are very rigid and the arrangement of molecules fixed. Above the melting temperature (0°C) the energy of thermal motion is sufficient to break the hydrogen bonds and to allow the water molecules to change their nearest neighbors continually. Even in the liquid form, however, at a given instant most water molecules are bound by four strong hydrogen bonds.

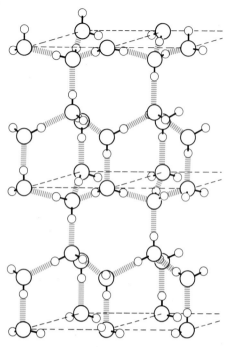

Figure 4–10
Schematic diagram of a lattice formed by H₂O molecules. The energy gained by forming specific hydrogen bonds (‖‖‖‖) between H₂O molecules favors the arrangement of the molecules in adjacent tetrahedrons. Oxygen atoms are indicated by large circles, and hydrogen atoms by small circles. Although the rigidity of the arrangement depends upon the temperature of the molecules, the pictured structure is, nevertheless, predominant in water as well as in ice. (Redrawn from L. Pauling, *The Nature of the Chemical Bond*, 3rd ed., Cornell Univ. Press, Ithaca, N.Y., 1960, p. 465, with permission.)

WEAK BONDS BETWEEN MOLECULES IN AQUEOUS SOLUTIONS

The average energy of a secondary bond, though small compared to that of a covalent bond, is nonetheless strong enough compared to heat energy to ensure that most molecules in aqueous solution will form secondary bonds to other molecules. The proportion of bonded to nonbonded arrangements is given by Eq. (4–4), corrected to take into account the high concentration of molecules in a liquid. It tells us that interaction energies as low as 2 to 3 kcal/mole are sufficient at physiological temperatures to force most molecules to form the maximum number of good secondary bonds.

The specific structure of a solution at a given instant is markedly influenced by which solute molecules are present, not only because molecules have specific shapes, but also because molecules differ in which types of secondary bonds they can form. These differences mean that a molecule will tend to move until it is next to a molecule with which it can form the strongest possible bond.

Solutions of course are not static. Because of the disruptive influence of heat, the specific configuration of a solution is constantly changing from one arrangement to another of approximately the same energy content. Equally important in biological systems is the fact that metabolism is continually transforming one molecule into another and so automatically changing the nature of the secondary bonds that can be formed. The solution structure of cells is thus constantly disrupted not only by heat motion, but also by the metabolic transformations of the cell's solute molecules.

ORGANIC MOLECULES THAT TEND TO FORM HYDROGEN BONDS ARE WATER SOLUBLE

The energy of hydrogen bonds per atomic group is much greater than that of van der Waals contacts. Thus those molecules that can form hydrogen bonds will form them in preference to van der Waals contacts. For example, if we try to mix water with a compound which cannot form hydrogen bonds, such as benzene, the water and benzene molecules rapidly separate from each other, the water molecules forming hydrogen bonds among themselves, while the benzene molecules attach to each other by van der Waals bonds. Thus it is effectively impossible to insert a non-hydrogen-bonding organic molecule into water.

On the other hand, polar molecules such as glucose and pyruvate, which contain a large number of groups that form excellent hydrogen bonds (e.g., $=O$ or $-OH$), are somewhat soluble in water (hydrophilic as opposed to hydrophobic). This effect occurs because, while the insertion of such groups into a water lattice breaks water–water hydrogen bonds, it results simultaneously in hydrogen bonds between glucose and water. These alternative arrangements, however, are not usually as energetically satisfactory as the water–water arrangements, so that even the most polar molecules ordinarily have only limited solubility.

97

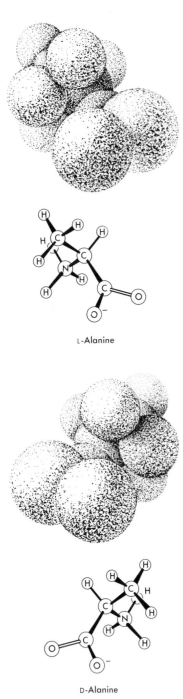

L-Alanine

D-Alanine

Figure 4–11
The two stereoisomers of the amino acid alanine. (Redrawn from *General Chemistry*, 2nd ed. by Linus Pauling. W. H. Freeman and Company, copyright © 1953, p. 598, with permission.)

Thus almost all the molecules which cells are constantly acquiring, either through food intake or through biosynthesis, are somewhat insoluble in water. These molecules, by their thermal movements, randomly collide with other molecules until they find complementary molecular surfaces on which to attach and thereby release water molecules for water–water interactions.

THE UNIQUENESS OF MOLECULAR SHAPES; THE CONCEPT OF SELECTIVE STICKINESS

Even though most cellular molecules are built up from only a small number of groups, such as OH, NH$_2$, and CH$_3$, great specificity exists as to which molecules tend to lie next to each other. This is because each molecule has unique bonding properties. One very clear demonstration comes from the specificity of stereoisomers. For example, proteins are always constructed from L-amino acids, never from their mirror images, the D-amino acids (Figure 4–11). Though the D- and L-amino acids have identical covalent bonds, their binding properties to asymmetric molecules are often very different. Thus most enzymes are specific for L-amino acids. If an L-amino acid is able to attach to a specific enzyme, the D-amino acid is unable to bind.

The general rule exists that most molecules in cells can make good "weak" bonds with only a small number of other molecules. This is partly because all molecules in biological systems exist in an aqueous environment. The formation of a bond in a cell depends not only upon whether two molecules bind well to each other, but also upon whether the bond will permit their water solvent to form the maximum number of good hydrogen bonds.

The strong tendency of water to exclude nonpolar groups is frequently referred to as *hydrophobic bonding*. Some chemists like to call all the bonds between nonpolar groups *in a water solution* hydrophobic bonds (Figure 4–12). In a sense this term is a confusing misnomer, for the phenomenon that it seeks to emphasize is the absence, not the presence, of bonds. (The bonds that tend to form between the nonpolar groups are due to van der Waals attractive forces.) On the other hand, the term hydrophobic bond is often useful, since it emphasizes the fact that nonpolar groups will try to arrange themselves so that they are not in contact with water molecules.

Consider, for example, the energy difference between the binding in water of the amino acids alanine and glycine to a third molecule which has a surface complementary to alanine. Alanine differs from glycine by the pres-

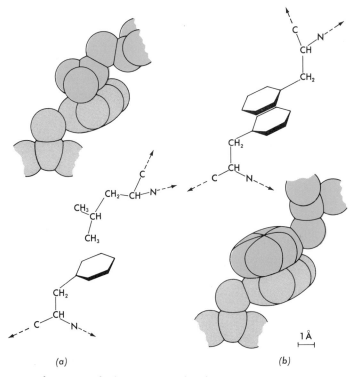

(a)   (b)

Figure 4–12
Illustrative examples of van der Waals (hydrophobic) bonds between the nonpolar side groups of amino acids. The hydrogens are not indicated individually. For the sake of clarity, the van der Waals radii are reduced by 20 per cent. The structural formulas adjacent to each space-filling drawing indicate the arrangement of the atoms; (a) phenylalanine–leucine bond; (b) phenylalanine–phenylalanine bond. (Redrawn from H. A. Scheraga, in *The Proteins,* H. Neurath (ed.), 2nd ed., Vol. I, Academic Press, New York, 1963, p. 527, with permission.)

ence of one methyl group in the former. When alanine is bound to the third molecule, the van der Waals contacts around the methyl group yield 1 kcal/mole of energy, which is not released when glycine is bound instead. This small energy difference alone would give [using Eq. (4–4)] only a factor of 6 between the binding of alanine and glycine. This calculation does not take into consideration, however, the fact that water is trying to exclude alanine much more than glycine. The presence of alanine's CH_3 group upsets the water lattice much more seriously than does the hydrogen atom side group of glycine. At present it is still difficult to predict how large a correction factor must be introduced for this disruption of the water lattice by the hydrophobic side groups. A current guess is that the water tends to exclude alanine, thrusting it toward a third molecule with a hydrophobic force 2 to 3 kcal/mole larger than the force excluding glycine.

We thus arrive at the important conclusion that the energetic difference between the binding of even the most similar molecules to a third molecule (when the difference involves a nonpolar group) is at least 2 to 3 kcal/mole greater in the aqueous interior of cells than under non-aqueous conditions. Frequently, the energetic difference is 3 to 4 kcal/mole, since the molecules involved often contain polar groups which can form hydrogen bonds.

THE ADVANTAGE OF ΔG'S BETWEEN 2 AND 5 KCAL/MOLE

We have seen that the energy of just one secondary bond (2 to 5 kcal/mole) is often sufficient to ensure that a molecule preferentially binds to a selected group of molecules. Equally important, these energy differences are not so large that rigid lattice arrangements develop within a cell—the interior of a cell never crystallizes as it would if the energy of secondary bonds were several times greater. Larger energy differences would mean that the secondary bonds break only seldom, resulting in low diffusion rates incompatible with cellular existence.

WEAK BONDS ATTACH ENZYMES TO SUBSTRATES

Secondary forces are necessarily the basis by which enzymes and their substrates initially combine with each other. Enzymes do not indiscriminately bind all molecules but, in general, have noticeable affinity only for their own substrates.

Since enzymes catalyze both directions of a chemical reaction, they must have specific affinities for both sets of reacting molecules. In some special cases it is possible to calculate an equilibrium constant for the binding of an enzyme and one of its substrates, which consequently enables us [Eq. (4–4)] to calculate the ΔG upon binding. This calculation in turn hints at which types of bonds may be involved. ΔG values ranging between 5 and 10 kcal/mole suggest that from one to several good secondary bonds are the basis of specific enzyme–substrate interactions. Also worth noting is that the ΔG of binding is never exceptionally high; thus enzyme–substrate complexes can be both made and broken apart rapidly as a result of random thermal movement. This fact explains why enzymes can function so quickly, sometimes as often as 10^6 times per second. If enzymes were bound to their substrates by more powerful bonds, they would act much more slowly.

MOST MOLECULAR SHAPES DETERMINED BY WEAK BONDS

The shapes of numerous molecules are automatically given by the distribution of covalent bonds. This inflexibility occurs when groups are attached by covalent bonds about which free rotation is impossible. Rotation is possible only when atoms are attached by single bonds. (For example, the methyl groups of ethane, H_3C—CH_3, rotate about the carbon–carbon bond.) When more than one electron pair is involved in a bond, rotation does not occur; in fact, the atoms involved must lie in the same plane. Thus the aromatic purine and pyrimidine rings are planar molecules 3.4 Å thick. There is no uncertainty about the shape of any aromatic molecules. They are almost always flat, independent of their surrounding environment.

On the contrary, for molecules containing single bonds, the possibility of rotation around the bond suggests that a covalently bonded molecule exists in a large variety of shapes. This theoretical possibility, however, is seldom in fact realized, because the various possible 3-D configurations differ in the number of good weak bonds which can be formed. Generally, there is one configuration that has significantly less free energy than any of the other geometric arrangements.

Two classes of secondary bonds may be important in determining 3-D shapes. One class is internal, forming between atoms connected by a chain of covalent bonds. Internal bonds often cause a linear molecule to bend back upon itself, allowing contacts between atoms separated by a large number of covalent bonds. In such molecules the final shape is usually compact (globular). The other class of bonds is external, forming between atoms not connected by covalent bonds. In cases in which the optimal 3-D configuration is achieved by forming external bonds, molecules most often have extended (fibrous) configurations.

It is no simple matter to guess the correct shape of a large molecule from its covalent-bond structure; although one configuration for a protein or a nucleic acid may be energetically much more suitable than any other, we cannot yet derive it from our knowledge of bond energies. There are two reasons for this difficulty. One is purely logistical. The number of possible configurations of a nucleic acid or even a small protein molecule is immense. Given present techniques for building molecular models, a single person (or even a small group of people) cannot rapidly calculate the sum of the energies of the weak bonds for each possible configuration; years would be required. Future work using electronic computers, however, should simplify some of this task. The second reason for our present inability to derive protein and nucleic acid structures is that our knowledge about the nature of weak bonds is still very incomplete; in many cases, we are not sure about either the exact bond energies or the possible angles they form to each other.

Today, protein and nucleic acid shapes can be revealed only by x-ray diffraction analysis. Fortunately these experimental structure determinations are beginning to reveal some general rules that tell us which weak chemical interactions are most important in governing the molecular shapes of large molecules. In particular, these rules emphasize the vital importance of interactions of the macromolecules with water, by far the most common molecule in all cells.

POLYMERIC MOLECULES ARE SOMETIMES HELICAL

Earlier we emphasized that polymeric molecules, like proteins and nucleic acids, have regular linear backbones in which specific groups (e.g.,—CO—NH—) repeat over and over along the molecule.

Often these regular groups are arranged in helical configurations held together by secondary bonds. The helix is a natural conformation for regular linear polymers,

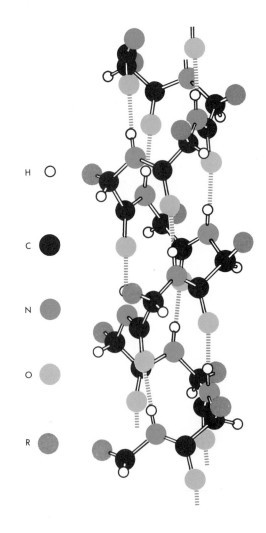

Figure 4–13
A polypeptide chain folded into a helical configuration called the α-helix. All the backbone atoms have identical orientations within the molecule. It may be looked at as a spiral staircase in which the steps are formed by amino acids. There is an amino acid every 1.5 Å along the helical axis. The distance along the axis required for one turn is 5.3 Å, giving 3.6 amino acids per turn. The helix is held together by hydrogen bonds between the carbonyl group of one residue and the imino group of the fourth residue down along the chain. (Redrawn from L. Pauling, *The Nature of the Chemical Bond*, 3rd ed., Cornell Univ. Press, Ithaca, N.Y., 1960, p. 500, with permission.)

since it places each monomer group in an identical orientation within the molecule (Figure 4–13). Each monomer therby forms the same group of secondary bonds as every other monomer. On the contrary, when a regular linear polymer has a nonhelical arrangement, different monomers must form different secondary bonds. Clearly, an unstable state occurs if any one set of secondary bonds is much stronger than any of the other arrangements. Thus, helical symmetry does not evolve from the particular shape of the monomer, but is instead the natural consequence of the existence of a unique monomer arrangement which is significantly more stable than all other arrangements.

It is important to remember that most biopolymers are not regular polymers containing identical monomers. Instead they often have irregular side groups attached to a regular backbone. When this happens, as it does in both nucleic acids and proteins, we need not necessarily expect a helical structure: A 3-D arrangement that is energetically very satisfactory for the backbone groups often produces very unsatisfactory bonding of the side groups. The 3-D structure of many irregular polymers is thus a compromise between the tendency of regular backbones to form a regular helix and the tendency of the side groups to twist the backbone into a configuration that maximizes the strength of the secondary bonds formed by the side groups.

PROTEIN STRUCTURES ARE USUALLY IRREGULAR

In the case of proteins, the compromise between the side groups and the backbone groups is usually decided in favor of the side groups. Thus, as we shall show in much greater detail in Chapter 6, most amino acids in proteins are not part of regular helices. This is because almost one-half of the side groups are nonpolar and can be placed in contact with water only by a considerable input of free energy. This conclusion was at first a surprise to many chemists, who were influenced by the fact that backbone groups could form strong internal hydrogen bonds, whereas the nonpolar groups could form only the much weaker van der Waals bonds. Their past reasoning was faulty, however, because it did not consider either the fact that the polar backbone can form almost as strong external hydrogen bonds to water, or the equally important fact that a significant amount of energy is necessary to push nonpolar side groups into a hydrogen-bonded water lattice.

This argument leads to the interesting prediction that in aqueous solutions macromolecules containing a large number of nonpolar side groups will tend to be more stable than molecules containing mostly polar groups. If we disrupt a polar molecule held together by a large number of internal hydrogen bonds, the decrease in free energy is often small, since the polar groups can then hydrogen bond to water. On the other hand, when we disrupt molecules having many nonpolar groups, there is usually a much greater loss in free energy, because the disruption necessarily inserts nonpolar groups into water.

DNA CAN FORM A REGULAR HELIX

At first glance, DNA looks even more unlikely to form a regular helix than does an irregular polypeptide chain. DNA not only has an irregular sequence of side groups, but in addition, all its side groups are hydrophobic. Both the purines (adenine and guanine) and the pyrimidines (thymine and cytosine), even though they contain polar

Figure 4–14
The hydrogen bonded base pairs in DNA. Adenine is always attached to thymine by two hydrogen bonds, whereas guanine always bonds to cytosine by three hydrogen bonds. The obligatory pairing of the smaller pyrimidine with the larger purine allows the two sugar–phosphate backbones to have identical helical configurations. All the hydrogen bonds in both base pairs are strong, since each hydrogen atom points directly at its acceptor atom (nitrogen or oxygen).

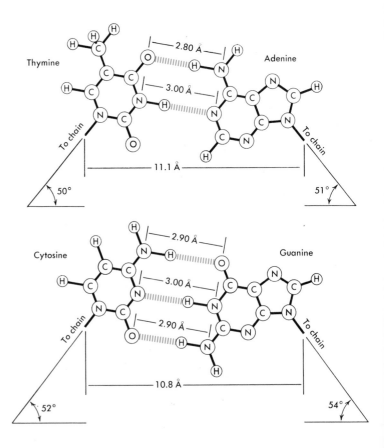

$C=O$ and NH_2 groups, are quite insoluble in water because their flat sides are completely hydrophobic.

Nonetheless, DNA molecules usually have regular helical configurations. This is because most DNA molecules contain two polynucleotide strands that have complementary structures (see Chapter 9 for more details). Both internal and external secondary bonds stabilize the structure. The two strands are held together by hydrogen bonds between pairs of complementary purines and pyrimidines (Figure 4–14). Adenine (amino) is always hydrogen bonded to thymine (keto), whereas guanine (keto) is hydrogen bonded to cytosine (amino). In addition, virtually all the surface atoms in the sugar and phosphate groups form bonds to water molecules.

The purine–pyrimidine base pairs are found in the center of the DNA molecule. This arrangement allows their flat surfaces to stack on top of each other and so limits their contact with water. This stacking arrangement would be much less satisfactory if only one chain were present. A single chain could not have a regular backbone because its pyrimidines are smaller than the purines, and so the angle of helical rotation would have to vary with the sequence of bases. The presence of complementary base pairs in double-helical DNA makes a regular structure possible, since each base pair is of the same size.

DNA MOLECULES ARE STABLE AT PHYSIOLOGICAL TEMPERATURES

The double-helical DNA molecule is very stable at physiological temperatures, for two reasons. First, disruption of the double helix breaks the regular hydrogen bonds and brings the hydrophobic purines and pyrimidines into contact with water. Second, individual DNA molecules have a *very large number of weak bonds,* arranged so that most of them cannot break without the simultaneous breaking of many others. Even though thermal motion is constantly breaking apart the terminal purine–pyrimidine pairs at the ends of each molecule, the two chains do not usually fall apart, because the hydrogen bonds in the middle are still intact (Figure 4–15). Once a break occurs, the most likely next event is the reforming of the same hydrogen bonds to restore the original molecular configuration. Sometimes, of course, the first breakage is followed by a second one, and so forth. Such multiple breaks, however, are quite rare, so that double helices held together by more than ten nucleotide pairs are very stable at room temperature.

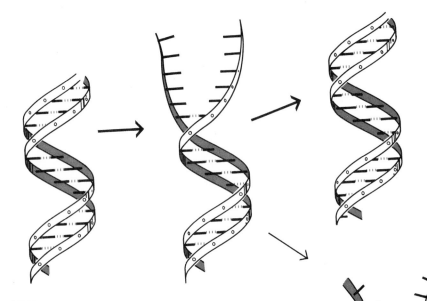

Figure 4–15
The breaking of terminal hydrogen bonds in DNA by random thermal motion. Because the internal hydrogen bonds continue to hold the two chains together, the immediate reforming of the broken bonds is highly probable. Also shown is the very rare alternative: the breaking of further hydrogen bonds, and the consequent disentangling of the chains.

The same principle also governs the stability of most protein molecules. Stable protein shapes are never due to the presence of just one or two weak bonds, but must always represent the cooperative result of a number of weak bonds.

Ordered collections of hydrogen bonds become less and less stable as their temperature is raised above physiological temperatures. At physiologically abnormally high temperatures, the simultaneous breakage of several weak bonds is more frequent. After a significant number have broken, a molecule usually loses its original form (the process of denaturation) and assumes an inactive (denatured) configuration.

MOST MEDIUM SIZE AND ALMOST ALL
LARGE PROTEIN MOLECULES ARE AGGREGATES
OF SMALLER POLYPEPTIDE CHAINS

Earlier we pointed out how the realization that macromolecules are all polymers constructed from small regular monomers, such as the amino acids, greatly simplified the problem of solving macromolecule structure. It has recently become clear that most of the very large proteins are regular aggregates of much smaller polypeptide chains, containing up to 400 amino acids apiece. For example, the protein ferritin, which functions in mammals to store iron atoms, has a molecular weight of about 480,000. It contains, however, not just one long polypeptide chain of 4000 amino acids, but instead 20 identical smaller polypeptide chains of about 200 amino acids each. Similarly, the protein component of tobacco mosaic virus was originally thought to have the horrendous molecular weight of 36,000,000. Most fortunately, it was subsequently discovered (see Chapter 15) that each TMV protein contains 2150 identical smaller protein molecules, each containing 158 amino acids. Even much smaller protein molecules are frequently constructed from a number of polypeptide chains. Hemoglobin, which has a molecular weight of only 64,500, contains four polypeptide chains, 2 α chains, and 2 β chains, each of which has a molecular weight of about 16,000.

In all three examples, as with most other protein aggregates, the smaller units are held together by secondary bonds. This fact is known because they can be dispersed by the addition of reagents (e.g., urea), which tend to break secondary bonds but not covalent bonds. But weak bonds are not the only force holding macromolecular units together. In some cases, for example, the protein insulin, disulfide bonds (S—S) between cysteine residues are also a binding force.

SUBUNITS ARE ECONOMICAL

Both the construction of polymers from monomers and the use of polymeric molecules themselves as subunits to build still larger molecules reflect a general building principle applicable to all complex structures, nonliving as well as living. This principle states that it is much easier to reduce the impact of construction mistakes if we can discard them before they are incorporated into the final product. For example, let us consider two alternative ways of constructing a molecule with 1,000,000 atoms. In scheme (a) we build the structure atom by atom; in scheme (b) we first build 1000 smaller units, each with 1000 atoms, and subsequently put the subunits together into the 1,000,000-atom product. Now consider that our building process randomly makes mistakes, inserting the wrong atom with a frequency of 10^{-5}. Let us assume that each mistake results in a nonfunctional product.

Under scheme (a), each molecule will contain, on the average, 10 wrong atoms, and so almost no good products will be synthesized. Under scheme (b), however, mistakes will occur in only 1 percent of the subunits. If there is a device to reject the bad subunits, then good products can be easily made and the cell will hardly be bothered by the presence of the 1 percent of nonfunctional subunits. This concept is the basis of the assembly line in which complicated industrial products such as radios and automobiles are constructed. At each stage of assembly there are devices to throw away bad subunits. In industrial assembly lines, mistakes were initially removed by human hands; now automation often replaces manual control. In cells, mistakes are sometimes controlled by the specificity of enzymes: if a monomeric subunit is wrongly put together, it usually will not be recognized by the polymer-making specific enzyme, and hence will not be incorporated into a macromolecule. In other cases, faulty substances are rejected because they are unable spontaneously to become part of the stable molecular aggregates.

THE PRINCIPLE OF SELF-ASSEMBLY

ΔG values of 1 to 5 kcal/mole mean not only that single weak bonds will be spontaneously made, but also that structures held together by several weak bonds will be spontaneously formed. For example, an unstable, unfolded polypeptide chain tends to assume a large number of random configurations as a result of thermal movements. Most of these conformations are thermodynamically unstable. Inevitably, however, thermal movements bring together groups that can form good weak bonds. These groups tend to stay together, because more free energy is lost when they form than can be regained by their breakage. Thus, by a random series of movements, the polypeptide chain gradually assumes a configuration in which most of, if not all, the atoms have fixed positions within the molecule.

Aggregation of separate molecules also occurs spontaneously. The protein hemoglobin furnishes a clear example (Figure 4–16). It can be broken apart by the addition

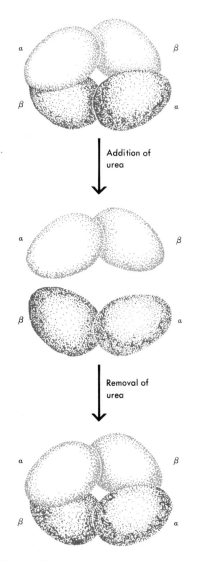

Figure 4–16
Formation of an active hemoglobin molecule from two half-molecules. Each hemoglobin molecule contains two α and two β chains. When placed in urea (a reagent which destabilizes weak bonds), the native molecule falls apart to two halves, containing one α and one β chain. Upon removal of the urea, the halves reassociate to form the complete molecule.

of reagents such as urea, which break secondary bonds to yield half-molecules of MW = 32,000. If, however, the urea is removed, the half-molecules quickly aggregate to form functional hemoglobin molecules. The surface structure of the half-molecules is very specific, for they bind only to each other and not with any other cellular molecules.

This same general principle of self-assembly operates to build even larger and more complicated structures, like the cell membrane and the cell wall. Both are mosaic surfaces containing large numbers of various molecules, some large, like proteins, and others much smaller, like lipids. At present, practically nothing is known about the precise arrangement of the molecules in these very large, complicated structures. Nonetheless, there is every reason to believe that the constituent molecules form stable contacts only with other molecules in the cell membrane (or wall). This situation is easy to visualize in the case of lipids, which are extremely insoluble in water because of their long, nonpolar hydrocarbon chains. Newly synthesized lipids have a stronger tendency to attach to other lipids in the cell membrane or cell wall by van der Waals forces than to enter some other more polar area, such as the aqueous (polar) interior of the cell.

SUMMARY

Many important chemical events in cells do not involve the making or breaking of covalent bonds. The cellular location of most molecules depends on the existence of "weak" (secondary) attractive or repulsive forces. In addition, weak bonds are important in determining the shape of many molecules, especially very large ones. The most important of these "weak" forces are: hydrogen bonding, van der Waals interactions, and ionic bonds. Even though these forces are relatively weak, they are still large enough to ensure that the right molecules (groups) interact with each other. For example, the surface of an enzyme is uniquely shaped to allow specific attraction of its substrates.

The formation of all chemical bonds, weak interactions as well as strong covalent bonds, proceeds according to the laws of thermodynamics. A bond tends to form when the result would be a release of free energy (ΔG negative). In order for the bond to be broken, this same amount of free energy must be supplied. Because the formation of covalent bonds between atoms usually involves a very large negative ΔG, covalently bound atoms almost never separate spontaneously. In contrast, the ΔG values accompanying the formation of weak bonds are only several times larger than the average thermal energy of molecules at physiological temperatures. Single weak bonds are thus frequently being made and broken in living cells.

Molecules having polar (charged) groups interact quite differently from nonpolar molecules (in which the charge is symmetrically distributed). Polar molecules can form good hydrogen bonds, whereas nonpolar molecules can form only van der Waals bonds. The most important polar molecule is water. Each water molecule can form four good hydrogen bonds to other water molecules. Although polar molecules tend to be soluble in water (to various degrees), nonpolar molecules are insoluble, because they cannot form hydrogen bonds with water molecules.

Every distinct molecule has a unique molecular shape that restricts the number of molecules with which it can form good secondary bonds. Strong secondary interactions demand both a complementary (lock-and-key) relationship between the two bonding surfaces, and the involvement of many atoms. Although molecules bound together by only one or two secondary bonds frequently fall apart, a collection of these weak bonds can result in a stable aggregate. The fact that double-helical DNA never falls apart spontaneously demonstrates the extreme stability possible in such an aggregate. The formation of such aggregates can proceed spontaneously, with the correct bonds forming in a step-by-step fashion (the principle of self-assembly).

The shape of polymeric molecules is determined by secondary bonds. All biological polymers contain single bonds about which free rotation is possible. They do not, however, exist in a variety of shapes as might be expected, because the formation of one of the possible configurations generally involves a maximum decrease in free energy. This energetically preferred configuration thus is formed exclusively. Some polymeric molecules have regular helical backbones held in shape by sets of regular internal secondary bonds between backbone groups. Regular helical structures cannot be formed, however, if they place the specific side groups in positions in which they cannot form favorable weak bonds. This situation occurs in many proteins where an irregular distribution of nonpolar side groups forces the backbone into a highly irregular conformation, permitting the nonpolar groups to form van der Waals bonds with each other. Irregularly distributed side groups do not always lead, however, to nonhelical molecules. In the DNA molecule, for example, the specific pairing of purines with pyrimidines in a double-stranded helix allows the nonpolar aromatic groups to stack on top of each other in the center of the molecule.

REFERENCES

LEHNINGER, A. L., *Bioenergetics,* 2nd ed., Benjamin, Menlo Park, California, 1971. A concise description of the laws of thermodynamics written for the beginning biology student appears in the first several chapters.

BLUM, H. F., *Time's Arrow and Evolution,* Harper, New York, 1962. Chapter 3 provides an exceptionally clear introduction to the thermodynamics applicable to biological systems.

KLOTZ, I. M. *Energy Changes in Biochemical Reactions,* Academic, New York, 1967. A somewhat advanced discussion of thermodynamics as it relates to biochemical reactions.

PAULING, L., *The Nature of the Chemical Bond,* 3rd ed., Cornell University Press, Ithaca, New York, 1960. One of the great classics of all chemical literature; a treatment of structural chemistry with considerable emphasis on the hydrogen bond.

MOROWITZ, H. J., *Entropy for Biologists,* Academic, New York, 1970.

Coupled Reactions and Group Transfers

In the previous chapter we looked at the formation of weak bonds from the thermodynamic viewpoint. Each time a potential weak bond was considered, the question was posed, "Does its formation involve a gain or a loss of free energy?", because only when ΔG is negative does the thermodynamic equilibrium favor a reaction. This same approach holds with equal validity for covalent bonds. The fact that enzymes are usually involved in the making or breaking of a covalent bond does not in any sense alter the requirement of a negative ΔG.

CHAPTER 5

On superficial examination, however, many of the important covalent bonds in cells appear to be formed in violation of the laws of thermodynamics, particularly those bonds joining small molecules together to form large polymeric molecules. The formation of such bonds involves an increase in free energy. Originally this fact suggested to some people that cells had the unique property of working somehow in violation of thermodynamics, and that this property was, in fact, the real "secret of life."

Now, however, it is clear that these biosynthetic processes do not violate thermodynamics but, instead, that they are based upon different reactions from those originally postulated. Nucleic acids, for example, do not form by the condensation of nucleoside phosphates; glycogen is not formed directly from glucose residues; proteins are not formed by the union of amino acids. Instead, the monomeric precursors, using energy present in ATP, are first

113

converted to high-energy "activated" precursors, which then spontaneously (with the help of specific enzymes) unite to form larger molecules. In this chapter we shall illustrate these ideas by concentrating on the thermodynamics of peptide (protein) and phosphodiester (nucleic acid) bonds. First, however, we must briefly look at some general thermodynamic properties of covalent bonds.

FOOD MOLECULES ARE THERMODYNAMICALLY UNSTABLE

There is great variation in the amount of free energy possessed by specific molecules. This is a consequence of the fact that covalent bonds do not all have the same bond energy. As an example, the covalent bond between oxygen and hydrogen is considerably stronger than the bonds between hydrogen and hydrogen or oxygen and oxygen. The formation of an O—H bond at the expense of O—O or H—H will thus release energy. Energetic considerations tell us that a sufficiently concentrated mixture of oxygen and hydrogen will be transformed into water.

A molecule thus possesses a larger amount of free energy if linked together by weak covalent bonds than if it is linked together by strong bonds. This idea seems almost paradoxical at first glance, since it means that the stronger the bond the less energy it can give off. But the notion automatically makes sense when we realize that an atom that has formed a very strong bond has already lost a large amount of free energy in this process. Therefore, the best food molecules (molecules which donate energy) are those molecules that contain weak covalent bonds and are, thereby, thermodynamically unstable.

For example, glucose is an excellent food molecule, since there is a great decrease in free energy when it is oxidized by O_2 to yield CO_2 and H_2O. On the contrary, CO_2 is not a food molecule in animals, since, in the absence of the energy donor ATP, it cannot spontaneously be transformed to more complex organic molecules even with the help of specific enzymes. CO_2 can be used as a primary source of carbon in plants only because the energy supplied by light quanta during photosynthesis results in the formation of ATP.

DISTINCTION BETWEEN DIRECTION AND RATE OF A REACTION

The chemical reactions by which molecules are transformed into other molecules which contain less free energy do not occur at significant rates at physiological temperatures in the absence of a catalyst. This is because even a "weak covalent bond" is, in reality, very strong and is only rarely broken by thermal motion within a cell. In order for a covalent bond to be broken in the absence of a catalyst, energy must be supplied to push apart the bonded atoms. When the atoms are partially apart, they can recombine with new partners to form stronger bonds. In the process of recombination, the energy released is the sum of the free energy supplied to break the old bond plus the difference in free energy between the old and the new bond (Figure 5–1).

The energy that must be supplied to break the old covalent bond in a molecular transformation is called the *activation energy*. The activation energy is usually less than the energy of the original bond because molecular rearrangements generally do not involve the production of completely free atoms. Instead, a collision between the two reacting molecules is required, followed by the temporary formation of a molecular complex (the *activated state*). In the activated state, the close proximity of the two molecules makes each other's bonds more labile, so that less energy is needed to break a bond than when the bond is present in a free molecule.

Most reactions of covalent bonds in cells are, therefore, described by

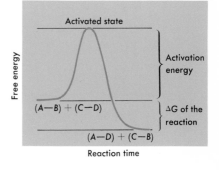

Figure 5–1
The energy of activation of a chemical reaction $(A-B) + (C-D) \rightarrow (A-D) + (C-B)$.

$$A\text{–}B + C\text{–}D \rightarrow A\text{–}D + C\text{–}B - \Delta G \qquad (5\text{–}1)$$

The mass action expression for such reaction is

$$K_{eq} = \frac{\text{conc}^{A\text{-}D} \times \text{conc}^{C\text{-}B}}{\text{conc}^{A\text{-}B} \times \text{conc}^{C\text{-}D}} \qquad (5\text{–}2)$$

where $\text{conc}^{A\text{-}B}$, $\text{conc}^{C\text{-}D}$, etc., are the concentrations of the several reactants in moles per liter. Here also, the value of K_{eq} is related to ΔG by Eq. (4–4).

Since energies of activation are generally between 20 and 30 kcal/mole, activated states practically never occur at physiological temperatures. High activation energies should thus be considered barriers preventing spontaneous rearrangements of cellular covalent bonds.

These barriers are enormously important. Life would be impossible if they did not exist, for all atoms would be in the state of least possible energy. There would be no way to temporarily store energy for future work. On the other hand, life would also be impossible if means were not found to selectively lower the activation energies of certain specific reactions. This also must happen if cell growth is to occur at a rate sufficiently fast so as not to be seriously impeded by random destructive forces, such as ionizing or ultraviolet radiation.

ENZYMES LOWER ACTIVATION ENERGIES

Enzymes are absolutely necessary for life because they lower activation energies. The function of enzymes is to speed up the rate of the chemical reactions requisite to cellular existence by lowering the activation energies of molecular rearrangements to values that can be supplied by the heat of motion. When a specific enzyme is present, there is no longer an effective barrier preventing the rapid formation of the reactants possessing the lowest amounts of free energy. Enzymes never affect the nature of an equilibrium: They merely speed up the rate at which it is reached. Thus, if the thermodynamic equilibrium is unfavorable for the formation of a molecule, the presence of an enzyme can in no way bring about its accumulation.

The need for enzymes to catalyze essentially every cellular molecular rearrangement means that knowledge of the free energy of various molecules cannot by itself tell us whether an energetically feasible rearrangement will, in fact, occur. The rate of the reactions must always be considered. Only if a cell possesses a suitable enzyme will the corresponding reaction be important.

A METABOLIC PATHWAY IS CHARACTERIZED BY A DECREASE IN FREE ENERGY

Thermodynamics tells us that all biochemical pathways must be characterized by a decrease in free energy. This is obviously the case for degradative pathways, in which thermodynamically unstable food molecules are converted to more stable compounds, such as CO_2 and H_2O, with the evolution of heat. All degradative pathways have two primary purposes: (1) to produce the small organic fragments necessary as building blocks for larger organic molecules, and (2) to conserve a significant fraction of the free energy of the original food molecule in a form that can do work, by coupling some of the steps in degradative pathways with the simultaneous formation of molecules that can store free energy (high-energy molecules) like ATP.

Not all the free energy of a food molecule is converted into the free energy of high-energy molecules. If this were the case, a degradative pathway would not be characterized by a decrease in free energy. No driving force would exist to favor the breakdown of food molecules. Instead, we find that all degradative pathways are characterized by a conversion of at least one-half the free energy of the food molecule into heat or entropy. For example, it is now believed that, in cells, approximately 40 percent of the free energy of glucose is used to make new high-energy compounds, the remainder being dissipated into heat energy and entropy.

HIGH-ENERGY BONDS HYDROLYZE WITH LARGE NEGATIVE ΔG's

A high-energy molecule contains a bond(s) whose breakdown by water (hydrolysis) is accompanied by a large decrease in free energy (5 kcal/mole or more). The specific bonds whose hydrolysis yields these large negative ΔG's are called *high-energy bonds*. Both these terms are, in a real sense, misleading, since it is not the bond energy but the free energy of hydrolysis that is high. Nonetheless, the term high-energy bond is generally employed, and, for convenience, we shall continue this usage by marking high-energy bonds with the symbol $\sim$.

The energy of hydrolysis of the average high-energy bond (7 kcal/mole) is very much smaller than the amount of energy that would be released if a glucose molecule were to be completely degraded in one step (688 kcal/mole). A one-step breakdown of glucose would be inefficient in making high-energy bonds. This is undoubtedly the reason why biological glucose degradation requires so many steps. In this way, the amount of energy released per degradative step is of the same order of magnitude as the free energy of hydrolysis of a high-energy bond.

The most important high-energy compound is ATP. It is formed from inorganic phosphate $\textcircled{P}$ and ADP, using energy obtained either from degradative reactions (some of which are shown in Chapters 2 and 3) or from the sun (photosynthesis). There are, however, many other important high-energy compounds. Some are directly formed during degradative reactions; others are formed using some of the free energy of ATP. Table 5–1 lists the most important types of high-energy bonds. All involve either phosphate or sulfur atoms. The high-energy pyrophosphate bonds of ATP arise from the union of phosphate groups. The pyrophosphate linkage ($\textcircled{P}\sim\textcircled{P}$) is not, however, the only kind of high-energy phosphate bond: The attachment of a phosphate group to the oxygen atom of a carboxyl group creates a high-energy acyl bond. It is now clear that high-energy bonds involving sulfur atoms play almost as important a role in energy metabolism as those involving phosphorus. The most important molecule containing a high-energy sulfur bond is acetyl-CoA. This

117

Table 5–1 Important Classes of High-Energy Bonds

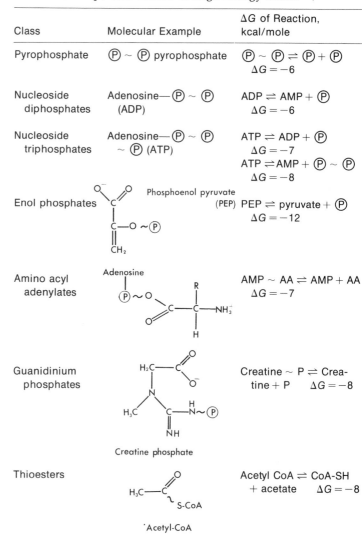

Class	Molecular Example	ΔG of Reaction, kcal/mole
Pyrophosphate	$\text{P} \sim \text{P}$ pyrophosphate	$\text{P} \sim \text{P} \rightleftharpoons \text{P} + \text{P}$ $\Delta G = -6$
Nucleoside diphosphates	Adenosine—$\text{P} \sim \text{P}$ (ADP)	$\text{ADP} \rightleftharpoons \text{AMP} + \text{P}$ $\Delta G = -6$
Nucleoside triphosphates	Adenosine—$\text{P} \sim \text{P}$ $\sim \text{P}$ (ATP)	$\text{ATP} \rightleftharpoons \text{ADP} + \text{P}$ $\Delta G = -7$ $\text{ATP} \rightleftharpoons \text{AMP} + \text{P} \sim \text{P}$ $\Delta G = -8$
Enol phosphates	Phosphoenol pyruvate (PEP)	$\text{PEP} \rightleftharpoons \text{pyruvate} + \text{P}$ $\Delta G = -12$
Amino acyl adenylates		$\text{AMP} \sim \text{AA} \rightleftharpoons \text{AMP} + \text{AA}$ $\Delta G = -7$
Guanidinium phosphates	Creatine phosphate	$\text{Creatine} \sim \text{P} \rightleftharpoons \text{Crea-}$ $\text{tine} + \text{P}$ $\Delta G = -8$
Thioesters	Acetyl-CoA	$\text{Acetyl CoA} \rightleftharpoons \text{CoA-SH}$ $+ \text{ acetate}$ $\Delta G = -8$

bond is the main source of energy for fatty acid biosynthesis.

The wide range of ΔG values of high-energy bonds (Table 5–1) means that calling a bond "high-energy" is sometimes arbitrary. The usual criterion is whether its hydrolysis can be coupled with another reaction to effect an important biosynthesis. For example, the negative ΔG accompanying the hydrolysis of glucose-6-P is 3 to 4 kcal/mole. This ΔG is not sufficient for efficient synthesis of peptide bonds, for example, so this phosphate ester bond is not included among high-energy bonds.

HIGH-ENERGY BONDS NECESSARY
FOR BIOSYNTHETIC REACTIONS

Often the construction of a large molecule from smaller building blocks requires the input of free energy. Yet a biosynthetic pathway, like a degradative pathway, would not exist if it were not characterized by a net decrease in free energy. This means that many biosynthetic pathways demand the existence of an external source of free energy. These free-energy sources are the "high-energy compounds." The making of many biosynthetic bonds is coupled with the breakdown of a high-energy bond, so that the net change of free energy is always negative. High-energy bonds in cells, therefore, generally have a very short life. Almost as soon as they are formed during a degradative reaction, they are enzymatically broken down to yield the energy needed to drive another reaction to completion.

Not all the steps in a biosynthetic pathway require the breakdown of a high-energy bond. Often only one or two steps involve such a bond. Sometimes this is because the ΔG, even in the absence of an externally added high-energy bond, favors the biosynthetic direction. In other cases, ΔG is effectively zero, or in some cases may even be slightly positive. These small positive ΔG's, however, are not significant so long as they are followed by a reaction characterized by the hydrolysis of a high-energy bond. Instead, it is the *sum* of all the free-energy changes in a pathway that is significant. It does not really matter that the K_{eq} of a specific biosynthetic step is slightly (80:20) in favor of degradation, if the K_{eq} of the succeeding step is 100:1 in favor of the forward biosynthetic direction.

Likewise, not all the steps in a degradative pathway generate high-energy bonds. For example, only two steps in the lengthy glycolytic (Embden–Meyerhof) breakdown of glucose generate ATP. Moreover, there are many degradative pathways that have one or more steps requiring the breakdown of a high-energy bond. The glycolytic breakdown of glucose is again an example. It uses up two molecules of ATP for every four that it generates. Here, of course, as in every energy-yielding degradative process, more high-energy bonds must be made than consumed.

PEPTIDE BONDS HYDROLYZE SPONTANEOUSLY

The formation of a dipeptide and a water molecule from two amino acids requires a ΔG of 1 to 4 kcal/mole, depending upon which amino acids are being bound. This

119

ΔG value decreases progressively if amino acids are added to longer polypeptide chains; for an infinitely long chain the ΔG is reduced to ~0.5 kcal/mole. This decrease reflects the fact that the free, charged NH_3^+ and COO^- groups at the chain ends favor the hydrolysis (breakdown accompanied by the uptake of a water molecule) of nearby peptide bonds.

These positive ΔG values by themselves tell us that polypeptide chains cannot form from free amino acids. In addition, we must take into account the fact that water molecules have a much, much higher concentration (generally > 100) than any other cellular molecules. All equilibrium reactions in which water participates are thus strongly pushed in the direction that consumes water molecules. This is easily seen in the definition of equilibrium constants. For example, the reaction forming a dipeptide,

$$\text{amino acid}(a) + \text{amino acid}(b) \rightarrow \text{dipeptide}(a\text{-}b) + H_2O \qquad (5\text{--}3)$$

has the following equilibrium constant:

$$K_{eq} = \frac{\text{conc }(a) \times \text{conc }(b)}{\text{conc }(a\text{-}b) \times \text{conc }(H_2O)} \qquad (5\text{--}4)$$

where concentrations are given in moles/liter. Thus, for a given K_{eq} value (related to ΔG by the formula $\Delta G = -RT \ln K$) a much greater concentration of H_2O means a correspondingly smaller concentration of the dipeptide. The relative concentrations are, therefore, very important. In fact, a simple calculation shows that hydrolysis may often proceed spontaneously even when the ΔG for the nonhydrolytic reaction is -3 kcal/mole.

Thus, in theory, proteins are unstable and, given sufficient time, will spontaneously degrade to free amino acids. On the other hand, in the absence of specific enzymes, these spontaneous rates are too slow to have a significant effect on cellular metabolism. That is, once a protein is made, it remains stable unless its degradation is catalyzed by a specific enzyme.

COUPLING OF NEGATIVE WITH POSITIVE ΔG

Free energy must be added to amino acids before they can be united to form proteins. How this could happen became clear with the discovery of the fundamental role of ATP as an energy donor. ATP contains three phosphate groups attached to an adenosine molecule (adenosine —O—Ⓟ~Ⓟ~Ⓟ). When one or two of the terminal

$\sim\!\textcircled{P}$ groups are broken off by hydrolysis, there is a significant decrease of free energy.

adenosine —O—$\textcircled{P}$ ~ $\textcircled{P}$ ~ $\textcircled{P}$ + H$_2$O →

    adenosine —O —$\textcircled{P}$ ~ $\textcircled{P}$ + $\textcircled{P}$ ($\Delta G = -7$ kcal/mole)   (5–5)
    (ADP)

adenosine —O—$\textcircled{P}$ ~ $\textcircled{P}$ ~ $\textcircled{P}$ + H$_2$O →

    adenosine —O—$\textcircled{P}$ + $\textcircled{P}$ ~ $\textcircled{P}$ ($\Delta G = -8$ kcal/mole)   (5–6)
    (AMP)

adenosine —O—$\textcircled{P}$ ~ $\textcircled{P}$ + H$_2$O →

    adenosine —O—$\textcircled{P}$ + $\textcircled{P}$ ($\Delta G = -6$ kcal/mole)   (5–7)
    (AMP)

All these breakdown reactions have negative ΔG values considerably greater in absolute value (numerical value without regard to sign) than the positive ΔG values accompanying the formation of polymeric molecules from their monomeric building blocks. The essential trick underlying those biosynthetic reactions, which by themselves have a positive ΔG, is that they are coupled with breakdown reactions characterized by negative ΔG of greater absolute value. Thus, during protein synthesis, the formation of each peptide bond ($\Delta G = +0.5$ kcal/mole) is coupled with the breakdown of ATP to AMP and pyrophosphate, which has a ΔG of -8 kcal/mole. This results in a net ΔG of -7.5 kcal/mole, more than sufficient to ensure that the equilibrium favors protein synthesis rather than breakdown.

ACTIVATION THROUGH GROUP TRANSFER

When ATP is hydrolyzed to ADP and $\textcircled{P}$, most of the free energy is liberated as heat. Since heat energy cannot be used to make covalent bonds, a coupled reaction cannot be the result of two completely separate reactions, one with a positive, the other with a negative, ΔG. Instead, a coupled reaction is achieved by two or more successive reactions. These are always *group-transfer* reactions: reactions, not involving oxidations or reductions, in which molecules exchange functional groups. The enzymes that catalyze these reactions are called transferases.

$$A\text{–}X + B\text{–}Y \rightarrow A\text{–}B + X\text{–}Y \qquad (5\text{–}8)$$

121

In this example, groups X and Y are exchanged with components A and B. Group-transfer reactions are arbitrarily defined to exclude H_2O as a participant. When H_2O is involved,

$$A{-}B + H{-}OH \rightarrow A{-}OH + BH \qquad (5\text{-}9)$$

the reaction is called a hydrolysis, and the enzymes involved, hydrolases.

The group-transfer reactions which interest us here are those involving groups attached by high-energy bonds (high-energy groups). When a high-energy group is transferred to an appropriate acceptor molecule, it becomes attached to the acceptor by a high-energy bond. Group transfer thus allows the transfer of high-energy bonds from one molecule to another. For example, Eqs. (5–10) and (5–11) show how energy present in ATP is transferred to form GTP, one of the precursors used in RNA synthesis:

adenosine —(P) ~ (P) ~ (P) + guanosine —(P) →
  (ATP)   (GMP)

        adenosine —(P) ~ (P) + guanosine —(P) ~ (P)   (5-10)
          (ADP)   (GDP)

adenosine —(P) ~ (P) ~ (P) + guanosine —(P) ~ (P) →
  (ATP)   (GDP)

        adenosine —(P) ~ (P) + guanosine —(P) ~ (P) ~ (P)   (5-11)
          (ADP)   (GTP)

The high-energy (P)~(P) group on GTP allows it to unite spontaneously with another molecule. GTP is thus an example of what is called an *activated molecule;* correspondingly, the process of transferring a high-energy group is called *group activation.*

ATP VERSATILITY IN GROUP TRANSFER

In Chapter 2 we emphasized the key role of ATP synthesis in the controlled trapping of the energy of food molecules. In both oxidative and photosynthetic phosphorylations, energy is used to synthesize ATP from ADP and P:

adenosine – (P)~(P) + (P) + energy → adenosine – (P)~(P)~(P)   (5-12)

Since ATP is, thus, the original biological recipient of high-energy groups, it must be the starting point of a variety of reactions in which high-energy groups are transferred to low-energy molecules to give them the potential to react spontaneously. ATP's central role utilizes the fact

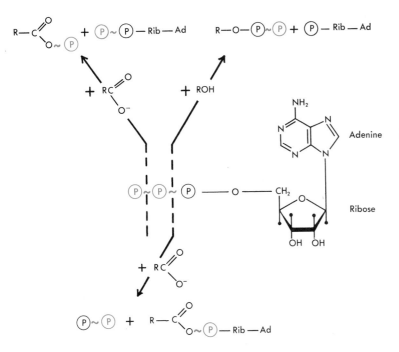

Adenine

Ribose

Figure 5–2
Important group transfers involving ATP.

that it contains two high-energy bonds whose splitting releases specific groups. This is shown in Figure 5–2, which shows three important groups arising from ATP: (1) $\textcircled{P}\sim\textcircled{P}$, a pyrophosphate group, (2) $\sim$AMP, an adenosyl monophosphate group, and (3) $\sim\textcircled{P}$, a phosphate group. It is important to notice that these high-energy groups retain their high-energy quality only when transferred to an appropriate acceptor molecule. For example, although the transfer of a $\sim\textcircled{P}$ group to a COO$^-$ group yields a high-energy COO$\sim\textcircled{P}$ acyl-phosphate group, the transfer of the same group to a sugar hydroxyl group (—C—OH), as for example in the formation of glucose-6-$\textcircled{P}$, gives rise to a low-energy bond (<5 kcal/mole decrease in ΔG upon hydrolysis).

ACTIVATION OF AMINO ACIDS BY ATTACHMENT OF AMP

The activation of an amino acid is achieved by transfer of an AMP group from ATP to the COO$^-$ group of the amino acid:

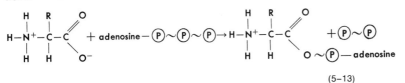

(5-13)

(R represents the specific side group of the amino acid.) The specific enzymes that catalyze this type of reaction are called amino-acyl synthetases. Upon activation, an amino acid is thermodynamically capable of being efficiently used for protein synthesis. Nonetheless, the AA~AMP complexes are not the direct precursors of proteins. Instead, for a reason which we shall explain in Chapter 11, a second group transfer must occur to transfer the amino acid, still activated at its carboxyl group, to the end of a tRNA molecule:

$$AA \sim AMP + tRNA \rightarrow AA \sim tRNA + AMP \qquad (5\text{--}14)$$

A peptide bond then forms by the condensation of the AA ~ tRNA molecule onto the end of a growing polypeptide chain:

AA ~ tRNA + growing polypeptide chain (of *n* amino acids) →

tRNA + growing polypeptide chain (of *n* + 1 amino acids) (5–15)

Thus the final step of this "coupled reaction," like that of all other coupled reactions, necessarily involves the removal of the activating group and the conversion of a high-energy bond into one with a lower free energy of hydrolysis. This is the source of the negative ΔG which drives the reaction in the direction of protein synthesis.

NUCLEIC ACID PRECURSORS ACTIVATED BY PRESENCE OF Ⓟ~Ⓟ

Both types of nucleic acid, DNA and RNA, are built up of mononucleotide monomers (nucleoside Ⓟ). Mononucleotides, however, are thermodynamically even less likely to combine than amino acids. This is because the phosphodiester bonds which link the former together release considerable free energy upon hydrolysis (-6 kcal/mole). This means that nucleic acids will spontaneously hydrolyze, at a slow rate, to mononucleotides. Thus it is even more important that activated precursors be used in their synthesis than in that of proteins.

Recently it has been found that the immediate precursors for both DNA and RNA are the nucleoside-5'-triphosphates. For DNA these are dATP, dGTP, dCTP, and dTTP (d stands for deoxy); for RNA the precursors are ATP, GTP, CTP, and UTP. ATP thus not only serves as the main source of high-energy groups in group-transfer reactions, but in addition is itself a direct precursor for RNA. The other three RNA precursors all arise by group-transfer reactions like those described in Eqs. (5–10) and (5–11). The deoxytriphosphates are formed in basically the same way: After the deoxymononucleotides have been synthesized, they are transformed to the triphosphate form by group transfer from ATP:

$$\text{deoxynucleoside} - \text{(P)} + \text{ATP} \rightarrow \text{deoxynucleoside} - \text{(P)} \sim \text{(P)} + \text{ADP} \qquad (5\text{--}16)$$

$$\text{deoxynucleoside} - \text{(P)} \sim \text{(P)} + \text{ATP} \rightarrow$$
$$\text{deoxynucleoside} - \text{(P)} \sim \text{(P)} \sim \text{(P)} + \text{ADP} \qquad (5\text{--}17)$$

These triphosphates can then unite to form polynucleotides held together by phosphodiester bonds. In this process (a group-transfer reaction), a pyrophosphate bond is broken and a pyrophosphate group released:

$$\text{deoxynucleoside} - \text{(P)} \sim \text{(P)} \sim \text{(P)}$$
$$+ \text{ growing polynucleotide chain (of } n \text{ nucleotides)} \rightarrow$$
$$\text{(P)} \sim \text{(P)} + \text{ growing polynucleotide chain } (n + 1 \text{ nucleotides)} \qquad (5\text{--}18)$$

This reaction, unlike that which forms peptide bonds, does not have a negative ΔG. In fact, the ΔG is slightly positive (~ 0.5 kcal/mole). This immediately poses the question, since polynucleotides obviously form: What is the source of the necessary free energy?

VALUE OF (P)~(P) RELEASE IN NUCLEIC ACID SYNTHESIS

The needed free energy arises from the splitting of the high-energy pyrophosphate group which is formed simultaneously with the high-energy phosphodiester bond. All cells contain a powerful enzyme, pyrophosphatase, which breaks down pyrophosphate molecules almost as soon as they are formed:

$$\text{(P)} \sim \text{(P)} \rightarrow 2\,\text{(P)} \qquad (\Delta G = -7 \text{ kcal/mole}) \qquad (5\text{--}19)$$

The large negative ΔG means that the reaction is effectively irreversible: This means that once (P)~(P) is broken down it never reforms.

The union of the nucleoside monophosphate group [Eq. (5–16)], coupled with the splitting of the pyrophosphate groups [Eq. (5–19)], has an equilibrium constant determined by the combined ΔG values of the two reactions: (0.5 kcal/mole) + (-7 kcal/mole). The resulting value ($\Delta G = -6.5$ kcal/mole) tells us that nucleic acids almost never break down to reform their nucleoside triphosphate precursors.

Here we see a powerful example of the fact that often it is the free-energy change accompanying a *group of reactions* that determines whether a reaction in the group will take place. Reactions with small, positive ΔG values, which by themselves would never take place, are often part of important metabolic pathways in which they are followed by reactions with large negative ΔG's. At all times we must remember that a single reaction (or even a single pathway) never occurs in isolation, but rather that the nature of the equilibrium is constantly being changed through the addition and through the removal of metabolites.

Ⓟ~Ⓟ SPLITS CHARACTERIZE MOST BIOSYNTHETIC REACTIONS

The synthesis of nucleic acids is not the only reaction where direction is determined by the release and splitting of Ⓟ~Ⓟ. In fact, the generalization is emerging that essentially all biosynthetic reactions are characterized by one or more steps that release pyrophosphate groups. Consider, for example, the activation of an amino acid by the attachment of AMP. By itself, the transfer of a high-energy bond from ATP to the AA~AMP complex has a slightly positive ΔG. Therefore, it is the release and splitting of ATP's terminal pyrophosphate group that provides the negative ΔG that is necessary to drive the reaction.

The great utility of the pyrophosphate split is neatly demonstrated by considering the problems that would arise if a cell attempted to synthesize nucleic acid from nucleoside diphosphates rather than triphosphates. Phosphate, rather than pyrophosphate, would be liberated as the backbone phosphodiester linkages were made. The phosphodiester linkages, however, are not stable in the presence of significant quantities of phosphate, since they are formed without a significant release of free energy. Thus, the biosynthetic reaction would be easily reversible; as soon as phosphate began to accumulate, the reaction would begin to move in the direction of nucleic acid breakdown (mass-action law). Moreover, it is not possible for a cell to remove the phosphate groups as soon as they are generated (thus preventing this reverse reaction), since all cells need a significant internal level of phosphate in order to grow. Thus the use of nucleoside triphosphates as precursors of nucleic acids is not a matter of chance.

This same type of argument tells us why ATP, and not ADP, is the key donor of high-energy groups in all cells. At first this preference seemed arbitrary to biochemists. Now, however, we see that many reactions using ADP as an energy donor would occur equally well in both directions.

SUMMARY

The biosynthesis of many molecules appears, at a superficial glance, to violate the thermodynamic law that spontaneous reactions always involve a decrease in free energy (ΔG is negative). For example, the formation of proteins from amino acids has a positive ΔG. This paradox is removed when we realize that the biosynthetic reactions do not proceed as initially postulated. Proteins, for example, are not formed from free amino acids. Instead, the precursors are first enzymatically converted to high-energy activated molecules, which, in the presence of a specific enzyme, spontaneously unite to form the desired biosynthetic product.

Many biosynthetic processes are thus the result of "coupled" reactions, the first of which supplies the energy that allows the spontaneous occurrence of the second reaction. The primary energy source in cells is ATP. It is formed from ADP and inorganic phosphate, either during degradative reactions (e.g., fermentation or respiration) or during photosynthesis. ATP contains several (high-energy) bonds whose hydrolysis has a large negative ΔG. Groups linked by high-energy bonds are called high-energy groups. High-energy groups can be transferred to other molecules by group-transfer reactions, thereby creating new high-energy compounds. These derivative high-energy molecules are then the immediate precursors for many biosynthetic steps.

Amino acids are activated by the addition of an AMP group, originating from ATP, to form an AA~AMP molecule. The energy of the high-energy bond in the AA~AMP molecule is similar to that of a high-energy bond of ATP. Nonetheless, the group-transfer reaction proceeds to completion because the high-energy Ⓟ↝Ⓟ molecule, created when the AA~AMP molecule is formed, is broken down by the enzyme pyrophosphatase to low-energy groups. Thus, the reverse reaction, Ⓟ↝Ⓟ + AA~AMP → ATP + AA, cannot occur.

The general rule exists that Ⓟ↝Ⓟ is released in almost all biosynthetic reactions. Almost as soon as it is made, it is enzymatically broken down to 2Ⓟ, thereby making impossible a reversal of the biosynthetic reaction. The great utility of the Ⓟ↝Ⓟ split provides an explanation for why ATP, not ADP, is the primary energy donor. ADP cannot transfer a high-energy group and, at the same time, produce Ⓟ↝Ⓟ groups as a by-product.

REFERENCES

MAHLER, H. R., AND E. H. CORDES, *Biological Chemistry*, 2nd ed., Harper and Row, New York, 1971. A very complete advanced text that covers all of biochemistry.

LEHNINGER, A. L., *Biochemistry*, 2nd ed., 1975 Worth Publishing Company, New York, 1970. A superb general biochemistry text containing an abundance of clear illustrations.

STREYER, H., *Biochemistry*, Freeman, San Francisco, 1975. A very new and excellent text that emphasizes general principles—contains multicolored illustrations.

WHITE, A., P. HANDLER, AND E. L. SMITH, *Principles of Biochemistry*, 5th ed., McGraw-Hill, New York, 1971. A survey of cell biochemistry, strongest in its emphasis on intermediary metabolism.

WOOD, W. B., J. H. WILSON, R. M. BENBOW, AND L. HOOD, *Biochemistry: A Problems Approach*, Benjamin, Menlo Park, California, 1974. A most useful, very up-to-date, text based upon solving problems.

KREBS, H. A., AND H. L. KORNBERG, "A Survey of the Energy Transformation in Living Material," *Ergeb. Physiol. Biol. Chem. Exptl. Pharmakol.*, **49**: 212, 1957. This comprehensive review is one of the few classics in biochemistry. Though many of its facts have been modified by subsequent research, it can still be used with great profit.

KORNBERG, A., "On the Metabolic Significance of Phosphorolytic and Pyrophosphorolytic Reactions," in M. Kasha and B. Pullman (eds.), *Horizons in Biochemistry*, pp. 251–264, Academic Press, New York, 1962. In this short article are found some detailed arguments about the importance of the release of pyrophosphate.

The Concept of Template Surfaces

By now our chemist knows that there are several "key secrets of life" upon which the ability of a cell to grow and divide depends. First, there must exist a highly organized surface membrane capable of maintaining, through selective permeability, a high concentration of internal molecules. Second, enzymes that catalyze the movement of the atoms from food molecules into new cellular building blocks must exist. Third, useful energy must be derived from food molecules or the sun to ensure that the thermodynamic equilibria favor biosynthetic rather than degradative reactions.

CHAPTER 6

All these properties depend intimately upon the existence of proteins. Only these very large molecules with their 20 different building blocks possess sufficient specificity to build selectively permeable membranes or to catalyze highly specific chemical transformations. We must thus add to the list of key secrets of life the ability to synthesize the physiologically correct amounts of specific proteins. This requirement at first might seem to fall under the more general prerequisite of enzyme-catalyzed biosynthesis. But as we shall soon learn, the synthesis of a protein does not proceed according to rules governing the synthesis of small molecules. This point becomes clear when we look at the way enzymes are used to construct increasingly larger molecules.

SYNTHESIS OF SMALL MOLECULES

Let us first look at how the amino acid serine is normally put together in *E. coli* cells growing upon glucose as their sole energy and carbon source. Figure 6–1 illustrates how serine is formed in three steps from 3-phosphoglyceric acid, a key metabolite in the normal degradation of glucose (Figure 2–6). Serine can be further broken down in several more steps (whose exact chemistry has yet to be worked out) to give the simplest amino acid, glycine. Each of these steps requires a specific enzyme with a characteristic surface capable of combining only with its correct substrate. Each of the other 18 amino acids is synthesized according to the same principle. In every case, a metabolite derived from glucose serves as the starting point for a series of specific enzymatically mediated reactions leading finally to an amino acid. Likewise, the purine and pyrimidine nucleotides, the building blocks from which the nucleic acids

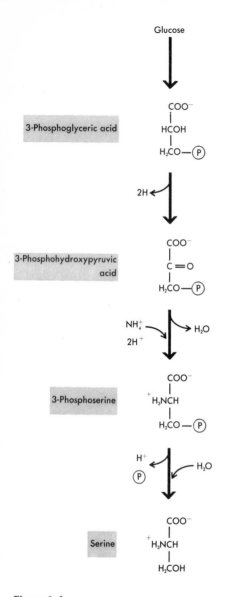

Figure 6–1
Serine biosynthesis.

Figure 6–2
The biosynthesis of uridine-5'-phosphate. ▶

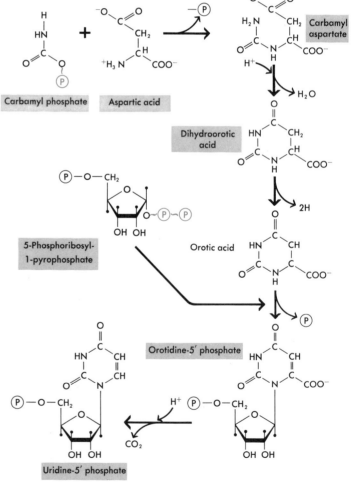

130

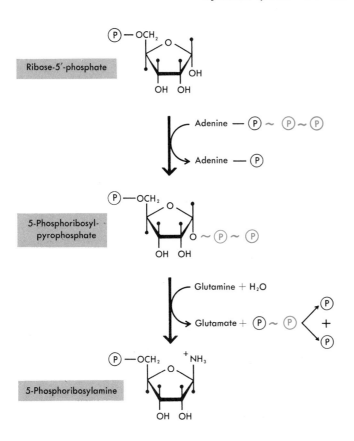

Figure 6–3
Initial steps in purine formation.

DNA and RNA are constructed, are synthesized by a series of consecutive reactions starting with smaller molecular units whose carbon atoms are derived from glucose molecules.

Some of the reactions leading to the synthesis of the pyrimidine nucleotide, uridine-5'-phosphate, are seen in Figure 6–2. The synthesis of the larger purine nucleotides requires more steps, since more covalent bonds must be built. Again, however, the same basic principles govern: (1) each reaction requires a different specific enzyme and (2) the sum of the reactions results in a release of free energy.

This energy release (usually as heat) means that the thermodynamic equilibrium favors the generation of the biosynthetic reaction products necessary for cell growth. It is often accomplished by having one of the substrates react with the energy-rich molecule ATP to form an activated substrate in which a phosphate ($\sim$ⓟ), pyrophosphate ($\sim$ⓟ$\sim$ⓟ), or adenylic acid ($\sim$AMP) group is attached to an atom involved in the formation of the desired biosynthetic bond. A typical ATP-driven synthesis is the transformation of ribose-5-ⓟ into 5-phosphoribosylamine (PRA) (Figure 6–3). This transformation, one of the initial

131

steps in purine nucleotide formation, occurs in two enzymatic steps. In the first, ribose-5-Ⓟ and ATP combine to form AMP and 5-phosphoribosylpyrophosphate (PRPP). The second step involves the reaction of PRPP with glutamine to yield PRA, Ⓟ~Ⓟ, and glutamic acid. The equilibrium of the first reaction favors PRPP synthesis because there is more energy in an ATP-pyrophosphate bond than in the phosphate ester (C—O—Ⓟ) bond attaching Ⓟ~Ⓟ to ribose-5-Ⓟ. Likewise, the second equilibrium favors PRA formation because the Ⓟ~Ⓟ product is broken down by pyrophosphatase to 2Ⓟ.

Both biosynthetic steps are thus accompanied by the release of energy as heat. In contrast there is little energy difference between the initial C—O bond of ribose-5-Ⓟ and the final C—N linkage. Hence, activation by an energy donor is a necessary prerequisite for this biosynthetic step. Activation is not, however, an obligatory feature of all biosyntheses. Sometimes the relevant covalent bonds in a necessary cell constituent have significantly less free energy than the bonds in the metabolites from which they are derived.

SYNTHESIS OF A LARGE "SMALL MOLECULE"

The construction of chlorophyll (Figure 6–4) is a good example. Here is a molecule (MW = 892) whose total laboratory synthesis has just recently been achieved, and which still looks very complex even to a first-rate organic chemist. It contains the complicated porphyrin ring, to which is attached a long unbranched alcohol (phytol). As yet, only the broad outlines of its biosynthesis are known. The porphyrin and phytol components are most likely synthesized separately and later joined together. Most of what is now known about its synthesis concerns the putting together of the porphyrin ring (Figure 6–5). Here a very large number of different enzymes are used to rearrange the C, N, O, and H atoms found initially in the much

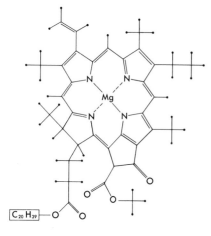

Figure 6–4
The structure of chlorophyll.

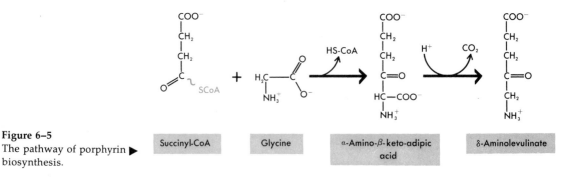

Figure 6–5
The pathway of porphyrin ▶
biosynthesis.

| Succinyl-CoA | Glycine | α-Amino-β-keto-adipic acid | δ-Aminolevulinate |

132

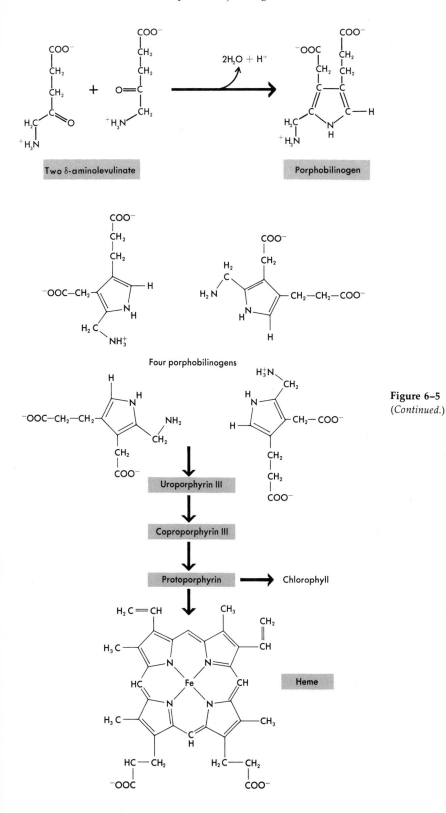

Two δ-aminolevulinate

Porphobilinogen

Four porphobilinogens

Uroporphyrin III

Coproporphyrin III

Protoporphyrin → Chlorophyll

Heme

Figure 6–5
(*Continued.*)

smaller glycine and succinyl~CoA precursors. No new qualitative features thus appear to distinguish the synthesis of molecules with chlorophyll-like complexity from the construction of small organic molecules. In both cases specific enzymes and favorable thermodynamic equilibria are necessary. There is only the quantitative difference that the biosynthesis of large complex molecules needs more different enzymes and usually more externally added energy.

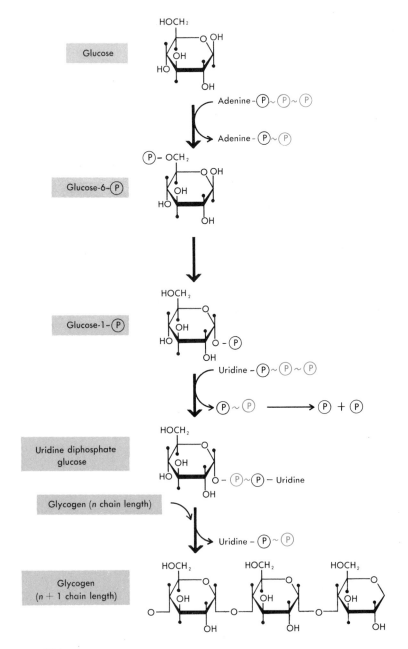

Figure 6–6
The biosynthesis of glycogen from glucose.

SYNTHESIS OF A REGULAR, VERY LARGE POLYMERIC MOLECULE

Glycogen is a macromolecule whose molecular weight is often above a million. Nonetheless, only four different enzymes are necessary to derive glycogen from glucose, because glycogen is a polymeric molecule built up by the repetitive linking together of glucose units. Figure 6–6 shows the specific chemical steps by which glucose is activated at its number 1 carbon atom and then polymerized. Only one enzyme is required for the final polymerization because each polymerization step makes the same type of chemical bond. Almost all the linkages are glucosidic bonds (C—O—C) between carbon atoms numbers 1 and 4. Infrequently, another enzyme catalyzes the formation of 1–6 glucosidic bonds. Thus glycogen is often branched.

We thus see that the number of enzymes necessary to synthesize a molecule is related not necessarily to its size, but to its chemical complexity. Thus glycogen, which is an easy molecule for the organic chemist to understand, also poses no fundamental problems to the biochemist.

A DEEPER LOOK INTO PROTEIN STRUCTURE

Before we go into the problems involved in protein synthesis, we must first look more closely into protein structure. Proteins are immensely complex macromolecules since they are polymers built up from 20 different building blocks (the amino acids). Thus the organic chemist must determine both how the amino acids are linked together and what their order is within a given linear polypeptide chain. Likewise, the biochemist wishes to know both how the backbone linkages are connected and what trick is used to order the amino acids during synthesis. In both types of work, the questions involving sequence have proved to be the more difficult questions to answer.

It was, in fact, not until 1953 that the first complete amino acid sequence became known. The protein studied was the hormone insulin, a relatively small protein containing 51 amino acids (Figure 6–7). More recently, the

Figure 6–7
The amino acid sequence of bovine insulin.
▼

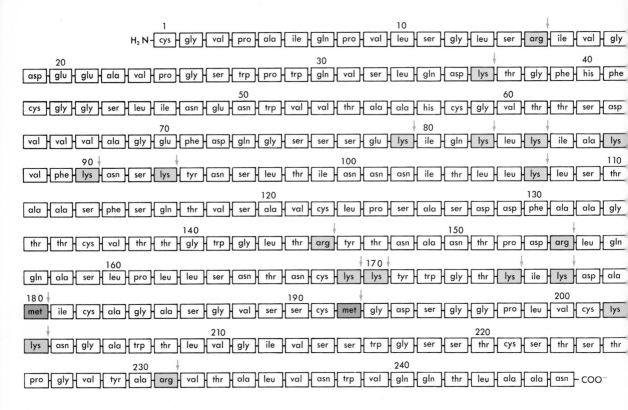

Figure 6–8

The amino acid sequence of the protein chymotrypsinogen, key to the establishment of such protein sequences in the isolation of smaller, well defined, polypeptide fragments. The enzyme trypsin, for example, cleaves on the carboxyl side of arginine and lysine residues; and the so-called "tryptic fragments" have proved invaluable in the working out of almost all protein sequences. Larger fragments generally are made by treatment with the reagent cyanogen bromide, which cleaves the peptide bond on the carboxyl side of the much less commonly employed amino acid methionine.

sequences of a large number of additional proteins have been solved. One of the larger, containing 246 amino acids, is the protein chymotrypsinogen, which, after conversion to chymotrypsin, catalyzes the hydrolysis of certain peptide bonds. The determination of its sequence (Figure 6–8) required almost 15 man years of work by several talented chemists. Now, new experimental techniques make sequence determinations much easier. Today, with luck, less than a year may be sufficient to solve the structure of a relatively small protein.

Aside from the question of sequence, there is also the problem of how polypeptide chains assume their final 3-D configurations. The correct functioning of almost all proteins depends not only upon possession of the correct amino acid sequence but also upon their exact arrangement in space. As we pointed out in Chapter 4, however, the polypeptide backbone is not completely rigid, for many of its atoms can freely rotate and assume different relative locations. Nonetheless, the tendency to form optimal weak bonds favors a unique conformation for a given protein. Very good indirect evidence indicates that in a given environmental situation all protein molecules with identical sequences have the same "native" 3-D form.

This belief receives direct support from the complete 3-D structural determination of the oxygen-carrying protein myoglobin. In Figure 6–9 is shown its structure as revealed by x-ray diffraction analysis. Though the molecule is immensely complex, detailed inspection shows an important, simplifying structural characteristic: the chain is folded to bring together atomic groupings that attract each other.

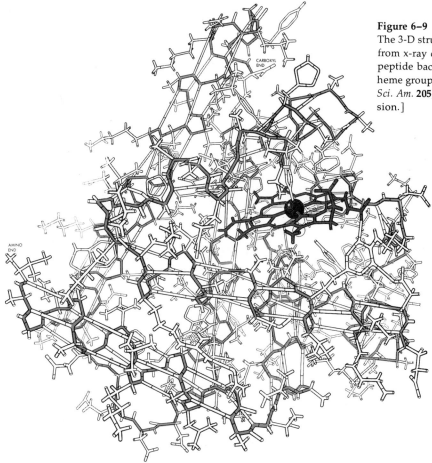

CARBOXYL END

AMINO END

Figure 6–9
The 3-D structure of myoglobin as derived from x-ray diffraction analysis. The polypeptide backbone is shown in color, the heme group in gray. [From J. C. Kendrew, *Sci. Am.* **205,** 100–101 (1961), with permission.]

THE PRIMARY STRUCTURES OF PROTEINS

Myoglobin, which has 153 amino acids, is one example of the many proteins that contain only one polypeptide chain. Many other proteins have two or more chains. For example, there are four polypeptide chains in the hemoglobin molecule, which has a MW of 64,500. The number of chains and the sequence of residues within them consti-

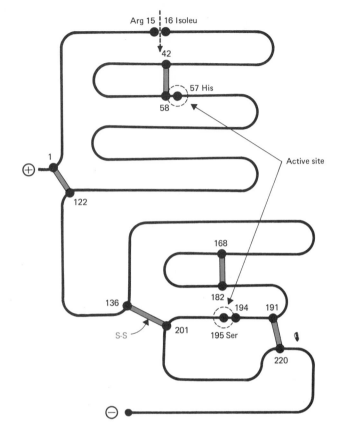

Figure 6–10

The arrangement of S—S bonds in chymotrypsinogen. Intact chymotrypsinogen molecules are enzymatically inactive. They become active by the enzymatic splitting of the peptide bond between amino acids 15 and 16 and subsequent excision of residues 14 and 15. The active split product is called chymotrypsin. The 3-D structure of chymotrypsin has been solved through x-ray analysis and its mode of action is known. Its active center involves the serine residue of position 195, the aspartic acid residue at position 102, and the histidine residue of position 57, and so the chain folds to bring these three amino acids near to each other.

tute the *primary structure of proteins.* When several polypeptide chains are present in the same molecule, they are often held together by secondary forces. In other cases, disulfide bonds (S—S) between cysteine side groups keep them together; they are what hold together the two chains of the insulin molecule (Figure 6–7). Disulfide bonds are important also in helping a single chain to maintain a rigid shape. In chymotrypsinogen there are 5 disulfide bridges, each linking specific cysteine residues (Figure 6–10).

In addition, a number of proteins have attached to them nonprotein (prosthetic) groups that play a vital role in their functional activity. They are often metal–organic compounds. Both myoglobin and hemoglobin contain the prosthetic group heme, a metal–organic compound closely related to the porphyrin component of chlorophyll. Heme combines with O_2 and gives to hemoglobin and myoglobin the ability to bind O_2.

A characteristic feature of prosthetic groups is that they possess very little functional activity unless they are attached to a polypeptide partner. Heme by itself, for example, combines with O_2 in an effectively irreversible fashion. Only when heme is attached to either myoglobin or hemoglobin does it possess the quality of reversibly binding oxygen. Then it can release bound oxygen when it is needed under conditions of oxygen scarcity.

SECONDARY STRUCTURES OF PROTEINS MAY BE SHEETS OR HELICES

The term *protein secondary structure* refers to the regular configurations of the polypeptide backbone. One class of these regular arrangements contains hydrogen bonds between groups on different polypeptide chains. These configurations, collectively called β structures, use polypeptide chains fully extended to form sheet-like structures held together by N—H $\cdot \cdot \cdot$ O=C hydrogen bonds (Figure 6–11). β structures are favored by the presence of large numbers of glycine and alanine residues. In nature they occur chiefly in silk proteins.

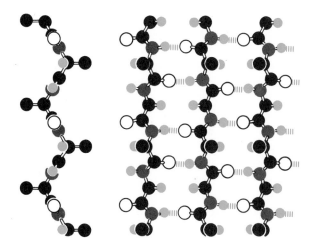

Figure 6–11
An example of extended polypeptide chains held together in sheets by hydrogen bonds (β configuration). (Redrawn from L. Pauling, *The Nature of the Chemical Bond*, 3rd ed., Cornell Univ. Press, Ithaca, N.Y., p. 501, with permission.)

The most important regular arrangement of the polypeptide chain, however, is brought about by hydrogen bonding between groups on the same chain. This bonding results in the twisting of the polypeptide backbone to form a helix. The most important polypeptide helix is the α-helix, which we have shown in Chapter 4 (Figure 4–13) to illustrate helical symmetry. X-ray diffraction analysis tells us that large sections of the polypeptide backbone of myoglobin are folded into α-helices. There is much suggestive evidence for helices in a large variety of proteins.

TERTIARY STRUCTURES OF PROTEINS ARE EXCEEDINGLY IRREGULAR

The *tertiary structure* of a protein is its 3-D form. It is, in many cases, very irregular. Practically no proteins exist in the form of a simple helix: Instead, many proteins contain both helical and nonhelical regions. Some, in fact, seem to have almost no helical regions. There are a number of stereochemical reasons why the α-helix or another regular arrangement is not found more extensively in spite of almost perfectly regular hydrogen bonding in the backbone. One reason is that the amino acid proline does not contain an amino group, and so where it occurs the regular hydrogen bonding must be interrupted. Another reason is the formation of disulfide (S—S) bridges between cysteine residues. When these cysteine residues are located on the same polypeptide chain, the helix necessarily undergoes distortion.

Perhaps, however, the most important reason for irregularity in protein structures arises from the diverse chemical nature of the amino acid side groups. Each of these side groups will tend to make the energetically most favorable secondary interactions with other atomic groups. As an example, the free hydroxyl group on tyrosine will tend to assume a position where it can form a hydrogen bond. The considerable energy of the bond would be lost if, for example, it were next to a hydrophobic isoleucine side group.

Furthermore, the side groups of several amino acids, like valine and leucine, are very insoluble in water, whereas others, like those of glutamic acid or lysine, are highly water-soluble. It thus makes chemical sense that the water-insoluble side groups are found stacked next to each other in the interior of myoglobin, and the external surface contains groups that mix easily with water. The 3-dimensional configuration represents the energetically most favorable arrangement of the polypeptide chain. Each specific sequence of amino acids takes up the particular "native" arrangement that makes possible a maximum number of favorable atomic contacts between it and its normal environment. This view is strongly supported by very striking experiments in which high temperature or some other unnatural condition breaks down the native 3-D form (*denaturation*) to give randomly oriented, biologically inactive polypeptide chains. When the denatured chains are carefully returned to their normal environment, some of them can then resume their native conformation (*renaturation*) with full biological activity.

S—S BONDS FORM SPONTANEOUSLY BETWEEN CORRECT PARTNERS

In many cases, renaturation of a disordered protein to an active form involves not only the formation of the thermodynamically favorable weak bonds, but also the making of specific disulfide (S—S) bridges. This was first shown by experiments with the enzyme ribonuclease, a protein constructed from one polypeptide chain of 124 amino acids, cross-linked by four specific S—S bonds. The native, active configuration of the enzyme can be destroyed by reducing the S—S groups to sulfhydryl (SH) groups in the presence of the denaturing agent 8 *M* urea and the reducing agent mercaptoethanol (Figure 6–12). When the urea is removed, the SH bonds are reoxidized in air to yield S—S bonds identical to those found in the original molecule. A given SH group reassociates, not randomly with any of the other seven SH groups in the molecule, but rather with a specific SH group brought into close contact with it by the folding of the polypeptide chain. Thus S—S bridges are not a primary reason for the peculiar folding of the chain. They

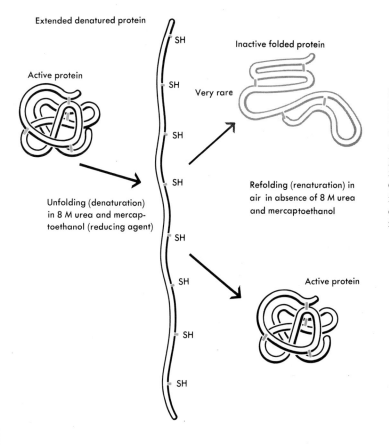

Extended denatured protein

SH

SH

SH

SH

SH

SH

SH

Active protein

Unfolding (denaturation) in 8 M urea and mercaptoethanol (reducing agent)

Very rare

Inactive folded protein

Refolding (renaturation) in air in absence of 8 M urea and mercaptoethanol

Active protein

Figure 6–12
Schematic illustration of the fate of S—S bonds during protein denaturation and renaturation. When the denaturing agents are removed, most of the polypeptide chains resume the native configuration with the original S—S bonds. Only a few polypeptide chains fold up in an inactive form characterized by a different set of S—S bonds than those found in the native molecules.

might be better viewed as a device for increasing the stability of an already stable configuration. The presence or absence of S—S bonds does not affect the argument that the final structure of a protein is determined by the amino acid sequence.

ENZYMES CANNOT BE USED TO ORDER AMINO ACIDS IN PROTEINS

We have seen that the sequence of amino acids in the polypeptide chains largely determines the 3-D structure of a protein. Now we come back to the ordering dilemma with the realization that it is the heart of the matter of protein synthesis. In comparison, the problem of how the connecting links form is minor, for this connective process involves the synthesis of only one type of covalent bond (the peptide bond), a fact that hints a need of only one enzyme or, at most, several enzymes.

On the other hand, the ordering itself cannot be accomplished by recourse to enzymes specific for each amino acid in a protein for the following reason. Such a device would require as many ordering enzymes as there are amino acids in the protein; but since all known enzymes are themselves proteins, still additional ordering enzymes would be necessary to synthesize the enzymes, and so on. This is clearly a paradox, unless we assume a fantastically interrelated series of syntheses in which a given protein can alternatively have many different enzymatic specificities. With such an assumption it might be just possible (and then with great difficulty) to visualize a workable cell. It does not seem likely, however, that most proteins really do have more than one task. All our knowledge, in fact, points toward the opposite general conclusion of one protein, one function.

It is, therefore, necessary to throw out the idea of ordering proteins with enzymes and to predict instead the existence of a specific surface, the *template* (Figure 6–13), that attracts the amino acids (or their activated derivatives) and lines them up in the correct order. Then a specific enzyme common to all protein synthesis could make the peptide bonds. It is, furthermore, necessary to assume that the templates must also have the capacity of serving either directly or indirectly as templates for themselves (self-duplication). That is, in some way their specific surfaces must be exactly copied to give new templates. Again we cannot invoke the help of specific enzymes, for this immediately leads us back to the "enzyme cannot make enzyme" paradox.

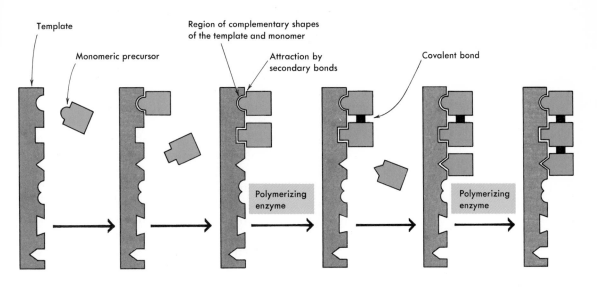

TEMPLATE INTERACTIONS ARE BASED ON RELATIVELY WEAK BONDS

The existence of proteins thus simultaneously demands the coexistence of highly specific template molecules. Moreover, the templates themselves must be macromolecules, at least as large as their polypeptide products. This is clear when we examine the rules that govern the selective binding of small molecules to their templates. We first see that the binding is not done by using strong covalent bonds. Instead the attraction is based on relatively weak bonds that can form without enzymes. These are (1) ionic bonds, (2) hydrogen bonds in which an electropositive hydrogen atom is attracted to electronegative atoms, such as oxygen or nitrogen, and (3) van der Waals forces.

Since all these forces operate only over very short distances (<5 Å), templates can order small molecules only when they are in close contact on the atomic level. Thus, it is to be expected that the specific (attracting) regions of the template will be in the same size range as the amino acid side groups in the protein product.

ATTRACTION OF OPPOSITES VS. SELF-ATTRACTION

Here we pose the obvious question: Can a polypeptide chain serve as a template for its own synthesis? If it could, it would make possible a great reduction in the chemical prerequisites for life. Then the problems of protein synthesis and template replication would be the same, and the additional biochemical complexity required to maintain a special class of template molecules would be unnecessary.

▲
Figure 6–13
Diagrammatic view of the formation of a specific polymeric molecule upon a template surface.

This conceptual possibility finds no support, however, from close inspection of the amino acid side groups. There is no chemical reason why, for example, the occurrence of valine on a template should preferentially attract the specific side group of another valine molecule. In fact, none of the amino acid side groups have specific affinities for themselves. Instead it is much easier to imagine molecules with opposite or complementary features attracting each other. Negative charges obviously attract positive groups, and hydrogen atoms can form hydrogen bonds only to electronegative atoms. Similarly, molecules can specifically attract by van der Waals forces only when they possess complementary shapes, to allow a cavity in one molecule to be filled with a protruding group of another molecule.

A formal way remains, however, to save the possibility of protein templates. We might imagine the existence of 20 different specific molecules that we could call connectors. Each would possess two identical surfaces complementary in charge and/or shape to a given amino acid. The intervention of these connector molecules would then make possible the lining up of amino acids in a sequence identical to that of the template polypeptide chain. No evidence exists for such molecules, however. Instead, as we shall shortly show, a specific template class (the nucleic acids) does in fact exist.

A CHEMICAL ARGUMENT AGAINST THE EXISTENCE OF PROTEIN TEMPLATES

The failure of proteins ever to evolve a template role may originate in the composition of the amino acid side groups. The argument can be made that no template whose specificity depends upon the side groups of closely related amino acids, like valine or alanine, could ever be copied with the accuracy demanded for efficient cellular existence. This follows from the fact that some amino acids are chemically similar. For example, valine and isoleucine differ only by the presence of an additional methyl group in isoleucine. Likewise, glycine and alanine also differ by only one methyl group. This close chemical similarity immediately poses the question whether any copying process can be sufficiently accurate to distinguish between such closely related molecules. Our answer depends in part upon what we mean by "sufficiently accurate." A good speculative guess is that each amino acid in an hereditary molecule would have to be copied with an accuracy of not more than one error in 10^8. On the other hand, a semirigorous chemical argument can be made that no chemical reaction could distinguish between molecules differing by only one methyl group with an accuracy of better than one in 10^6. Moreover, when we look at the accuracy of protein synthesis itself, we observe that some amino acids can be inserted into polypeptide chains with no greater than a 99.9 percent accuracy. Thus proteins do not have the "smell" of an hereditary molecule.

SUMMARY

The frequent occurrence of most chemical reactions within cells depends both on the presence of specific enzymes and on the availability of an external energy supply (the sun or food molecules) to make energy-rich molecules like ATP. There is a different enzyme for almost every specific reaction involved in the synthesis of a small molecule. This rule holds even when the "small molecule" is as large as chlorophyll. The problem arises, however, whether the same general scheme can hold for the biosynthesis of the enzymes themselves. Are a number of specific enzymes used to synthesize each enzyme involved in the metabolism of small molecules? This number would need to be very large, since all enzymes are proteins, themselves very large molecules, constructed by the linear linking together in a definite order of the 20 amino acids. The average protein contains about 300 to 500 amino acids, and so an equivalent number of enzymes would be necessary if enzymes are used to specify amino acid sequences.

This is clearly an unworkable scheme, and the ordering of amino acids in proteins is instead accomplished by template molecules. The templates for protein are also macromolecules. They have surfaces that specifically attract and thereby line up the amino acids in the correct sequence. There is a specific template for each specific protein. No enzymes are involved in the attraction of the amino acid residues to the templates. Attraction is accomplished by weak secondary forces. Specific regions of the template specifically attract one of the 20 different amino acid residues.

When a cell grows and divides, the number of protein template molecules must also double. Templates must in some way also be templates for their own highly exact synthesis. The templates are not protein molecules; there are chemical arguments why proteins should not be highly accurate templates. Instead, all cells contain a special class of molecules specifically devoted to being templates for protein synthesis.

REFERENCES

KORNBERG, A., *DNA Synthesis,* Freeman, San Francisco, 1974. An excellent summary of purine and pyrimidine nucleotide biosynthesis is found in Chapter 2.

DICKERSON, R. E., AND I. GEIS, *The Structure and Action of Proteins,* Harper and Row, New York, 1969. An extraordinarily well-illustrated introduction to the ways polypeptide chains fold up.

WOOD, W. B., J. H. WILSON, R. M. BENBOW, AND L. HOOD, *Biochemistry: A Problems Approach,* Benjamin, Menlo Park, California, 1974. The first third of this most useful "problems text" is about protein structure and function.

BERNHARD, S., *Enzymes: Structure and Function,* Benjamin, Menlo Park, 1968. A short but excellent introduction to proteins as enzymes.

"Structure and Function of Proteins at the Three-Dimensional Level," *Cold Spring Harbor Symposium on Quantitative Biology* **XXXVI,** 1972. The most complete review of the protein structure field now in press, containing many drawings specifically prepared for three-dimensional viewing.

NEURATH, H., AND R. L. HILL, (eds.), *The Proteins,* Volume I, 3rd ed., Academic Press, New York, 1975. The first volume of a new collection of advanced articles on all aspects of protein structure and function.

PERUTZ, M. F., "The Hemoglobin Molecule," *Scientific American,* November, 1964, pp. 64–76. Relates much of what is currently known about the architecture of the hemoglobin molecule, in the presence or absence of oxygen.

PHILLIPS, D., "The Three-Dimensional Structure of an Enzyme Molecule," in R. H. Haynes and P. C. Hanawalt (eds.), *The Molecular Basis of Life,* Freeman, San Francisco, 1968, pp. 52–64. A description of the beautiful work that solved the structure of lysozyme, the first enzyme for which it was possible to correlate structure and function.

STROUD, R. M., "A Family of Protein-Cutting Proteins," *Scientific American,* July, 1974. An exceptionally well illustrated article describing the structure and functioning of trypsin and chymotrypsin-like enzymes.

DICKERSON, R. E., "The Structure and History of an Ancient Protein," *Scientific American,* August 1972. A superbly illustrated introduction to protein structure, as shown by evolution of the cytochrome c structure.

Muller, H. J., "The Gene," *Proc. Roy. Soc.* (London), B134, 1–37, 1947. A lecture given in 1945, in which a distinguished geneticist traces the history of the gene concept and speculates about how it might function as a template.

The Arrangement of Genes on Chromosomes

Our chemical intuition tells us that proteins are unlikely to serve as the templates necessary to order amino acid sequences in proteins. Instead, we must look for a class of molecules capable of both the protein template function and self-replication. Here the direction of our search is completely dictated by the results of modern genetics. This flourishing science has shown in amazing detail how the chromosomes are responsible for the perpetuation of heredity: it is by means of genes, located on the chromosomes, that daughter cells come to resemble parental cells. The major task of the geneticists has been to show how this resemblance occurs.

Parallel to their work in mapping the location of genes, geneticists began to ask the fundamental question of how the genes chemically controlled specific cellular processes. Usually, however, the mutations they studied were not easily analyzed. In the 1920's and the 1930's, as even today, virtually nothing was known of the biochemical basis of development.

Fortunately, however, the mutations affecting color in flowers and eyes were open to a chemical approach. For example, mutations in many genes change the color of the eyes of *Drosophila*. Here biochemical analysis was possible because it was known that eye color is directly related to the presence of definite colored molecules called pigments. It could thus be asked how a gene difference could convert the color of fruit-fly eyes from red to white. The obvious and correct answer is that no pigment is found in the eyes of flies thus altered. This in turn hints that an enzyme necessary for its synthesis is absent, a suggestion soon extended to the general hypothesis that genes directly control the synthesis of all proteins, whether or not they are enzymes.

As this way of thinking became generally accepted (about 1946), geneticists began to deal with the deeper problem of how a gene dictates which particular protein is present. Little progress could be made until protein chemists showed unambiguously (in the early 1950's) that proteins were linear collections of the 20 amino acids. It was then a simple matter for the more theoretically inclined geneticists to hypothesize that the chromosomes carry the genetic information that orders amino acid sequences, and to predict that the study of the structure of genes might lead to the elucidation of the molecular basis of the templates that order amino acid sequences. This was, in fact, what did happen. But before we can examine the problem more deeply, some genetic concepts must first be explained.

MUCH REMAINS TO BE LEARNED ABOUT THE MOLECULAR ASPECTS OF CHROMOSOME STRUCTURE

Even today, our knowledge of the molecular structure of chromosomes is very incomplete. This is especially true for the more complex chromosomes of higher plants and animals. In bacteria and viruses, as well as in higher cells, there is evidence (which we shall later relate) that the principal genetic component is deoxyribonucleic acid (DNA). The chromosomes of higher organisms, however, also contain a significant fraction (as great as 65 percent in higher plants and animals) of protein.

Much of this protein in higher plants and animals belongs to a class of protein molecules called *histones*. All histones are basic (have a net positive charge), and it is believed that they neutralize part of the negative charge of the acid DNA molecules. Before 1943 many biologists believed that histones carried genetic information, but their structural invariance throughout much of evolution now argues against the assignment of a fundamental genetic role. Instead there is a growing belief that they have essentially structural functions and play key roles in the coiling and uncoiling of chromosomes during the cell cycle.

Up to now, electron microscopy has provided no useful insight into the exact structure of chromosomes of higher organisms. This failure is in striking contrast with the success of electron microscopy in examining very thin sections of muscle and nerve fibers. The failure arises from the *irregular* shape not only of the highly extended chromosomes found during interphase, but even of the contracted metaphase chromosomes. At the molecular level the various muscle proteins are nicely lined up in parallel array. In contrast, the path of a chromosome through a cell is excessively irregular: When a thin section is observed, it has so far been impossible even to follow the contour, much less to observe details of molecular structure. Thus, morphological examination by itself has given us no useful information about the chromosomal arrangement of genes; instead genetic crosses are the only way of attacking this problem. Fortunately, as we shall soon see, the powers of resolution of this method are indeed very great.

THE GENETIC CROSS

A variety of devices exist in nature that bring together genetic material from different organisms (the genetic cross) for the purpose of achieving genetic recombination. These devices, collectively known as *sexual processes*, greatly speed up the rate of evolution, since they bring collections of favorable mutations into one cell much faster than successive cycles of favorable mutations could by themselves. Here, however, we are not at all concerned with the evolutionary advantages of the various sexual processes. Our interest in them arises instead from the variety of tools that their existence has provided for finding the location of genes along chromosomes.

The essential trick of locating genes through genetic crosses involves the determination of whether the genes donated by a given parent remain together when haploid segregants are produced. When genes are located on different chromosomes they will assort randomly, and there will be a 50–50 chance that they will be found together in a given haploid segregant. When, on the contrary, they are located on the same chromosome, they will tend to segregate together, unless they have been separated by crossing over.

Crossing over occurs at the stage of meiosis where two homologous chromosomes specifically attract each other to form pairs. The mechanism of this attraction (*pairing*) remains a great mystery. It is clearly a very specific process, since it occurs only between chromosomes containing the same genes. Following the formation of pairs, both chromosomes occasionally break at the same point and rejoin crossways. This allows the formation of recombinant chromosomes containing some genes derived from the paternal chromosome and some from the maternal one. Crossing over greatly increases the amount of genetic recombination and, except in highly specialized cases, is universally observed. The frequency of crossing over, however, varies greatly with the particular species involved. On the average, one to several crossovers occur every time chromosomes pair.

All our early knowledge of gene locations came from the study of gene segregation after conventional meiotic divisions. The organisms studied were those in which there is a regular fusion of the male and the female cells to produce diploid cells half of whose chromosomes are derived from the male parent and half from the female. When meiosis occurs, the diploid chromosome number is regularly reduced to the haploid number. Crossing over always takes place after each of the parental chromosomes has split to form two chromatids held together by a single centromere. Two chromatids are involved in each crossover event, so that each crossover produces two recombinant chromatids and leaves the two parental chromosomes intact (Figure 7–1). (This does not mean that half the chromatids produced during meiosis have the parental genotype. Each chromatid in a pair has an equal chance of crossing over, so that, even though a particular chromatid

Figure 7–1

Crossing over between homologous chromatids during meiosis.

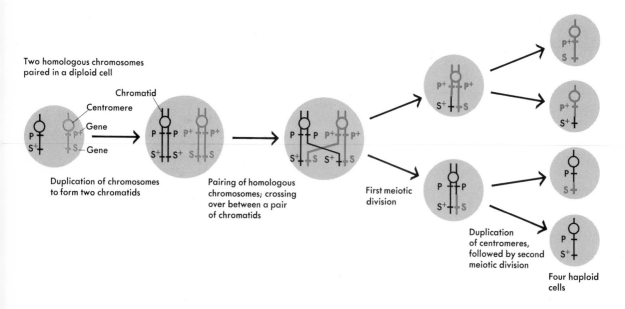

Two homologous chromosomes paired in a diploid cell

Chromatid

Centromere

Gene

Gene

Duplication of chromosomes to form two chromatids

Pairing of homologous chromosomes; crossing over between a pair of chromatids

First meiotic division

Duplication of centromeres, followed by second meiotic division

Four haploid cells

is not involved in a given crossover, it may still participate in another.)

CHROMOSOME MAPPING

The existence of crossing over provides a means of locating genes along chromosomes. Crossing over occurs randomly throughout the length of many chromosomes, so that the farther apart two genes are, the greater the probability that a break will occur between them to cause genetic recombination. The way in which the frequencies of the various recombinant classes are used to locate genes is straightforward. It is especially easy when the progeny are haploid, and the question of dominance versus recessiveness is irrelevant. Consider the segregation pattern of three genes all located on the same chromosome of a haploid organism. The arrangement of the genes can be determined by means of three crosses, in each of which two genes are followed (two-factor crosses). A cross X between a^+b^+ and ab yields four progeny types: the two parental genotypes (a^+b^+ and ab) and two recombinant genotypes (a^+b and ab^+). The cross Y between a^+c^+ and ac similarly gives the two parental combinations as well as the a^+c and ac^+ recombinants, whereas the cross Z between b^+c^+ and bc produces the parental types, and the recombinants b^+c and cb^+. Each cross will produce a specific ratio of parental to recombinant progeny. Consider, for example, the result that cross X gives 30 per cent recombinants, cross Y, 10 per cent, and cross Z, 25 per cent. This hints that genes a and c are closer together than a and b or b and c, and that the genetic distances between a and b and b and c are more similar. The gene arrangement which best fits this data is acb (Figure 7–2).

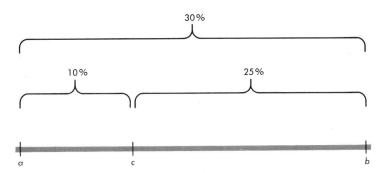

Figure 7–2
Assignment of the tentative order of three genes on a basis of 3 two-factor crosses (see text for details).

The correctness of gene orders suggested by crosses of two gene factors can usually be unambiguously confirmed by three-factor crosses. When the three genes used

Figure 7–3
The use of three-factor crosses to assign gene order. The least frequent pair of reciprocal recombinants must arise from a double crossover. The percentages listed for the various classes are the theoretical values expected for an infinitely large sample. When finite numbers of progeny are recorded, the exact values will be subject to random statistical fluctuations.

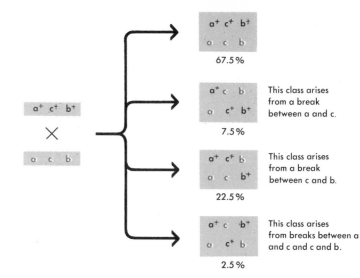

$a^+ c^+ b^+$
$a \quad c \quad b$
67.5%

$a^+ c \quad b$
$a \quad c^+ b^+$
7.5%
This class arises from a break between a and c.

$a^+ c^+ b$
$a \quad c \quad b^+$
22.5%
This class arises from a break between c and b.

$a^+ c \quad b^+$
$a \quad c^+ b$
2.5%
This class arises from breaks between a and c and c and b.

in the above example are followed in the cross $a^+b^+c^+ \times abc$, six recombinant genotypes are found (Figure 7–3). They fall into three groups of reciprocal pairs. The rarest of these groups arises from a double crossover. By looking for the least frequent class, it is often possible instantly to confirm (or deny) a postulated arrangement. The results in Figure 7–3 immediately confirm the order hinted by the two-factor crosses. Only if the order is acb does the fact that the rare recombinants are a^+cb^+ and ac^+b make sense.

The existence of multiple crossovers means that the amount of recombination (ab) between the outside markers a and b is usually less than the sum of the recombination frequencies (ac and cb) between a and c and c and b. To obtain a more accurate approximation of the distance between the outside markers, we calculate the probability (ac × cb) that, when a crossover occurs between c and b, an ac crossover also occurs, and vice versa (cb × ac). This probability subtracted from the sum of the frequencies expresses more accurately the amount of recombination. This gives the simple formula

$$ab = ac + cb - 2(ac)(cb)$$

It is applicable in all cases where the occurrence of one crossover does not affect the probability of another crossover. Unfortunately, accurate mapping is often disturbed by *interference* phenomena, which can either increase or decrease the probability of correlated crossovers.

The results of a very large number of such crosses have led to an important genetic conclusion: All the genes on a chromosome can be located on a line. The gene

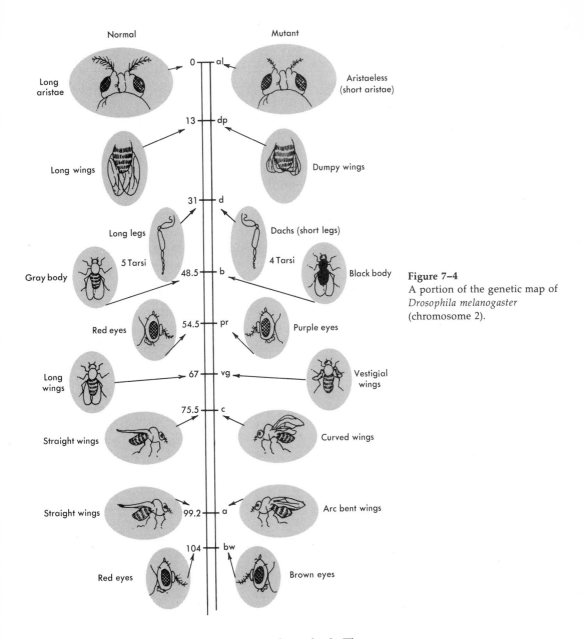

Figure 7–4
A portion of the genetic map of
Drosophila melanogaster
(chromosome 2).

arrangement is strictly linear, and never branched. Thus chromosomes are linear, not only in shape, but also in gene arrangement. The arrangement of genes on a particular chromosome is called a *genetic map,* and the locating of genes on a chromosome is often referred to as mapping a gene. Figure 7–4 shows the genetic map of one of the chromosomes of *Drosophila.* Distances between genes on a map are usually measured in map units, which are related to the frequency of recombination between the genes: Thus if the frequency of recombination between two genes is found to be 5 percent, the genes are said to be separated by five map units. Because of the occurrence of double crossovers, the assignment of map units can be considered accurate only if recombination between closely spaced genes is followed.

Even when two genes are at the far ends of a very long chromosome, they will show not less than 50 percent linkage (assort together at least 50 percent of the time), because of multiple crossovers. Genes will be separated if 1,3,5,7 . . . crossovers occur between them; they will end up together if 2,4,6,8 . . . occur between them. Thus in the beginning of the genetic analysis of an organism, it is practically impossible to determine immediately whether two genes are on different chromosomes or are at the opposite ends of one long chromosome. After a large number of genes have been mapped, we often find that two genes thought to be on different chromosomes are, in fact, on the same one. For example, the genes of the phage T4 were thought for many years to lie on three separate chromosomes, until further mapping, using newly discovered genes, revealed that they all lie on a single chromosome.

It is important to remember that a genetic map derived from recombinant frequencies gives only the relative physical distances between mutable sites. It would give the actual physical distances only if the probability of crossing over were the same throughout the length of a chromosome. Thus, just as soon as geneticists made maps, they sought methods that could relate mutational sites to true physical locations. Now a number of tricks, some too complicated to be explained here, tell us that often, *but not always*, genetic maps are a good reflection of the actual chromosome structure. One of these tricks will be described in a subsequent section, in which we discuss the genetic map of *E. coli*.

IMPORTANCE OF WORK WITH MICROORGANISMS

Most of our initial ideas about genes arose from work with large, multicellular plants and animals. Now, however, unless there is an economic or social need for information about a particular species (e.g., the corn plant, man), microorganisms may be much more preferable for study. Several important advantages favor work with microorganisms. First, they are usually haploid, so that the ease of genetic analysis does not depend upon whether a mutation is dominant or recessive. Since most mutations are recessive, they cannot be detected when the normal wild-type gene is also present. In work with diploids, several generations of genetic crosses must often be carried out to detect the presence of a particular mutant gene; in a haploid organism the mutant gene can express itself almost immediately. Second, microorganisms multiply very rapidly. There is enormous advantage to working with an organism, such as a bacterium, which has a new cell generation every 20 minutes, rather than with a plant like corn, where at best only two generations per year can be studied, even in tropical environments.

Sixty years ago, *Drosophila,* with its average life cycle of 14 days, looked very attractive. Today only with great difficulty can it be used to answer the most fundamental questions being asked at the molecular level about what genes are and how they act. Instead the most exciting materials for pure genetics are the yeasts, molds, bacteria, and viruses. Genetic work with these effectively began some 35 years ago. Until then their small size was considered an enormous disadvantage. Microorganisms do not have easily recognized morphological features, such as red eyes, and so it was very difficult to know when they contained mutations. Until 1945, it was generally believed that some did not have chromosomes and some biologists even suspected they might not have genes.

THE VALUE OF MUTAGENS

Most mutant genes studied by the early Mendelian geneticists arose spontaneously. Now there is increasing use of mutations specifically induced by external agents, such as ionizing radiation, ultraviolet light, and certain specific chemicals. These agents, collectively called mutagens, greatly increase the rate at which geneticists can isolate mutant genes. For many years the various forms of radiation were the most powerful mutagens known. Now chemical mutagens are more often used, because they produce a much higher fraction of mutated genes. Treatment of bacteria with the highly reactive compound nitrosoguanidine can produce viable mutations in almost 1 percent of the bacterial genes.

Mutagens act quite indiscriminately. No presently known mutagen increases the probability of mutating a given gene without also increasing the probability of mutating all other genes. Until recently, the mechanisms of mutagenesis were completely unclear. Now, as we shall point out in Chapter 9, the realization that DNA is the primary genetic material allows the development of precise hypotheses about how several chemical mutagens act.

BACTERIAL MUTATIONS: THE USE OF GROWTH FACTORS

The essential breakthrough in the use of bacteria as genetic material came in 1944 with the realization that mutations could be obtained affecting the ability of bacteria to synthesize essential metabolites. For example, *E. coli* ordinarily grows well with only glucose as a carbon source. But as a result of specific mutations, there now exist mutant *E. coli* strains that will grow only when their normal medium

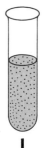

Treatment of *E. coli* cells
with a mutagen—nitrosoguanidine

Figure 7–5
The isolation of mutant *E. coli* cells with specific growth factor requirements.

The treated cells are placed on a Petri dish filled with a rich nutrient solid agar medium containing the 20 amino acids, the various purines and pyrimidines, all known vitamins, etc. Many of the treated cells fail to multiply because they are killed by the mutagen. The remaining survivors multiply to form distinct colonies on the solid agar surface.

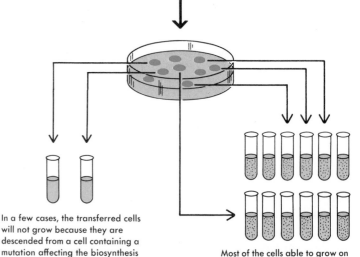

In a few cases, the transferred cells will not grow because they are descended from a cell containing a mutation affecting the biosynthesis of a compound present in the rich medium. The identity of the needed growth factor can be found quickly by selective addition of the various compounds present in the rich medium.

Most of the cells able to grow on the rich medium can grow also when transferred to a minimal medium containing only glucose and inorganic salts.

is supplemented with a specific metabolite (*growth factor*). These types of mutation had been described just a few years earlier (1941) in the haploid lower plant *Neurospora* (a mold). Such mutations are very easy to work with: To test for their presence one need merely grow a suspected mutant both in the presence and in the absence of a metabolite, for example, the amino acid arginine (Figure 7–5). If a mutation-inhibiting arginine biosynthesis has occurred, the bacteria will grow only in the presence of arginine. The use of this approach quickly led (with the help of mutagens) to the isolation of a large number of different gene mutations affecting the synthesis of specific molecules.

Another important type of mutation involves the resistance of bacteria to poisonous compounds such as antibiotics. For example, most *E. coli* cells are rapidly killed by small amounts of streptomycin. Very rarely, however, there occur mutations (Strep^R), which make the cells resistant to certain amounts of the drug. Mutations also can occur to make cells resistant to the growth of viruses. One of the most useful mutations in *E. coli* strain B confers resistance to the phage T1; these mutant cells are designated B/1. Correspondingly, *E. coli* strain B cells resistant to phages T2 and T4 are called B/2 and B/4, respectively. Still other mutations affect the ability of *E. coli* cells to grow upon sugars such as lactose, galactose, or maltose; a specific mutation can cause the loss of the ability of *E. coli* to use one of these sugars as a sole carbon source.

The isolation of growth factor, antibiotic resistance, and phage resistance mutations was quickly followed by experiments demonstrating existence of genetic recombination (hence a sexual process) in *E. coli* (Figure 7–6). Twofold use was made of these mutant genes. First, they were used as conventional genetic markers, their segregation patterns revealing the chromosomal arrangement of genes. Second, they provided a method of detecting a genetic recombination process occurring in only a very small fraction of the population at a given time. In *E. coli,* for example, simple morphological examination of bacterial cells gave no clues that cell fusion and genetic recombination existed: To detect recombination it was necessary to devise an experiment in which only the recombinant cells would be able to multiply. This was done by using parental strains with specific growth requirements such that they could not multiply in minimal media.

In these experiments, two strains of bacteria, each possessing specific growth requirements, were mixed together. Neither strain alone was able to grow in the ab-

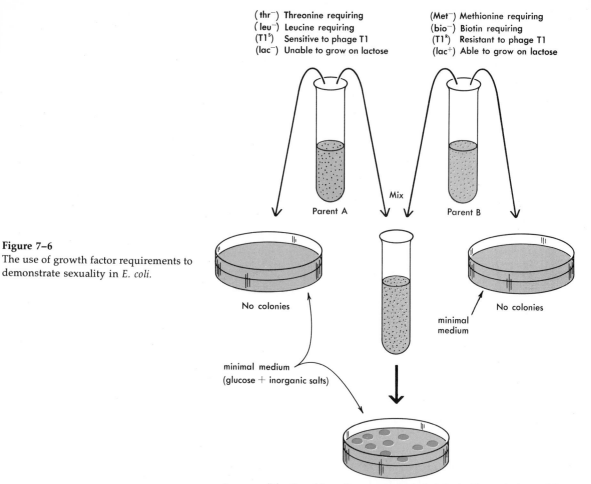

(thr⁻) Threonine requiring (Met⁻) Methionine requiring
(leu⁻) Leucine requiring   (bio⁻) Biotin requiring
(T1ˢ) Sensitive to phage T1 (T1ᴿ) Resistant to phage T1
(lac⁻) Unable to grow on lactose (lac⁺) Able to grow on lactose

Figure 7–6

The use of growth factor requirements to demonstrate sexuality in *E. coli*.

A very small fraction of the cells are met⁺, bio⁺, thr⁺, leu⁺. They arise by genetic recombination shown by examination of the lac and T1 markers. In addition to the parental lac⁻T1ˢ and lac⁺T1ᴿ genotypes, there are found lac⁻T1ᴿ and lac⁺T1ˢ cells.

sence of specific metabolites or growth factors (that is, the amino acids threonine and leucine were required by one strain, the vitamin biotin and the amino acid methionine by the other). After the two strains had been mixed together, a small number of cells were able to grow without any growth factors. This meant that they had somehow acquired good copies of each of their mutant genes. This result strongly suggested that *E. coli* has a sexual phase that can bring together the chromosomes of two different cells. Crossing over could then place in one chromosome good copies of all its necessary genes. Further genetic analysis confirmed this hypothesis, and within the past 10 years *E. coli* has become one of the genetically best known of all organisms.

VIRUSES ALSO CONTAIN CHROMOSOMES

Chromosomal control of heredity even extends to viruses. These disease-causing particles, much smaller than bacteria, can enter (infect) cells and multiply to form large numbers of new virus particles. The common cold, influenza, and poliomyelitis are among the many diseases caused by viruses. The relation of viruses to their host cells is very intimate, since they are able to increase in numbers only after they have entered a cell; outside cells they are completely inert. There exist viruses active on most, if not all, plants and animals. Viruses can even multiply in bacteria (bacterial viruses are usually called bacteriophages or phages). The replication of many new virus particles within a single cell usually kills the host cell—hence their disease-causing property.

Our knowledge of the genetics of several bacterial viruses has shown an expansion similar to and simultaneous with the expansion of our knowledge of bacteria. Before 1940, almost no one thought about the genetics of viruses. To most people viruses seemed much too small to be studied unless they caused a disease that we wished to control. They could not be seen in the light microscope, and were generally detected only by their property of killing cells. Several factors changed this outlook. First, it was realized that the bacterial viruses were very easy to experiment with and that the phage–bacterium system was ideally suited to the study of the general problem of how genes multiply and work. Second, it was found that phage mutations were as easy to obtain as mutations in bacteria, if not easier.

Chemically, viruses are extremely heterogeneous both in size and in variety of molecular constituents (Figure 7–7). For many years their biological significance was obscure and the question often asked was, "Are viruses living?" Now we realize that they are small pieces of genetic material, each enclosed within a protective coat, rich in protein, which allows it to be transported from one cell to another. Progeny virus particles resemble their parents because they contain identical chromosomes. We also see that they are no more "alive" than isolated chromosomes; both the chromosomes of cells and those of viruses can duplicate only in the complex environment of a living cell. The study of viruses has been of immense value to the understanding of how cells live: Viruses are almost unique in affording convenient systems for quickly studying the consequences of the sudden introduction of new genetic material into a cell.

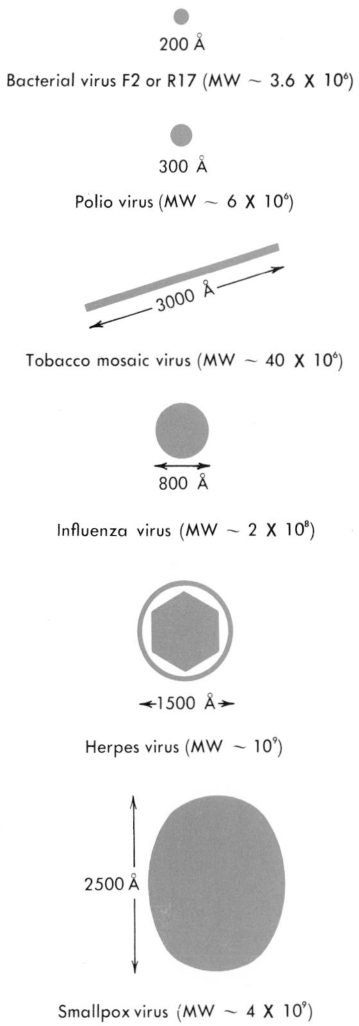

200 Å
Bacterial virus F2 or R17 (MW ~ 3.6 X 10⁶)

300 Å
Polio virus (MW ~ 6 X 10⁶)

3000 Å
Tobacco mosaic virus (MW ~ 40 X 10⁶)

800 Å
Influenza virus (MW ~ 2 X 10⁸)

◄1500 Å►
Herpes virus (MW ~ 10⁹)

2500 Å
Smallpox virus (MW ~ 4 X 10⁹)

Figure 7–7
Variation in size and shape of a number of viruses.

VIRUSES DO NOT GROW BY GRADUAL INCREASE IN SIZE

The life cycle of an average cell involves gradual increase to twice its initial size, followed by a division process (mitosis) producing two identical daughter cells. Viruses, however, do not multiply in this fashion. They are not produced by the fission of a large preexisting particle—all the virus particles of a given variety have approximately similar (in some cases identical) masses. During viral multiplication there is a temporary disappearance (the eclipse period) of the original parental particle, because the parental particle breaks down upon infection and releases its chromosome from the protective outer shell. Then the free chromosome serves as a template to direct the synthesis of new viral components. This breakdown process is an obligatory feature of viral multiplication, for as long as the chromosome is tightly enclosed within the protein-containing shell, it cannot be a template for the synthesis of either new chromosomes or new shell protein molecules.

VIRUSES ARE PARASITES AT THE GENETIC LEVEL

In most cases, the viral chromosomes (which we shall later show always to be constructed from nucleic acids) and proteins are constructed from the same four main nucleotides and 20 amino acids used in normal cells. All these precursors are usually synthesized by host cell enzymes. Likewise, the energy needed to push the various chemical reactions in the biosynthetic direction usually comes from ATP produced by food molecule degradation controlled by host enzymes.

The parasitism of viruses is thus obligatory. It is impossible to imagine viruses reproducing outside cells, on which they are completely dependent for supply of both the necessary precursors and the ribosomes, the structural machinery for making proteins. The essential aspects of viruses is thus not really their small size, but the fact that they do not possess the capacity to independently construct proteins. Thus for their replication, they must insert their nucleic acid into a functional cell. This fact enables us to distinguish the larger viruses, like smallpox, from very small cellular organisms, like the *Rickettsiae*. Even though the *Rickettsiae* are obligatory parasites, they contain both DNA and RNA and grow by increasing in size and splitting into two smaller cells. At no time does their cell membrane break down; moreover, all their protein is synthesized upon their own protein-synthesizing machinery.

BACTERIAL VIRUSES (PHAGES) ARE OFTEN EASY TO STUDY

As mentioned before, the most exciting viruses from the viewpoint of the geneticist are the bacterial viruses (phages). Their discovery in 1914 produced great excitement, for it was hoped that they might afford an effective and simple way to combat bacterial diseases. Phages were unfortunately never medically useful, because their bacterial hosts mutate readily to forms resistant to viral growth. Thus almost everyone lost interest in the phage, and almost no one studied how it multiplied until the late 1930's. Then a small group of biologists and physicists, intrigued by the problem of gene replication, began to investigate the reproduction of several phages which multiplied in *E. coli.* They chose to work with phages rather than with plant or animal viruses because, under laboratory conditions, it is immensely easier to grow bacterial cells than plant or animal cells.

Almost all work with phages has concentrated on several particular phages, arbitrarily given names like T1, T2, P1, F2, or λ (Figure 7–8). Among the best known are

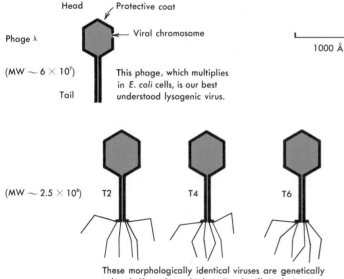

Head / Protective coat

Phage λ ←— Viral chromosome

1000 Å

(MW ∼ 6 × 10⁷)

Tail

This phage, which multiplies in *E. coli* cells, is our best understood lysogenic virus.

(MW ∼ 2.5 × 10⁸) T2 T4 T6

These morphologically identical viruses are genetically related. They also multiply in *E. coli* cells and are now among the best understood of all genetic objects.

Figure 7–8
Some bacterial viruses that have been important in the study of the chemistry of genetics.

F2 R17 MS2 (MW ∼ 3.6 × 10⁶)

These are the smallest known group of *E. coli* viruses. Even though they have been known only for a few years, they are quickly becoming some of the most intensively studied of all viruses.

the closely related strains T2, T4, T6, and λ. These strains reproduce in essentially the same way. The growth cycle starts when a phage particle collides with a sensitive bacterium and the phage tail specifically attaches to the bacterial wall. An enzyme in the phage tail then breaks down a small portion of the cell wall, creating a small hole through which the viral chromosome enters the cell. The viral chromosome duplicates, and the daughter chromosomes continue to duplicate, to form eventually 100 to 1000 new chromosomes, which become encapsulated with newly synthesized protective coats, to form a large number of new bacteriophage particles. The growth cycle is complete when the bacterial cell wall breaks open (lyses) and releases the progeny particles into the surrounding medium.

PHAGES FORM PLAQUES

The presence of viable phage particles can be quickly demonstrated by adding the virus-containing solution to the surface of a nutrient agar plate, on which bacteria susceptible to this virus are rapidly multiplying. If no virus particles are present, the rapidly dividing bacteria will form a uniform surface layer of bacteria. But if even one virus particle is present, it will attach to a bacterium and multiply to form several hundred new progeny virus particles, which are then suddenly released by dissolution (lysis) of the cell wall, some 15 to 60 minutes after the start of phage infection. Each of these several hundred progeny particles can then attach to a new bacterium and multiply. After several such cycles of attachment, multiplication, and release, all the bacteria in the immediate region of the original virus particle are killed. These regions of killed bacteria appear as circular holes (plaques) in the lawn of healthy bacteria (Figure 7–9).

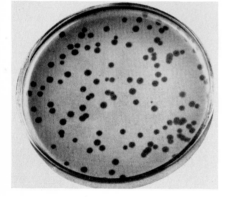

Figure 7–9
Photograph of phage T2 plaques on a lawn of *E. coli* bacteria growing in a Petri plate. (From G. S. Stent, *Molecular Biology of Bacterial Viruses*, Freeman, San Francisco, 1963, p. 41, with permission.)

VIRUS CHROMOSOMES ARE SOMETIMES INSERTED INTO THE CHROMOSOMES OF THEIR HOST CELLS

Some bacterial viruses (e.g., phage λ) do not always multiply upon entering a host cell. Instead their chromosome sometimes becomes inserted into a specific section of a host chromosome. Then the viral chromosome is, for all practical purposes, an integral part of its host chromosome and is duplicated, like the bacterial chromosome, just once every cell generation (Figure 7–10). The virus chromosome when it is integrated into a host chromosome is called the *prophage;* those bacteria containing prophages are called

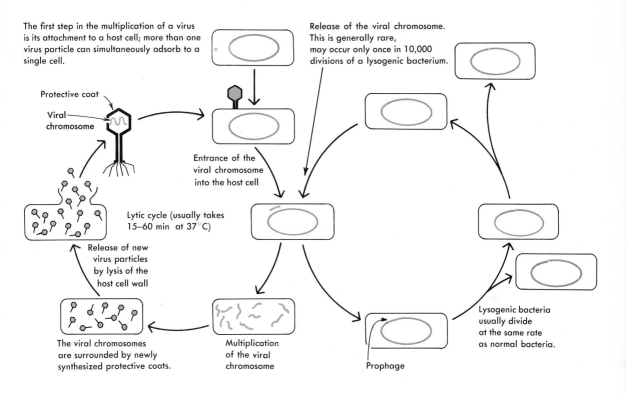

The first step in the multiplication of a virus is its attachment to a host cell; more than one virus particle can simultaneously adsorb to a single cell.

Protective coat

Viral chromosome

Entrance of the viral chromosome into the host cell

Release of the viral chromosome. This is generally rare, may occur only once in 10,000 divisions of a lysogenic bacterium.

Lytic cycle (usually takes 15–60 min at 37°C)

Release of new virus particles by lysis of the host cell wall

The viral chromosomes are surrounded by newly synthesized protective coats.

Multiplication of the viral chromosome

Prophage

Lysogenic bacteria usually divide at the same rate as normal bacteria.

lysogenic bacteria; and those types of virus whose chromosomes can become prophages are known as *lysogenic viruses.* In contrast, those viruses (e.g., T2) that always multiply when they enter a host cell are called *lytic viruses.* It is often difficult to know when a bacterium is lysogenic. We can be sure only when the virus chromosome is released from the host chromosome and the multiplication process which forms new progeny particles commences. Why only certain viruses (lysogenic viruses) form lysogenic associations and what advantage they receive from this association is still an open question.

How the viral chromosome is transformed into prophage was until recently very mysterious. Now there is good genetic evidence that the integration of the viral chromosome is achieved by crossing over between the host chromosome and a *circular form* of the viral chromosome. Prior to integration, the viral chromosome forms a circle and attaches to a specific region of the host chromosome. Both the host chromosome and the viral chromosome then break and rejoin in such a way that the broken ends of the viral chromosomes join to the broken ends of the bacterial chromosome instead of to each other, thereby inserting the prophage into the host chromosome (Figure

Figure 7–10
The life cycle of a lysogenic bacterial virus. We see that, after its chromosome enters a host cell, it sometimes immediately multiplies like a lytic virus and at other times becomes transformed into prophage. The lytic phase of its life cycle is identical to the complete life cycle of a lytic (nonlysogenic) virus. Lytic bacterial viruses are so called because their multiplication results in the rupture (lysis) of the bacteria.

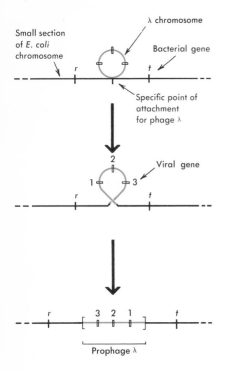

Small section of *E. coli* chromosome

λ chromosome

Bacterial gene

Specific point of attachment for phage λ

Viral gene

Prophage λ

Figure 7–11
Insertion of the chromosome of phage λ into the *E. coli* chromosome by crossing over.

7–11). The prophage detaches from the host chromosome by the reverse process: The two ends of the prophage pair prior to a crossover event which ejects the viral chromosome. The now free viral genome then can begin to multiply as if it were the chromosome of a lytic virus.

BACTERIAL CHROMOSOME MAPPING BY MATING

Many organisms now of greatest use in revealing what genes are and how they work do not have a conventional meiosis. When sexuality in bacteria was first discovered, it seemed simplest to believe that a conventional cycle of cell fusion followed by meiotic segregation occurred. Now we know, however, that the *E. coli* cycle has distinctive factors complicating conventional genetic analysis. The complications arise from the nature of the mating process. As in higher organisms, there exist male and female cells. The sexual cycle starts when the male and female cells attach to each other by a narrow bridge (Figure 7–12). A male chromosome then begins to move through the bridge to the female cell. Usually the transfer is incomplete, and only part of the male chromosome enters the female cell before the mating cells separate. Thus only rarely are complete diploid cells formed, partially diploid cells usually occurring instead. Crossing over then occurs between the female parent and the male chromosome (or fragment), fol-

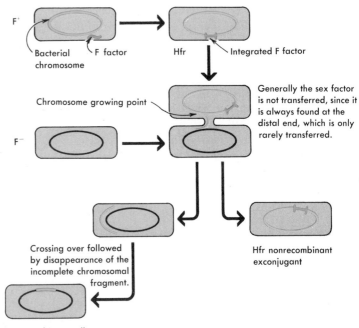

F+

Bacterial chromosome

F factor

Hfr

Integrated F factor

Chromosome growing point

Generally the sex factor is not transferred, since it is always found at the distal end, which is only rarely transferred.

F−

Crossing over followed by disappearance of the incomplete chromosomal fragment.

Hfr nonrecombinant exconjugant

F− recombinant cell

lowed by a segregation process, which yields haploid progeny cells.

The sex difference between the male and female cells is determined by the presence of a specific genetic factor, called the *F(ertility) factor*. When it is present in cells (F^+) the cells are male, and when it is absent (F^-) they are female. The F factor can exist in two alternative states, either as part of the *E. coli* chromosome or as a very small free circular chromosome that multiplies once per cell division. In the latter case, the F^+ cells are only potentially male, since they cannot transfer their genes to F^- cells; only when the F factor is part of the chromosome can the male cells mate. F^+ cells containing integrated sex factors are called *Hfr* (high frequency of recombination).

A variety of different Hfr strains exist, each containing an F factor integrated into a different region of the chromosome. Until recently, there was no good hypothesis about how the F agent becomes part of the host chromosome. Now there are strong hints that the F agent, like the chromosomes of lysogenic phages, has a circular shape and becomes integrated by crossing over.

Both the F factor and the lysogenic phage chromosomes are called *episomes*. An episome is defined as a genetic particle that can exist either free or as a part of a normal cellular chromosome. Now we have good evidence for the existence of episomes only in bacteria; there are hints, however, that they exist in higher plants and animals as well.

At first the fact that only part of the male chromosome enters the female cell made genetic analysis more difficult. Then it was realized that a fixed end of the male chromosome, specific for a given Hfr strain, always enters

◀ **Figure 7–12**
Sexuality in *E. coli*. Chromosomal transfer from male to female cells provides a primitive sexuality in *E. coli*. When the F agent has become attached to the male chromosome, the male cell, now Hfr, is competent to transfer a chromosome to a female cell. When the Hfr and F^- cells join together by a narrow bridge, the Hfr chromosome breaks and begins to duplicate at the site where the F factor has been inserted. A free end of one of the daughter Hfr chromosome segments then begins to move into the F^- cell. At 37°C, the transfer of a complete chromosome requires about 90 minutes. Generally, however, only part of the Hfr chromosome moves to the F^- cell before the cells separate; crossing over then occurs between the donor genetic material and the F^- chromosome. The haploid condition is reestablished by elimination of the supernumerary genes that have not become part of complete chromosomes. The recombinant cells can then be screened for the presence of various Hfr and F^- genes, thereby allowing the construction of a genetic map.

the female cell first, and that the relative frequency with which male genes are incorporated into the recombinant chromosome is a measure of how close they are to the entering end. Moreover it is possible to break apart the male and female cells artificially by violent agitation (this was first done in a mixing machine called the Waring Blendor); matings can be made and the couples violently agitated at fixed times during the process (the Blendor experiments). If the pairs are disrupted soon after mating, only the genes very close to the forward end will have entered the cell. It is thus possible to obtain the *E. coli* gene positions merely by observing the time intervals at which various male alleles have entered the female cells.

BACTERIAL CHROMOSOMES ARE CIRCULAR

The genetic map obtained through analysis of interrupted matings is the same as that arrived at by analysis of frequencies of various recombinant classes. As in the chromosomal maps of higher organisms, the bacterial genes are arranged on an unbranched line. However, one important distinction exists: The genetic map of *E. coli* is a circle (Figure 7–13); the male chromosome must break at a certain point before a free end can move into a female cell. If the point of breakage were not always the same point, we would not observe that in a given bacterial strain some genes tend to be transferred before others. The place where the break occurs, however, is not the same in all strains. This is because the break always occurs at the point where the *F* factor is integrated. There is one Hfr strain, for example, in which some of the genes involved in the synthesis of threonine and leucine are transferred soon after mating, whereas in another, a gene involved in methionine synthesis is among the first to enter the female cell. The existence of many strains, with various breakage points, has been very important in assigning gene locations. If only one entering point existed, it would be nearly impossible to assign even a rough order to those genes that would always enter last, since they enter the female cell only rarely.

We are still not sure what force drives the male chromosome into the female cell. There are hints that the transfer may be connected to the process of chromosome duplication; that is, the male chromosome does not merely break and one end begin to be transferred. Instead it looks as if the breakage initiates a cycle of chromosome replication, and that as the Hfr chromosome splits to form two chromosomes, one of the progeny chromosomes moves

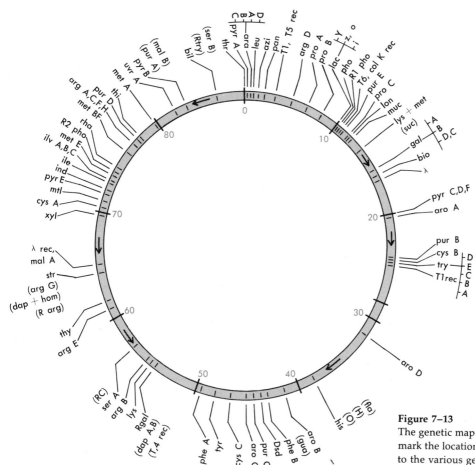

Figure 7–13
The genetic map of *E. coli*. The symbols mark the locations of genes. A key to the various gene abbreviations is found in Table 7–1. Those genes whose locations are only approximately known are shown in parentheses. The numbers divide the map into time intervals corresponding to the time in minutes which it takes each male chromosomal segment to move into a female cell. Thus 89 minutes are now thought to be required for complete transfer. The arrows mark the points at which various Hfr chromosomes break prior to transfer into a female cell; the direction of the arrows indicates transfer direction. [Redrawn from A. L. Taylor and M. S. Thoman, *Genetics*, **50**, 667 (1964), with permission.]

into the female cell. Chromosome transfer may thus accompany the synthesis of a new chromosome or chromosome segment.

The biological significance of circular genetic maps is surrounded with mystery. Circular maps also exist for many viral chromosomes, and so the *E. coli* form should not be viewed as a strange exception. We shall return to a discussion of circles when we look at the precise chemistry of the chromosome.

PLASMIDS

In addition to the sex factors which carry genes directly involved in mating, most bacteria contain other freely replicating circular chromosomal elements. Collectively they are all called *plasmids*, with the term *episome* reserved for those DNA elements like the sex factors which regularly move

Table 7–1 Key to the Genes of the *E. coli* Chromosome[a]

Genetic Symbols	Mutant Character	Enzyme or Reaction Affected
araD araA araB araC	Cannot use the sugar arabinose as a carbon source	L-Ribulose-5-phosphate-4-epimerase L-Arabinose isomerase L-Ribulokinase
argB argC argH argG argA argD argE argF	Requires the amino acid arginine for growth	N-Acetylglutamate synthetase N-Acetyl-γ-glutamokinase N-Acetylglutamic-γ-semialdehyde dehydrogenase Acetylornithine-*d*-transaminase Acetylornithinase Ornithine transcarbamylase Argininosuccinic acid synthetase Argininosuccinase
aroA, B, C aroD	Requires several aromatic amino acids and vitamins for growth	Shikimic acid to 3-enolpyruvyl-shikimate-5-phosphate Biosynthesis of shikimic acid
azi	Resistant to sodium azide	
bio	Requires the vitamin biotin for growth	
cysA cysB cysC	Requires the amino acid cysteine for growth	3-Phosphoadenosine-5-phosphosulfate to sulfide Sulfate to sulfide; 4 known enzymes
dapA dapB	Requires the cell wall component diaminopimelic acid	Dihydrodipicolinic acid synthetase N-Succinyl-diaminopimelic acid deacylase
dap + hom	Requires the amino acid precursor homoserine and the cell-wall component diaminopimelic acid for growth	Aspartic semialdehyde dehydrogenase
Dsd	Cannot use the amino acid D-serine as a nitrogen source	D-Serine deaminase
fla	Flagella are absent	
galA galB galD	Cannot use the sugar galactose as a carbon source	Galactokinase Galactose-1-phosphate uridyl transferase Uridine-diphosphogalactose-4-epimerase
gua	Requires the purine guanine for growth	
H	The H antigen is present	
his	Requires the amino acid histidine for growth	10 known enzymes[b]
ile	Requires the amino acid isoleucine for growth	Threonine deaminase

Table 7–1 *(continued)*

Genetic Symbols	Mutant Character	Enzyme or Reaction Affected
ilvA ilvB ilvC	Requires the amino acids isoleucine and valine for growth	α-Hydroxy-β-keto acid rectoisomerase α,β-dihydroxyisovaleric dehydrase[b] Transaminase B
ind (indole)	Cannot grow on tryptophan as a carbon source	Tryptophanase
λ	Chromosomal location where prophage λ is normally inserted	
lac Y	Unable to concentrate β-galactosides	Galactoside permease
lac Z	Cannot use the sugar lactose as a carbon source	β-Galactosidase
lac O	Constitutive synthesis of lactose operon proteins (see Chapter 14)	Defective operator
leu	Requires the amino acid leucine for growth	3 known enzymes[b]
lon (long form)	Filament formation and radiation sensitivity are affected	
lys	Requires the amino acid lysine for growth	Diaminopimelic acid decarboxylase
lys + met	Requires the amino acids lysine and methionine for growth	
λ rec, malA	Resistant to phage λ and cannot use the sugar maltose	Phage λ receptor, and maltose permease
malB	Cannot use the sugar maltose as a carbon source	Amylomaltase(?)
metA metB metF metE	Requires the amino acid methionine for growth	Synthesis of succinic ester of homoserine Succinic ester of homoserine + cysteine to cystathionine 5,10-Methylene tetrahydrofolate reductase
mtl	Cannot use the sugar mannitol as a carbon source	Mannitol dehydrogenase (?)
muc	Forms mucoid colonies	Regulation of capsular polysaccharide synthesis
O	The O antigen is present	

(Continued)

[a] Each known gene or gene cluster is listed by its symbol and with the character caused by a mutation in the gene or gene cluster. The enzyme affected or reaction prevented is listed where known.
[b] Denotes enzymes controlled by the homologous gene loci of *Salmonella typhimurium.*

Table 7–1 (*continued*)

Genetic Symbols	Mutant Character	Enzyme or Reaction Affected
pan	Requires the vitamin pantothenic acid for growth	
phe A, B	Requires the amino acid phenylalanine for growth	
pho	Cannot use phosphate esters	Alkaline phosphatase
pil	Has filaments (pili) attached to the cell wall	
proA proB proC	Requires the amino acid proline for growth	
purA purB purC, E pur D	Requires certain purines for growth	Adenylosuccinate synthetase Adenylosuccinase 5-Aminoimidazole ribotide (AIR) to 5-amino-imidazole-4-(N-succino carboximide) ribotide Biosynthesis of AIR
pyrA	Requires the pyrimidine uracil and the amino acid arginine for growth	Carbamate kinase
pyrB pyrC pyrD pyrE pyrF	Requires the pyrimidine uracil for growth	Aspartate transcarbamylase Dihydroorotase Dihydroorotic acid dehydrogenase Orotidylic acid pyrophosphorylase Orotidylic acid decarboxylase
R arg	Constitutive synthesis of arginine (see Chapter 14)	Repressor for enzymes involved in arginine synthesis
R gal	Constitutive production of galactose	Repressor for enzymes involved in galactose production
R1 pho, R2 pho	Constitutive synthesis of phosphatase	Alkaline phosphatase repressor
R try	Constitutive synthesis of tryptophan	Repressor for enzymes involved in tryptophan synthesis
RC (RNA control)	Uncontrolled synthesis of RNA	
rha	Cannot use the sugar rhamnose as a carbon source	
serA serB	Requires the amino acid serine for growth	3-Phosphoglycerate dehydrogenase Phosphoserine phosphatase
str	Resistant to or dependent on streptomycin	
suc	Requires succinic acid	

Table 7–1 *(continued)*

Genetic Symbols	Mutant Character	Enzyme or Reaction Affected
T1, T5 rec	Resistant to phages T1 and T5 (mutants called B/1,5)	T1, T5 receptor sites absent
T1 rec	Resistant to phage T1 (mutants called B/1)	T1 receptor site absent
T6, colK rec	Resistant to phage T6 and colicine K	T6 and colicine receptor sites absent
T4 rec	Resistant to phage T4 (mutants called B/4)	T4 receptor site absent
thi	Requires the vitamin thiamine for growth	
thr	Requires the amino acid threonine for growth	
thy	Requires the pyrimidine thymine for growth	Thymidylate synthetase
trpA		Tryptophan synthetase, A protein
trpB	Requires the amino acid tryptophan for growth	Tryptophan synthetase, B protein
trpC		Indole-3-glycerolphosphate synthetase
trpD		Phosphoribosyl anthranilate transferase
trpE		Anthranilate synthetase
tyr	Requires the amino acid tyrosine for growth	
uvrA	Resistant to ultraviolet radiation	Ultraviolet-induced lesions in DNA are reactivated
xyl	Cannot use the sugar xylose as a carbon source	

[a] Each known gene or gene cluster is listed by its symbol and with the character caused by a mutation in the gene or gene cluster. The enzyme affected or reaction prevented is listed where known.
[b] Denotes enzymes controlled by the homologous gene loci of *Salmonella typhimurium*.

on and off the main bacterial chromosome. Plasmid sizes vary tremendously; some are not much larger than 10^6 MW and carry only 1 to 3 genes, while others have sizes up to 10 to 20 percent of the main chromosome. Among the most common nonepisomal plasmids are many of the *resistance transfer factors* (RTF) that confer simultaneous multiple resistance to many different antibiotics (e.g., ampicillin, streptomycin, and tetracycline). Many such plasmids are present in multiple copies (10 to 50) and duplicate independently of the main chromosomal element. Conceivably, autonomously replicating plasmids may provide the quickest way to increase the relative number of genes that confer antibiotic resistance.

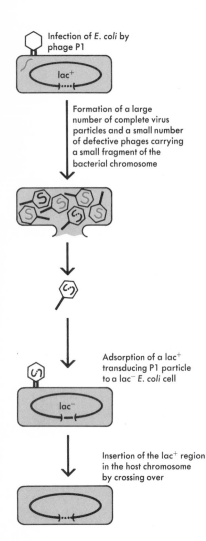

Infection of *E. coli* by phage P1

lac⁺

Formation of a large number of complete virus particles and a small number of defective phages carrying a small fragment of the bacterial chromosome

Adsorption of a lac⁺ transducing P1 particle to a lac⁻ *E. coli* cell

lac⁻

Insertion of the lac⁺ region in the host chromosome by crossing over

Figure 7–14
Passive transfer of genetic material from one bacterium to another by means of carrier phage particles (transduction).

PHAGES OCCASIONALLY CARRY BACTERIAL GENES

Not only can bacterial genes be transferred in mating, but they can also be passively carried from one bacterium to another by phage particles (*transduction*). This happens when a virus particle is formed that accidentally contains a very small portion (usually less than 1 to 2 percent) of its host chromosome (a *transducing phage*). When this virus particle (usually biologically inactive because its viral chromosome is incomplete or totally missing) attaches to a host cell, the fragment of bacterial chromosome is injected into the cell. It then can engage in crossing over with the host chromosome; if the transducing phage has been grown on a bacterial strain genetically different from the strain subsequently infected with the phage, a genetically altered bacterium may be produced (Figure 7–14). For example, a suspension of phage particles P1 grown on a strain of *E. coli* that is able to grow on lactose contains a small number of particles carrying the gene (lac⁺) involved in lactose metabolism. Addition of these phages to an *E. coli* strain unable to use lactose (lac⁻) transforms a small number of the lac⁻ bacteria to the lac⁺ form, by means of genetic recombination.

Because transduction is generally very rare, it might be guessed that it would not be a useful tool for probing chromosome structure. In fact, it has been most helpful in telling us whether two genes are located close to each other and what their exact order is. This is because the number of bacterial genes carried by a single transducing particle is small, so that only genes located very close to each other will be enclosed in the same transducing particle. Thus, by determining the frequencies with which groups of genes can be transduced by the same phage particle, we can very accurately establish their relative locations.

TRANSFER OF PURIFIED CHROMOSOME FRAGMENTS

Transformation is the name given to genetic recombination brought about by the introduction of purified chromosomes (DNA). It has provided the crucial biological system for the chemical identification of the genetic material (DNA). Transformation was originally discovered in 1928, when the observation was made that the addition of heat-killed cells of a pathogenic strain of *Diplococcus pneumoniae* to a suspension of live, nonpathogenic pneumonia cells caused a small fraction of the live bacteria to become pathogenic. The hereditary nature of this transformation was shown by using descendants of the newly pathogenic strain to transform still other nonpathogenic bacteria. This suggested that when the pathogenic cells are killed by heat, their chromosomes (now known to be DNA) are undamaged, and free chromosomal material, liberated somehow from the heat-killed cells, can pass through the cell wall of the living cells and subsequently undergo genetic recombination with the host chromosome (Figure 7–15).

Subsequent experiments have confirmed the genetic interpretation of the transformation phenomenon. The pathogenic character is caused by a gene S (smooth), which affects the chemistry of the bacterial cell wall and causes the formation of a carbohydrate capsule. When the R (rough) allele of this gene is present instead, no capsule is formed, and the cell is not pathogenic.

Although the first transformation experiments involved only changes in capsule chemistry, it is now clear that all genes can be transformed by means of the addition of extracted chromosomes. Because only small chromosomal fragments are generally transformed in this way, this process can also reveal which genes are located close to each other. The fact that transformation of most types of bacteria is very inefficient has, up to now, severely restricted its general applicability to genetic problems. Thus transduction rather than transformation is our most efficient means of determining the precise order of *E. coli* genes. Transformation has, nevertheless, been very useful in locating the genes of the bacteria *Bacillus subtilis*, for which a useful transducing phage is not yet known.

Ever since transformation was first demonstrated, much speculation has existed whether it is possible in organisms larger than bacteria. Particular attention has been focused on the possibility of transforming mammals, especially man. So far, however, all results have been negative, except for some special cases where viral chromosomes are able to transform normal cells into cancer cells. Here, however, we may be dealing not with the change of an existing gene, but rather with the introduction of entirely new genetic material. In any case, though, too few good experiments have yet been performed to give us a feeling for whether this type of genetic analysis can be extended to higher organisms.

PHAGES ALSO MUTATE

The plaques formed by a given type of phage are quite characteristic and can often be distinguished easily from those of genetically distinct phages. For example, the plaques of phage T2 can easily be separated from those made by phage λ or by phage F2. More significantly, mu-

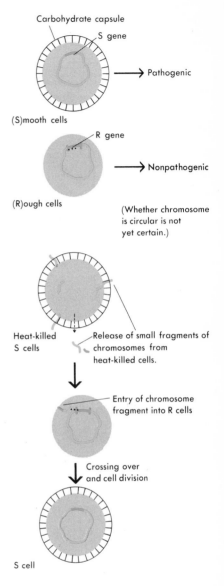

Figure 7–15

Transformation of the genetic character of a bacterial cell (*Diplococcus pneumoniae*) by addition of heat-killed cells of a genetically different strain. Here we show an R cell receiving a chromosomal fragment containing the S gene. Most R cells, however, receive other chromosome fragments, and so the efficiency of transformation for a given gene is usually less than 1 percent.

tations occur that change the morphology of phage plaques. We do not usually know the biochemical basis of these plaque differences, but this does not really matter. The important fact is that these differences are usually reproducible and simple to score. It is easy to look at the morphology of thousands of plaques to see if any differ from the plaques made by the wild-type phage. In this way a large number of different plaque-type mutations were found (Figure 7–16).

Another class of mutations changes the ability of phage to adsorb to bacteria. For example, wild-type T2 cannot multiply on *E. coli* strain B/2, because the B/2 mutation changes the cell surface, thereby preventing the attachment of T2. Mutant T2 particles can, however, multiply on B/2: They are called T2h, and are able to adsorb because they possess altered tail fibers.

Another very large and important class of mutants exists which have the ability to multiply at 25°C, but cannot multiply at 42°C (temperature-sensitive *conditional lethals*). Although we cannot now pinpoint the exact reason why these mutant phages do not multiply at the higher temperature, we suspect that the high temperature destroys the 3-D structure of a protein necessary for their reproduction. This type of mutation has proved useful because a large number of viral genes mutate to a temperature-sensitive form. When we isolate a mutant unable to multiply at the higher temperature, the mutation may be located in a large variety of different genes.

The existence of conditional lethal mutations has allowed phage geneticists to find mutations in essentially all the genes of phages T4 and λ. Despite the fact that genetic recombination between mutant viruses was not discovered until 1945, the phage T4 is now the best and most completely characterized genetic object.

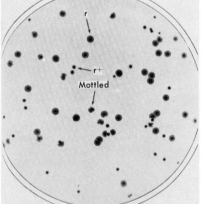

Figure 7–16
Photograph of mutant phage plaques. Shown are a mixture of T2r⁺ (wild-type) plaques and T2r (a rapid lysis mutant) plaques. The mottled plaques arise from the simultaneous growth of both r and r⁺ phages in the same plaque. (From *Molecular Biology of Bacterial Viruses* by Gunther S. Stent. W. H. Freeman and Company, copyright © 1963, p. 177, with permission.)

PHAGE CROSSES

More than one phage particle at a time can grow in a single bacterium. If several particles adsorb at once, the chromosomes from all of them enter the cell and duplicate to form large numbers of new copies. So long as the chromosomes exist free (unenclosed by a protective coat) they can cross over with similar chromosomes (Figure 7–17). This is shown by infecting cells with two or more genetically distinct phage particles and finding recombinant genetic types among the progeny particles. For example, it is easy to obtain mutant T4 particles differing from the

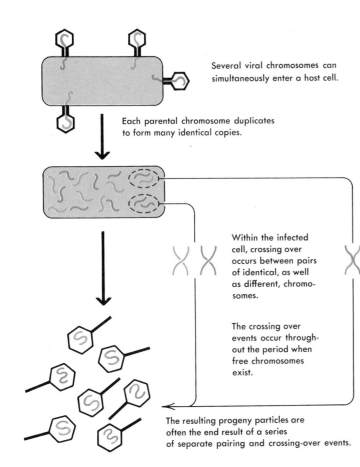

Several viral chromosomes can simultaneously enter a host cell.

Each parental chromosome duplicates to form many identical copies.

Within the infected cell, crossing over occurs between pairs of identical, as well as different, chromosomes.

The crossing over events occur throughout the period when free chromosomes exist.

The resulting progeny particles are often the end result of a series of separate pairing and crossing-over events.

◀ **Figure 7–17**
Genetic recombination following infection of a bacterium with several genetically distinct phage particles.

Figure 7–18
Plaques found after infecting bacteria with T2hr and T2h⁺r⁺ phages. The technique used to see all four progeny types (hr, h⁺r⁺, h⁺r, hr⁺) is to look for plaques on a mixture of strain B and strain B/2. Only phages possessing the h gene can kill both B and B/2 cells. Phages with the h⁺ gene kill only B cells, and their plaques look turbid because of the presence of live B/2 cells. (From *Molecular Biology of Bacterial Viruses* by Gunther S. Stent. W. H. Freeman and Company, copyright © 1963, p. 185, with permission.)
▼

wild type by two mutations, one in an h gene, which allows them to grow on *E. coli* strain B/4 (a strain resistant to wild-type T4 particles) and the other in an r gene, which causes them to form larger and clearer plaques than wild-type T4 particles. These double mutant phages are designated T4hr, and the wild type is called T4h⁺r⁺. When an *E. coli* cell is infected simultaneously with a T4h⁺r⁺ and a T4hr phage, four types of progeny particles are found: the parental genotypes h⁺r⁺ and hr, and the recombinant genotypes hr⁺ and h⁺r (Figure 7–18).

The frequency with which recombinants are found depends upon the particular mutants used in the cross. Crosses between some pairs of markers give almost 50 percent recombinant phage; crosses between others give somewhat lower recombinant values, and sometimes almost no recombinants are found. This immediately suggests that viruses also have unbranched genetic maps, a suggestion now completely confirmed by intense analysis of a large number of independently isolated mutations.

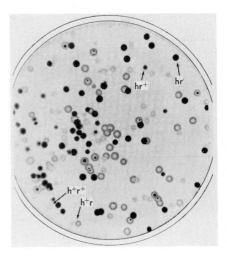

Figure 7–19

The genetic map of T4. The numbers refer to a collection of 56 genes in which conditional lethal mutations have been located. Their existence as distinct genes has been shown by complementation studies (see Chapter 8). The minimal length is shown for some of the genes (color segments). The length of other genes (black segments) has not yet been worked out. The boxes show either deficiencies in synthesis associated with some mutant genes or incomplete viral components seen by EM investigation of infected cells. The genes in the upper left of the circle control functions in the first half of the life cycle. The remaining genes function in the last half of the cycle. Nothing is known about how this differential timing is accomplished. (From "The Genetics of a Bacterial Virus," R. S. Edgar and R. H. Epstein; *Sci. Amer.* (February 1965); p. 76. Copyright 1965 by Scientific American, Inc. All rights reserved.)

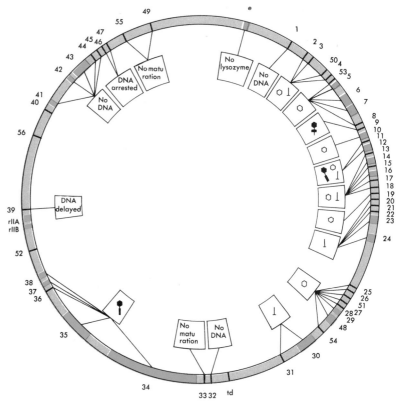

At present our best known map is that of phage T4 (Figure 7–19), a circular map, like that of *E. coli.* We do not know whether all viruses will have circular maps; hints now exist that several viruses previously thought to have strictly linear maps will be found to have circular ones.

VIRAL CROSSES INVOLVE MULTIPLE PAIRINGS

Thus the genetic structure of viral chromosomes appears to be essentially similar to that of cellular chromosomes. Nonetheless, it is worth pointing out a distinct feature of viral crosses. In conventional meiosis each chromosome pairs just once, whereas in a viral cross, pairing and crossing over may occur repeatedly throughout the period when free chromosomes are present. Thus a given chromosome may participate in several pairing and crossing-over events. This point is simply demonstrated by infecting a cell with three virus particles, each with a distinct genetic marker, and finding single progeny particles that have derived chromosomal regions from all three particles. The products of a phage cross are essentially, then, the products of a large number of distinct pairings and crossings over. There is great variation among viruses in the amount of crossing over; for example, a chromosome of phage T4 crosses over, on the average, 5 to 10 times, and a phage λ chromosome only 0.5 times, during each growth cycle. The existence of many distinct crossovers does not, however, restrict our ability to map the genetic markers, since the general rule still holds that genes located close together seldom recombine.

Another way in which a viral cross differs from conventional meiosis involving cell fusion is that viruses are not separated into male and female particles. Differentiation into sexes can be considered a device to bring about cell fusion between genetically distinct organisms. Virus particles, however, do not have to fuse, since genetic recombination occurs when two viral chromosomes of different genotypes enter a single host cell. Moreover, in cells infected with several genetically distinct viruses, crossing over can occur between both genetically identical and different chromosomes; thus only a fraction of the crossovers in a cell infected with different viruses result in recombinant chromosomes of a new genotype. Evidence also exists that crossing over occurs after infection with a single virus particle. This phenomenon cannot be revealed by genetic analysis, since all the progeny chromosomes are genetically identical, and so experiments with isotopically labeled virus (see Chapter 9) are necessary for its demonstration.

SUMMARY

Chromosomes control the hereditary properties of all cells and are linear collections of specific genetic factors called genes. Each gene can affect the character of a cell in a highly specific way. This is shown by the striking cellular effects of hereditary changes in gene structures (mutations): Various mutations can alter, for example, the eye color or body size of an organism. Although spontaneous mutations occur only rarely, it is possible to increase mutation rates by applying specific chemicals or radiation (mutagens).

Most of the early work in genetics was with large and complicated diploid plants and animals. Now, however, the most favorable objects for use in studying what a gene is and how it functions are the haploid microorganisms like the bacteria and their viruses, the bacteriophages (phages). They have the advantage of very short life cycles and ease of growth under controllable laboratory conditions.

Bacterial mutations involving growth factors and resistance to specific antibiotics and viruses are particularly useful because of the ease of separating wild-type and mutant particles. The same is true for phage mutations involving host range and temperature requirements.

The location of genes on chromosomes is revealed by a study of the segregation of genes in genetic crosses. Recombination of alleles can be caused by both random assortment of chromosomes and crossing over of homologous chromosomes. We can determine whether given genes are located on different or the same chromosomes by whether they assort randomly, as chromosomes do, or whether they exhibit (more than 50 percent) linkage. Crossing over of homologous chromosomes provides a basis for determining the relative positions of genes on chromosomes (chromosome mapping). The more frequently crossovers occur between two genes, the farther apart the genes must be. When three genes are considered at once (three-factor cross), the least frequent recombinant type results from a double crossover; the determination of this recombinant type provides a check for the results of several two-factor crosses dealing with the same three factors.

Most genetic crosses designed to reveal what genes are are now performed with bacteria and phages, where typical meiosis does not occur. Bacterial crosses usually involve crossovers between a chromosome fragment and an intact chromosome. Genetic recombination occurs in bacteria as a result of mating between male (Hfr) and female cells (conjugation), attachment of phage particles containing the genes of former bacterial hosts (transduction), and the introduction of foreign chromosomal (DNA) extracts (transformation). Phage crosses take place when a bacterium is infected with two or more genetically distinct phages; they involve many cycles of pairing and crossing over. Despite the difficulties which these phenomena present to the study of phage and bacterial genetics, the chromosomes of the bacterium E. coli and its various phages are quickly becoming the best understood of all genetic material.

REFERENCES

HAYES, W., *The Genetics of Bacteria and Their Viruses: Studies in Basic Genetics and Molecular Biology,* 2nd ed., Wiley, New York, 1969. A comprehensive description of the genetic systems of viruses and bacteria, showing how they have been used to elucidate many of the fundamental principles of molecular genetics.

STENT, G. S., *Molecular Biology of Bacterial Viruses,* Freeman, San Francisco, 1963. Though somewhat outdated, still a beautiful introduction to the world of bacteriophages.

ADELBERG, E. A. (ed.), *Papers on Bacterial Genetics,* 2nd ed., Little, Brown, Boston, 1966. A collection of significant reprints on the development of bacterial genetics.

STENT, G. S. (ed.), *Papers on Bacterial Viruses,* 2nd ed., Little, Brown, Boston, 1965. A collection of many of the papers that shaped the development of current research with bacterial viruses.

WHITEHOUSE, H. L. K., *Towards an Understanding of the Mechanism of Heredity,* 2nd ed., Edward Arnold, London, 1969. A good general genetics text, with much emphasis on crossing over.

HERSKOWITZ, I. H., *Principles of Genetics,* MacMillan, New York, 1973. An excellent up-to-date text.

GOODENOUGH, U., AND R. P. LEVINE, *Genetics,* Holt, Rinehart, and Winston, New York, 1974. A very well illustrated comprehensive text that features molecular genetics.

MILLER, J., *Experiments in Molecular Genetics,* Cold Spring Harbor Laboratory, New York, 1972. A superior laboratory manual which doubles as an advanced text for bacterial genetics.

CLOWES, R., "The Molecule of Infectious Drug Resistance," *Scientific American,* April 1973. The molecular biology of a medically important plasmid.

MEYNELL, G. G., *Bacterial Plasmids,* Macmillan Press, London, MIT Press, Cambridge, 1973. An advanced text that emphasizes genetic and molecular aspects.

Gene
Structure
and Function

For many years it was generally thought that crossing over occurred between genes, not within the genes themselves. The chromosome was viewed as a linear collection of genes held together by some nongenetic material, somewhat like a string of pearls. Now, however, we realize that this viewpoint is completely wrong and the exact opposite may be true—all crossing over may occur by breakage and reunion of the genetic molecules themselves. The original impression arose from the fact that crossing over between two regions is much easier to detect if the regions are far apart on a chromosome. If they are very close, recombination is extremely rare and can be detected only by examining a very large number of progeny, too large to make study of intragenic recombination practicable when genetic work was restricted to higher organisms. Even with the intensively studied fruit fly *Drosophila*, it is difficult to look at more than 50,000 progeny from a single cross. Newer techniques, however, permit rapid screening of millions of the progeny from crosses between genetically different molds, yeast, bacteria, or viruses. With these organisms it has been simple to show that crossing over can cause recombination of material within a gene. Each gene contains a number of different sites at which mutations occur and between which crossing over occurs. This result provides a powerful method for investigating topological structures of genes.

CHAPTER 8

RECOMBINATION WITHIN GENES ALLOWS CONSTRUCTION OF A GENE MAP

Up to now, the most striking results on the genetic structure of the gene itself have come from the work with the rIIA and rIIB genes of the bacterial virus T4. These are two adjacent genes which influence the length of the T4 life cycle; mutations in both thereby affect the size of plaques produced in a bacterial layer growing on an agar plate. The presence of either an rIIA or an rIIB mutation can cause a shorter life cycle of T4 phage within an *E. coli* cell. T4-infected cells on an agar plate normally do not break open and release new progeny phage until several hours after they have been infected. Cells infected with rII mutants, however, always break open more rapidly, hence the designation r (II stands for the fact that there do exist other genes which cause rapid cell lysis). Thus rII mutants produce larger plaques than wild-type phage. The rII mutants were chosen to work with because of the possibility of detecting a very small number of wild-type particles among a

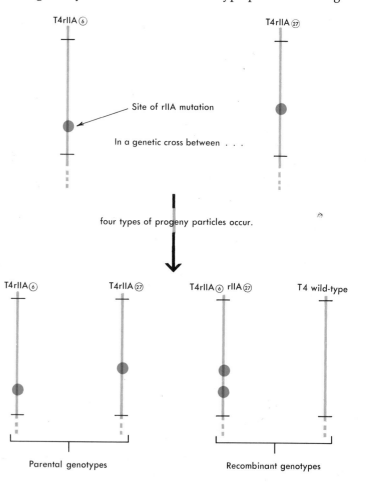

Figure 8–1
The use of T4rII mutations in the demonstration of crossing over within the gene. Equal numbers of wild-type and double r recombinants occur. The wild-type recombinants are easily found, because they are the only progeny genotype which will form plaques on K12(λ). It is much harder to identify the double $rIIA_6rIIA_7$ recombinants, inasmuch as their plaques are indistinguishable from single r plaques. To detect them, it is necessary to isolate a large number of progeny r viruses with the r phenotype and use them for new genetic crosses with both parent r strains. The double r mutants will not produce wild-type recombinants with either of the single r mutants.

184

very large number of mutants. Although the wild-type and the rII mutants grow equally well on *E. coli* strain B, there is another strain, *E. coli* K(λ), on which only the wild type can multiply. Thus when the progeny of a genetic cross between two different rII mutants are added to K12(λ), only the wild type recombinants form plaques. Even as few as one wild type recombinant per 10^6 progeny is easily detected.

Over two thousand independent mutations in the rIIA and rIIB genes have been isolated and used in breeding experiments. In a typical cross, *E. coli* strain B bacteria were infected with two phage particles, each bearing an independently isolated rIIA (or rIIB) mutation. As the virus particles multiplied, genetic recombination occurred. The progeny were then grown on *E. coli* K(λ) to test for the wild type. Normal particles were found in a very large fraction of the crosses, indicating recombination within the gene (Figure 8–1). If recombination occurred only *between* genes it would be impossible to produce wild-type recombinants by crossing two phage particles with mutations in the same gene. A large spectrum of recombination values was found, just as in crosses between mutants for separate genes. This spread of values indicates that some mutations occur closer together than others, and allows for the construction of genetic maps for the two rII genes. Examination of these genetic maps yields the following striking conclusions:

1. A large number of different sites of mutation (mutable sites) occur within the gene. This number is on the order of 1000 to 1500 altogether for the rIIA and rIIB genes (Figure 8–2).
2. The rIIA and rIIB genetic maps are unambiguously linear, strongly hinting that the gene itself has a linear construction.
3. Most mutations are changes at only one mutable site. Genes containing such mutations are able to be restored to the original wild-type gene structure by the process of undergoing a second (reverse) mutation at the same site as the first mutation.
4. Other mutations cause the deletion of significant fractions of the genetic map. These are the result of a physical deletion of part of the rII gene (Figure 8–3). Mutations deleting more than one mutable site are highly unlikely to mutate back to the original gene form.

The genetic fine structure of a number of other viral and bacterial genes has also been extensively mapped. The

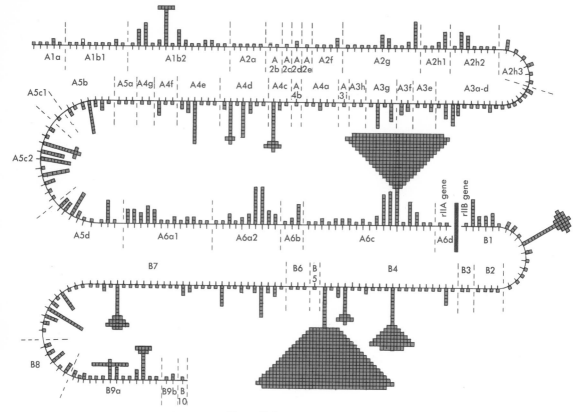

Figure 8–2

The genetic map of the rIIA and rIIB genes of phage T4. This map shows the assignment of a large number of different mutations in specific regions (A1a, A1b₁, etc.) of the gene. In most cases, the order of the mutations within a given region has not yet been determined. Each square corresponds to an independent occurrence of a mutation at a given site on the chromosome. Thus some regions ("hot spots") mutate much more frequently than others. (From S. Benzer, *Proc. Natl. Acad. Sci. U.S.* **47**, 410 (1961), with permission.)

Figure 8–3

Deletion mutations within the rII region of T4. About 10 percent of the spontaneous rII mutations do not map at a distinct point. They are due to deletions of a large number of adjacent mutable sites. Some deletions, for example 1272, involve both the rIIA and rIIB genes. The small rectangles indicate that the deletion most likely extends into the adjacent gene. The existence of deletion mutations has considerably facilitated genetic mapping. By crossing a newly isolated rII mutant with a number of deletion mutations covering increasingly larger regions, it is quickly possible to assign an approximate map location to the new mutant. In this map the size of the rIIB gene has been arbitrarily reduced.

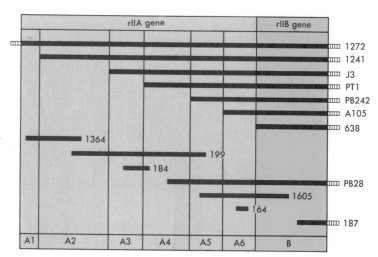

lengths of these maps vary from gene to gene, suggesting that some gene products are larger than others. Though no other study has been as extensive as the rII work, each points to the same conclusions—that all genes have a very large number of sites at which mutation can occur, and that these mutable sites are arranged in a strictly linear order. The geneticist's view of a gene is thus: *a discrete chromosomal region which (1) is responsible for a specific cellular product and (2) consists of a linear collection of potentially mutable units (mutable sites), each of which can exist in several alternative forms and between which crossing over can occur.*

THE COMPLEMENTATION TEST DETERMINES IF TWO MUTATIONS ARE IN THE SAME GENE

Since mutations in both the rIIA and rIIB genes result in a larger sized plaque, it is natural to ask why they are considered two genes: Would it not be simpler to consider them parts of the same gene? Our answer is straightforward. If *E. coli* strain K12(λ) is infected simultaneously with a T4rIIA and a T4rIIB mutant, the chromosomes of both viruses multiply, and progeny virus is produced (Figure 8–4). In contrast, simultaneous infection of the K12(λ)

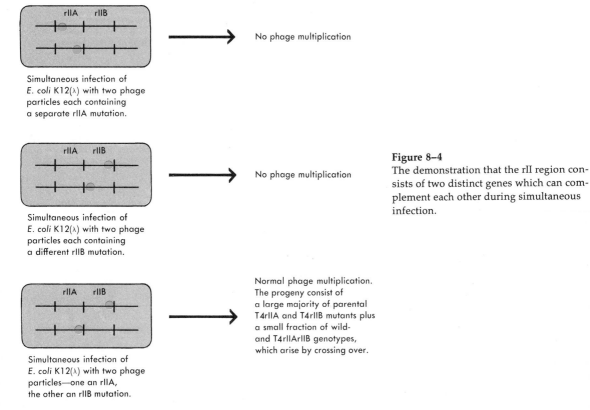

No phage multiplication

Simultaneous infection of
E. coli K12(λ) with two phage
particles each containing
a separate rIIA mutation.

No phage multiplication

Simultaneous infection of
E. coli K12(λ) with two phage
particles each containing
a different rIIB mutation.

Normal phage multiplication.
The progeny consist of
a large majority of parental
T4rIIA and T4rIIB mutants plus
a small fraction of wild-
and T4rIIArIIB genotypes,
which arise by crossing over.

Simultaneous infection of
E. coli K12(λ) with two phage
particles—one an rIIA,
the other an rIIB mutation.

Figure 8–4
The demonstration that the rII region consists of two distinct genes which can complement each other during simultaneous infection.

187

with either two different T4rIIA or two different T4rIIB mutants results in no virus multiplication. This demonstrates that rIIA and rIIB genes carry out two different functions, each necessary for multiplication on K12(λ).

In infection with a single T4rIIA mutant, the rIIB gene functions normally, but no active form of the rIIA product is available, so that no viral multiplication is possible. Likewise, in infection with a single rIIB mutant, an active rIIA product results, but only an inactive form of the rIIB product arises. Virus multiplication occurs only when we simultaneously infect a bacterium with an rIIA phage mutant and an rIIB phage mutant, because each mutant chromosome is able to produce the gene product that its infecting partner is unable to make. Two chromosomes thus can *complement* each other when the mutations are present in distinct genes—the *complementation test*. (This experiment is not affected by the possibility of intragenic crossing over between mutants to produce wild-type phage, because the number of such recombinants is very small; in contrast, complementation between different mutant genes yields a normal number of progeny particles.)

It is easy to perform complementation tests with phage mutants. All that is necessary is to infect a bacterium with two different mutants at once. This automatically creates a cell containing one copy of each mutant chromosome. It is more difficult to carry out complementation tests with normally haploid cells, like *E. coli*. But fortunately, genetic tricks too complicated to be described here enable special strains to be constructed with some chromosome sections present twice (partially diploid strains). These have often been useful in telling us that a chromosomal region thought to contain only one gene actually produces several gene products, and so must contain a corresponding number of genes.

GENETIC CONTROL OF PROTEIN FUNCTION

A direct relationship between genes and enzymes was hypothesized as early as 1909 from a study of the metabolism of phenylalanine in patients suffering from certain hereditary diseases; but it was not until the decade following 1941, when a variety of growth factor mutants became available in *Neurospora* and *E. coli*, that the one gene–one enzyme hypothesis became an established fact. One of the first proofs involved the biosynthesis of the amino acid arginine. Its pathway starts with glutamic acid and proceeds by eight chemical reactions, each catalyzed by a distinct enzyme (Figure 8–5). For each of these eight steps

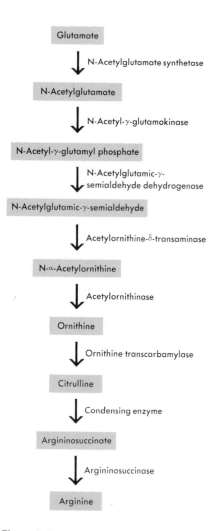

Figure 8–5
Pathway of arginine biosynthesis in *E. coli*.

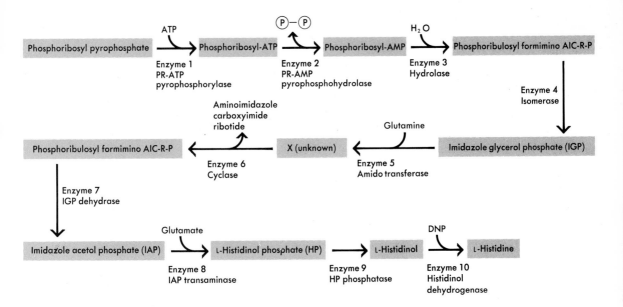

Figure 8–6
The pathway for histidine biosynthesis in *Salmonella typhimurium*. This bacterium, which is closely related to *E. coli*, appears to have a chromosome with a similar gene arrangement.

mutant cells have been isolated that fail to carry out that specific enzymatic reaction. This suggests that a separate gene controls the presence of each enzyme, a hypothesis confirmed by the absence of the specific active enzyme in cell extracts prepared from such mutant cells. The biosynthesis of histidine (Figure 8–6) provides another beautiful example of this relationship: There exist specific mutations resulting in the absence of each of the ten enzymes necessary for this biosynthesis.

Enzymes are not the only proteins directly controlled by genes. Each specific protein is controlled by a gene unique for that protein. One of the first clean proofs of this idea came from the study of the hemoglobin present in people suffering from sickle-cell anemia. This is a disease whose genetic basis is well worked out. If the sickle (s) gene is present in both homologous chromosomes, a severe anemia results, characterized by the red blood cells having a sickle shape. If only one s gene is present, and the allele in the homologous chromosome is normal ($+$), the anemia is less severe, and the red blood cells almost normal in shape. The type of hemoglobin in red blood cells is likewise correlated with the genetic pattern. In the ss case, all the hemoglobin is of an abnormal type, characterized by a solubility different from that of normal hemoglobin, whereas in the $+s$ condition, half the hemoglobin is normal and half sickle.

Even now it is often hard to identify the protein product of a given gene. One of the most annoying cases involves the rII region of T4. We still do not know the

189

biochemical changes that underlie the larger plaques produced by rII mutant phages and, despite much effort, no one has yet found a specific protein absent in *E. coli* cells infected with T4rII mutants and present in cells infected with wild-type phage. Thus there is now a strong tendency to concentrate further genetic analysis on those genes for which we already know how to isolate the corresponding protein.

ONE GENE–ONE POLYPEPTIDE CHAIN

Until recently, the above ideas were stated by the slogan "one gene–one protein (enzyme)." Now we realize that a more correct statement is "one gene–one polypeptide chain." When a protein contains more than one polypeptide chain, each chain is made separately. Only after their synthesis do they aggregate to form the final protein. Generally, the two or more polypeptide chains needed to form a functional protein are controlled by adjacent genes. This is, however, not always the case. The α and β hemoglobin chains are not controlled by linked genes. The complementation test thus tells us not whether two genes control different proteins, but whether they control two different polypeptide chains.

RECESSIVE GENES FREQUENTLY DO NOT PRODUCE FUNCTIONAL PRODUCTS

Most mutant genes are recessive with respect to wild-type genes. This fact, puzzling to early geneticists, is now partially understood in terms of the gene–enzyme relation. The recessive phenotype often results from the failure of mutant genes to produce *any* functional protein (enzyme). In heterozygotes, however, there is always present one "good" gene, and correspondingly, a number of good gene products. Because the wild-type gene is present only once in heterozygotes, the possibility exists that there are always fewer good copies of the relevant protein in heterozygotes than in individuals with two wild-type genes. If this were the case, we might guess that the heterozygous phenotype would tend to be intermediate between the two homozygous phenotypes. Usually, however, this does not happen, for one of two reasons. Either there are still enough good enzyme molecules to catalyze the metabolic reaction even though the total number is reduced, or the recessive gene is not noticeable, because control mechanisms cause the wild-type gene in a heterozygote to produce more gene products than does each wild-type gene in a homozygote. In Chapter 14 we shall discuss how the rate at which a gene acts may be controlled.

GENES WITH RELATED FUNCTIONS
ARE OFTEN ADJACENT

Until ten years ago, geneticists believed that the chromosomal location of genes was purely random; there seemed to be no tendency for genes with related effects to be located near each other. Now, however, there are strong indications from viral and bacterial genetics that a sizable fraction of genes are situated in groups carrying out related functions. There are two main reasons for this change.

First, many geneticists now study mutations directly affecting the biosynthesis of known cellular molecules. Until the advent of work with microorganisms, most genetic markers involved very complex characters, like eye color or wing shape. The development of wings or eyes is, from a chemical viewpoint, fantastically complicated; clearly there is a large variety of *unrelated chemical reactions* whose absence might lead to a misshapen wing. It is thus not surprising that genes affecting the wings occur in many places on all *Drosophila's* four different chromosomes. In contrast, a mutant character that shows itself as the inability to synthesize a relatively simple molecule like the amino acid serine is most likely due to the absence of one of a much smaller number of *related chemical reactions.* If related genes are, indeed, next to each other on a chromosome, we are more likely to observe this phenomenon when we are studying mutations that we can assign immediately to a particular category of chemical upset.

The second main reason why we now frequently find adjacent genes with related functions is the availability of the complementation test. This often tells us that a chemical phenomenon is more complicated than originally guessed. The splitting of the rII region into the rIIA and rIIB genes is a typical case. Numerous situations now exist where a region previously thought to contain a single gene has been shown by complementation tests to perform a number of different, yet related, tasks. Subsequent biochemical investigations have then revealed that several chemical reactions (enzymes) are involved. One of the most spectacular examples involves the ability of *E. coli* to synthesize histidine. Thirty years ago we would have guessed that a series of mutations blocking histidine synthesis and

Figure 8–7

Clustering of the genes involved in the biosynthesis of histidine by the bacterium *Salmonella typhimurium.* The gene order was determined by transduction experiments. Each gene is responsible for the synthesis of one of the 10 enzymes needed to transform phosphoribosyl pyrophosphate into histidine. Here the genes are designated by numbers 1–10. Enzyme 1 is responsible for catalyzing the first reaction in the biosynthesis, enzyme 2 for the second step, etc. The names of these enzymes are given in Figure 8–6.

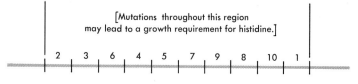

Histidine region

[Mutations throughout this region may lead to a growth requirement for histidine.]

2 3 6 4 5 7 9 8 10 1

mapping in the same region all fell in the same gene. Today, however, ten different genes in this region have been identified (Figure 8–7), each concerned with a different enzyme in the biosynthesis of histidine starting from phosphoribosyl pyrophosphate. Likewise, the region concerned with tryptophan biosynthesis contains a cluster of five different genes each concerned with a distinct step (enzyme) in the biosynthesis of tryptophan.

At first it was thought that the order of genes in the cluster corresponded to the order of their respective enzymes in the biosynthetic pathway. Now, however, there are several known exceptions to the rule. For example, the first two enzymes in histidine biosynthesis are located at opposite ends of the cluster.

Clusters of related genes are not always connected with the biosynthesis of essential metabolites. Degradation of many specific food molecules, such as the sugars galactose and lactose, also involves several consecutive chemical reactions; these also tend to be controlled by adjacent genes. Galactose breakdown requires three specific chemical steps, and lactose utilization at least two, probably three, different genes. Another striking example of adjacent related genes concerns genetic control of various structural proteins found in the protective coat of phage T4: The genes affecting the synthesis of head proteins are in one region, and those affecting the tail fibers in another.

The grouping together of related genes is connected with the fact, which we shall examine in detail in Chapter 14, that not all genes function at the same time. Mechanisms exist that tell genes whether or not to work. For example, the genes controlling lactose metabolism function only when a cell is growing on lactose; when lactose is absent, there is no need for these genes to work. The switching on and off of the genes is controlled by a specific molecule, the lactose repressor. As we shall see in Chapter 14, the ability of these repressor molecules to simultaneously control the synthesis of several proteins is dependent upon the fact that the corresponding genes are physically adjacent.

PROOF THAT GENES CONTROL AMINO ACID SEQUENCES IN PROTEINS

The first experimental demonstration that genes control amino acid sequences involved sickle-cell hemoglobin. Wild-type hemoglobin molecules are constructed from two different kinds of polypeptide chains: α chains and β chains. Each chain has a molecular weight of about 16,100. Two α chains and two β chains are present in each molecule, giving hemoglobin a molecular weight of about 64,500. The α and β chains are controlled by two distinct genes, so a single mutation will affect either the α or the β chain but not both. Sickle hemoglobin differs from normal hemoglobin by the change of one amino acid in the β chain: at position 6 (Figure 8–8), the glutamic acid residue found in the wild type hemoglobin is replaced by valine. Except for this one change the entire amino acid sequence is identical in normal and mutant hemoglobin peptides.

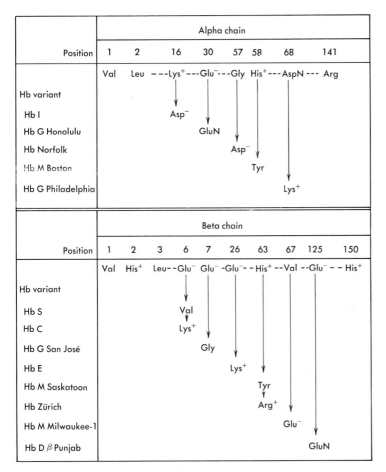

	Alpha chain							
Position	1	2	16	30	57	58	68	141
	Val	Leu	---Lys$^+$---	Glu$^-$---	Gly	His$^+$---	AspN ---	Arg
Hb variant								
Hb I			Asp$^-$					
Hb G Honolulu				GluN				
Hb Norfolk					Asp$^-$			
Hb M Boston						Tyr		
Hb G Philadelphia							Lys$^+$	

	Beta chain									
Position	1	2	3	6	7	26	63	67	125	150
	Val	His$^+$	Leu--	Glu$^-$	Glu$^-$	-Glu$^-$--	His$^+$	--Val	--Glu$^-$ – –	His$^+$
Hb variant										
Hb S				Val						
Hb C				Lys$^+$						
Hb G San José					Gly					
Hb E						Lys$^+$				
Hb M Saskatoon							Tyr			
Hb Zürich							Arg$^+$			
Hb M Milwaukee-1								Glu$^-$		
Hb D β Punjab									GluN	

Figure 8–8
A summary of the established amino acid substitutions in human hemoglobin variants.

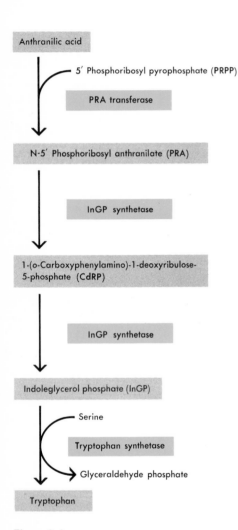

Figure 8–9
Last steps in the pathway of tryptophan biosynthesis.

This shows that a mutation in a gene results in a specific change in the template for hemoglobin, and strongly hints that all the information required to order hemoglobin amino acid sequences is present in the genes. Strongly supporting this belief are the analyses of amino acid sequences in hemoglobin isolated from persons suffering from other forms of anemia; here sequence analysis shows that each specific anemia is characterized by a single amino acid replacement at a unique site along the polypeptide chain.

The impracticality of large-scale breeding experiments with mammals makes it impossible to correlate the changes in hemoglobin sequences with the location of the various mutations along the genetic map. It is, however, possible to do this form of analysis with altered enzymes found in microorganisms whose genetics are well known.

COLINEARITY OF THE GENE AND ITS POLYPEPTIDE PRODUCT

The best understood example of the relationship between the order of the mutable sites in a gene and the order of their corresponding amino acid replacements involves the *E. coli* enzyme tryptophan synthetase, one of the several enzymes involved in tryptophan synthesis (Figure 8–9). This enzyme consists of two easily separable polypeptide chains, A and B, neither of which is enzymatically active by itself. A large number of mutants unable to synthesize tryptophan have been isolated; they lack a functional A chain and so are enzymatically inactive. When these mutants were genetically analyzed, it was found that changes at a large number of different mutable sites can give rise to inactive A chains. Accurate mapping of these mutants revealed that they all could be unambiguously located on the linear genetic map shown in Figure 8–10. It was possible to isolate the inactive A chains from many of these mutants and to begin to compare their amino acid sequences with the sequence of the wild-type A chain, which contains 267 amino acids. This sequence allows us to see how the location of a mutation within a gene is correlated with the location of amino acid replacements in its polypeptide chain product. Since both genes and polypeptide chains are linear, the simplest hypothesis is that amino acid replacements are in the same relative order as the mutationally altered sites in the corresponding mutant genes. This was most pleasingly demonstrated in 1964. The location of each specific amino acid replacement is exactly

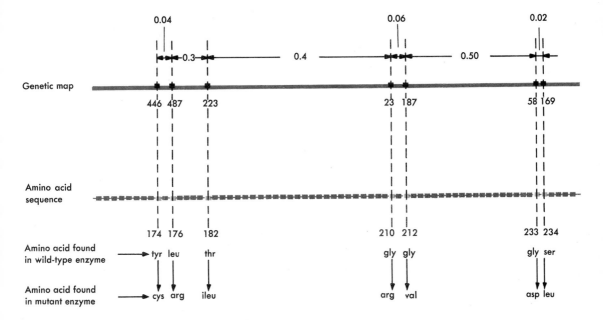

Figure 8–10

correlated (colinearity) with its location along the genetic map (shown in Figure 8–10). Thus each amino acid in a polypeptide chain is controlled (coded) by a specific region of the gene.

A MUTABLE SITE CAN EXIST IN SEVERAL ALTERNATIVE FORMS

Enzymatically inactive tryptophan synthetase molecules resulting from different mutations of the *same* mutable site (as shown by failure to give wild-type recombinants) do not always contain the same amino acid replacement. For example, depending upon the exact mutant strain examined, a change at the same mutable site will result in glycine being replaced by either glutamic acid or valine. This result means that a mutable site can exist in at least three alternative forms. In the next chapter, we shall discuss other evidence which tells us that the mutable sites are the deoxynucleotide building blocks from which the genes (regions of DNA molecules) are constructed. Since only four types of deoxynucleotides exist, this leads us to expect that genetic evidence will never show more than four alternative configurations for a mutable site.

Colinearity of the gene and its protein product. Here is illustrated the genetic map for one-fourth of the gene coding for the amino acid sequences in the *E. coli* protein tryptophan synthetase A. The symbols ←0.04→, etc., refer to map distances (frequencies of recombination) between the various tryptophan synthetase mutations A446, A487, etc. The numbers in the amino acid sequence refer to their position in the 267 residues of the A protein. Following convention, the amino terminal end of the segment is on the left.

SINGLE AMINO ACIDS ARE SPECIFIED BY SEVERAL ADJACENT MUTABLE SITES

A one-to-one relationship between mutable sites and specific amino acids does not exist. Instead, there is genetic evidence showing that some, if not all, amino acids are jointly specified by several adjacent sites. The relevant evidence comes from the study of residue 47 of the tryptophan synthetase fragment illustrated in Figure 8–10. Treatment with a mutagen of the wild strain has given rise to mutant A23, in which arginine replaces glycine, and the mutant A46, in which glutamic acid replaces glycine. The difference between A23 and A46 does not involve changes to alternative forms of the same mutable site, since a genetic cross between A23 and A46 yields a number of wild-type (glycine in position 210) recombinants (Figure 8–11). If these changes were at the same mutable site, no wild-type recombinants could be produced. The very low frequency of the wild-type recombination suggests that the mutable sites involved may be adjacent to each other.

Another type of evidence suggesting that several mutable sites code for a single amino acid comes from observing how A23 and A46 themselves mutate upon treat-

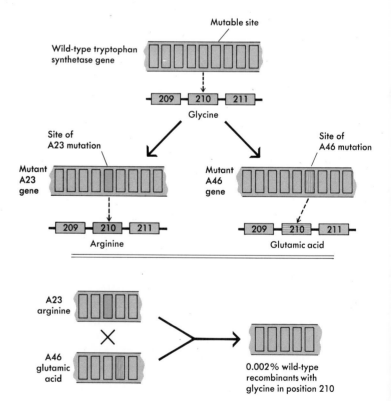

Figure 8–11
Demonstration that a single amino acid is specified by more than one mutable site (see text for details).

ment with mutagens. After exposure to a mutagen, both strains can give rise to new strains containing tryptophan synthetase molecules with glycine in position 210. These reverse mutations most likely involve changing the altered mutable sites back to the original wild-type configuration. However, strains also arise in which the amino acid in position 210 is replaced by another amino acid. Most significantly, the type of possible replacement differs for strains A23 and A46. Strain A23, besides back-mutating to glycine, mutates to threonine and serine, whereas A46 mutates to alanine and valine, in addition to glycine. The failure of A23 ever to give rise to alanine or valine, whereas A46 never mutates to threonine or serine, is very difficult to explain if their differences from wild types are based on the possession of alternative configurations of the same mutable site. Instead, if the changes are at two different sites, then a change from arginine to valine might require changes at both sites and so occur at too low a level to be detected under ordinary mutagen treatment.

These experiments by themselves place only the lower limit (two) of the number of mutable sites coding for a single amino acid. Much more extensive results would be necessary before an upper limit could be assigned on the basis of these types of experimentation. In the next chapter we shall talk about other evidence which tells us that three mutable sites (nucleotides) specify a given amino acid.

UNIQUE AMINO ACID SEQUENCES ARE NOT REQUIRED FOR ENZYME ACTIVITY

The ability of a chain to be enzymatically active does not demand a unique amino acid sequence. This is shown by examination of the new mutant strains obtained by treating strains A23 and A46 with mutagens. The fact that the possession of either glycine or serine in position 210 yields a fully active enzyme (Figure 8–12), whereas threonine in the same position yields an enzyme with reduced activity, demonstrates that the activity of an enzyme does not demand a unique amino acid sequence. In fact, a variety of evidence now indicates that amino acid replacements in many parts of a polypeptide chain can occur without seriously modifying catalytic activity. Most likely, however, one sequence is best suited to a cell's particular needs, and it is this sequence that is coded for by the wild-type allele. Even though other sequences are almost as good, they will tend to be selected against in evolution unless they are equally functional.

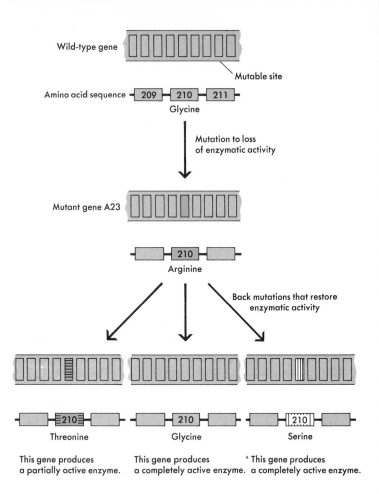

Figure 8–12
Evidence that many amino acid replacements do not result in loss of enzymatic activity (see text for details). We have arbitrarily shown the mutations occurring at the same mutable site.

"REVERSE" MUTATIONS SOMETIMES CAUSE A SECOND AMINO ACID REPLACEMENT

The conclusion that a unique amino acid sequence is not necessary for enzyme activity is extended by the finding that some mutations, which convert inactive mutant enzyme to an active form, work by causing a second amino acid replacement in the mutant enzyme. If we start with cells of mutant A46, which produces inactive tryptophan synthetase because of the substitution of arginine for glycine, distant second-site mutations that result in active enzyme occasionally emerge. For example, the second-site mutation A446 is located one-tenth of a gene length away from the first (Figure 8–13). The double mutant A46A446 produces active enzyme molecules containing two amino acid replacements: the original glycine-to-arginine shift, and a tyrosine-to-cysteine shift located 36 amino acids away.

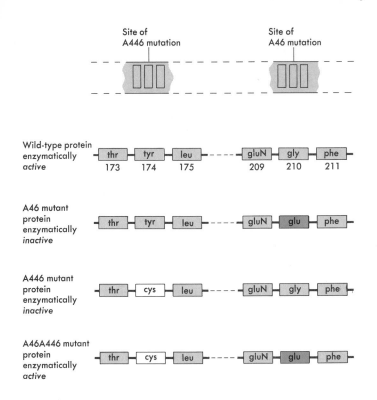

Figure 8–13
Reversal (suppression) of mutant pheno-type by a second mutation at a second site in the same gene.

The second shift can be studied independently of the first by obtaining recombinant cells with only the A446 mutation. Most interestingly, the A446 change when present alone also results in an inactive enzyme. We thus see that a combination of two wrong amino acids can produce an enzyme with an active 3-D configuration. Only sometimes do two wrong amino acids, however, cancel out each other's faults. For example, double mutants containing A446 and A23, or A446 and A187, do not produce active enzyme. It does not now seem wise to speculate on how the various amino acid residues are folded together in the 3-D configuration, and why only some combinations are enzymatically active. This kind of analysis must await the establishment of the 3-D structure of tryptophan synthetase.

SUMMARY

Crossing over occurs within the gene, thereby making possible the mapping of mutations within the gene. Like the order of genes on a chromosome, the arrangement of mutable sites in a gene is strictly linear. There appear to be many mutation sites within an average gene (500 to 1500). Sometimes it is initially difficult to determine whether a chromosomal region contains more than one gene; this usually can be resolved by introducing more than one chromosome (or chromosomal fragment) mutant for that region into a cell. If two mutants can complement each other, their mutations must be in different genes (the complementation test).

Genes control a cell's phenotype by determining which proteins the cell can synthesize. Each gene is responsible for the synthesis of a specific polypeptide. Genes work by controlling the sequence of amino acids in proteins. This is clearly shown by the discovery that many mutations cause the production of protein molecules that differ from normal protein by a single amino acid replacement. Mutant genes are usually recessive to the wild type because they most often express themselves by the failure to produce a protein product; in such heterozygotes, the wild-type gene dominates by producing enough of its product to bring about the wild-type phenotype.

Good evidence is accumulating that the gene and its polypeptide product are colinear: Mutations that map at an end of a gene affect the amino acid sequence at an end of the polypeptide, and so forth. Genes controlling a series of related biochemical reactions are often adjacent to each other. Various mutants of tryptophan synthetase A demonstrate that a mutable site can exist in at least three alternative states (we shall see in the next chapter that the number is thought to be exactly four). Experiments with these mutants also reveal that each amino acid is under the joint control of more than one (as we shall see, three) mutable site, and that an enzyme does not require a unique amino acid sequence in order to be active.

REFERENCES

PONTECORVO, G., *Trends in Genetic Analysis,* Columbia University Press, New York, 1958. A very readable early essay on the fine-structure analysis of the gene.

BENZER, S., "The Fine Structure of the Gene," *Scientific American,* January, 1962, pp. 70–84. A lucid presentation of the author's classic investigations of rII gene of phage T4.

HARTMAN, P. E., AND S. R. SUSKIND, *Gene Action,* 2nd ed., Prentice-Hall, Englewood Cliffs, New Jersey, 1969. A very clear description of how genes work; introduces many of the experimental details that led to the general principles.

INGRAM, V. M., *The Hemoglobins in Genetics and Evolution,* Columbia University Press, New York, 1963. A description of the variety of ways in which hemoglobin has been used to establish important biological principles. Particularly relevant to this chapter is the discussion of the abnormal hemoglobin molecules.

YANOFSKY, C., "Gene Structure and Protein Structure," in Hanawalt, P. C., and R. H. Haynes (eds.), *The Chemical Basis of Life*, Freeman, San Francisco, 1973. An introductory description of the author's classic work on the tryptophan synthetase system.

The Replication of DNA

Genetics tells us that the information that directs amino acid sequences in proteins is carried by long-chain polymeric molecules, the deoxyribonucleic acids (DNA).[1] Though DNA was first recognized in chromosomes about 70 years ago, it was definitely identified as having a genetic function only 30 years ago, and even then many geneticists thought that some information might also reside in the protein component of chromosomes. Now, however, there is no reason to believe that any genetic information is carried in other than nucleic acid molecules.

This poses the chemical question of how the information is transferred at the molecular level. In addition, we must ask how the DNA molecules are exactly copied during chromosome duplication. These questions could not be immediately attacked in 1943, when the genetic role of DNA was established. At that time, only fragmentary information concerning its structure existed. It was known to be a very long, large molecule (Figure 9–1), but the extent of its complexity was unclear. Consequently, there was much apprehension not only that DNA structure would vary from one gene to another but that, even in the simplest case, the solution of its structure would pose almost insuperable problems. Moreover, there existed an

---

[1] As we shall see in Chapter 15, the chromosomes of some viruses do not contain DNA but instead are composed of the chemically very similar compound ribonucleic acid (RNA).

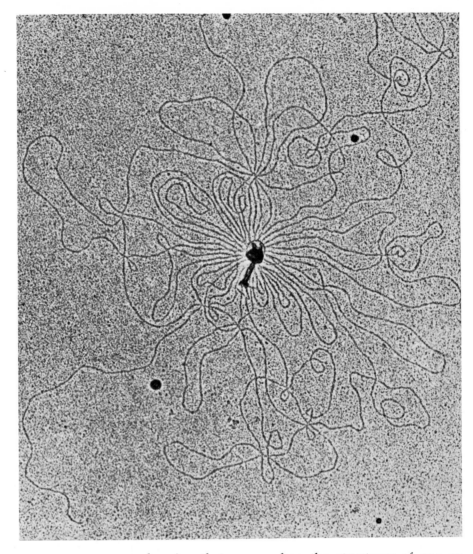

Figure 9–1
Electron micrograph of T2 DNA × 100,000.
[Reproduced from A. K. Kleinschmidt *et al., Biochem. Biophys. Acta,* **61,** 857 (1962), with permission.]

apprehension that, even when the structures of one or more DNA molecules were known, we would not be presented with any obvious clues about how it acted as a template for its self-replication or how it could control the sequence of amino acids in proteins. Fortunately, these fears were unfounded. Only ten years elapsed between the identification of DNA as the transforming substance and the 1953 elucidation of the double-helical structure of most DNA molecules. Moreover, the basic features of the double helix were simple and immediately told how DNA stores genetic information; even more pleasing, they suggested a chemical mechanism for the self-replication of DNA. From this moment on, the way in which geneticists investigated the gene entered a completely new phase. Their hypotheses no longer needed to be based only on genetic crosses.

Instead, it made sense always to ask how the structure of DNA affects the interpretation of their genetic crosses. As a consequence, many subjects (e.g., the action of mutagens) previously refractory to a systematic treatment became open to rational analysis.

THE GENE IS (ALMOST ALWAYS) DNA

The first serious assignment of the primary genetic role to DNA arose from experiments (discussed in Chapter 7) in-

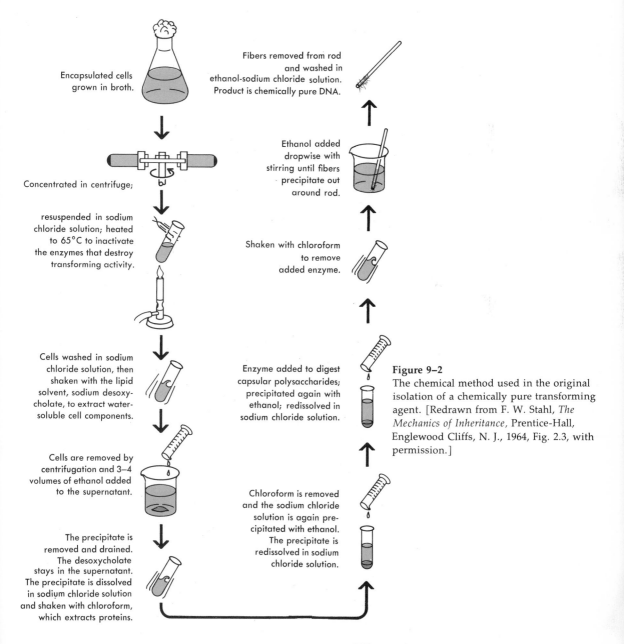

Encapsulated cells grown in broth.

Concentrated in centrifuge;

resuspended in sodium chloride solution; heated to 65°C to inactivate the enzymes that destroy transforming activity.

Cells washed in sodium chloride solution, then shaken with the lipid solvent, sodium desoxycholate, to extract water-soluble cell components.

Cells are removed by centrifugation and 3–4 volumes of ethanol added to the supernatant.

The precipitate is removed and drained. The desoxycholate stays in the supernatant. The precipitate is dissolved in sodium chloride solution and shaken with chloroform, which extracts proteins.

Enzyme added to digest capsular polysaccharides; precipitated again with ethanol; redissolved in sodium chloride solution.

Chloroform is removed and the sodium chloride solution is again precipitated with ethanol. The precipitate is redissolved in sodium chloride solution.

Shaken with chloroform to remove added enzyme.

Ethanol added dropwise with stirring until fibers precipitate out around rod.

Fibers removed from rod and washed in ethanol-sodium chloride solution. Product is chemically pure DNA.

Figure 9–2
The chemical method used in the original isolation of a chemically pure transforming agent. [Redrawn from F. W. Stahl, *The Mechanics of Inheritance*, Prentice-Hall, Englewood Cliffs, N. J., 1964, Fig. 2.3, with permission.]

volving the genetic transformation of pneumonia bacteria. Soon after the discovery that cell extracts were effective in transformation, careful chemical studies were begun to determine which type of molecule was responsible. Since virtually everybody believed that genes were proteins, there was initially great surprise when Avery and his co-workers reported in 1943 that the active genetic principal was deoxyribonucleic acid (DNA) (Figure 9–2). A major experimental result prompting their conclusion was the observation that the transforming activity of the extract was destroyed by deoxyribonuclease, an enzyme which specifically degrades DNA molecules to their nucleotide building blocks and has no effect on the integrity of protein molecules or ribonucleic acid (RNA). The addition of either the enzyme ribonuclease (which degrades RNA) or various proteolytic (protein-destroying) enzymes had no influence on transforming activity.

THE AMOUNT OF CHROMOSOMAL DNA IS CONSTANT

Even though the transformation results were clear-cut, there was initially great skepticism about their general applicability; people doubted that they would be found relevant to anything but certain strains of bacteria. Thus the momentous nature of Avery's discovery was only gradually appreciated.

One important confirmation came from studies on the chemical nature of chromosomes. DNA was found to be located almost exclusively in the nucleus, and essentially never where detectable chromosomes were absent. Moreover, the amount of DNA per diploid set of chromosomes was constant for a given organism and equal to twice the amount present in the haploid sperm cells. Another type of evidence that favored DNA as the genetic molecule was the observation that it is metabolically stable. It is not rapidly made and broken down like many other cellular molecules: Once atoms are incorporated into DNA they do not leave it as long as healthy cell growth is maintained.

VIRAL GENES ARE ALSO NUCLEIC ACIDS

Even more important confirmatory evidence came from chemical studies with viruses and virus-infected cells. It was possible by 1950 to obtain a number of essentially pure viruses and to determine which types of molecules were present in them. This work led to the very important generalization that all viruses contain nucleic acid. Since there was, at that time, a growing realization that viruses contain genetic material, the question immediately arose as to whether the nucleic acid component was the viral chromosome. The first crucial experimental test of the question came from isotopic study of the multiplication of T2, a virus containing a DNA core and a protective shell built up by the aggregation of a number of different protein molecules. In these experiments the protein coat was labeled with a radioactive isotope S^{35}, and the DNA with the radioactive isotope P^{32}. The labeled virus was then used for following the fates of the phage protein and nucleic acid as virus multiplication proceeded, particularly to see which labeled atoms from the parental virus entered the host cell and later appeared in the progeny phage.

Clear-cut results emerged from these experiments (Figure 9–3): Much of the parental nucleic acid and none of the parental protein was detected in the progeny phage. Moreover, it was possible to show that little of the parental protein ever enters the bacteria—instead it stays attached

Figure 9–3

Demonstration that only the DNA component of T2 carries genetic information and that the protein coat functions as a protective shell which facilitates DNA transfer to new host cells.

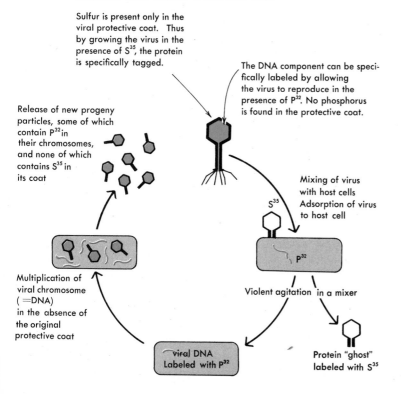

Sulfur is present only in the viral protective coat. Thus by growing the virus in the presence of S^{35}, the protein is specifically tagged.

The DNA component can be specifically labeled by allowing the virus to reproduce in the presence of P^{32}. No phosphorus is found in the protective coat.

Release of new progeny particles, some of which contain P^{32} in their chromosomes, and none of which contains S^{35} in its coat

Mixing of virus with host cells Adsorption of virus to host cell

S^{35}

P^{32}

Multiplication of viral chromosome (=DNA) in the absence of the original protective coat

Violent agitation in a mixer

viral DNA Labeled with P^{32}

Protein "ghost" labeled with S^{35}

to the outside of the bacterial cell, performing no function after the DNA component has passed in. This point was neatly shown by violently agitating infected bacteria after the entrance of the DNA: The protein coats were shaken off without affecting the ability of the bacteria to form new virus particles.

With some viruses it is now possible to do an even more convincing experiment. For example, purified DNA from the mouse virus polyoma can enter mouse cells and initiate a cycle of viral multiplication producing many thousands of new polyoma particles. The primary function of viral protein is thus to protect its genetic nucleic acid component in its movement from one cell to another. Thus no reason exists for the assignment of any genetic role to protein molecules.

DNA IS USUALLY A DOUBLE HELIX

The most important feature of DNA is that it usually consists of two complementary polymeric chains twisted about each other in the form of a regular double helix (Figures 9–4 and 9–5). The diameter of the helix is about 20 Å and each chain makes a complete turn every 34 Å. Each chain is a polynucleotide (Figure 3–10), a regular polymeric collection of nucleotides in which the sugar of each nucleotide is linked by a phosphate group to the sugar of the adjacent nucleotide. There are 10 nucleotides on each chain every turn of the helix. The distance per nucleotide base is thus 3.4 Å. Four main nucleotides exist, each of them containing a deoxyribose residue, a phosphate group, and a purine or pyrimidine base (Figure 3–9). There are two pyrimidines, thymine (T) and cytosine (C), and two purines, adenine (A) and guanine (G). In the polynucleotide chain the sugar and phosphate junction always involves the same chemical groups. Hence, this part of the molecule, called the backbone, is very regular. In contrast, the order of the purine and pyrimidine residues along the chain is highly irregular and varies from one molecule to another. Both the purine and pyrimidine bases are flat, relatively water-insoluble molecules which tend to stack above each other perpendicular to the direction of the helical axis.

The two chains are joined together by hydrogen bonds between pairs of bases. Adenine is always paired with thymine and guanine with cytosine (Figure 4–14). Only these arrangements are possible, for two purines would occupy too much space to allow a regular helix and, correspondingly, two pyrimidines would occupy too little.

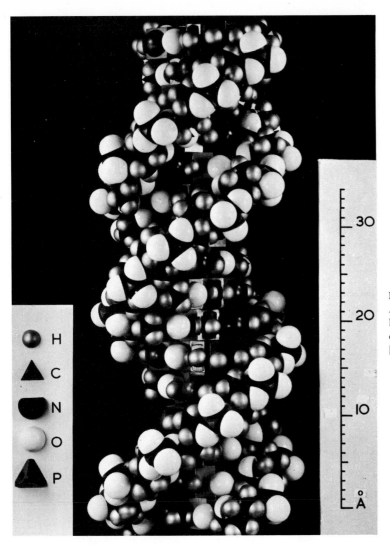

Figure 9–4
A space-filling model of double-helical DNA. The size of the circles reflects the van der Waals radii of the different atoms. [Courtesy of M. H. F. Wilkins.]

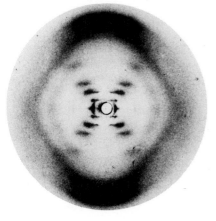

Figure 9–5
The key x-ray photograph involved in the elucidation of the DNA structure. This photograph, taken at Kings College, London, in the winter of 1952–1953, by Rosalind Franklin, experimentally confirmed the then current guesses that DNA was helical. The helical form is indicated by the crossways pattern of x-ray reflections (photographically measured by darkening of the x-ray film) in the center of the photograph. The very heavy black regions at the top and bottom tell that the 3.4-Å thick purine and pyrimidine bases are regularly stacked next to each other, perpendicular to the helical axis. [Reproduced from R. E. Franklin and R. Gosling, *Nature,* **171,** 740 (1953), with permission.]

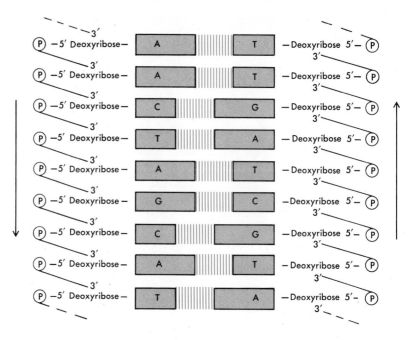

Figure 9–6
Diagram of the DNA double helix showing the specific pairing of the bases, the unrestricted sequence of bases along any one chain, and the reversed direction of the 3'-5'-phosphodiester linkages of the two chains. The heavy lines indicate the bases, and the interrupted lines the hydrogen bonds between base pairs (A, adenine; G, guanine; T, thymine; C, cytosine). [Redrawn from W. Hayes, *The Genetics of Bacteria and Their Viruses*, Blackwell, Oxford, 1964, p. 229, with permission.]

The strictness of these pairing rules results in a complementary relation between the sequences of bases on the two intertwined chains. For example, if we have a sequence ATGTC on one chain, the opposite chain must have the sequence TACAG. A sterochemical consequence of the formation of the A····T and G····C base pairs is that the two polynucleotide chains run in opposite directions. Thus, if the helix is inverted by 180°, it superficially looks the same (Figure 9–6).

An important chemical feature of the structure is the position of the hydrogen atoms in the purine and pyrimidine bases. Before 1953, many people thought that some of the hydrogen atoms were highly mobile, randomly moving from one ring nitrogen or oxygen atom to another, and so could not be assigned a fixed location. Now we realize that, although these movements (called tautomeric shifts) do occur, they are generally quite rare, and most of the time the H atoms are found at precise locations. The N atoms attached to the purine and pyrimidine rings are usually in the *amino* (NH_2) form and only very, very rarely assume the *imino* (NH) configuration. Likewise, the oxygen atoms attached to the C-6 atoms of guanine and thymine normally have the *keto* (C=O) form and only rarely take up the *enol* (COH) configuration. As we shall see, these relatively stable locations are essential to the biological function of DNA. For if the H atoms had no fixed locations, adenine could often pair with cytosine, and guanine with thymine.

210

In contrast to the ratios of A/T and G/C, which must always be 1 to satisfy the base-pairing rules, there are wide variations in the $(A + T)/(G + C)$ contents of different DNA molecules (Table 9–1). Higher plants and animals all have an excess of $A + T$ over $G + C$ in their DNA, whereas among the viruses, bacteria, and lower plants, there is much more variation, and both $(A + T)$-rich and $(G + C)$-rich species occur. These variations, however, are not purely random, and the base ratios of taxonomically related organisms are quite similar. No one yet knows the reason for the wide base-ratio spread. It may be a consequence of extensive evolution but certainly not a prerequisite. Witness extreme differences between higher plants and animals despite roughly similar percentages of the four main bases. This fact tells us that variation in the sequences of the bases is, by itself, sufficient to produce the gene differences between plants and animals.

DNA molecules with a high $G + C$ content are more resistant to thermal melting than $(A + T)$-rich molecules. When double-helical DNA molecules are heated above physiological temperatures (to near 100°C), their hydrogen bonds break and the complementary strands often separate from each other (DNA denaturation). Because the G····C base pair is held together by three hydrogen bonds, higher temperatures are necessary to separate $(G + C)$-rich strands than to break apart $(A + T)$-rich molecules. At intermediate temperatures or in the presence of destabilizing agents such as alkali or formamide, a DNA molecule can be partially denatured. In this case, those regions of the double helix relatively rich in A····T base pairs melt apart, while regions rich in G····C base pairs retain their double-helical character (see Figure 9–16).

Even complete denaturation is not necessarily an irreversible phenomenon. If a heated DNA solution is slowly cooled, a single strand can often meet its complementary strand and reform a regular double-helical molecule. This ability to renature DNA molecules permits us to show that artificial hybrid DNA molecules can be formed by slowly cooling mixtures of denatured DNA from two different species. For example, hybrid molecules can be formed containing one strand from a man and one from a mouse. Only a fraction (25 percent) of the DNA strands from a man can form hybrids with mouse DNA. This is, of course, not surprising since it merely means that some genes of man are very similar to those of a mouse, whereas others have quite different nucleotide sequences. It now appears that this molecular technique may be quite useful in establishing the genetic similarity of the various taxonomic groups.

Table 9–1 Examples of the Spread of $A + T/G + C$ Ratios in the DNA Molecules of Taxonomically Diverse Organisms

Source of DNA	$\dfrac{A + T}{G + C}$
Pseudomonas aeruginosa (a bacterium)	.51
Escherichia coli	.97
Bacillus megaterium (a bacterium)	1.66
Mycobacterium tuberculosis (a bacterium)	.60
Saccharomyces cerevisiae (a yeast)	1.80
Aspergillus niger (a fungus)	1.00
Scenedesmus quadricauda (an alga)	.57
Rhabdonema adriaticum (an alga = diatom)	1.71
Wheat	1.22
Paracentrotus lividos (sea urchin)	1.86
Locusta migratoria (an insect)	1.41
Trout	1.34
Domestic chicken	1.36
Horse	1.33
Man	1.40
Phage T2	1.84
Phage λ	1.06

211

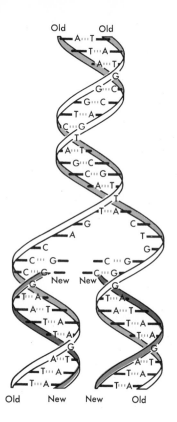

Figure 9–7
The replication of DNA.

THE COMPLEMENTARY SHAPE IMMEDIATELY SUGGESTS SELF-REPLICATION

Earlier, in the discussion of how templates must act, the point was emphasized that, in general, two identical surfaces will not attract each other and that it is instead easier to visualize the attraction of oppositely shaped or charged groups. Thus, without any detailed structural knowledge, we might guess that a molecule as complicated as the gene could not be directly copied. Instead, replication would involve the formation of a molecule complementary in shape, and this in turn would serve as a template to make a replica of the original template. Some geneticists, in the days before detailed knowledge of protein or nucleic acid structure existed, wondered whether DNA serves as a template for a specific protein that in turn serves as a template for a corresponding DNA molecule.

The realization that it is possible to form a DNA molecule in which the specific genetic surfaces (the purine and pyrimidine bases) of the two polynucleotide strands are complementary in shape and charge immediately tells us to reject the possibility, which we already suspected to be chemically tricky if not impossible, of having specific protein formation as the essential aspect of DNA replication. Instead, it is immensely simpler to imagine that DNA replication involves strand separation and formation of complementary molecules on each of the free single strands (Figure 9–7). Under this scheme, each of the two strands of every DNA molecule would serve as the template for the formation of its complement.

No difficulty arises from the need to break the hydrogen bonds joining the base pairs in the parental templates. Though they are highly specific, they are at the same time relatively weak and no enzymes are needed either to make or to break them. Thus they are the ideal means of specifically holding together a template and its complementary replica. Nor does the need to untwist a section of DNA to separate the two intertwined strands present a real problem. The DNA molecule is very thin (20 Å), and rotation about its axis involves almost no energy.

BASE PAIRING SHOULD PERMIT VERY ACCURATE REPLICATION

Earlier we stated that the chemistry of the specific amino acid side groups argued against their employment as accurate templates. Just the opposite, however, is true of the purine and pyrimidine bases, for each of them can form several hydrogen bonds. These bonds are ideal template forces, for they are highly specific, unlike the van der Waals interactions, whose attractive forces are both weaker and virtually independent of specific chemical groupings.

The average energy of a hydrogen bond is about 4 kcal/mole, or eight times the average energy of thermal motion of molecules at room temperature. This value allows us to estimate the percentage of time that a given group (e.g., amino) will be hydrogen bonded. Under ordinary cellular conditions the ratio of bonding to nonbonding is about 10^4. This means that two molecules held together by several hydrogen bonds are almost never found in the nonbonded form. For example, the frequency with which adenine will attach to cytosine during DNA replication might be as low as 10^{-8} times the frequency of its bonding to thymine ($10^{-4} \times 10^{-4}$ since two hydrogen bonds are involved). Since the G$\cdots$C pair is held together by three hydrogen bonds, its replication should be even more accurate.

This argument is not affected by the fact that the bases also can form almost equally strong hydrogen bonds with the hydrogen and oxygen atoms of water. It does not matter if, say, the potential thymine site is temporarily filled with several water molecules. They cannot be chemically joined to the growing polynucleotide chain, and they will soon diffuse away and be replaced by the suitable hydrogen bonding nucleotide.

It is not yet possible to give a similar quantitative argument for why two purines or two pyrimidines never accidentally bond to each other. Though both arrangements seriously distort the sugar–phosphate backbone and clearly are energetically unfavorable, the large number of atoms involved makes the precise calculation of energy differences impossible now. Part of the difficulty arises from the fact that the precise locations of the atoms in DNA are not known at the 0.1-Å level. Furthermore, it is unclear what effect this distortion would have on the binding of the enzyme that catalyzes the formation of the internucleotide covalent bonds.

DNA CARRIES ALL THE SPECIFICITY INVOLVED IN ITS SELF-REPLICATION

Proof for this statement came from the 1956 discovery that DNA synthesis could be observed in a cell-free system, prepared from *E. coli* cells. The enzyme involved, DNA polymerase I, was the first to be found involved in the synthesis of polydeoxynucleotides. It links the nucleotide precursors by 3'-to-5' phosphodiester bonds (Figure 9–8). Furthermore, it works only in the presence of DNA, which is needed to order the four nucleotides in the polynucleotide product.

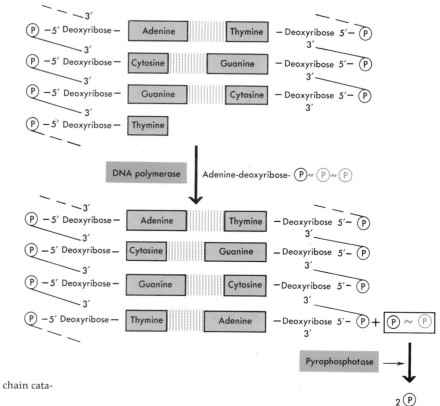

Figure 9–8
Enzymatic synthesis of a DNA chain catalyzed by DNA polymerase.

DNA polymerase I recognizes only the regular sugar–phosphate portion of the nucleotide precursors and so cannot determine sequence specificity. This is neatly demonstrated by allowing the enzyme to work in the presence of DNA molecules that contain varying amounts of A····T and G···C base pairs. In every case, the enzymatically synthesized product has the base ratios of the template DNA (Table 9–2). During this cell-free synthesis, no synthesis of protein or any other molecular class occurs, unambiguously eliminating any non-DNA compounds as intermediate carriers of genetic specificity. There is thus no doubt that DNA is the direct template for its own formation. As we might expect, the enzymatic product, like the primer, has a double-helical structure.

Table 9–2 A Comparison of the Base Composition of
Enzymatically Synthesized DNA and Their DNA Templates

Source of DNA Template	Base Composition of the Enzymatic Product				$\dfrac{A+T}{G+C}$ in product	$\dfrac{A+T}{G+C}$ in template
	Ade-nine	Thy-mine	Gua-nine	Cyto-sine		
Micrococcus lysodeiticus (a bacterium)	0.15	0.15	0.35	0.35	0.41	0.39
Aerobacter aerogenes (a bacterium)	0.22	0.22	0.28	0.28	0.80	0.82
Escherichia coli	0.25	0.25	0.25	0.25	1.00	0.97
Calf thymus	0.29	0.28	0.21	0.22	1.32	1.35
Phage T2	0.32	0.32	0.18	0.18	1.78	1.84

Exactly how DNA polymerase I works at the molecular level is not known since little has been determined about its 3-D structure. Because it is present in cells in only limited amounts, purification in large amounts is a horrendous task. It is known, however, to consist of a single polypeptide built from about 1000 amino acids. Eventually, crystallization should be achieved, opening up the prospect of x-ray diffraction studies of its three-dimensional shape.

The ability of DNA polymerase I to make complementary copies of DNA chains immediately suggested that this might be the major enzyme in *E. coli* involved in linking up nucleotides during DNA replication. But as we shall see later, DNA replication is generally a complex process involving several different polymerizing enzymes. Discovery of the first DNA polymerase, however, was of immense importance in showing that the assembly of complementary nucleotide sequences was amenable to relatively straightforward test-tube analysis.

SOLID EVIDENCE IN FAVOR OF DNA STRAND SEPARATION

To prove this point, methods had to be found for the physical separation of parental and daughter DNA molecules. This was first accomplished through use of heavy isotopes such as N^{15}. Bacteria grown in a medium containing the heavy isotope N^{15} have denser DNA than bacteria grown under normal conditions with N^{14}. Heavy DNA can be separated from light DNA by equilibrium centrifugation in concentrated solutions of heavy salts such as cesium chloride. When high centrifugal forces are applied, the

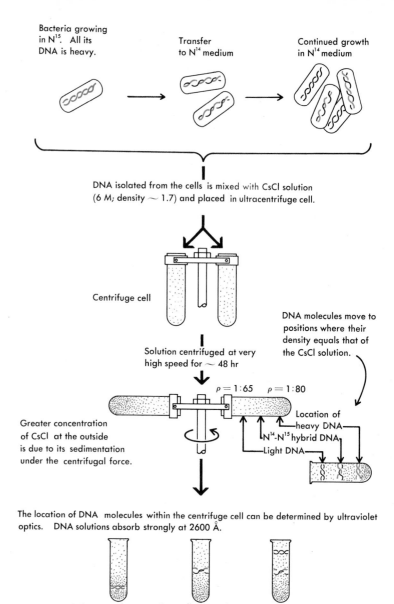

Figure 9–9
Use of a cesium chloride density gradient for the demonstration of the
separation of complementary strands during DNA replication.

solution becomes more dense at the outside of the centrifuge cell. If the correct initial solution density is chosen,
the individual DNA molecules will move to the central region of the centrifuge cell where their density equals that
of the salt solution. Thus the heavy molecules will band at

a higher density than the light molecules. If bacteria containing heavy DNA are transferred to a light medium (containing N^{14}) and allowed to grow, the precursor nucleotides available for use in DNA synthesis will be light; hence, DNA synthesized after transfer will be distinguishable from DNA made before transfer.

If DNA replication involves strand separation, definite predictions can be made about the density of the DNA molecules found after various growth intervals in a light medium. After one generation of growth, all the DNA molecules should contain one heavy and one light strand and thus be of intermediate hybrid density. This result is exactly what is observed. Likewise, after two generations of growth, half the DNA molecules are light and half hybrid (Figure 9–9), just as strand separation predicts.

SINGLE-STRANDED DNA ALSO IS REPLICATED BY BASE PAIRING

At first it was thought that all DNA molecules are double-stranded except during replication, when part is temporarily in a nonhydrogen-bonded, single-stranded form. Many people were surprised, therefore, when conclusive experiments revealed that the DNA of several groups of small bacterial viruses exists normally as circular single-stranded molecules in which $A \neq T$ and $G \neq C$. Among these single-stranded DNA phages are the viruses ϕX174, s13, F1, and M13.

This discovery of single-stranded DNA immediately posed the question of whether an additional copying mechanism exists in which a single DNA strand serves as a template for an identical copy. This type of replication cannot be accomplished by the formation of a double helix in which identical bases hydrogen bond to each other, because such a structure is stereochemically impossible. If self-replication were to exist, it would have to use connector molecules of the type discussed when we considered whether a polypeptide chain could be the template for an identical copy. Four different connectors would have to exist, each having two identical surfaces that could make several hydrogen bonds with a specific base. The connectors could line up and connect the nucleotides to form a new polynucleotide chain identical to the first.

There is no evidence, however, for the existence of such connectors. Instead we find that as soon as the single-stranded viral DNA (which we shall call the "+" strand) enters its host cell, the strand serves as a template for the formation of a complementary "–" strand (Figure

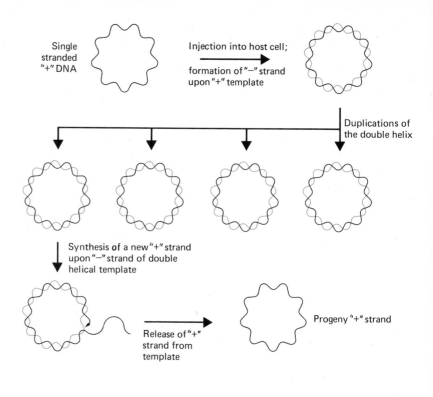

Figure 9–10
The replication of a single-stranded DNA molecule. Each double helix generally serves as the template for the formation of a large number of new "+" strands.

9–10). The resulting double helix in turn serves as a template for the formation of new single "+" strands, which then become incorporated into new virus particles. Thus, the fundamental mechanism for ordering nucleotides during synthesis of single-stranded DNA is basically the same as that used for double-helical DNA. Nucleotide selection always occurs by attraction of the complementary base. The difference is that with the single-stranded viruses, only one of the two strands in the double helix (the "−") functions as a template for progeny particle DNA.

Despite the existence of single-stranded DNA in some viruses, most cellular DNA must be double-stranded. This follows from the facts that (1) the two complementary chains do not have identical base sequences and hence, if both used, would code for entirely different amino acid arrangements, and (2) in bacteria (and probably in many higher cells) there is usually only one copy of a given gene.

Thus, if the single-stranded form were the usual form and the double helix existed only briefly during DNA duplication, then following mitosis the two daughter cells would contain completely different sets of genetic information.

THE CHROMOSOMES OF VIRUSES AND *E. COLI* ARE SINGLE DNA MOLECULES

Early estimates of the molecular weights of DNA centered at about a million, a size that we shall see is that of the average gene. Therefore, it was natural to equate single DNA molecules with single genes. But these first reports were misleading because of DNA breakage during its isolation and study. Now it is clear that virtually all undegraded DNA molecules contain the information of several genes. Our most convincing molecular weight values come from DNA-containing viruses. Here no matter whether the DNA content is relatively small or large, each virus particle contains a single DNA molecule. For example, all the DNA of the small monkey virus SV40 is present within a single molecule of MW 3×10^6, while the DNA molecule of the large bacterial virus T2 has a MW 1.2×10^8 (Figure 9–11). In these cases we must equate the viral chromosomes, not their genes, with single DNA molecules. This must also be done for the single chromosome of an *E. coli* cell; a single DNA molecule whose molecular weight is about 2×10^9 and whose extended length is almost a millimeter. Exactly how many DNA molecules are contained in the individual chromosomes of higher plants and animals is not generally known, though there are hints that here again the number will always be one. In any case, some DNA molecules are larger by several powers of 10 than any other biological molecule.

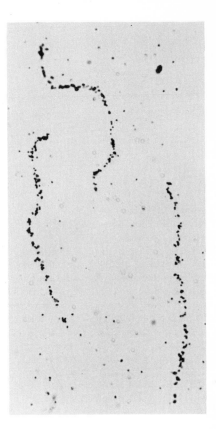

Figure 9–11
Autoradiograph of several T2 chromosomes. The total length is 52 μ. [Courtesy of J. Cairns.]

CIRCULAR VS. LINEAR DNA MOLECULES

Initially our evidence, largely from autoradiography and electron microscopy, suggested that all DNA molecules were linear and had two free ends. As, however, it becomes possible to look more easily at undegraded DNA molecules, we find that many DNA molecules have circular forms. For instance, the SV40 chromosome is present within the virus particle as a duplex ring and, as mentioned earlier, certain single-stranded DNA molecules (e.g., that isolated from the phage ϕX174) have a ring shape. Not only small DNA's are circular, for most, if not all, bacterial chromosomes also have circular configurations.

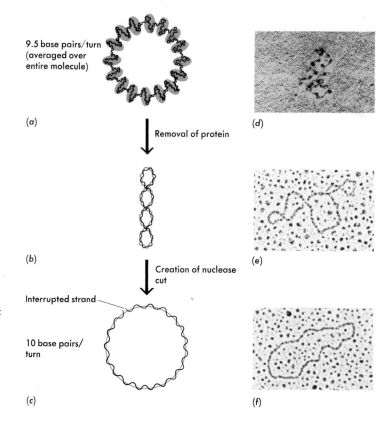

9.5 base pairs/turn
(averaged over
entire molecule)

(a)

Removal of protein

(b)

Creation of nuclease
cut

Interrupted strand

10 base pairs/
turn

(c)

(d)

(e)

(f)

Figure 9–12

Different conformations of SV40 DNA. a) Complexed with protein to form a mini-chromosome characterized by repeating packets of folded double-helical DNA and protein. b) Purified DNA that takes up a superhelical conformation. c) Relaxed DNA created by a cut in one strand. d) EM view of a minichromosome. Note that it does not have a superhelical form [from Germond, *et al.*, *PNAS* **72**, 1975]. e) EM photo of covalently closed superhelix. [Courtesy E. Daniell.] f) EM photo of relaxed form created by mild nuclease treatment. [Courtesy E. Daniell.]

Circular DNA molecules are often twisted upon themselves to form *supercoils* (Figure 9–12). Such supercoils are the result of a change in the average number of base pairs per turn of the double helix which occur when the protein components of chromosomes are removed in the process of preparing pure DNA. Once two intertwined chains are linked together by covalent bonds, the absolute number of turns cannot change. Any subsequent change in the number of base pairs per helical turn must be compensated by the formation of a supercoil in the opposite direction. The fact that virtually all purified circular DNA's contain negative supercoils suggests that they are all coated with protein during the DNA duplication process. The finding that the occasional circular DNA molecule lacks the supercoiled form does not negate this hypothesis. All such simple circles contain one or more breaks in their polynucleotide backbone that automatically allow the formation of a complete double helix without supercoiling.

INTERCONVERSION BETWEEN LINEAR AND CIRCULAR FORMS

Sometimes a DNA molecule that is linear when isolated from a virus particle (e.g., phage λ) is found as a circle inside the host cell. This tells us that the linear and circular forms of such DNA molecules are interconvertible. A molecular mechanism which converts a DNA circle to the rod form (or vice versa) is shown in Figure 9–13. Starting with a closed circle, a specific enzyme(s) introduce(s) two breaks, one in the "+" strand and one in the "−" strand. As these breaks are very close to each other, the intervening hydrogen bonds occasionally are broken by thermal agitation and then the circle unfolds. Most importantly, the resulting rod contains single-stranded ends which have complementary nucleotide sequences ("sticky ends"). Thus at a later time the rod can, by base pairing, resume a circular configuration. If the missing phosphodiester bonds are replaced, then the covalent circle is regenerated.

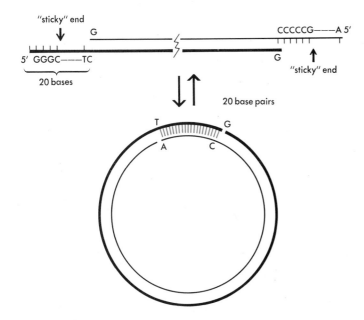

Figure 9–13
Interconversion of the linear and circular forms of λ DNA.

Solid evidence exists that λ DNA interconverts between its circular and rod forms, using precisely this mechanism. Later we shall see that the duplication of such "sticky" DNA's involves circular replicative intermediates. So the rod form most likely is an adaptation for injection of the viral chromosome through the narrow phage tail. It is important that many other linear DNA molecules, for example that of T7, also have potentially sticky ends but never assume circular forms *in vivo*. So circularity, *per se*, is not required for DNA replication.

GENERATION OF UNIQUE FRAGMENTS BY RESTRICTION ENZYMES

Enzymes which cut the phosphodiester bonds of polynucleotide chains are called *nucleases*. Those nucleases which preferentially break internal bonds are known as *endonucleases*, while those which clip off terminal nucleotides are referred to as *exonucleases* (Figure 9–14).

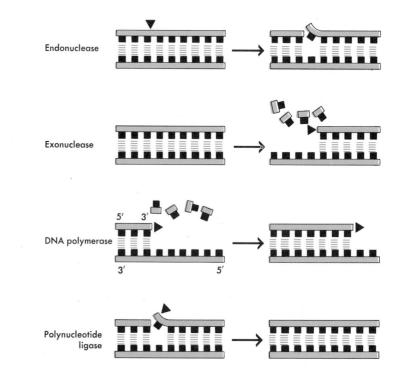

Figure 9–14
Schematic representation of the action of four different enzymes concerned with making and breaking the backbone bonds of DNA.

For a long time it was thought that all endonucleases were relatively unspecific, and cut polynucleotide bonds surrounded by many different nucleotide sequences. Over the past several years, however, a new class of endonucleases of microbial origin has been discovered whose action is limited to very specific nucleotide sequences. These specific nucleases, which collectively are referred to as *restriction enzymes,* make breaks only within sequences which exhibit twofold symmetry around a given point, e.g.,

```
AAG CTT
TTC GAA
```

Because the same sequence (running in opposite directions) is found on both strands, such enzymes always create double-stranded breaks. A large number of restriction enzymes with different specificity have already been characterized (Table 9–3), giving us the possibility of cutting up given DNA molecules into many specific classes of smaller fragments. For example, the restriction enzyme Eco RI recognizes a sequence which is present once in SV40 DNA (3×10^6 MW), while the restriction enzyme Hind II breaks SV40 at five different sites (Figure 20–11). More-

Table 9–3 Recognition Sequences of Several Restriction Endonucleases

Microbial Origin	Enzyme	Recognition Site	Number of Recognition Sites per Viral Genome		
			SV40[1]	λ[2]	Adenovirus 2[3]
Escherichia coli KY13	Eco RI	Axis of symmetry Cut bond 5′ ↓ GAA | TTC CTT | AAG ↑	1	5	5
Hemophilus influenzae Rd	Hind II	5′ ↘ GTPy | PuAC CAPu | PyTG ↖	5	34	720
	Hind III	5′ ↓ AAG | CTT TTC | GAA ↑	6	6	11
Hemophilus parainfluenzae	Hpa I	5′ ↘ GTT | AAC CAA | TTG ↖	3	11	6
	Hpa II	5′ ↓ CC | GG GG | CC ↑	1	750	750
Hemophilus aegyptius	Hae III	5′ ↘ GG | CC CC | GG ↖	17	750	750

[1] MW = 3×10^6.
[2] MW = 32×10^6.
[3] MW = 25×10^6.

over, sequences present in one form of DNA are frequently absent in other DNA's. For instance, phage T7 (MW 25×10^6) does not contain even one example of the

GAA TTC
CTT AAG

sequence recognized by Eco R1.

Restriction enzymes are now a *very* important tool in studying DNA. Not only do they allow the possibility of converting circular molecules into linear rods with well defined ends, but they also permit the isolation of fragments sufficiently short for direct determination of their nucleotide sequences. Until two years ago even the smallest DNA molecules seemed much too large for serious attempts at sequencing. Now, however, it has suddenly become possible to almost routinely work out stretches of 50 to 100 nucleotides (Figure 9–15).

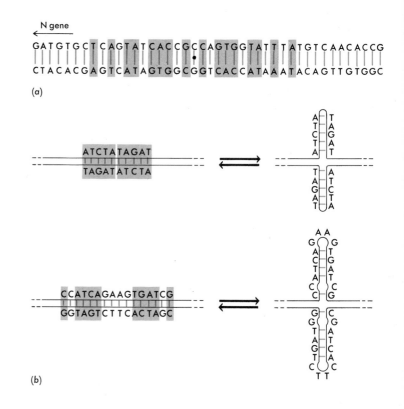

(a)

(b)

Figure 9–15
a) A 45 base-pair length of λ DNA from the operator (O$_L$) involved in the control of gene expression. In blocks are illustrated an imperfect palindrome centered about the black dot. b) How inverted repetitions (palindromes) can produce hairpin loops. Note in the lower example how an imperfect palindrome leads to imperfect loops.

PALINDROMES

An unexpected consequence of the working out of these DNA nucleotide sequences has been the discovery of the widespread occurrence of inverted repetitious sequences (*palindromes*) in which the same (or almost the same) sequences run in opposite directions if one switches from one strand to its complement at the central axis of symmetry $\left(\text{e.g., } \begin{smallmatrix} ..\text{GTATCCGGATAC}.. \\ ..\text{CATAGGCCTATG}.. \end{smallmatrix}\right)$. Some palindromes, like those recognized by certain restriction enzymes (Table 9–3), are relatively short (3 to 10 bases in a given direction). Others are much longer, comprising hundreds of base pairs. Many of the shorter palindromes most likely serve as DNA recognition sites for *multimeric proteins* composed of identical subunits. The significance of the longer palindromes, however, remains very unclear. Conceivably they sometimes undergo a steric rearrangement in which the complementary sequences on the same chain base pair to form cruciform (hairpin) loops extending out from the main helical axis (Figure 9–15 (b)). Such loops may sometimes be involved in the recognition of DNA binding proteins.

In any case, those sections of RNA chains transcribed off palindromic regions will tend to form hydrogen-bonded hairpin loops, the stability of which will be directly related to how perfectly symmetrical the palindromic sequences are. Many palindromes may thus have evolved to endow structural compactness to their RNA products.

PARTIAL DENATURATION MAPS

Brief exposure of linear DNA molecules to denaturing conditions (e.g., high temperature, high or low pH, hydrogen bond breaking reagents) preferentially unwinds those sections which are rich in $A \cdots T$ base pairs, leaving intact regions largely held together by the much stronger $G \cdots C$ base pairs. When such partially denatured molecules are treated with formaldehyde, these single-stranded regions become stabilized, since formaldehyde combines with the free amino groups of adenine and prevents the formation of the $A \cdots T$ base pairs. Visualization of such treated molecules in the EM reveals the melted regions as single-stranded bubbles that reproducibly occur at fixed positions along a given type of DNA molecule (Figure 9–16). Such "denaturation maps" now provide an easy way to distinguish the ends of linear DNA molecules, and so to precisely define the location(s) where DNA replication starts.

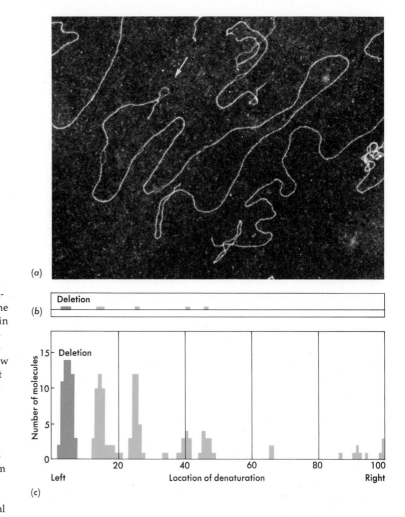

(a)

(b)

(c)

Figure 9–16
a) Electron micrograph of a partially denatured T7 DNA molecule. b) Mapping of the location of the denatured regions shown in (a). c) A histogram summarizing the location of single-stranded regions in a large number of denatured molecules. The arrow in (a) points to a single-stranded loop-out that is present in the DNA molecule even before partial denaturation. The loop-out occurs because this was not just an ordinary T7 DNA molecule but a "heteroduplex" molecule (see Chapter 10) derived by reannealing separated strands from wild-type T7 DNA with strands from T7 DNA molecules that contained a deletion known by genetic mapping to be located near the left end. It is this additional physical landmark that allows us to say that the AT-rich, partial denaturation sites are located in the left end of the bacteriophage chromosome. [Courtesy of Drs. John Wolfson and David Dressler.]

VISUALIZATION OF THE REPLICATION OF A LINEAR DNA MOLECULE

Now it is easy to isolate DNA in the process of replication and visualize it by electron microscope techniques. When this was first done for the linear T7 DNA, it was expected that replication would commence at one or both ends since, superficially, strand separation is more easily imagined for terminal as opposed to internal sections of a double helix. Surprisingly, however, replication was found to commence internally, always to a point some 17% along the molecule from the left end. Once DNA synthesis starts, it moves in both directions generating —O— (eye)-shaped intermediates which convert to Y-shaped molecules when the left-hand replicating fork reaches its respective terminal (Figure 9–17). These structures clearly rule out a two-step replication process in which the parental strands first completely unwind and only later act as templates to form new

226

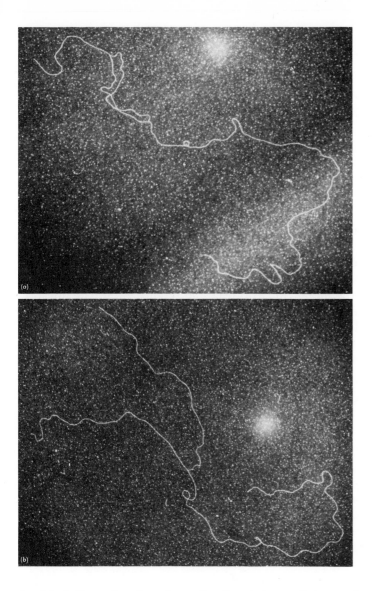

Figure 9–17
a) Electron micrograph of a T7 DNA molecule that has just begun replication and has a small replicating bubble centered about a point 17% from the left end of the genetic map. b) A more advanced replicating molecule whose Y shape is the result of the left replicating fork having run off its respective end. [Courtesy of Drs. John Wolfson and David Dressler.]

double helices. In all such replication intermediates, only traces of unwound single-stranded regions can be seen. Thus daughter polynucleotide strands are synthesized almost as soon as the parental strands separate.

OVERALL CHAIN GROWTH IN BOTH 5' TO 3' AND 3' TO 5' DIRECTIONS

The opposing chain directions (5' to 3' and 3' to 5') of the two strands of a double helix mean that the two daughter strands being synthesized in each replicating fork must also run in opposite directions. So the overall direction of chain growth must be 5' to 3' for one daughter strand and 3' to 5' for the other daughter strand (Figure 9–18). Paradoxically, however, all known forms of DNA polymerase,

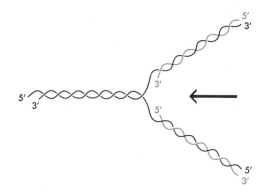

Figure 9–18
Overall direction of progeny chain growth at a replicating fork.

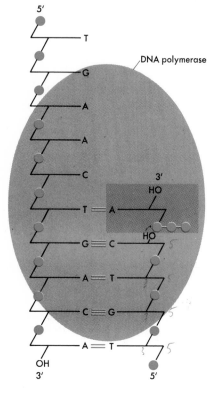

Figure 9–19
Reaction of a nucleoside triphosphate with the 3'OH of a growing DNA chain.

the only type of enzyme which can add nucleotide precursors to DNA, extend chains only in the 5' to 3' direction. The chemical reaction that it catalyzes allows a nucleoside triphosphate to react with the free 3' end of a growing polynucleotide strand (Figure 9–19). The question thus arises whether other replicating enzymes exist, particularly those which would add nucleotides onto free 5' ends. Now we would guess not for, despite very extensive searches, no enzyme with such specificity has been found.

SMALL DNA FRAGMENTS AS PRECURSORS OF LONG CHAINS

Resolution of the apparently paradoxical 3' to 5' direction of synthesis came from the discovery that most deoxynucleotides are not added directly to very long daughter polynucleotides. Instead they are first found as parts of much shorter DNA chains (~1000 bases long) that later link up to the main daughter strands through the action of the polynucleotide joining enzyme DNA ligase. This finding opened up the possibility that a chain whose final direction of growth is 3' to 5' might in fact be formed by the linkup of smaller chains, each of which grew in the conventional 5' to 3' manner. In this way the same set of enzymes could be responsible for making all progeny strands. Proof that this indeed is the case comes from direct studies of the ends of growing fragments to which precursor nucleotides have recently been added. All such experiments show growth exclusively at the 3' end.

At first it was believed that small fragments were precursors only to the chain with overall 3' to 5' growth. But now there is evidence that both daughter chains at a given replicating fork form by the linking up of smaller units (Figure 9–20). This means that unraveling of the parental strands at replicating forks is not the direct consequence of nucleotide incorporation but that some positive force creates extended sections of single-stranded

DNA in front of the sites of actual chain growth. Most likely this force comes from attachment of the *unwinding proteins*. These molecules bind exclusively to single-stranded DNA and, when present in large quantities, tend to destabilize double helices into their strand components. Some 200 molecules of unwinding protein are found on each fork with some 8 to 10 nucleotides complexed to each such protein. Each arm of the replicating fork thus may have some 2000 unpaired bases upon which the short DNA fragments may be initiated.

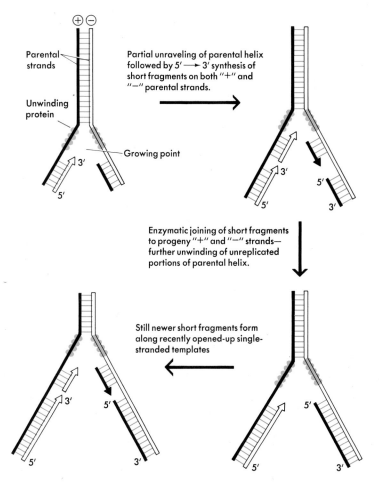

Figure 9–20
Hypothesis that replication of both progeny strands involves joining together of short fragments each synthesized in the 5'-to-3' direction.

THREE FORMS OF DNA POLYMERASE

The exact ways in which enzymes are used to make complete DNA chains are now realized to be much more complicated than initially suspected when the original DNA polymerase (DNA polymerase I) was first worked with. Now at least three different DNA polymerases have been identified within *E. coli;* and it will be surprising if each is not shown to have some distinctive role (Table 9–4). While DNA polymerase I was long thought to be the major enzyme that joins together deoxynucleotides, we now realize that DNA polymerase III, the most recently discovered form, is the main polymerizing agent. In contrast, DNA polymerase I is largely used in the filling in of gaps between the small fragment precursors (Figure 9–20). No specific task has been found for DNA polymerase II and, since mutant cells exist that have very few molecules, it may not be indispensable and finding out why it exists may be tricky.

Table 9–4 Properties of DNA Polymerases from *E. coli**

	Pol I	Pol II	Pol III
Functions			
5′ → 3′ polymerization	+	+	+
5′ → 3′ exonuclease	+	−	−
3′ → 5′ exonuclease	+	+	+
Activity			
Inhibition by high salt	−	−	+
Affinity for triphosphate precursors	low	low	high
Inhibition by SH blocking agents	−	+	+
General	Single chains	Single chains	Double chains 140,000 + 40,000
Molecular weight	109,000	120,000	180,000
Molecules/cell	400	100	10
Nucleotides polymerized at 37°/minute/molecule	∼1,000	∼50	∼15,000

* Data from *DNA Synthesis*, A. Kornberg (Freeman, 1974).

CORRECTION OF MISTAKES BY 3′ TO 5′ EXONUCLEASE ACTION

Extensive purification of all three forms of DNA polymerase reveals that each such enzyme has a 3′ to 5′ exonuclease activity as well as a polymerizing ability. Possession of this exonuclease specificity means that each polymerase molecule has the ability to cut back nascent polynucleotides as well as to extend them. In the presence of even moderate levels of the deoxynucleoside-triphosphate precursors, synthesis is overwhelmingly favored over degradation. The question arose why this specific degradative potential should be obligatorily coupled to polymerizing action. At first sight, possession of a specific degradative action makes no sense. But the reason for its presence immediately became obvious when it was shown that it acted preferentially on incorrectly paired bases. If by chance the wrong base becomes added onto a growing chain, it has a very high probability of being removed before the next base is added. The 3' to 5' exonuclease activity thus provides a proofreading activity which gives DNA replication much higher fidelity than it would have if selection were the result of only one base-pairing selection step.

A constant need for proofreading may be the reason no enzyme has evolved which adds on deoxynucleotides in the 3' to 5' direction. If chain growth were to occur this way, the growing ends would of necessity be terminated by triphosphate groups whose high-energy bonds are necessary for the addition of the next nucleotide (Figure 9–21). Removal of deoxynucleotides from such triphosphate-terminated molecules for proofreading purposes would thus create 5' monophosphate-ended chains energetically incapable of being extended unless a source of energy were to be provided. Some form of ligase-like reaction could do the job, but such a step might so intensely slow down DNA replication that its evolution has never made sense.

Figure 9–21
Advantage of a 5' → 3' chain growth mechanism for a replication system which requires proofreading.

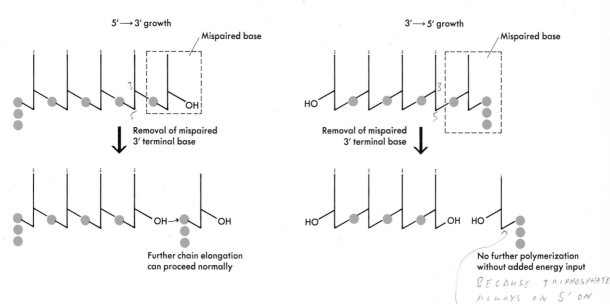

INITIATION OF DNA CHAINS BY RNA PRIMERS

As the exact specificity of the several DNA polymerases became established, the unexpected fact emerged that each such enzyme could only add on nucleotides to preexisting polynucleotide chains. None can initiate new DNA chains, thereby raising the question whether still undiscovered DNA polymerases exist whose sole role might be the starting of new DNA chains. Now we guess not, since short stretches of RNA recently have been found attached to the 5' ends of many newly synthesized DNA chains.

The starting points for DNA replication may therefore be recognized not by a DNA polymerase, but by a form of *RNA polymerase,* the group of enzymes whose role is the growth of complementary RNA chains upon DNA templates (see Chapter 11). Following binding of a starter RNA polymerase, a number of ribonucleotides are sequentially linked together until a stop signal is read. The enzyme then detaches leaving a short RNA chain still bound to its DNA template. This RNA chain then serves as a *primer* onto which DNA polymerase III adds deoxynucleotides (Figure 9–22).

Removal of the RNA primer occurs later through the action of other enzyme(s) which can digest away the RNA component of RNA/DNA hybrids. One enzyme that has this capacity is DNA polymerase I, whose primary task may be to close the gaps between RNA-primed DNA fragments. In doing so, it adds nucleotides to the 3' end of the fragment at its back and simultaneously cuts away the RNA primer that lies ahead. The snipping off of ribonucleotides is catalyzed by a completely different active site

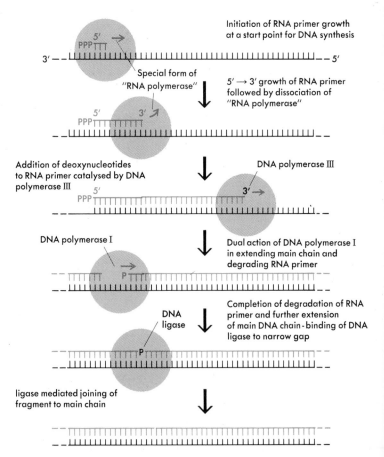

Figure 9–22
Use of an RNA primer in the initiation of DNA synthesis.

than that involved in its 3′ to 5′ proofreading capability (Figure 9–23). This 5′ to 3′ exonuclease activity can remove deoxyribonucleotides as well as ribonucleotides, and so its use is not exclusively restricted to primer removal. As we will show later, DNA polymerase I is also employed to remove sections of DNA damaged by radiation attack.

Not all RNA primers become excised by DNA polymerase I. Those which become linked on both sides to DNA are digested away by specific endonucleases like Ribonuclease H, which specifically cuts out the RNA component of hybrid regions.

The nature of the nucleotide sequence(s) signifying "start DNA replication" has not yet been worked upon. Soon, however, it should be possible to sequence viral DNA fragments that contain "origins" of replication and, hopefully, their sequences may have common features. Also just under investigation is (are) the form(s) of RNA polymerase that make the RNA primers. Hints exist that some primers are made by modified RNA polymerases containing a "starting factor," but what this component is at the molecular level is only just being studied. So it may be some time before we understand how RNA polymerase plays key roles in both DNA and RNA synthesis. In particular, attention must be paid to the need for RNA primers not only to start the vast numbers of small fragments involved in the progressive movement of replicating forks, but also to initiate DNA synthesis at the origin of replication.

Hopefully at some step in these investigations we may find we understand why RNA primers seem to be so universally needed. Now the only hypothesis that has a ring of possible truth argues that they serve to prevent an unlivable number of mistakes at the starting points of DNA chains. Conceivably the laying down of the first few nucleotides along a DNA template is inherently less accurate than when nucleotides add onto already formed long sections of double helix. So there may be great selective advantage for an initiation mechanism which allows easy discrimination and subsequent removal of the starting nucleotide sequences.

COMPLETING THE ENDS OF LINEAR DNA MOLECULES

Realization that the RNA primers are ordinarily removed raises the question of how the extreme ends of linear DNA chains are completed. Not only would a small RNA fragment appear to have to initiate at the terminal 3′ nucleotide of the template strand, but then some way would have to be found to fill in the gaps created by ribonucleotide ex-

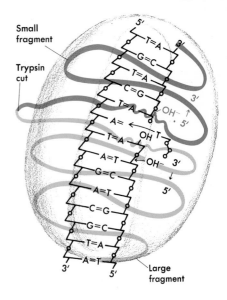

Figure 9–23
Diagrammatic view of DNA polymerase I, showing its single polypeptide chain. Addition of trypsin splits the chain into a "small fragment" which carries the 5′ → 3′ exonuclease activity and a large fragment that has both the polymerase and the 3′ → 5′ exonuclease activities. [Redrawn from A. Kornberg, *DNA Synthesis*, 1974.]

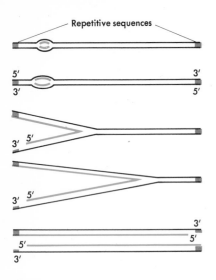

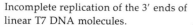

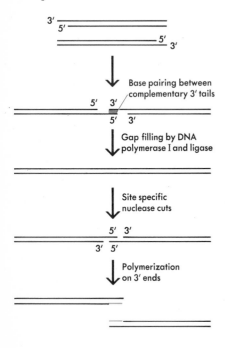

Figure 9–24
Incomplete replication of the 3′ ends of linear T7 DNA molecules.

Figure 9–25
Employment of concatemers to complete the replication of linear T7 DNA.

cision (Figure 9–24). No such terminal gaps exist, however, in T7, all of whose terminal deoxynucleotides form conventional base pairs. Resolution of this dilemma comes from the tying together of two unexpected properties that characterize most linear forms of DNA. Firstly, linear DNA's tend to have redundant ends—the sequence found at the left end is exactly repeated at the right end. Secondly, linear molecules replicate to immediately produce not unit length progeny molecules but, instead very long concatemers containing the genome sequence repeated over and over.

Why linear DNA should pass through a concatemer stage can be seen through consideration of the incompleteness of the two daughter molecules produced by the initial replication of a linear double helix. One end of each daughter molecule will be completely replicated while the other end will have a 3′-ended single-stranded tail. Because of the redundant terminal sequences, the two tails will have complementary sequences that can base pair to form a two-unit length concatemer (Figure 9–25). Following gap-filling and ligase action, it can be replicated to again reproduce unfinished progeny that will tend to stick together to produce a 4-mer, and so forth. Unit length molecules can be produced from such concatemers by the action of specific endonucleases that act to create staggered nicks. These can easily be filled in by a DNA polymerase to yield double helices identical to those present in the original parental particle.

No such dilemma exists in the replication of circular DNA. There the gaps created by removal of the RNA primers can eventually be filled in by extension of the 3′ termini growing around the circle.

θ-SHAPED INTERMEDIATES IN THE REPLICATION OF CIRCULAR DNA

Duplication of circular DNA need not involve the temporary creation of linear DNA. Visualization of the replicating intermediates of many ring chromosomes reveals that the parental strands maintain a circular form throughout replication (Figure 9–26). They always possess a θ-like shape that comes into existence by the initiation of a replicating bubble at some fixed point (the origin) (Figure 9–27). Circular λ DNA, for example, has its replication origin near gene 0 and synthesis proceeds outwardly in both the clockwise and counterclockwise directions (Figure 9–28). The normal origin of DNA replication on the *E. coli* chromosome is near the ilv locus at 74 minutes on the standard genetic map. Here again replication proceeds in both directions.

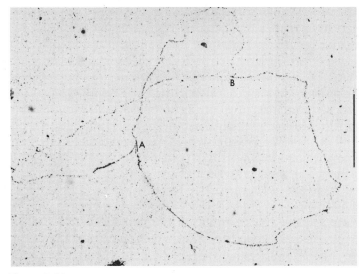

Figure 9–26
Autoradiograph showing the Y-shaped growing points *A* and *B* of an *E. coli* chromosome. The DNA was labeled with H³-thymidine for two generations of DNA replication. The *θ*-shaped appearance is the result of looking at a circular chromosome which has been two-thirds duplicated. The scale shows 100 *μ*. [Reproduced from J. Cairns, *Cold Spring Harbor Symp. Quant. Biol.* **XXVIII**, 44 (1963), with permission.]

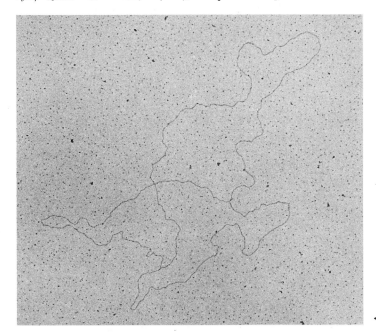

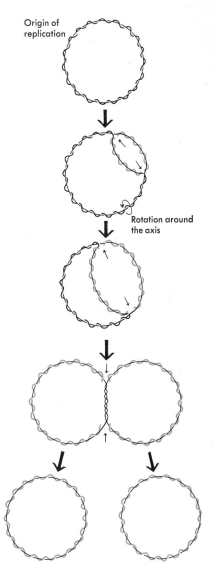

Figure 9–27
Simplified version of the bidirectional replication of circular DNA molecules. Only a fraction of the replication intermediates are so visualized in the EM, since those molecules which possess intact parental strands take up supercoiled configurations.

◄ **Figure 9–28**
θ-shaped replicating form of λ DNA. [Courtesy of Dr. David Dressler, Harvard University.]

From the moment of its discovery, circular DNA posed the question of how two covalently closed intertwined strands could unravel during replication. As long as each chain remains intact, even minor untwisting of one section of a circular double helix demands the cre-

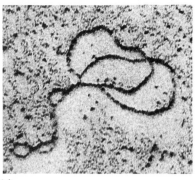

(a)

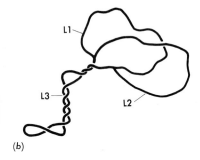

(b)

Figure 9–29
Twisted replicating molecules of circular
SV40 DNA: a) electron micrograph of an
early phase of replication. b) a diagrammatic
representation indicating details of the unreplicated, supercoiled parental strands—L1
and L2 indicate portions already replicated,
while L3 indicates the unreplicated portion
which takes up a supercoiled configuration.
[From N. P. Salzman *et al., Cold Spring
Harbor Symposia on Quantitative Biology,*
XXXVIII, 257 (1973).]

ation of supercoils in the opposing direction. The semiconservative replication of such DNA thus obligatorily demands temporary cuts in one or both of its polynucleotide
backbones. The question thus becomes how and when are
these breaks made? And are they the result of accidental
nuclease breaks, or do there exist specific cutting enzymes
whose principal function is to make possible the untwisting of parental DNA strands?

At first the possibility was considered that some permanent molecular swivel existed around which the
parental strands untwisted. Now it seems probable that
such breaks usually persist for very short intervals and
may involve cuts in only one of the two parental strands. A
simple break (nick) in one of the polynucleotide strands
would in fact generate such a swivel, since free rotation is
possible about most of the single bonds which make up
the polynucleotide backbone. But this scheme, at least in
its simplest form, cannot work since, when a replicating
fork passes through a nicked region, replication should
stop on the broken template strand.

This potential dilemma has in a sense been lessened
by the finding that *both* parental strands remain intact
throughout most of the replication cycle, with the unreplicated portion of the θ-shaped intermediates becoming
periodically supercoiled. So the cuts that allow untwisting
must have only a fleeting existence. In fact the replication
of the circular DNA may start by the growth of a very
small replicating bubble within completely covalently
closed DNA (Figure 9–29). Chain elongation within this
tiny bubble goes on to create supercoiling until further
growth ceases, due to the steric impossibility of creating
more supercoiled turns. Such halted bubbles can resume
growth only when a specific enzyme (*swivelase*) creates a
single-stranded cut that relaxes the supercoiled molecule
into an untwisted circular molecule. After repair of the
cut, further chain growth occurs, leading again to the generation of supercoils which are relaxed by the unswiveling
enzyme.

Swivelases now have been isolated from both bacterial and higher animal cells. The same enzyme both makes
the single-stranded cuts and then sews them together
again. Surprisingly, no external energy seems to be
required to remake the polynucleotide bonds. This may be
the consequence of a catalytic mechanism which attaches
one of the cut ends to the swivelase by a covalent bond
that preserves the energy of the backbone phosphodiester
bond. This energy in turn is used to reform the cut bond
after supercoil relaxation. Still very unclear is whether the

swivelase acts on *any* polynucleotide bonds or whether it works only at very specific sites (e.g., origin or termination of replication).

Still also to be elucidated are the events which permit separation of two parental strands at the end of replication. One of the two strands has to break, but how this occurs without generating an incomplete linear daughter helix remains an intriguing dilemma. Most likely, the parental strands become untwisted by unwinding proteins considerably in advance of the laying down of new daughter strands. Creation of a cut by the swivelase at the termini could thus lead to separation of the daughter molecules before they become converted into covalently closed double helices. This scheme by itself, however, does not explain how the now free ends of the broken parental strand become rejoined. Conceivably the "separating cut" occurs in regions where the spearated single strands can form hairpin loops like those known to occur in single-stranded RNA. If so, the nicked loops could be enzymatically closed by a ligase (Figure 9–30).

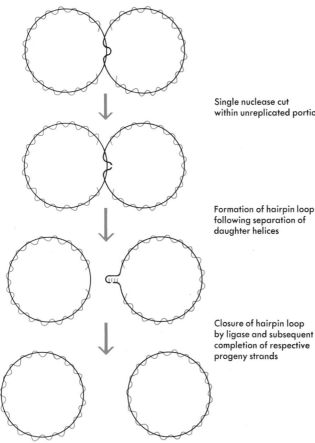

Single nuclease cut
within unreplicated portion

Formation of hairpin loop
following separation of
daughter helices

Closure of hairpin loop
by ligase and subsequent
completion of respective
progeny strands

Figure 9–30
A mechanism for the separation of the circular DNA progeny that employs the formation of a hairpin-like loop to bring together the cut ends of a parental template strand.

The Replication of DNA

ROLLING CIRCLE REPLICATION

An alternative way to replicate circular DNA is the *Rolling Circle* mechanism. This replication scheme explains the multiplication of many viral DNA's, bacterial mating, and gene amplification (see page 533). According to this model, replication starts with a specific cut in one specific strand of the parental duplex circle, thereby converting this strand to a polynucleotide with two chemically distinct ends (Figure 9–31). One terminal nucleotide contains a free 3' OH group on its sugar moiety while the other end displays a 5' phosphate group. The latter may be the binding site for the enzyme that makes the cut at the specific "starting sequence." DNA synthesis starts by the displacement of the enzyme-bound 5' end into solution, thereby permitting DNA polymerase to add deoxynucleotides to the free 3' OH end. As replication proceeds, the 5' end of the open strand is rolled out as a free tail of increasing length. This replicating structure is called a rolling circle

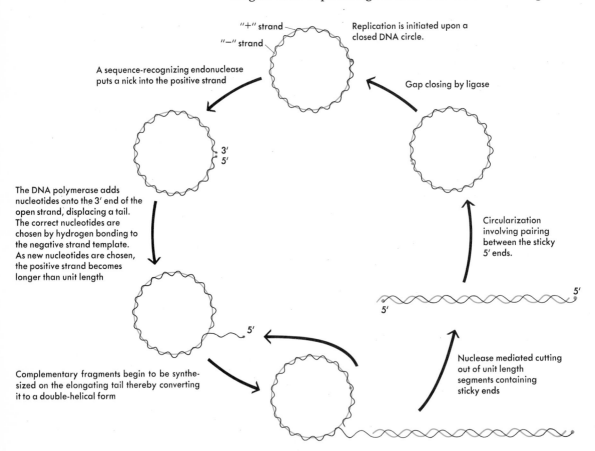

Figure 9–31
Duplication of circular DNA using the rolling circle mechanism.

since the unraveling of the free single strand is accompanied by rotation of the double-helical template about its axis. Such unraveling does not create any topological problems since the circle is held together by only one intact polynucleotide strand which will freely rotate about many of the covalent bonds comprising its backbone.

When this device is used to replicate double-stranded DNA, the 5′-ended tails serve as templates for the synthesis of small DNA fragments which eventually are laced together by the joining enzyme DNA ligase. Such growing tails thus have a double-stranded character soon after their formation (Figure 9–32). Elongation of such tails sometimes goes on to produce tails many times the contour length of the original circle. How these tails become converted into

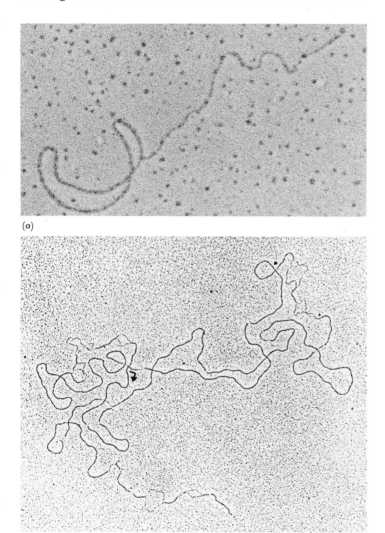

(a)

(b)

Figure 9–32
Electron micrographs of the rolling circle intermediates in (a) single-stranded φX174 DNA. [Courtesy of David Dressler, Harvard University.] and (b) in the replication of double-stranded phage P2 DNA. Here partial denaturing conditions were used to show sequence homologies between circle and tail sections. [Courtesy of R. B. Inman, University of Wisconsin.]

progeny DNA molecules is not yet established. The best guess is that the tails are cut by specific endonucleases of the form that can convert λ circles into λ rods (see page 221). The resulting unit length progeny rods could remain linear and become encapsulated into mature virus particles (see Chapter 15). Alternatively they could use their single-stranded complementary ends to form new circular molecules capable of becoming new rolling circles.

Rolling circles can also be the replicating intermediates for the synthesis of circular single-stranded DNA (Figure 9–33). Such molecules result when the nascent tails become covered with the viral structural proteins almost simultaneously with their synthesis. Being so covered,

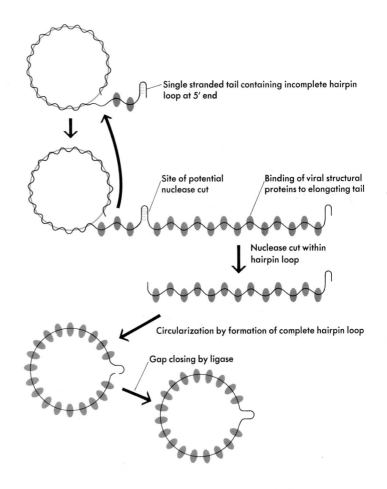

Figure 9–33
A rolling circle mechanism for the production of circular single-stranded DNA. Here is represented the process thought involved in the production of the single-stranded DNA circles of phages like φX174 and fd.

they can never be converted into double helices. How such protein-covered, single-stranded tails are converted into unit length circles is not known, though here, as for the formation of closed circular duplexes, the best guess is that hairpin-like loops are involved. If so, we might imagine that a specific cutting enzyme exists which cuts the tail within the hairpins to create unit length molecules. Subsequently the free ends must find each other to form an intramolecular hairpin loop within which the cut can be sealed by ligase action.

SYNTHESIS AND TRANSFER OF SINGLE-STRANDED DNA DURING BACTERIAL MATING

When bacteria mate, the DNA that the male transfers into the female is single-stranded, not double-stranded. A rolling circle is used for its synthesis with the process starting by the nicking of one specific strand of the male chromosome. This strand is then elongated at its 3' end, leading to the displacement of growing 5'-ended single strands. Most likely the enzyme that makes the specific cut, like that which starts off rolling circles, remains attached to the 5' end and somehow guides it into the female cell (Figure 9–34). Once the male DNA is inside the recipient it is converted to normal double-helical DNA. It may then by genetic recombination be exchanged for a portion of the original female chromosome. Transfer in this way couples DNA replication and mating in such a way as never to lead to genetically deficient male parents.

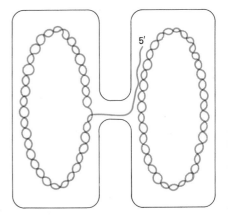

Figure 9–34
Transfer of single-stranded DNA from a male bacteria into the female via a rolling circle mechanism.

MUTATIONS WHICH BLOCK DNA SYNTHESIS

Mutations at a large number of different sites around the *E. coli* chromosome have been found to specifically block DNA synthesis (Figure 9–35). Some of these mutations (dna A, dna C) specifically block formation of new replication forks but have no effect on the growth of preexisting forks. Other mutations quickly stop the synthesis in preexisting replicating forks, either by loss of the ability to initiate further small fragments (e.g., dna G) or by the loss of enzymes involved in the polymerization process itself (dna E = Pol III). Certainly some of the genes involved here control the synthesis of the proteins involved in making RNA primers. Mutations also have been found which block the polymerizing actions of DNA polymerase I (Pol I) and II (Pol II) as well as the joining action of DNA ligase. These mutations do not stop synthesis, but lead to the buildup of unusually large numbers of unsealed DNA frag-

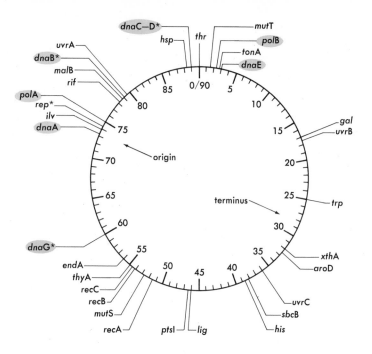

Figure 9–35

The location on the *E. coli* chromosome of genes involved in DNA replication (for a more complete map, see Figure 7–13). [Redrawn from *DNA Synthesis* by Arthur Kornberg. W. H. Freeman and Company. Copyright © 1974.]

Locus	Minutes	Comments
mut T[a]	1.5	Mutator, AT → CG transversions
dna E[b]	2.5	pol. III (pol. III*)
pol B[b]	2.5	pol. II
uvr B[a]	17.5	Thymine dimer excision
terminus[c]	~25	Terminus of replication
xth A[d]	32	Exonuclease III
uvr C[a]	36	Thymine dimer excision
sbc B[e]	38	Exonuclease I
lig[f]	46	Polynucleotide ligase
rec A[a]	51.5	Recombination
mut S[a]	52.5	Mutator
rec B[a]	54.5	*rec* BC nuclease
rec C[a]	54.5	*rec* BC nuclease
end A[g]	56	Endonuclease I
dna G[h]	60–65	DNA replication
dna A[h]	73	DNA replication
rep[a]	74	M13, φX174 RF replication
origin[c]	~74	Origin of replication
pol A[a]	75	pol. I
rif[a]	78.5	β subunit of RNA polymerase (rifampicin sensitivity)
dna B[i]	~79	DNA replication
uvr A[a]	79.5	Thymine dimer excision
hsp[a]	88.5	DNA restriction and modification endonuclease and methylase
dna C–D[h]	89	DNA replication

Reference loci[a] are: *thr, tonA, gal, trp, aroD, his, ctr, thyA, ilv,* and *malB*.

ments. We must anticipate that some genes code for replication enzymes, for whose existence we still have no direct clues. For example, DNA polymerase III binds to DNA in a reaction that requires ATP, a finding that suggests the existence of a still-to-be-understood energy-requiring step that must occur prior to polymerization.

TEST-TUBE REPLICATION OF COMPLETE DOUBLE HELICES

While initially biochemists were very pleased to find any test-tube incorporation of nucleotides into DNA-like molecules, now their objectives are much more precise. Not only do they want systems where DNA synthesis occurs rapidly, but they also aim for conditions which carefully reflect the processes by which *in vivo* DNA synthesis occurs. A clear goal is the synthesis of complete DNA molecules rather than the mere addition of nucleotides to growing chains of an undefined structure. Now after much jerky progress, conditions have been obtained that clearly mimic those present in growing cells. Not only does DNA synthesis now occur rapidly in cell-free extracts, but the products obtained are indistinguishable from intracellularly-produced DNA molecules. Extracts made from *E. coli* cells not only can make linear T7-length molecules of full length, but also carry out the complete semiconservative replication of circular plasmid DNA. In this latter case, replicating bubbles start *in vitro* at the same sites that are used as origins *in vivo*, and the θ-shaped intermediates grow until completion of replication and separation of circular double helices. Most importantly, these daughter circles can serve as templates for still additional rounds of semiconservative replication. Hopefully, these extracts can soon be fractionated to reveal the nature and mode of action of each enzyme involved in the replication process. Only then will our still all-too-speculative ideas about replication of DNA be replaced by a rigorous enzymological description.

Precise conditions now are also being worked out for the *in vitro* conversion of the single-stranded DNA of viruses like ϕX174 and M13 into their double-helical replicative intermediates. Surprisingly, the enzymology involved in converting single-stranded ϕX174 DNA into the duplex replicative form shows major differences from that involved in making M13 DNA duplex. In particular, several bacterial "DNA replication genes" are required for ϕX174 replication but not for that of M13 (Table 9–5). Moreover, while RNA polymerase is used to make the M13

Table 9–5 Protein Requirements for Synthesis of ϕX174 and M13 Double Helices*

	M13	ϕX174
Initiation		
RNA polymerase	+	−
Gene dna A product	−	+ (?)
Gene dna B product	−	+
Gene dna C product	−	+
Gene dna G product	−	+
Unwinding protein	+	+
Elongation		
DNA polymerase III	+	+
Unwinding protein	+	+
Termination		
DNA polymerase I	+	+
DNA ligase	+	+

* Data from Schekman *et al., Science,* **186,** 987 (1974).

243

RNA primers, (an)other enzyme(s) is (are) involved in making the φX174 primers. Since plasmid replication, but not that of the main chromosome DNA, is blocked by the RNA polymerase inhibitor rifampicin, the suspicion exists that M13 uses an enzyme complex normally involved in the replication of plasmid DNA while φX174 conversion employs the somewhat different set of enzymes used in the normal replication of the main *E. coli* chromosome. If so, we must face the complication that there exists no one unique way to replicate superficially quite similar DNA molecules.

REPAIR SYNTHESIS

For many years biochemists believed that incorporation of nucleotide precursors into DNA was necessarily a measure of the synthesis of new polynucleotide chains. Today, however, they realize that nucleotides can also be incorporated into full-length DNA chains by enzymatic reactions which repair damaged DNA. Creation of modified bases or chain breaks as a result of radiation action need not be lethal if the appropriate repair processes occur. Now several distinct types of repair processes are known to exist. In large part, they were first revealed by the discovery of mutants with increased sensitivity to radiation damage. Subsequent biochemical work then revealed the lack of specific repair processes in these mutants. Thus, in normal cells, much radiation damage is usually repaired before it has time to express itself.

The best understood case is the creation of thymine–thymine dimers by ultraviolet light. When ultraviolet light is absorbed by adjacent thymine molecules, they fuse together to form the structure shown in Figure 9–36. If unrepaired, this event is normally lethal: the fused thymine molecules do not act as faithful templates for the production of progeny DNA strands. Normal replication usually takes place only if the dimers are excised. This occurs in several stages, each mediated by a specific enzyme. In the first step, a specific endonuclease recognizes the damaged region, cutting the relevant polynucleotide strand on one side of the dimer. The second step involves the $5' \rightarrow 3'$-exonuclease activity of DNA polymerase I digesting away nucleotides adjacent to the cut. In the third step, 5'-nucleoside triphosphates bind to the resulting single-stranded region and subsequently become linked to the 3' end of the adjacent chain fragment by an enzyme, possibly DNA polymerase I. In the last step, the resulting gap is bridged by the joining enzyme polynucleotide ligase (Figure 9–37).

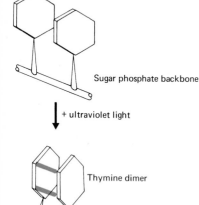

Figure 9–36
Formation of thymine dimers by uv irradiation.

Sugar phosphate backbone

+ ultraviolet light

Thymine dimer

244

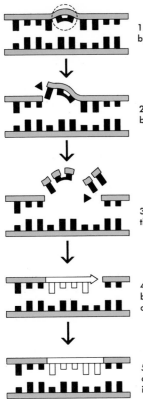

1. A distortion in the DNA molecule caused by a uv light-induced thymine dimer.

2. A specific endonuclease breaks the backbone of one chain near the dimer.

3. The excision of a small region containing the thymine dimer by an exonuclease.

4. 5'—3' synthesis of new strand. The correct bases are inserted by base pairing with those on the intact strand.

5. Polynucleotide ligase joins up the two ends of the strand and the "repaired" molecule is complete.

Figure 9–37
Some of the enzymatic steps involved in the repair of DNA molecules containing thymine dimers.

There is evidence for the existence of repair processes in virtually all cells examined, from bacteria to higher plants and animals. This is not surprising, since all cells are constantly subjected to various forms of radiation. During evolution, those cells possessing enzymatic systems for repairing damaged DNA molecules must have had an enormous selective advantage. In fact, evolution of repair systems might have been a necessary prerequisite for the development of organisms with larger and larger amounts of DNA. For the larger the amount of necessary DNA, the greater the probability that a given dose of radiation would cause a lethal event.

MEMBRANE INVOLVEMENT IN REPLICATION

The growth and division of a bacterial cell has to involve not only the accurate duplication of a circular DNA chromosome into two identical daughter chromosomes, but also their correct partitioning into the two progeny cells. So it has always seemed natural to postulate a connection between the cell membrane and some portion of the bacterial chromosome. Support for this supposition comes from electron micrographs which suggest chromosomal attachment to invaginated portions of the cell membrane called the *mesosome* (Figure 9–38). Moreover, when very gentle methods are used to isolate intact bacterial chromosomes, they frequently are found associated with membrane fragments.

Figure 9–38
Attachment of a *Bacillus subtilis* chromosome to an invagination (mesosome) of the bacterial membrane. (N) Nuclear chromosomal material; (M) Mesosome. [Photograph kindly supplied by A. Ryter, Institut Pasteur, Paris.]

Only replicating chromosomes seem to be membrane-bound, and interesting *in vivo* DNA transformation experiments suggest that both the origin and termini of DNA replication are preferentially bound to the membrane. So it is tempting to suggest that the binding substances (protein?) play an active role in the initiation (termination) of chromosomal replication. In contrast, there appear to be no firm bonds between the enzymes at the replicating fork and the cell membrane.

Unfortunately, firm proof for any of these latter conjectures may be very hard to obtain since there are dismally few facts about the protein components of any bacterial membrane. Conceivably more hard facts may come from study of the multiplication of the DNA of the small *E. coli* plasmids. They contain tightly bound protein which may be involved not only in creating single-stranded cuts, but also in attaching the 5' end of the cut strand to specific regions of the bacterial membrand. Intensive study of their function seems bound to yield concepts relevant not only to plasmid-membrane interactions, but also to the behavior of the main chromosome.

SUMMARY

The primary genetic material is DNA. It usually consists of two polynucleotide chains twisted about each other in a regular helix. Each chain contains a very large number of nucleotides. There are four main nucleotides and their sequence along a given chain is very irregular. The two chains are joined together by hydrogen bonds between pairs of bases. Adenine (purine) is always joined to thymine (pyrimidine) and guanine (purine) is always bonded to cytosine (pyrimidine). The existence of the base pairs means that the sequences of nucleotides along the two chains are not identical but complementary. If the sequence of one chain is known, that of its partner is automatically known.

Cellular duplication of DNA occurs with the two strands separated, allowing the single strands to act as templates for the formation of complementary strands. The strands do not completely separate before the synthesis of the new strands. Instead, duplication goes hand in hand with strand separation. The monomeric precursors for DNA synthesis are the deoxynucleoside-triphosphates, which are enzymatically joined together in the presence of a DNA template. Selection of the correct nucleotide by hydrogen bonding to the template is a very accurate process. The average probability of an error in the insertion of a new nucleotide under optimal conditions may be as low as 10^{-8} to 10^{-9}.

Individual DNA molecules may have either linear or circular shapes, with some rod-shaped molecules having the potential for conversion to the circular form. Except for DNA isolated from small viruses, most DNA molecules are very, very large. The E. coli chromosome, for example, is a circular DNA molecule of MW about 2×10^9. With bacteria and their viruses, the replication of both linear and circular forms of chromosomes usually starts at a fixed location (the origin) to produce a replicating bubble containing two replicating forks. Chain growth occurs at each replicating fork by the addition of precursor nucleotides to the 3' ends of short DNA fragments that, in turn, become linked up to the main daughter strands. While initially it was thought that only one to several enzymes might be needed to carry out DNA replication, now it appears that each cell requires many such enzymes. Initiation, elongation, and termination all require special enzymes, as does the creation of the single-stranded cuts required for the unraveling of the parental strands of circular DNA. Several different polymerizing enzymes (DNA polymerases) exist, with one specialized for chain elongation, and another for the filling in of the gaps between the small DNA fragments found at the replicating forks. No known DNA polymerase can initiate DNA chains, a process now thought to start by the synthesis of small

RNA primers catalyzed by a form of the enzyme RNA polymerase.

All DNA polymerases have the capacity to recognize incorrectly incorporated bases and to enzymatically cut them out before addition of the next precursor nucleotide (proofreading). Some of the enzymes involved in DNA duplication also function in the repair of damaged DNA molecules. There are also hints that one or more of the enzymes involved in initiating (terminating) DNA replication may associate with portions of the bacterial membrane. Correct positioning of progeny DNA molecules into their respective daughter cells may depend upon such interactions. Proof of this type of conjecture, however, may be difficult to obtain because of continued poverty of knowledge at the molecular level about bacterial membranes.

REFERENCES

STENT, G. S., *Molecular Genetics*. Freeman, San Francisco, 1971. A splendid narrative of the emergence of molecular genetics.

DAVIDSON, J. N., *The Biochemistry of Nucleic Acids*, 7th edition, Academic Press, 1972. A much enlarged and useful revision of one of the first texts about nucleic acids.

DNA Synthesis in vitro, edited by Welles, R. D., and R. B. Innan, University Park Press, Baltimore, Md., 1973. A collection of research articles arising from a July 1972 meeting.

KORNBERG, A., *DNA Synthesis*, Freeman, San Francisco, 1974. A superb advanced text that should be obligatory reading for all persons interested in DNA replication.

WATSON, J. D., AND F. H. C. CRICK, "Genetical Implications of the Structure of Deoxyribonucleic Acid," *Nature,* **177,** 964 (1953).

MESELSON, M., AND F. W. STAHL, "The Replication of DNA in *Escherichia coli,*" *Proc. Nat. Acad. Sci.,* **44,** 671 (1958). A classic experiment in molecular biology, showing that DNA replication involves the separation of the two complementary polynucleotide chains.

CAIRNS, J. "The Bacterial Chromosome," *Scientific American,* **214,** 36 (1966). A description of the elegant work that established the shape of the replicating *E. coli* chromosome.

SCHNÖS, M., AND R. D. INMAN, "Position of Branch Points in Replicating λ DNA," *J. Mol. Biol.,* **51,** 61 (1970). Demonstration of the bidirectional character of circular DNA replication.

SALZMAN, N. P., G. C. FAREED, E. D. SEBRING, AND M. M. THOREN, "The Mechanism of SV40 Replication," *Cold Spring Harbor Symposia on Quantitative Biology,* **XXXVIII,** 257 (1973). A summary of the experiments which showed the intactness of replicating circular DNA.

GILBERT, W., AND D. DRESSLER, "DNA Replication: The Rolling Circle Model," *Cold Spring Harbor Symposia on Quantative Biology,* **XXXIII,** 473 (1969). A description of the model plus early experiments in its favor.

WOLFSON, J. D., D. DRESSLER, AND M. MAGAZIN, "Bacteriophage T7 DNA Replication: A Linear Replicating Intermediate," *Proc. Nat. Acad. Sci.,* **69,** 998 (1972). The first experimental description of how linear DNA replicates.

WATSON, J. D., "The Origin of Concatemeric T7 DNA," *Nature New Biology,* **239,** 197 (1972). A solution to how the ends of linear DNA can be completed.

GROSS, J., "DNA Replication in Bacteria," *Current Topics in Microbiology and Immunology,* **57,** 39, Springer-Verlag, 1972. A comprehensive review emphasizing mutations which block DNA synthesis.

SCHECKMAN, R., A. WEINER, AND A. KORNBERG, "Multienzyme Systems of DNA Replication," *Science,* **186,** 987 (1974). A review article summarizing evidence that several distinct systems for DNA replication exist within the same *E. coli* cells.

MANIATIS, T., A. JEFFREY, AND D. G. KLEID, "Nucleotide Sequence of the Rightand Operator of Phage λ," *Proc. Nat. Acad. Sci.,* **72,** in press (1975). An elegant paper showing the power of new methods for sequencing DNA.

Papers in Biochemical Genetics, edited by Marmur, J., and G. Zubay; Holt, Rinehart and Winston, Inc. (1973). An excellent collection of original papers on DNA, RNA, and protein synthesis and regulation.

GETTER, M., "DNA Replication," *Ann. Rev. Biochem.* **44,** 45 (1975); and DRESSLER, D., "The Recent Excitement in the DNA Growing Point Problem," *Ann. Rev. Microbiology,* **29,** 525 (1975). Two up-to-date reviews emphasizing the enzymology of DNA replication.

The Genetic Organization of DNA

Even before the double helix was found, the simplest hypothesis imaginable was that the genetic information of DNA resided in the sequence of its various nucleotides. But until the structure was uncovered, it was possible that knowledge of the nucleotide order was not by itself sufficient. The real key might lie in some weird 3-D form not easily, if at all, deducible from the sequence of bases. Fortunately, the double helix tells us that all nucleotides are geometrically equivalent, unambiguously revealing that the genetic code resides in their linear sequences.

CHAPTER 10

THEORETICALLY, A VERY, VERY LARGE NUMBER OF DIFFERENT SEQUENCES CAN EXIST

Since the sugar-phosphate backbone is the same in all DNA molecules, it necessarily cannot carry any genetic information. The information must instead be carried in the sequence of the four (A, G, C, T) bases. This requirement, however, poses no real restriction on the effective amount of information in DNA. Since each molecule is very long, the number of sequence permutations is 4^n, where n is the number of nucleotides in a given molecule. A virtually infinite number of genetic messages can be coded with the four letters, A, T, C, and G of the nucleic acid alphabet. The possible number of different genes of $MW = 10^6$ is 4^{1500}, a value very much larger than the number of different genes that have existed in all the chromosomes present since the origin of life.

MUTATIONS ARE CHANGES IN THE SEQUENCE OF BASE PAIRS

The genetic mapping of the T4rIIA gene revealed at least 500 sites at which mutations can occur and between which genetic recombination (crossing-over) is possible. The magnitude of this number immediately tells us that these sites are the specific base pairs along the gene (Figure 10–1).

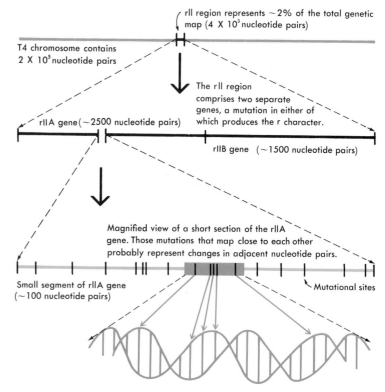

Figure 10–1

The relationship of mutations in the rII region of the chromosome of phage T4 to the structure of DNA.

Many mutations are base pair switches from, for example, AT to GC, CG, or TA (Figure 10–2). Thus, by studying the fine details of the genetic map very carefully, it is possible to obtain important information about the sequence of base pairs along the DNA molecule. One major reservation, however, exists. There is no *a priori* reason to believe that all changes in the genetic code will necessarily cause functional changes in the corresponding proteins. The number of observed mutable sites is, therefore, likely to be a serious underestimate of the number of nucleotide pairs.

Most single base switches are reversible, and often the rate of the "back" mutation to the normal nucleotide arrangement has an order of magnitude similar to that of the rate of change to the mutant arrangement. These mutations most likely reflect failures in the replication process.

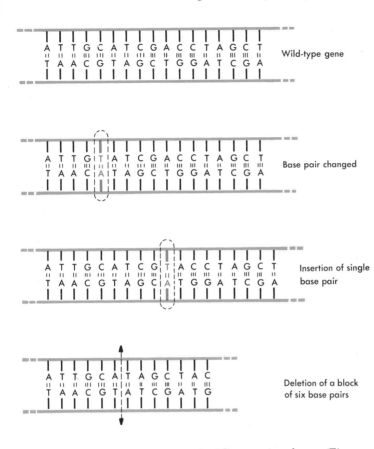

Figure 10–2

Three classes of mutations result from introducing defects in the sequence of bases (A, T, G, C) that are attached to the backbone of the DNA molecule. In one class, a base pair is simply changed from one into another (i.e., G–C to A–T). In the second class, a base pair is inserted (or deleted). In the third class, a group of base pairs is deleted (or inserted).

Either the wrong purine–pyrimidine pairs form (Figure 10–3) or an adenine mistakenly pairs with guanine (or cytosine with thymine). There are other rare locations where the rates of forward (backward) mutation greatly exceed the reverse step. Most of these highly mutable nucleotide

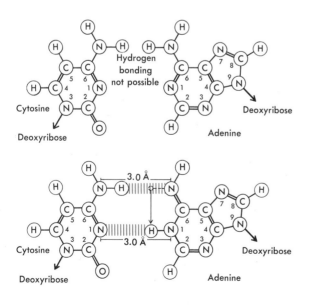

Figure 10–3

This demonstrates how the specificity of base pairing in DNA is determined by hydrogen bonding. The hydrogen atoms are indicated by solid circles, the bonds by ||||. (a) Shows why the pairing of cytosine with an adenine molecule, having the most stable distribution of its hydrogen atoms, cannot lead to hydrogen bonding. (b) Shows how the shift of a hydrogen atom in an adenine molecule, from the 6-amino group to the N_1 position, permits hydrogen bonding with cytosine. The normal position of the hydrogen atom is indicated by the small open circle. (The dimensions shown are only approximate.) [Redrawn from W. Hayes, *The Genetics of Bacteria and Their Viruses*, Blackwell, Oxford, 1964, p. 228, with permission.]

pairs—"hot spots"—probably do not arise by normal duplication, but as we shall see below, are connected with mispairing during crossing-over.

Other spontaneous mutations involve the loss (deletions) or gain (insertions) of nucleotides. Sometimes hundreds to thousands of nucleotides are involved in deletions, and in rare cases whole genes are lost. Reverse (back) mutation to the normal gene arrangement is clearly impossible for large deletions and insertions, and occurs only at low rates for simple one-nucleotide deletions (insertions).

ERROR LEVELS PER INCORPORATED NUCLEOTIDE RANGE FROM 10^{-6} to 10^{-9}

Measurements of mutation frequencies for a variety of bacterial genes suggest that detectable gene mutations spontaneously arise, on the average, about once every 10^6 gene duplications. This figure can be extended to the nucleotide level only if we can estimate the fraction of base pairs that must remain unchanged for the maintenance of normal gene function. If we make the assumption that all the base pairs are essential, then for a gene containing 1000 nucleotides the error at the nucleotide level is approximately 10^{-9} mistakes per incorporated nucleotide. Data from the two rII genes, however, suggest that many base pair changes do not lead to detectable mutation. Here the 500 mutable sites established reflect only one-fourth of the total number of base pairs found in these two genes. Further work is certain to reveal still more mutable sites, but it would be surprising if more than half the base pairs had to be present in their *current* form in order to retain biological activity. Detectable mutations in the rII gene, however, occur at a higher rate than in many other loci; and here the total mutation frequency (even if we exclude the hot spots) is about 10^{-4} to 10^{-5}. This may mean that in most other genes many fewer nucleotide changes lead to detectable mutations than in the rII genes. If so, the general mistake level may be as high as 10^{-7} per nucleotide replication.

CONTROL OF MUTATION LEVELS BY THE RELATIVE EFFICIENCIES OF FORWARD POLYMERIZING AND BACKWARD NUCLEASE ACTIVITIES

At first it was thought that these error frequencies were a direct reflection of the inherent accuracy of $A \cdots T$ and $G \cdots C$ base pair formation. If this indeed were the case, then only very, very rarely could any of the bases assume the "incorrect" tautomeric form (enol instead of keto, and imino instead of amino). Direct chemical measurements of the frequencies at which the various bases assume the "wrong" forms, however, gave values much higher than seemed compatible with known mutation rates. This led for many years to the chemically always questionable postulate that the electronic environment of the polymerizing site on a DNA polymerase molecule somehow affected the tautomeric form of its bound nucleotide. Fortunately, this dilemma vanished with the discovery of the *proofreading* capability of DNA polymerases. Though the *initial* error frequencies during base pair selection are consistent with measured tautomeric ratios, and so, much higher than anyone first guessed, almost all the wrongly inserted bases are later removed by the $3' \rightarrow 5'$-exonuclease component of the respective DNA polymerase. Every known DNA polymerase possesses this activity, a fact which initially seemed most bizarre; but now we realize it to be a necessary component of an accurate DNA replication machinery. In fact studies with mutant DNA polymerase molecules show that mutation levels may be controlled by the ratio of backwards ($3' \rightarrow 5'$)-nuclease activities to forward ($5' \rightarrow 3'$)-polymerizing capabilities. If the $3' \rightarrow 5'$ exonuclease is inefficient, then abnormally high mutation rates result. Correspondingly, a very efficient $3' \rightarrow 5'$-nuclease activity leads to a very low mutation rate.

We see that the spontaneous mutation level for a given species is itself determined by evolutionary forces. Too high a rate will so burden a species as to lead to its failure to produce viable progeny, while too low a rate will prevent the emergence of new variants necessary for survival in a changing environment.

PRECISE STATEMENTS ABOUT SOME CHEMICAL MUTAGENS

It is now possible to make intelligent statements about how some chemical mutagens produce changes in the genetic code. For example, nitrous acid (HNO_2) is a very powerful mutagen because it acts directly on the nucleic acids, replacing amino groups by keto groups. Thus, it directly alters the genetic code by converting one base into another (Figure 10–4). Other substances cause mutations as a result of their incorporation into the DNA molecule itself. These latter compounds are base analogues which, because of their structural similarity to the normal DNA bases, can be incorporated into DNA without destroying its capacity for replication. Their different structures, how-

255

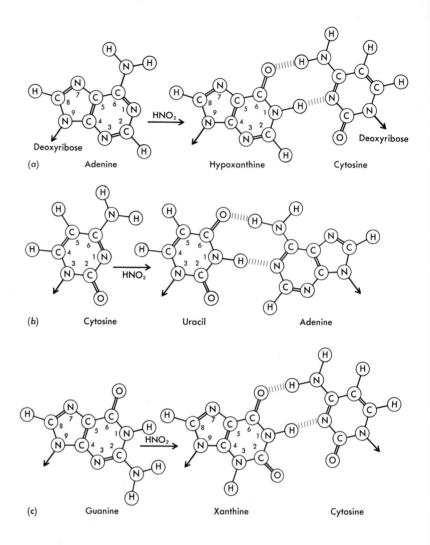

Figure 10–4
The oxidative deamination of DNA bases by nitrous acid, and its effects on subsequent base pairing. (a) Adenine is deaminated to hypoxanthine, which bonds to cytosine instead of to thymine. (b) Cytosine is deaminated to uracil, which bonds to adenine instead of to guanine. (c) Guanine is deaminated to xanthine, which continues to bond to cytosine, though with only two hydrogen bonds. Thymine, and the uracil of RNA, do not carry an amino group and so remain unaltered. [Redrawn from W. Hayes, *The Genetics of Bacteria and Their Viruses*, Blackwell, Oxford, 1964, p. 280, with permission.]

ever, often cause less accurate base pair formation than normal, leading to mistakes during the replication process. One of the most powerful base analogue mutagens is 5-bromouracil, an analogue of thymine. It is believed to cause mutations because its hydrogen atom at position 1 is not as firmly fixed as the corresponding hydrogen atom in thymine. Sometimes this hydrogen atom is bonded to the oxygen atom attached to carbon atom 6 (Figure 10–5). When this happens, the 5-bromouracil can pair with guanine.

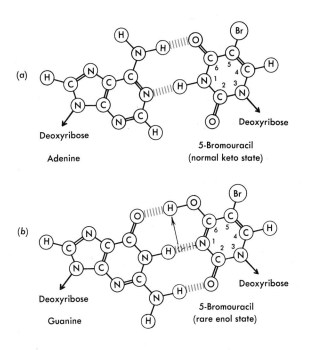

(a)

Deoxyribose

Adenine

5-Bromouracil
(normal keto state)

(b)

Deoxyribose

Guanine

Deoxyribose

5-Bromouracil
(rare enol state)

Figure 10–5
The base-pairing attributes of 5-bromouracil. (a) In the normal keto state, with a hydrogen atom in the N_1 position, bromouracil bonds to adenine. (b) In the rare enol state, a tautomeric shift of this hydrogen atom determines specific pairing with guanine. [Redrawn from W. Hayes, *The Genetics of Bacteria and Their Viruses*, Blackwell, Oxford, 1964, p. 278, with permission.]

GAPS BETWEEN GENES ARE RELATIVELY SHORT

Usually there is much less crossing-over between mutations located at adjacent ends of two contiguous genes than between two mutations at extreme ends of even the shortest genes. This suggests that adjacent genes often are separated by relatively few nucleotide pairs. Confirmation of this point will not be possible until the exact nucleotide sequences of a number of contiguous gene sets are determined chemically. Until recently, this task seemed virtually impossible, but new techniques are now beginning to reveal long stretches of nucleotide sequences of certain viral and bacterial chromosomes. As we shall see in Chapter 13, the first intergenic sequence data showed a group of 30 nucleotide pairs separating two viral genes. More recent data on two contiguous genes involved in tryptophan synthesis suggest that at most only a few nucleotides need separate two genes.

AGREEMENT OF A GENETIC MAP WITH THE CORRESPONDING DISTANCE ALONG A DNA MOLECULE

Since crossing-over analysis began, geneticists have wondered how closely their genetic maps corresponded with actual physical length along the chromosome. The visualization, in the 1930's, of *Drosophila's* highly extended salivary chromosomes permitted the exact mapping of the sites of mutations. Good correlations between the physical and genetic distances along some chromosomal stretches were shown. But these views of chromosomes in the light miscroscope tell us nothing about the arrangement of the constituent DNA molecules. Firm answers at the molecular level could come only when it became possible to locate the sites of specific mutations along a well defined viral chromosome (DNA molecule).

The chromosome best studied so far is that of phage λ. Several methods have been used to assign physical locations to the various mutations. The first utilized our ability to break DNA helices into fragments of defined size by stirring them in a blender. A certain stirring speed will break λ DNA into halves, a higher speed will produce quarters, and so on. Fragments which have different densities can then be separated from each other by equilibrium centrifugation in a cesium chloride gradient. As fragments rich in GC have a higher density than those rich in AT, separation becomes routinely possible if the various fragments are constituted from different AT/GC ratios. This is the case in λ, where the left-hand side is much richer in GC than the right-hand side (Figure 15–9). Purified left and right halves can then be used in DNA transformation experiments to see which genes are in each fragment. The results show good correlation with the left and right halves of the genetic map.

Much more precise data come from the mapping of well defined deletions of large numbers of nucleotides. This can be done using the technique of DNA–DNA hybridization. In this procedure, the two strands of the λ double helix are separated by heating the DNA to near 100°C. At that temperature virtually all the hydrogen bonds holding the molecule together break, and the resulting free single strands unwind from each other (DNA denaturation). The "+" and "−" strands then can be separated by their density differences in a CsCl solution. Reformation of the double helices from separated single strands (renaturation) occurs if the single strands are mixed together, heated to near 100°C, and then gradually cooled.

Under these conditions the original hydrogen bonds between the complementary chains reform, and the resulting molecules are indistinguishable in the electron microscope from molecules which have never been heated.

If, however, renaturation occurs between a normal "+" strand and a "−" strand containing a deletion, the "+" strand section complementary to the deleted "−" section cannot form hydrogen bonds. It exists as a single-stranded loop extended out from the predominantly double-stranded molecule (Figure 10–6). These loops are easily detectable in the electron microscope (Figure 10–7), allowing the location of a number of different deletions to be precisely spotted along the λ DNA molecule. Putting these data alongside the very extensively studied λ genetic map gives the picture shown in Figure 10–8. It reveals a much closer

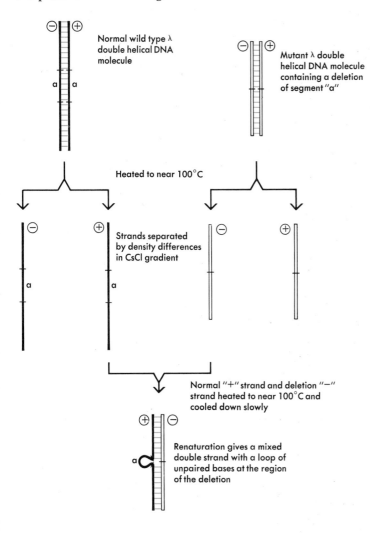

Normal wild type λ double helical DNA molecule

Mutant λ double helical DNA molecule containing a deletion of segment "a"

Heated to near 100°C

Strands separated by density differences in CsCl gradient

Normal "+" strand and deletion "−" strand heated to near 100°C and cooled down slowly

Renaturation gives a mixed double strand with a loop of unpaired bases at the region of the deletion

Figure 10–6
Mapping of deletions by electron microscope vizualization of renatured DNA molecules.

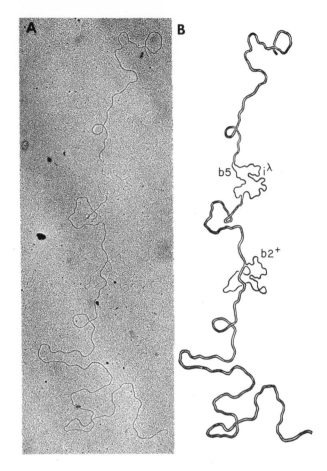

Figure 10–7
An electron micrograph (left) of a heteroduplex formed between strand 1 of λ and strand r of λb2b5 together with an interpretive drawing (right). λb2b5 contains a deletion (b2) and a nonhomologous section (b5) unable to pair with its λ counterpart. [Reproduced with permission from Westmoreland, Szybalski, and Ris, *Science*, **163**, 1343 (1969).]

Figure 10–8
A comparison of the λ (recombination) map with the true physical map. The recombination map was drawn using data of Amati and Meselson [*Genetics* **51**, 369 (1965)], while the physical map is redrawn from a recent map prepared by W. Szybalski (University of Wisconsin).

correspondence between the genetic map and the actual physical structure of the DNA than had been anticipated. Thus, at medium resolution, the probability of crossing-over is approximately equal throughout the DNA molecule. Exceptions to this point are obvious in Figure 10–8, and caution must be taken not to overinterpret genetic linkage data.

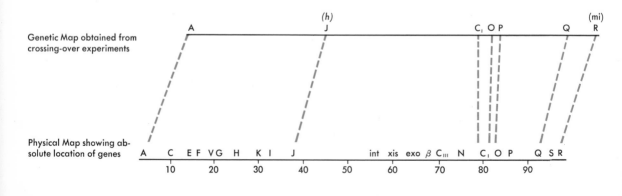

THE AVERAGE GENE CONTAINS ABOUT 900 TO 1500 NUCLEOTIDE PAIRS

There are several ways to reach this conclusion. The most direct one divides the number of nucleotides in a chromosome by the number of genes located along it. For example, over 100 genes in the bacterial virus T4 chromosome are already described. The chromosome has a molecular weight of 120 million, giving an average molecular weight of 1.2 million per gene. This is clearly an upper estimate, since a significant fraction of the T4 genes have probably not yet been discovered. In the more thoroughly studied bacterial virus λ, some 40 genes have been mapped. As the molecular weight of its DNA is 32 million, the upper limit of its average gene size is 0.8 million.

Similar size ranges are given by genetic mapping experiments. For example, the rIIA gene of T4 occupies about 1 percent of the total genetic map (Figure 10–1). If it is of average size, there are about 100 genes in T4. The validity of this method also depends upon the assumption that crossing-over occurs with approximately equal frequency in all regions along the chromosome. When this is not true, regions in which much crossing-over takes place will genetically seem much farther apart than regions of equal physical size that have limited recombination.

Our first argument is eventually the more rigorous, suffering only the complication that not all the genes in T4 and λ are known. Hence, we should look closely at the rates at which new genes are still being discovered. It now looks as if the number for both T4 and λ will be at most doubled (the rate of discovery, especially for λ, is rapidly slowing down). Thus, the average gene size may be between a half and one million, depending upon the organism under study. Since the molecular weight of a pair of bases is slightly over 600, this means that an average gene is a linear arrangement of approximately 900 to 1500 base pairs.

CROSSING-OVER IS DUE TO BREAKAGE AND REJOINING OF INTACT DNA MOLECULES

Until recently, not even a superficial understanding of the molecular basis of crossing-over existed. The classical picture of crossing-over, developed in the 1930's from cytological observations, hypothesized that during meiosis the paired, coiled chromosomes were sometimes physically broken at the chromatid level as a result of tension created by their contraction. The broken ends could then relieve the tension by crossways reunion, creating two reciprocally recombinant chromatids as well as two parental chromatids (Figure 10–9). According to this model, recombination occurs after chromosome duplication is complete—that is, at the four-strand stage. This hypothesis fell into disfavor about 1955, when geneticists found that crossing-over occurred within the gene, by then realized to be part of a DNA molecule. A seemingly unpleasant consequence was that the effective breakage points must necessarily lie between the same nucleotides in the two homologous chromatids. Otherwise recombination would generate new DNA molecules differing in length from the parental molecules.

(a) Breakage and reunion

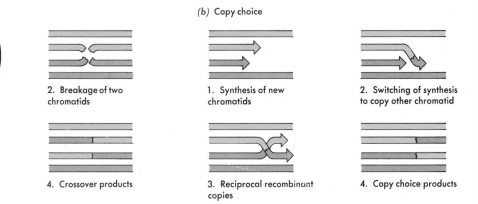

1. Pairing and coiling

2. Breakage of two chromatids

3. Crosswise reunion of broken chomatids

4. Crossover products

(b) Copy choice

1. Synthesis of new chromatids

2. Switching of synthesis to copy other chromatid

3. Reciprocal recombinant copies

4. Copy choice products

Figure 10–9
Diagrammatic representation of two possible mechanisms of crossing-over.

To avoid these dilemmas, enthusiasm developed for a hypothesis relating recombination to chromosome duplication. This alternative hypothesis proposed that during replication of the paired chromosomes, the new DNA strand being formed along the paternal chromosome (for example) switches to the maternal one that it thereafter copies. If the complementary replica of the maternal strand also switches templates when it reaches the same point, two reciprocally recombinant strands would be formed. This hypothetical process is called *copy choice*. A fundamental distinction between the two hypotheses lies in their prediction of the physical origin of recombinant chromosomes. Following breakage and reunion, the recombinant chromosomes inherit physical material from the two parental chromosomes. In contrast, the recombinant chromosomes produced by copy choice are synthesized from new material.

These alternative hypotheses were tested by experiments using isotopically heavy (C^{13}, N^{15}) parental phage λ particles (Figures 10–10 and 10–11). Here again the heavy isotopes were used to allow a cesium chloride gradient to distinguish between parental and daughter DNA strands.

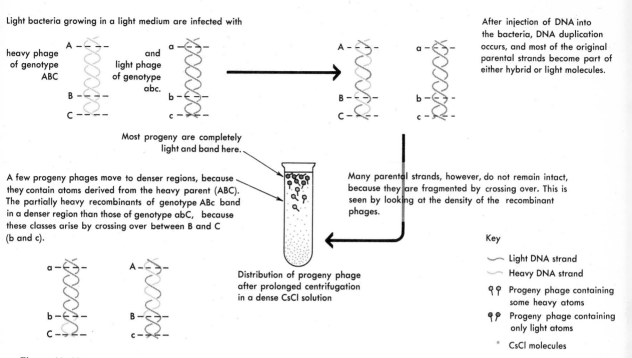

Light bacteria growing in a light medium are infected with

heavy phage of genotype ABC

and light phage of genotype abc.

After injection of DNA into the bacteria, DNA duplication occurs, and most of the original parental strands become part of either hybrid or light molecules.

Most progeny are completely light and band here.

A few progeny phages move to denser regions, because they contain atoms derived from the heavy parent (ABC). The partially heavy recombinants of genotype ABc band in a denser region than those of genotype abC, because these classes arise by crossing over between B and C (b and c).

Many parental strands, however, do not remain intact, because they are fragmented by crossing over. This is seen by looking at the density of the recombinant phages.

Distribution of progeny phage after prolonged centrifugation in a dense CsCl solution

Key

⌇ Light DNA strand

⌇ Heavy DNA strand

φ φ Progeny phage containing some heavy atoms

φ φ Progeny phage containing only light atoms

· CsCl molecules

Figure 10–10
The employment of heavy isotopes to study the mechanism of crossing-over.

Figure 10–11
An experimental demonstration that crossing-over and DNA duplication are independent phenomena.

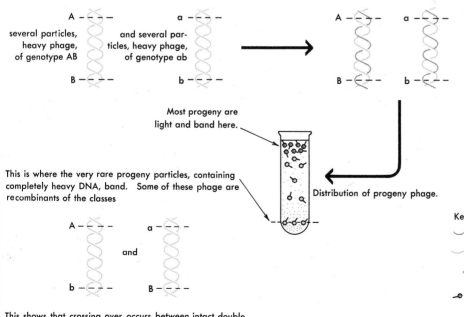

Light bacteria growing in a light medium are infected with

several particles, heavy phage, of genotype AB

and several particles, heavy phage, of genotype ab

After injection of the DNA molecules into the bacteria, most heavy strands become part of hybrid molecules. Rarely, however, an infecting molecule fails to duplicate and, when the new progeny particles are formed, these unreplicated molecules, in which the parental strands have never separated, become enclosed in new protein shells. This phenomenon enables us to ask whether the rare, completely heavy, DNA molecules are ever recombinants. Again use is made of a CsCl gradient to separate progeny particles of different density.

Most progeny are light and band here.

This is where the very rare progeny particles, containing completely heavy DNA, band. Some of these phage are recombinants of the classes

Distribution of progeny phage.

and

This shows that crossing over occurs between intact double helices and that extensive DNA synthesis is not involved in crossing over.

Key

⌇ Light DNA strand

⌇ Heavy DNA strand

· CsCl molecule

⌐φ φ Progeny phage containing only light atoms

φ φ Progeny phage containing some heavy atoms

Genetic crosses between heavy (or between heavy and light) phage particles were made in *E. coli* cells growing in a light (C^{12}, N^{14}) medium. Under these conditions, all of the newly synthesized viral DNA molecules are derived from light precursors; thus, if copy choice is the correct mechanism, all the recombinant particles should be light. On the contrary, if recombinants are derived by breakage and reunion, some of the recombinant phage particles will contain heavy atoms derived from the parental chromosomes. The progeny particles of these crosses were placed in dense CsCl solutions and rapidly centrifuged to separate particles of different density. Phage particles of varying density were then collected and genetically tested to see which were recombinants. The experimental results (Figure 10–10) were clear-cut and, to the surprise of most molecular biologists, showed that some recombinant particles contained heavy atoms. Moreover, further experiments (Figure 10–11) revealed that recombination can occur between nonreplicating DNA molecules. Breakage and reunion of intact double helices must, therefore, be the primary mechanism of crossing-over in bacteriophage. Crossing-over in both bacteria and higher organisms will most likely have a similar basis.

INVOLVEMENT OF BASE PAIRING IN CROSSING-OVER

Two assumptions underlie most current experiments designed to disclose the exact way(s) in which the strands recombine. The first assumption is that two homologous double-helical DNA molecules will not attract each other at specific points. No obvious pairing force can be imagined. The second assumption is that recognition involves hydrogen bond formation between complementary regions of single-stranded DNA.

One way by which single-stranded regions might arise starts with the production of single-stranded cuts by an endonuclease. This creates free ends at which DNA polymerases can add new nucleotides, thus displacing the preexisting strands to form a number of single-stranded tails (Figure 10–12). Random collisions of tails with complementary sequences should then lead to the formation of double-helical junctions between different DNA molecules. If subsequent endonuclease cuts occur, as shown in Figure 10–12, followed by gap filling, then a recombinant DNA molecule is produced, together with two partial molecules. Exonuclease attack on the remaining fragments will then produce the complementary single-stranded regions

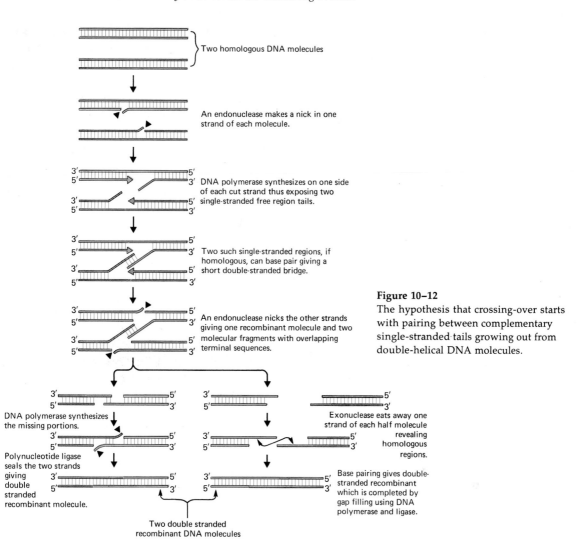

Two homologous DNA molecules

An endonuclease makes a nick in one strand of each molecule.

DNA polymerase synthesizes on one side of each cut strand thus exposing two single-stranded free region tails.

Two such single-stranded regions, if homologous, can base pair giving a short double-stranded bridge.

An endonuclease nicks the other strands giving one recombinant molecule and two molecular fragments with overlapping terminal sequences.

Figure 10–12
The hypothesis that crossing-over starts with pairing between complementary single-stranded tails growing out from double-helical DNA molecules.

DNA polymerase synthesizes the missing portions.

Polynucleotide ligase seals the two strands giving double stranded recombinant molecule.

Two double stranded recombinant DNA molecules

Exonuclease eats away one strand of each half molecule revealing homologous regions.

Base pairing gives double-stranded recombinant which is completed by gap filling using DNA polymerase and ligase.

(Figure 10–12) necessary to bring about their union. The gaps in the individual strands can then be filled by a DNA polymerase. The final result is a second recombinant molecule whose genetic structure is essentially reciprocal to the first recombinant.

STABILIZATION OF EXTENDED SINGLE-STRANDED TAILS BY RECOMBINATION-PROMOTING PROTEIN

Apparently at variance with the above mechanism is the strong tendency of all known DNA single strands to form hydrogen-bonded hairpin loops like those of single-stranded RNA chains (see Chapter 2). At 37°C outside of cells, the majority of bases in single-stranded DNA are part of hydrogen-bonded loops (Figure 11–11), which normally break only when the temperature is raised to above 50°C. Thus, many people found it very hard to believe that freely extended single-stranded DNA chains were a common intracellular feature. This dilemma vanished, however, with the recent discovery of the *unwinding protein*, molecules of which open up the hairpin loops by binding tightly to single-stranded DNA. The resulting extended polynucleotide chains easily form double helices when they collide with similarly extended strands of complementary sequence.

USE OF SPECIFIC ENZYMES TO PROMOTE CROSSING-OVER

Solid evidence implicating nucleases in crossing-over first came from the study of phages T4 and λ. When they multiply, much more crossing-over occurs than is observed in corresponding lengths of *E. coli* DNA. Simultaneously, several viral-specific nucleases appear. Each is coded by a specific gene on the viral chromosome (see Chapter 15). In T4 infection, both a viral-specific endonuclease and a new enzyme with DNA polymerase-like specificity have so far been discovered, while λ infection is marked by the appearance of new exonuclease and endonuclease activities. Most importantly, mutations which block the synthesis of the λ exonuclease lead to greatly reduced levels of recombination between infecting λ DNA molecules.

There is now also conclusive evidence linking nucleases to recombination in *E. coli* itself. Two closely linked genes (recB and recC), mutations in which result in much lower crossing-over rates, code for two polypeptide subunits of a powerful nuclease which attacks both single-stranded and double-stranded DNA.

STRAND EXCHANGES BETWEEN CLOSELY ALIGNED DOUBLE HELICES

A second potential mode of recombination involves direct exchange between parallel-aligned double helices, following cuts in two identically oriented strands. The free rotation of the backbone bonds permits a given strand to move from one double helix to another without loss of any potential base pairs or the creation of strained chemical bonds (Figure 10–13). Such cross-linked double helices may form following prior creation of single-stranded regions (tails) by nuclease action.

(a)

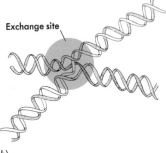

Exchange site

(b)

Figure 10–13
a) A photograph of a molecular model of sections of two double helices, showing that crossed strand connections can be made without disruption of the individual DNA molecules. [From N. Sigel and B. Alberts, *J. Mol. Biol.*, **71**, 789, (1972).] b) The mutual interchange of parallel strands between double helices (branch migration) as a consequence of right-handed axial rotation. [Redrawn from T. Broker, *J. Mol. Biol.*, **81**, 1, (1973).]

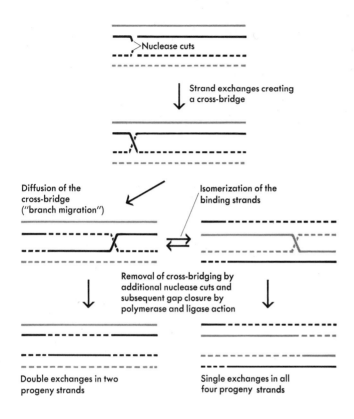

Double exchanges in two progeny strands

Single exchanges in all four progeny strands

Figure 10–14

An alternative hypothesis for crossing-over that involves parallel strand switches between two double helices. Subsequent isomerization of the bridging strands leads to an equal number of single-strand and double-strand switches.

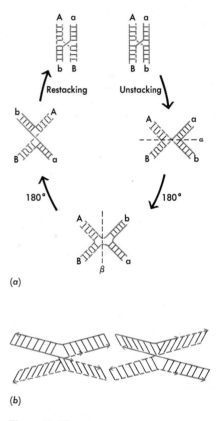

Figure 10–15

a) Steps in the isomerization of the connecting strands between cross-bridged double helices. [From Meselson and Radding, *Proc. Nat. Acad. Sci.* (1975).] b) A three-dimensional view of the equivalence of the isomeric forms. [Redrawn from Wood, Wilson, and Hood, *The Molecular Biology of Eucaryotic Cells,* W. A. Benjamin (1975).]

Once two helices are so joined, the cross connection can easily diffuse by a zipper-like action in which equivalent bases on the two original molecules exchange places. In this way, extensive sections of a given chain can move from one helix to another (Figure 10–14), frequently generating long regions of hybrid (heteroduplex) DNA in which the base sequences on the two chains may not be exactly complementary. Further rearrangements of parental strands can result when the cross-linked helices undergo a simple steric rearrangement which converts the bridging strand pair into outside strands and vice versa (Figure 10–15). Such "isomerizations" can occur very frequently, and so the further nuclease cuts which complete the crossing-over process do not necessarily occur in the strands which initiated the crossing-over (for example, in the "+" strands). Instead, they have an equal probability of rearranging the, until then, intact "−" pair. Crossing-over between paired helices can thus lead either to a double switch in two of the strands or to a single switch in all four strands (Figure 10–14).

DIRECT VISUALIZATION OF CROSSING-OVER

Direct proof that strand exchange can occur comes from electron microscope visualization of the crossing-over process. As crossing-over occurs very frequently during phage replication, the λ and T4 systems now offer the best opportunities for directly checking proposed molecular mechanisms. Our cleanest results so far use the λ system, where denaturation mapping shows that the cross bridges always link together homologous regions of the pairing partners (Figure 10–16). Equally important, the individual DNA strands can be visualized in the bridging region, showing unambiguously their passage from one double helix to another.

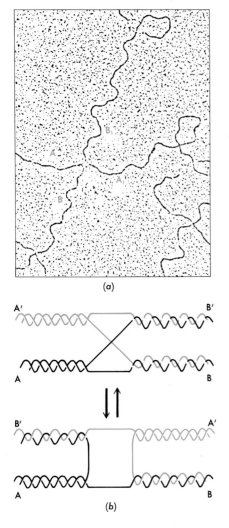

Figure 10–16
a) Electron micrograph visualization of crossing-over between two homologous regions of λ DNA, b) Schematic illustration to show how the box-shaped crossing-over diagram is generated by 180° rotation of the conventional criss-cross diagram. [Courtesy of R. Inman, University of Wisconsin.]

HETERODUPLEXES

Additional experimental support for the hypothesis that the fundamental recombination event involves pairing between regions of single-stranded DNA comes from the frequent generation of heteroduplex segments during crossing-over. These are regions on recombinant DNA molecules where the two strands are not exactly complementary. They arise when the primary pairing region, together with the adjacent region of branch migration, encompasses the site(s) of genetic differences between the two parental chromosomes. Evidence exists for heteroduplex regions in all viruses whose genetics have been extensively analyzed. On the average, one such region, several thousand nucleotides long, exists on each T4 molecule. Heteroduplexes have also been well characterized in phage λ and again are much longer than might be expected if they reflected only the primary pairing event. So most of their length probably arises by the branch migration transferring strands from one double helix to another.

Heteroduplex lifetime can be very short, for when the recombinant DNA molecules duplicate, the alternative alleles segregate out (Figure 10–17). This was first observed

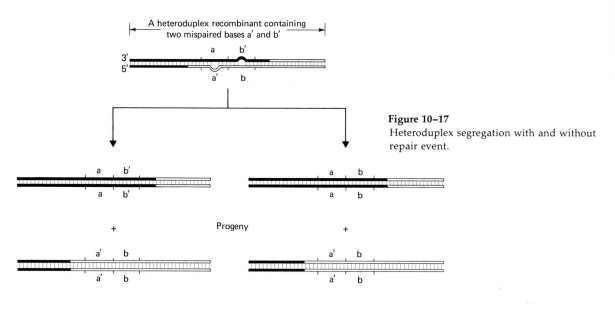

Figure 10–17
Heteroduplex segregation with and without repair event.

(a) Segregation following replication with no repair event

(b) Segregation following replication with a single repair event b′ → b

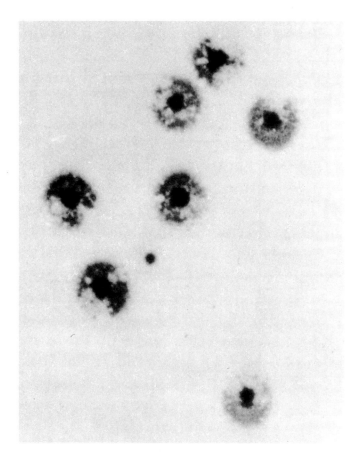

Figure 10–18
Photographs of several mottled plaques arising from segregation of a heteroduplex DNA molecule containing both rII⁺ and rII markers on opposing strands. [Kindly supplied by A. D. Hershey.]

in phage systems, where mere inspection of recombinant plaques reveals mixtures of two genetically distinct types. When the DNA of a parental T4 phage particle contains a heteroduplex rII region, the resulting plaque has a mottled appearance, with some sections characteristic of the rII phenotype and others of the wild type r$^+$ phenotype (Figure 10–18).

RECOMBINATION IS NOT ALWAYS RECIPROCAL AT THE SITE OF CROSSING-OVER

Early investigation of crossing-over between different genes revealed the seemingly obligatory occurrence of reciprocal recombinants (Figure 10–19). Exceptions to this feature, however, were discovered when the recombinants studied arose between nearby sites in the same gene. Then, nonreciprocal behavior was often observed. This phenomenon, called "gene conversion," is best studied in those organisms, like yeast or *Neurospora,* where all the products of a single meiotic event can be seen. Here instead of always observing equal (2:2) segregation of the entering alleles, cases of 3:1 segregation are found.

Figure 10–19
Reciprocal and nonreciprocal recombination following crossing-over.

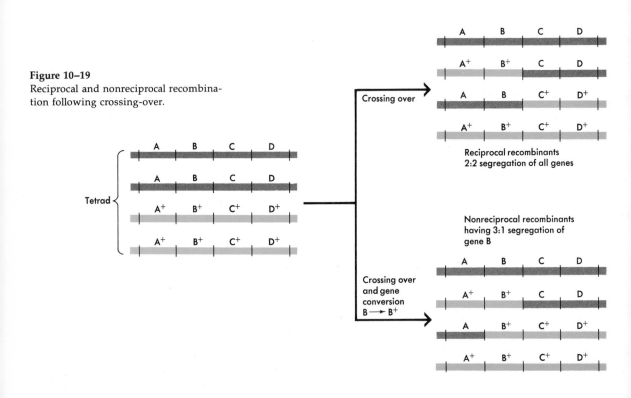

The crossing-over hypothesis outlined above permits such exceptions. This is shown in Figure 10–17 where the final segregation pattern of genes localized around the crossing-over site may be affected by repair processes which recognize heteroduplex distortions in the double helix and randomly remove one of a mismatched base pair. Depending on the bases removed, either 2:2 or 3:1 ratios will be found. We do not yet know which fraction of gene conversion is actually due to repair synthesis, for the complicated series of endonuclease and exonuclease cuts during recombination should also produce disturbed segregation patterns. Seen from a broad perspective, however, the most striking thing about recombination, even at the molecular level, is the prevalence of reciprocal recombinants.

INSERTIONS (DELETIONS) ARISING FROM ERRORS IN CROSSING-OVER

The extreme accuracy of most crossing-over events depends on the correct juxtaposition of the complementary single-stranded regions during reformation of the hydrogen bonds. This results from the uniqueness of most long nucleotide sequences. When a random chain of polynucleotides contains more than 12 nucleotides, it will virtually never have the same sequence as another fragment of similar length. Thus, as long as the single-stranded segments are relatively long, it is very, very unlikely that a fragment of one gene will mistakenly be linked up to the wrong part of its own gene or a different gene. In those rare cases, however, where two different regions have considerable homology, we should expect occasional misjoinings, leading to the insertion (deletion) of large blocks of nucleotides (Figure 10–20).

Deletions of single base pairs are the consequences of mispairing like that shown on the right in Figure 10–21. The resulting double helix has two mispaired bases which if uncorrected by any repair process will give rise to two progeny helices, each containing a different nucleotide pair deletion. In contrast, if pairing occurs as on the left in Figure 10–21, the final result will be two progeny helices, each with the same single base pair insertion.

With T4 it seems likely that most insertions (deletions) arise during crossing-over. In contrast, during normal *E. coli* replication, there is much less crossing-over, so the frequency of insertion (deletion) mutations is very much lower than with T4.

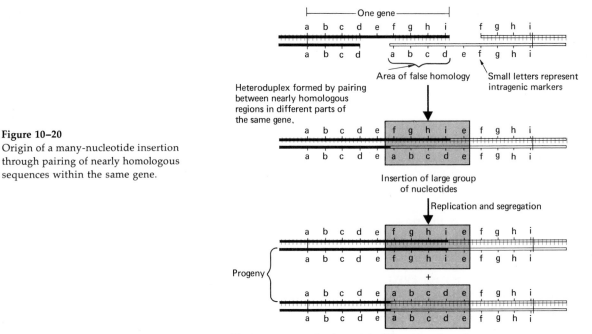

Figure 10–20
Origin of a many-nucleotide insertion through pairing of nearly homologous sequences within the same gene.

Figure 10–21
The possible origin of deletions and insertions during crossing-over by mispairing of regions containing stretches of identical bases.

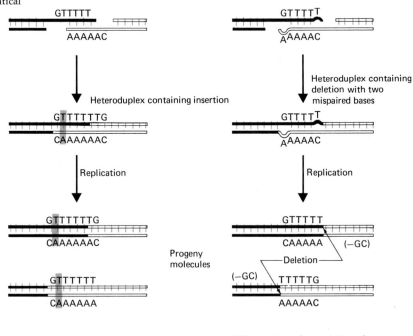

HOT SPOTS ARE OFTEN SITES OF MISMATCHING

The above model implies that the probability of an insertion (deletion) arising is dependent upon the base sequence around it. The longer the stretch of repeating bases, the higher the frequency of mutation. Evidence supporting this hypothesis came from studies on the T4-specific enzyme lysozyme, which is coded by the T4 DNA. Nucleotide sequence data, obtained from insertion and deletion experiments like those described in the next section, suggest that one highly mutable spot (hot spot) involves the deletion of an A from a stretch of 6 A residues. The reverse mutation, in which an additional A adds on to a 5A region, occurs at a 100-fold lower frequency. Thus, the frequency of the insertion (deletion) mutation most probably is a function of a very high power of the number of identical repeating bases (base doublets, etc.).

SITE-SPECIFIC RECOMBINATION

Until recently we suspected that the nucleases which mediate recombination must initially break only one strand of a given double helix. No one could imagine how such a completely severed helix could easily be rejoined. But now we suspect that certain forms of site-specific recombination are initiated by the staggered double breaks made by nucleases which cut only after binding to very specific nucleotide sequences. Following such cuts, the respective fragments can subsequently pair with new partners, thereby generating rearrangements of the original genetic material. For example, staggered cuts in two double helices, each bearing the same recognition site, open up the possibility of a crossing-over between the two helices (Figure 10–22).

Only those enzymes which recognize quite long sequences are likely to be employed in recombination since, if the recognition sequence were present at several sites, the original gene order might be completely rearranged. But if some ten base pairs were involved in recognition, then conceivably only one such site might randomly be found on a DNA molecule the length of the *E. coli* chromosme. Such an enzyme may make the cuts which insert a circular λ DNA molecule into the receptor site on the *E. coli* chromosome (Figure 10–23). In contrast, there seems no way to imagine staggered cut involvement in generalized bacterial recombination, since any cutting enzyme which recognized many different sites would quickly reshuffle the original gene arrangement.

The realization that nucleases which make specific staggered cuts can promote site-specific recombination is

273

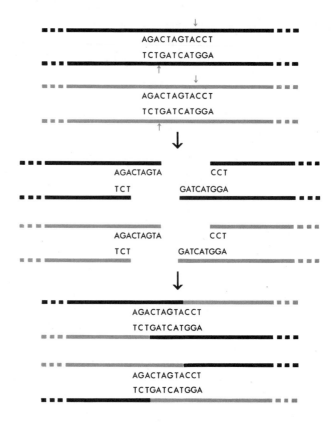

Figure 10–22 ▶

Crossing-over of sticky DNA fragments created by site-specific nuclease cuts.

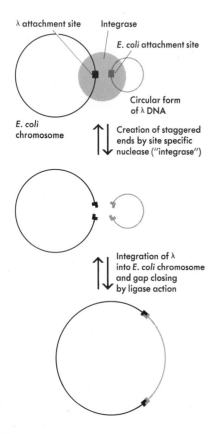

λ attachment site   Integrase

E. coli attachment site

Circular form
of λ DNA

E. coli
chromosome

Creation of staggered
ends by site specific
nuclease ("integrase")

Integration of λ
into E. coli chromosome
and gap closing
by ligase action

◀ **Figure 10–23**

A mechanism for the site-specific recombination of λ DNA into the *E. coli* chromosome.

now beginning to provide the basis for a form of *genetic engineering* in which specific chromosomal fragments can be transferred from one organism to another. For example, "sticky" fragments of eucaryotic DNA now can be easily inserted *in vitro* into circular viral (or plasmid) DNA molecules (Figure 10–24). Since modified molecules often can multiply within bacterial host cells, the functioning of eucaryotic genes within a bacterial environment now has become open for investigation. Whether such "foreign" genes will function normally in these abnormal environments is just beginning to be studied, and, hopefully, important data on the control of gene action will soon be obtained.

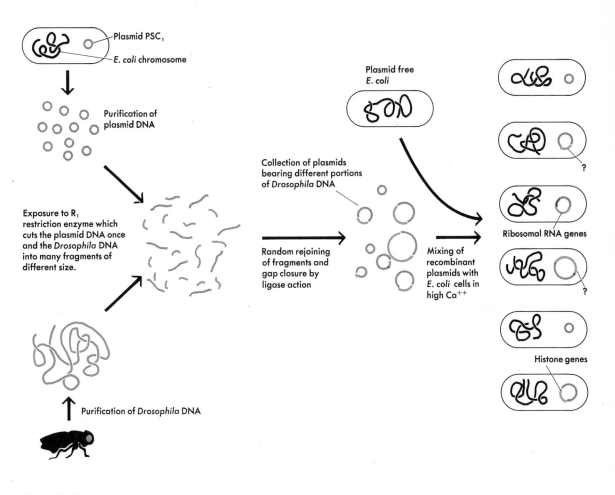

Figure 10–24
Use of specific staggered nuclease (EcoR₁) cuts to insert fragments of
Drosophila DNA into *E. coli* plasmid DNA.

THE GENETIC CODE IS READ IN GROUPS OF THREE

An obvious consequence of the fact that there are 20 amino
acids and only 4 bases is that each amino acid must be
coded for by groups of nucleotides. There cannot be a one-
to-one correspondence between the DNA bases and the
amino acids. Genetic evidence tells us that groups of three
nucleotides are fundamental units and that the code is
read linearly, starting from one end. These results arise
from crosses between mutants with deletions (or insertions) of one or more nucleotides. Deletion or insertion
mutations generally lead to completely nonfunctional
genes. In contrast, simple nucleotide switches often lead to
"leaky" genes, whose mutant proteins have partial enzymatic activity because of a single amino acid replace-

275

ment. The virtually complete absence of enzymatic activity in the deletion (insertion) mutants tells us that their protein product is completely changed.

This striking qualitative difference arises from the fact that during protein synthesis the reading of the genetic code starts from one end of the protein template and occurs in consecutive blocks of three bases. As a result, if a deletion or insertion occurs, the reading frame is completely upset (Figure 10–25). For example, if nor-

Figure 10–25

The effect of mutations that add or remove a base is to shift the reading of the genetic message.

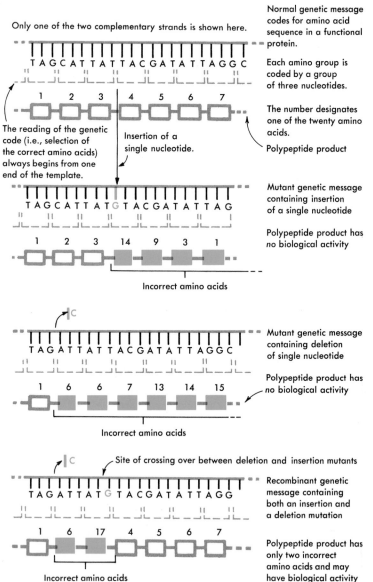

Only one of the two complementary strands is shown here.

Normal genetic message codes for amino acid sequence in a functional protein.

Each amino group is coded by a group of three nucleotides.

The number designates one of the twenty amino acids.

Polypeptide product

The reading of the genetic code (i.e., selection of the correct amino acids) always begins from one end of the template.

Insertion of a single nucleotide.

Mutant genetic message containing insertion of a single nucleotide

Polypeptide product has no biological activity

Incorrect amino acids

Mutant genetic message containing deletion of single nucleotide

Polypeptide product has no biological activity

Incorrect amino acids

Site of crossing over between deletion and insertion mutants

Recombinant genetic message containing both an insertion and a deletion mutation

Polypeptide product has only two incorrect amino acids and may have biological activity

Incorrect amino acids

mally the gene sequence ATTAGACAC . . . is read as (ATT), (AGA), (CAC), . . . , then the insertion of a new nucleotide ATTCAGACAC . . . leads to reading in the following groups: (ATT), (CAG), (ACA), (C . .). A similar consequence follows from a deletion. Crossing of two deletion (or two insertion) mutants yields double mutants in which the reading frame is still misplaced.

Partially active genes, however, can be produced by crossing-over between an insertion and a nearby deletion. Crossing-over between the deletion and insertion restores the correct reading frame except in the region between them (Figure 10–25). When the resulting protein product is normal except for several amino acid replacements, it may have some enzymatic activity. It is also sometimes possible to obtain functional genes by producing recombinants containing three closely spaced insertions (deletions). Recombinants containing four close insertions (deletions), however, produce only completely nonfunctional proteins. It is these latter two experiments that tell us that the reading group contains three nucleotides, since by combining three deletions (insertions), the reading frame is restored except for the deletion (insertion) region (Figure 10–26).

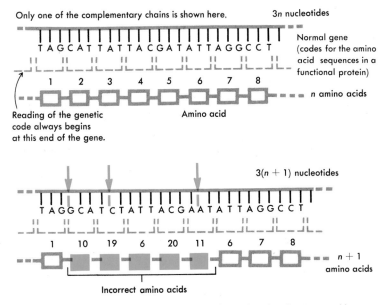

Figure 10–26
The effect of the addition of three nucleotide pairs on the reading of the genetic code. When the three nucleotides are added close together, the genetic message is scrambled only over a short region. The same type of result is achieved by the deletion of three nearby nucleotides.

Polypeptide chain contains five incorrect amino acids; its chain length is increased by one amino acid. It may have some biological activity depending upon how the five wrong amino acids influence its 3-D structure.

Thus an average gene containing 900 nucleotide pairs is subdivided into 300 reading units, each of which codes for a single amino acid. This corresponds very nicely with the average size of known proteins, somewhere around 30,000 (300 amino acids). Inasmuch as each estimate (average gene size and average protein size) is uncertain by 50 to 100 percent, this level of agreement is almost better than the experimental evidence demands.

SUMMARY

The genetic information of a DNA molecule resides in the linear sequence of the four bases. Theoretically, a very, very large number of different sequences can exist, since the number of possible permutations is 4^n, where n is the number of nucleotides along a DNA chain. All mutations involve changes in the sequence of base pairs. Many mutations are single base pair replacements at a definite location, while others involve insertions or deletions of one to many nucleotide pairs. Replacements of a single base pair by another often result from mistakes in normal hydrogen bonding during replication, while insertion or deletion events frequently are connected with mistakes in crossing-over.

Most nucleotides along a DNA molecule comprise genes coding for specific polypeptide chains. The distance between the end of one gene and the start of the adjacent one may be very short, sometimes only several nucleotides in length. Most DNA molecules contain a large number of genes, each of which usually contains between 600 and 1800 base pairs. As groups of three successive nucleotides code for an individual amino acid, most polypeptides are composed of 200 to 600 amino acids.

Crossing-over arises from breakage and rejoining of double-stranded DNA molecules. It is not directly connected with replication and must be mediated by specific cutting enzymes that expose long segments of single-stranded DNA. Formation of hydrogen bonds between these single-stranded regions of DNA may lead to the formation of recombinant molecules. Most molecular events involved in crossing-over remain unknown.

REFERENCES

BENZER, S., "The Fine Structure of the Gene." A lucid 1962 *Scientific American* article, reprinted in *The Molecular Basis of Life*, R. H. Haynes and P. C. Hanawalt (eds.), Freeman, San Francisco (1968), p. 130.

STAHL, F., *The Mechanics of Inheritance,* 2nd ed., Prentice-Hall, Englewood Cliffs, N.J., 1969. Mutation and recombination as seen by one of the more incisive phage workers. Particularly valuable are the questions accompanying each chapter.

DRAKE, J. W., *The Molecular Basis of Mutation,* Holden-Day, San Francisco, 1970. A new and extensive summary about how mutations occur.

SWANSON, C. P., T. MERZ AND W. J. YOUNG, *Cytogenetics,* Prentice-Hall, Englewood Cliffs, N.J., 1967. A brief introduction to many of the cytological facts necessary for interpreting recombination experiments in higher organisms.

MUZYCZKA, N., R. L. POLAND, AND M. J. BESSMAN, "Studies on the Biochemical Basis of Spontaneous Mutation," *J. Biol. Chem.* **247,** 7116. An elegant demonstration of proofreading during DNA replication.

WHITEHOUSE, H. K. L., *Toward an Understanding of the Mechanism of Heredity,* 2nd ed., Edward Arnold, London, 1969. The most complete treatment of the mechanism of crossing-over to be found in any text.

MESELSON, M., AND J. J. WEIGLE, "Chromosome Breakage Accompanying Genetic Reconstruction in Bacteriophage," *Proc. Natl. Acad. Sci. U.S.,* **47,** 857–868 (1961). A classic paper which told the molecular biologists that chromosomes break during crossing-over.

EMERSON, S., "Linkage and Recombination at the Chromosomal Level," in *Genetic Organization,* E. W. Caspari and A. W. Ravin (eds.), Academic, New York, 1969. The first thorough review that analyzes gene conversion in terms of DNA repair processes.

HOTCHKISS, R. D., "Models of Genetic Recombination," *Ann. Rev. Microbiol.,* **28,** 445 (1974). A most needed comparison of current molecular models for crossing-over.

SOBELL, H. M., "Symmetry in Protein–Nucleic Acid Interaction and Its Genetic Implications," *Adv. in Genetics,* **17,** 411 (1973). An attempt to tie in crossing-over with the existence of symmetrical nucleotide sequences.

SIGAL, N., AND B. ALBERTS, "Genetic Recombination: The Nature of a Crossed Stranded Exchange between Two Homologous DNA Molecules," *J. Mol. Biol.* **71,** 789 (1972). The important result that single strands can exchange between double helices without loss of any base pairs.

BROKER, T. R., AND I. R. LEHMAN, "Branched DNA Molecules: Intermediates in T4 Recombination," *J. Mol. Biol.* **60,** 131 (1971). Electron micrograph visualization of DNA in process of recombination.

CRICK, F. H. C., "The Genetic Code," *Sci. Am.,* October, 1962, pp. 66–74. A description of the original Crick–Brenner frameshift experiments with phage T4, revealing that the genetic code is read in groups of three nucleotides. Reprinted in *The Molecular Basis of Life,* R. H. Haynes and P. C. Hanawalt, (eds.), Freeman, San Francisco (1968), p. 198.

STREISINGER, G., Y. OKADA, J. EMRICH, J. NEWTON, A. TSUGITA, E. TERZAGHI, AND M. INOYE, "Frameshift Mutations and the Genetic Code," *Cold Spring Harbor Symp. Quant. Biol.* **XXXI,** 77 (1966). A most elegant proof of the frameshift hypothesis involving amino acid sequence analysis.

The Transcription of RNA Upon DNA Templates

We are now ready to approach the problem of how DNA controls the sequence of amino acids in proteins. Compared with DNA replication, we should anticipate this selection procedure to be a more chemically sophisticated task, since many of the amino acid side groups neither form specific hydrogen bonds nor have surfaces obviously complementary in shape to any nucleotide or nucleotide group. Yet somehow the correct amino acids are inserted into a given polypeptide chain with less than 1 error per 1000 amino acids. It is, then, not surprising that the solution of this problem (commonly known as the coding problem) has required much theoretical insight, produced many unanticipated results, and employed a greater variety of experimental approaches than those needed in understanding the mechanism of DNA replication.

THE CENTRAL DOGMA

We should first look at the evidence that DNA itself is not the direct template that orders amino acid sequences. Instead, the genetic information of DNA is transferred to another class of molecules which then serve as the protein templates. These intermediate templates are molecules of ribonucleic acid (RNA), large polymeric molecules chemically very similar to DNA. Their relation to DNA and protein is usually summarized by the *central dogma,* a flow scheme for genetic information first proposed some twenty years ago.

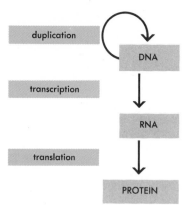

Here the arrows indicate the direction of transfer of the genetic information. The arrow encircling DNA signifies that it is the template for its self-replication; the arrow between DNA and RNA indicates that all cellular RNA molecules are made on DNA templates.[1] Correspondingly, all protein sequences are determined by RNA templates. Most importantly, the last two arrows are unidirectional, that is, RNA sequences are never copied on protein templates; likewise, RNA never acts as a template for DNA.[2]

PROTEIN SYNTHESIS IN ABSENCE OF DNA

Many experiments now exist which show that proteins can be constructed in the absence of DNA. The most obvious demonstration comes from nucleated cells where most protein synthesis occurs in the cytoplasm; almost all the DNA is found in the chromosomes within the nucleus. This observation unambiguously tells us that an intermediate template must carry genetic information to the cytoplasmic sites of synthesis. No such simple *in vivo* demonstration can exist for bacterial cells that lack a nucleus, but the use of *in vitro* cell-free systems (see below) shows that DNA's lack of direct participation is a general phenomenon.

The intermediate is clearly RNA. In the first place, there is evidence from many types of nucleated cells that all cellular RNA synthesis is restricted to the DNA-con-

[1] Although this statement holds for normal cellular RNA, it does not hold for cells infected with certain RNA viruses (See Chapter 13).
[2] In Chapter 20 we shall see that the RNA components of certain viruses serve as templates for synthesis of complementary DNA chains.

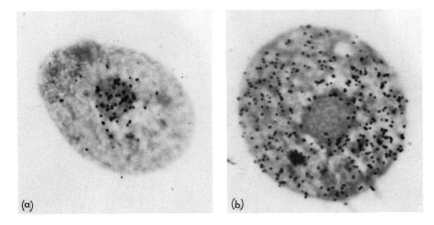

Figure 11–1
(a) Autoradiograph of a cell (*Tetrahymena*) exposed to radioactive cytidine for 15 min. Superimposed on a photograph of a thin section of the cell is a photograph of an exposed silver emulsion. Each dark spot represents the path of an electron emitted from an H³ (tritium) atom that has been incorporated into RNA. Almost all the newly made RNA is found within the nucleus. (b) An autoradiograph of a similar cell, exposed to radioactive cytidine for 12 min and then allowed to grow for 88 min in the presence of nonradioactive cytidine. Practically all the label incorporated into RNA in the first 12 min has left the nucleus and moved into the cytoplasm. [Photographs courtesy of D. M. Prescott, University of Colorado Medical School; reproduced from *Progr. Nucleic Acid Res.*, **III**, 35 (1964), with permission.]

taining nucleus (Figure 11–1).[3] No RNA strands are made in the cytoplasm, from which DNA is absent.[4] RNA is thus synthesized where it should be if it is made on DNA. After their synthesis, many of the RNA molecules move to the cytoplasm in which most protein synthesis is taking place.

In the second place, the amount of protein synthesized is directly related to the cellular content of RNA. Cells rich in RNA synthesize much protein, whereas little protein is made in RNA-poor cells. For example, the RNA-rich pancreas (Figure 11–2) synthesizes large quantities of proteolytic enzymes which are secreted into the digestive tract. Correspondingly, little RNA is found in muscle cells,

---

[3] This statement does not hold for certain virus-infected cells (see Chapter 15).

[4] This statement must now be qualified to take into account the facts that both the mitochondria and chloroplasts contain small amounts of DNA which serve as templates for some of the RNA molecules involved in the synthesis of specific mitochondria and chloroplast proteins.

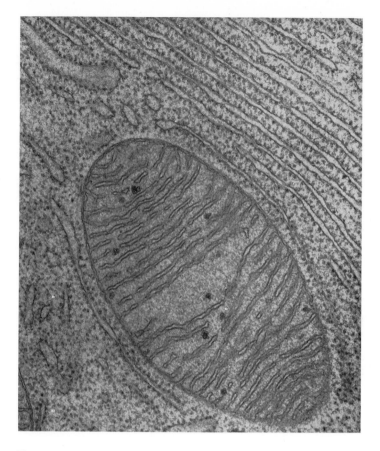

Figure 11–2
Electron micrograph (×105,000) of a portion of a cell in the pancreas of a bat, showing a mitochondrion and large numbers of ribosomes. Some ribosomes exist free; others (especially in the upper right) are attached to a membranous component, the endoplasmic reticulum. [Courtesy of K. R. Porter.]

in which little protein is made. The significance of this correlation is strengthened by isotopic experiments designed to locate accurately the synthetic sites within the cytoplasm. In these experiments cells are very briefly exposed to amino acids labeled with radioactive isotopes. During these short intervals ("pulses"), some radioactive amino acids become incorporated into proteins. The cells are then quickly broken open to see in which cellular fraction the newly made protein (identified by its possession of radioactive amino acids) is found. In all cells the results are the same: Newly synthesized polypeptide chains are

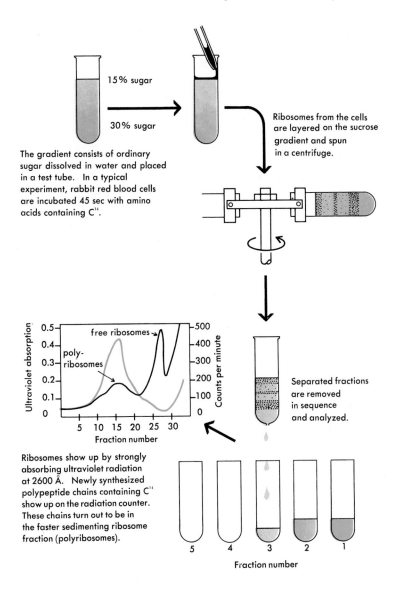

The gradient consists of ordinary sugar dissolved in water and placed in a test tube. In a typical experiment, rabbit red blood cells are incubated 45 sec with amino acids containing C¹⁴.

Ribosomes from the cells are layered on the sucrose gradient and spun in a centrifuge.

Separated fractions are removed in sequence and analyzed.

Ribosomes show up by strongly absorbing ultraviolet radiation at 2600 Å. Newly synthesized polypeptide chains containing C¹⁴ show up on the radiation counter. These chains turn out to be in the faster sedimenting ribosome fraction (polyribosomes).

Figure 11–3
Sucrose gradient demonstration that protein synthesis occurs on ribosomes. [Redrawn from A. Rich, *Sci. Am.,* **209**, 46–47 (December 1963).]

found associated with spherical RNA-containing particles, the ribosomes (Figure 11–3). Likewise, a small amount of protein can be made *in vitro* in carefully prepared extracts of cells, and here also use of radioactive isotopes shows attachment of much of this new protein to the RNA-rich ribosomes.

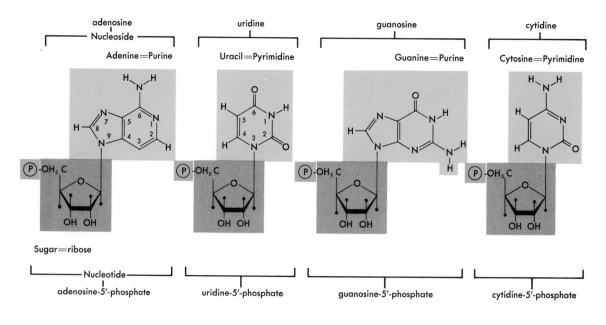

Figure 11–4
The nucleotide building blocks of RNA.

RNA IS CHEMICALLY VERY SIMILAR TO DNA

Mere inspection of RNA structure shows how it could be exactly synthesized on a DNA template. Chemically it is very similar to DNA. It is also a long, unbranched molecule containing four types of nucleotides (Figure 11–4) linked together by 3′–5′ phosphodiester bonds (Figure 11–5). Two differences in its chemical groups distinguish RNA from DNA. The first is a minor modification of the sugar component (Figure 11–6). The sugar of DNA is deoxyribose, whereas RNA contains ribose, identical to deoxyribose except for the presence of an additional OH (hydroxyl) group. The second difference is that RNA contains no thymine but instead contains the closely related pyrimidine uracil. Despite these differences, however, polyribonucleotides have the potential for forming complementary helices of the DNA type. Neither the additional hydroxyl group nor the absence of the methyl group found in thymine affects RNA's ability to form double-helical structures held together by hydrogen-bonded base pairs.

RNA IS USUALLY SINGLE-STRANDED

RNA molecules do not usually have complementary base ratios (Table 11–1). The amount of adenine does not often equal the amount of uracil, and the amounts of guanine and cytosine also usually differ from each other. This tells

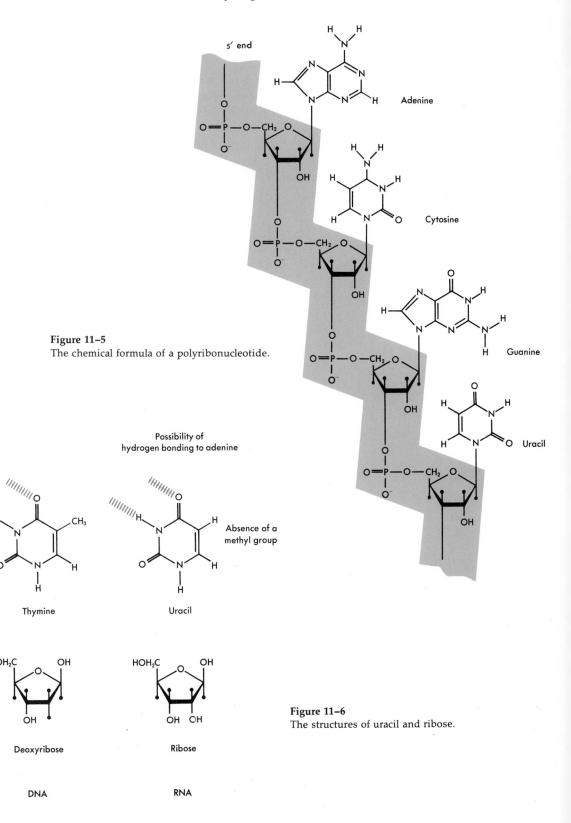

Figure 11–5
The chemical formula of a polyribonucleotide.

5′ end

Adenine

Cytosine

Guanine

Uracil

Possibility of
hydrogen bonding to adenine

Absence of a
methyl group

Thymine

Uracil

Figure 11–6
The structures of uracil and ribose.

HOH₂C

Deoxyribose

Ribose

DNA

RNA

287

us that most RNA does not possess a regular hydrogen-bonded structure but, unlike double-helical DNA, exists as single polyribonucleotide strands. Because of the absence of regular hydrogen bonding, these single-stranded molecules do not have a simple regular structure like DNA. This structural uncertainty initially caused much pessimism, for there was general belief that we should have to see the template before we could attack the problem of how it selected amino acids during protein synthesis. Fortunately, as we show later, this hunch was wrong.

Table 11–1 The Base Composition of RNA from Various Sources

RNA Source	Proportion of the Four Main Bases			
	Adenine	Uracil	Guanine	Cytosine
E. coli	24	22	32	22
Proteus vulgaris (a bacterium)	26	19	31	24
Euglena (an alga)	22	21	30	27
Turnip yellow mosaic virus	23	22	17	38
Poliomyelitis virus	30	25	25	20
Rat kidney	19	20	30	31

ENZYMATIC SYNTHESIS OF RNA UPON DNA TEMPLATES

The fact that RNA, like DNA, is a long, unbranched chain using four different nucleotides immediately suggests that the genetic information of DNA chains is transferred to a complementary sequence of RNA nucleotides. According to this hypothesis, the DNA strands at one or more stages in the cell cycle separate and function as templates onto which complementary ribonucleotides are attracted by DNA-like base pairing [adenine with thymine (uracil) and guanine with cytosine]. It further tells us that some control mechanism must dictate whether the separated DNA strands will function as templates for a complementary DNA strand or a complementary RNA strand.

Direct evidence for the hypothesis comes from the discovery of the appropriate enzyme, RNA polymerase, which exists in virtually all cells. This enzyme links together ribonucleotides by catalyzing the formation of the internucleotide 3′–5′ phosphodiester bonds that hold the RNA backbone together (Figure 11–7). It does so, however, only in the presence of DNA, a fact that suggests that DNA must line up the correct nucleotide precursors in order for

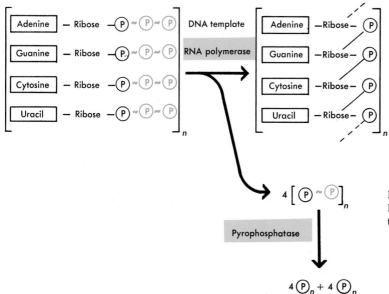

Figure 11–7
Enzymatic synthesis of RNA upon a DNA template.

RNA polymerase to work. Support for this idea comes from seeing how the RNA base composition varies with the addition of DNA molecules of different AT/GC ratios. In every enzymatic synthesis, the RNA AU/GC ratio is roughly similar to the DNA AT/GC ratio (Table 11–2).

Table 11–2 Comparison of the Base Composition of Enzymatically Synthesized RNA's with the Base Composition of Their Double-Helical DNA Templates

Source of DNA Template	Composition of the RNA Bases				$\frac{A+U}{G+C}$ Ob-served	$\frac{A+T}{G+C}$ in DNA
	Ade-nine	Uracil	Gua-nine	Cyto-sine		
T2	0.31	0.34	0.18	0.17	1.86	1.84
Calf thymus	0.31	0.29	0.19	0.21	1.50	1.35
E. coli	0.24	0.24	0.26	0.26	0.92	0.97
Micrococcus lyso-deikticus (a bacterium)	0.17	0.16	0.33	0.34	0.49	0.39

Further experiments directly demonstrate that the template is a single DNA strand. One of the clearest experiments uses DNA from the virus ϕX174. This virus belongs to one of the very special viral classes that contain single-stranded DNA instead of the customary double-helical form. Only one of the two possible complementary DNA strands is present, and when it is used as the tem-

Table 11-3 Base Composition of Enzymatically Synthesized RNA Using Single-Stranded φX174 DNA as Template

	φX174 DNA	Observed Values of RNA Product	Predicted RNA Composition
Adenine	0.25	0.32	0.33
Uracil	0.33 (thymine)	0.25	0.25
Guanine	0.24	0.20	0.18
Cytosine	0.18	0.23	0.24
Total	1.00	1.00	1.00

plate for RNA polymerase, the enzymatic product has a complementary base sequence (Table 11–3). Moreover, in this special case the RNA product remains attached to its DNA template, allowing the isolation of a hybrid DNA–RNA double helix. This result contrasts with experiments using double-helical DNA. Here the RNA product quickly detaches from its template and the two DNA strands again come together in specific register. Apparently the double helix made from two complementary DNA strands is energetically more stable than the hybrid DNA–RNA structure, so that the free single DNA strand quickly displaces the RNA product after RNA polymerase has moved over the corresponding template region (Figure 11–8).

Figure 11–8

Transcription of an RNA molecule upon a unique strand of its DNA template. The attachment of the enzyme RNA polymerase to a DNA molecule opens up a short section of the double helix, thereby allowing free bases on one of the DNA strands to base pair with the ribonucleoside —Ⓟ~Ⓟ~Ⓟ precursors. As RNA polymerase moves along the DNA template, the growing RNA strand peels off, allowing hydrogen bonds to reform between two complementary DNA strands. Thus almost immediately after the synthesis of an RNA strand commences, its front end becomes available to bind to a ribosome (see Chapter 12).

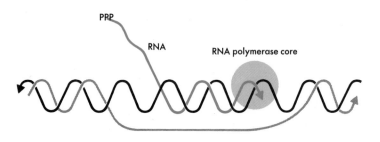

We thus see that the fundamental mechanism for the synthesis of RNA is very similar to that of DNA. In both cases, the immediate precursors are nucleoside triphosphates that use the energy in one of their pyrophosphate bonds to drive the reaction toward synthesis. Also in both cases, a single enzyme works on all four possible nucleotides, whose correct selection is dictated by the obligatory need to base pair with a polynucleotide template.

RNA synthesis is thus necessarily quite an accurate process. No evidence, however, exists for any proofreading, and so the overall precision of transcription does not approach the almost perfect copying characteristic of

DNA self-replication. But, since cellular RNA is not self-replicating, the rare mistakes which do occur are not genetically perpetuated.

ALONG EACH GENE ONLY ONE DNA STRAND ACTS AS AN RNA TEMPLATE

If each of the two DNA strands of a given gene served as an RNA template, each gene would produce two RNA products with complementary sequences which should code for two different proteins. Since our genetic evidence tells us that each gene controls only one protein, we must assume that either only one of the two possible RNA strands is made or, if both are synthesized, that only one is functional for some special reason. It appears that the former possibility is correct–*in vivo* only one of the two possible gene products is found. This can clearly be seen by examining the RNA synthesized *in vivo* under the direction of the virus SP8, which multiplies in the bacterium *Bacillus subtilis*. The two complementary DNA strands of the SP8, unlike those of most viruses, have quite different base compositions and can be relatively easily separated. It is thus possible to ask whether the RNA products have base sequences complementary to one or both the DNA strands.

To answer this question, use was made of our ability to form artificial DNA–RNA hybrid molecules by mixing RNA molecules with single-stranded DNA molecules formed by heating of double-helical DNA. Heating DNA molecules to temperatures just below 100°C breaks the hydrogen bonds holding the complementary strands together; they then quickly separate from each other (DNA denaturation). If the temperature is gradually lowered, the complementary strands again form the correct hydrogen bonds and the double-helical form is regained (renaturation of DNA). If, however, this gentle cooling is done in the presence of single-stranded RNA that has been synthesized on homologous DNA, then DNA–RNA hybrids form as well as renatured DNA double helices (Figure 11–9). These DNA–RNA hybrids are very specific and form only if some of the nucleotide sequences in the DNA are complementary to the RNA nucleotide sequences. This technique lets us ask whether the *in vivo* RNA products will form hybrids with only one or with both of the complementary SP8 DNA strands. A clear result emerges: Only one DNA strand is copied.

The copying of only one DNA strand within a given gene allows us to understand why the base ratios of RNA

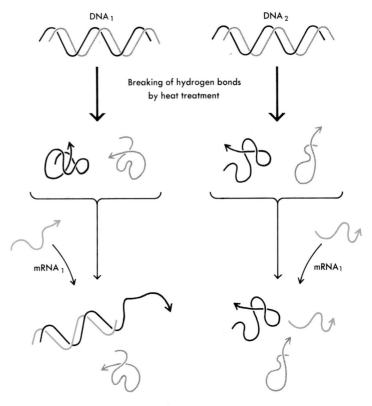

Figure 11–9
The use of DNA–RNA hybrids to show the complementarity in nucleotide sequences between an RNA molecule and one of the two strands of its DNA template. The left side of the diagram shows the formation of a hybrid molecule between an RNA molecule and one of the two strands of the template. The specificity of this method is shown on the right side of the diagram. Here the same RNA molecule is mixed with unrelated DNA. No hybrid molecules are formed.

need not be complementary even though RNA is made on a DNA template. In a given DNA strand, there is no reason why the A should equal T or G equal C. These ratios need be one-to-one only when the corresponding bases on the two complementary chains are added together. Thus, in general, we must assume that only in the exceptional DNA molecule will the single strands be found on close inspection to have the complementary ratios. Correspondingly, only rarely will single-stranded RNA molecules be found in which A equals U and G equals C.

Differential copying of the two strands of a DNA molecule can also be observed *in vitro*. This result depends on the use of DNA templates which have not been severely damaged. If, for example, T7 DNA has been denatured to produce single strands or if it has been broken to produce single-stranded regions, then both strands are copied. When, however, intact molecules of T7 DNA $(MW = 2.6 \times 10^7)$ are used as *in vitro* templates, the DNA strand which is copied is the same one copied during *in vivo* viral reproduction (Figure 11–10).

The transcription patterns along other chromosomes are often not as clear-cut as with T7 or SP8. Both strands of

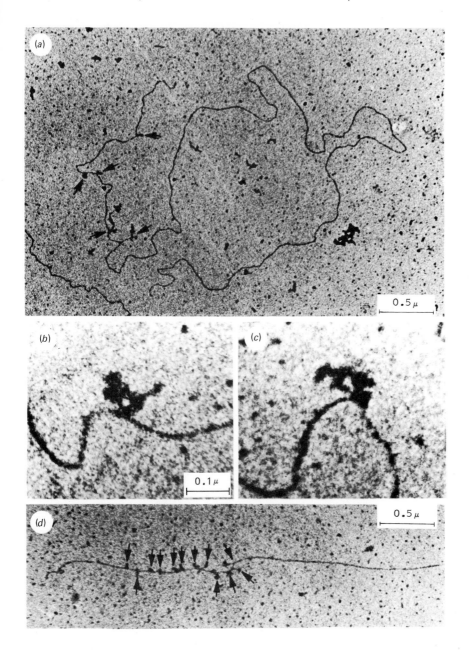

Figure 11–10
Electron micrographs of *in vitro* transcription of T7 DNA. In (a) several
growing RNA chains can be seen attached to the DNA template. All of
the RNA chains are transcribed off the same T7 strand, initiating their
growth at a unique site near the end visible at the right side. Total synthesis time is 5 min at 37°C. In (b) and (c) views of two attached RNA
chains are shown at higher magnification. In (d) synthesis was terminated after 2½ min at 37°C.—12 nascent chains all at one end can be seen.
[Kindly supplied by Dr. R. Davis, California Institute of Technology.]

the T4 and λ chromosomes are transcribed, one strand serving as the template for some genes (usually contiguous groups) while the remaining genes are transcribed along the other strand (Figure 15–24). The same picture holds for the *E. coli* chromosome, which can be transcribed both clockwise and counterclockwise. Many more genes (e.g., lactose, tryptophan, and galactose groups) are read counterclockwise than clockwise. Whether this has any real significance remains to be ascertained. Since the replicating bacterial chromosome is constantly rotating in a fixed direction, unwinding the nascent RNA chains might appear to be connected with the rotation of their templates. But this does not seem to be true. An example is known where the orientation of lactose genes within the *E. coli* chromosome is reversed so that they must be read clockwise instead of counterclockwise. As far as we can tell, their transcription rates are the same and so are independent of gene orientation.

RNA CHAINS ARE NOT CIRCULAR

All cellular RNA chains examined so far have a linear shape. No example of circular RNA has been found despite quite extensive searches.[5] The sizes of these linear molecules show much greater variation than those observed for

Figure 11–11
Schematic folding of an RNA chain showing several double-helical regions held together by hydrogen bonds.

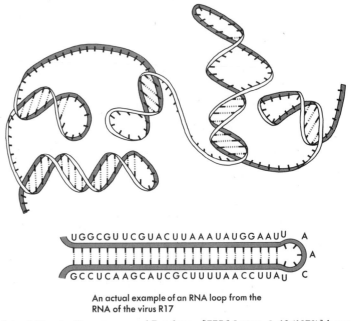

UGGCGUUCGUACUUAAAUAUGGAAUU A
|||||||||||||||||||||||||| A
GCCUCAAGCAUCGCUUUUAACCUUAU C

An actual example of an RNA loop from the
RNA of the virus R17

[5] Agol, Drygin, Romanova, and Bogdanov [*FEBS Letters*, **8**, 13 (1970)] have reported finding a circular replicating form of the RNA of the mouse encephalomyocarditis (EMC) virus. All other known forms of viral RNA are linear.

DNA chains. Some are built up from as few as 75 nucleotides, while others may contain over 10,000. Independent of their length, they generally do not have a completely random shape. A majority of the bases are hydrogen bonded to each other using DNA-type base pairing (A with U, G with C). This is accomplished by hairpin folds (Figure 11–11) that bring bases of the same chain into a DNA-like double-helical arrangement. Such configurations are possible when the number of adenine residues along a given chain section is approximately equal to the number of uracil residues, or when the guanine number is approximately equal to the cytosine number.

SYNTHESIS OF RNA CHAINS OCCURS IN A FIXED DIRECTION

Each RNA chain, like each DNA chain, has a direction defined by the orientation of the sugar–phosphate backbone. The chain end terminated by the 5′ carbon atom is called the 5′ end, while the end containing the 3′ carbon atom is called the 3′ end (Figure 11–5). Now it seems certain that all RNA chains grow in the 3′ to 5′ direction or vice versa. If they grow 5′ to 3′, then we expect the beginning nucleotide to possess a ⓟ~ⓟ~ⓟ group (Figure

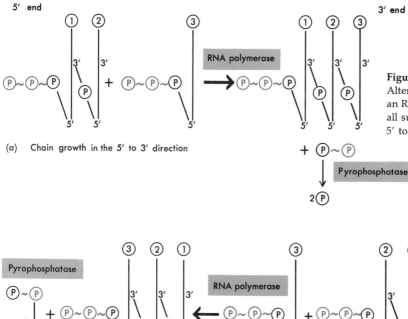

Figure 11–12
Alternative directions for the synthesis of an RNA chain. Many types of experiments all suggest that chains always grow in the 5′ to 3′ direction.

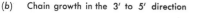

(b)   Chain growth in the 3′ to 5′ direction

11–12). But if the chains grow 3' to 5', then the nucleotide at the growing end would contain the $\textcircled{P}\sim\textcircled{P}\sim\textcircled{P}$ group. Firm evidence now shows that the direction of growth is 5' to 3'. Newly inserted nucleotides are found at the 3' ends, while the $\textcircled{P}\sim\textcircled{P}\sim\textcircled{P}$ groups are found attached to the nucleotides which commenced chain growth.

Work with the metabolic inhibitor 3'-deoxyadenosine confirms the 5' to 3' direction of growth. When added to cells, it is first phosphorylated to 3'-deoxyadenosine-$\textcircled{P}\sim\textcircled{P}\sim\textcircled{P}$ and then joined to the 3' growing end. Because it contains no 3'-OH group, further nucleoside-$\textcircled{P}\sim\textcircled{P}\sim\textcircled{P}$ cannot attach, and RNA synthesis stops.

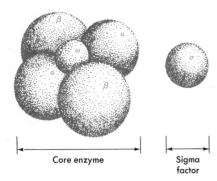
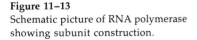

Figure 11–13
Schematic picture of RNA polymerase showing subunit construction.

CONSTRUCTION OF RNA POLYMERASE FROM SUBUNITS

In contrast to DNA polymerase I, which is built up from only a single polypeptide chain, RNA polymerase has a very complicated subunit structure. The active form, called the *holoenzyme*, sediments at about 15S, and contains 5 different polypeptide chains, β', β, σ, α, and ω, with respective molecular weights of 160,000, 150,000, 90,000, 40,000, and 10,000. No covalent bonds run between the various chains; aggregation results from the formation of secondary bonds. Each specific chain, with the exception of α, which appears twice, is present once within the active molecule, giving the complete holoenzyme ($\beta'\beta\alpha_2\omega\sigma$) a molecular weight slightly over a half million (Figure 11–13). The attachment of σ to the other chains is not very firm, so it is relatively easy to isolate a $\beta'\beta\alpha_2\omega$ aggregate. This specific grouping is known as the *core enzyme*, for it catalyzes the formation of the internucleotide phosphodiester bonds equally well in the presence or absence of σ.

RECOGNITION OF START SIGNALS

Thus, σ by itself has no catalytic function; its role is the recognition of start signals along the DNA molecule. *In vitro* experiments show that, in the absence of σ, the core enzyme will occasionally initiate RNA synthesis, but this is the result of starting mistakes. RNA chains made by the core enzyme alone are started randomly along both strands of a given gene. When σ is present, however, the correct strand is selected, indicating that the holoenzyme has the capacity to specifically bind to a "starting" sequence. Whether σ itself directly recognizes start signals or acts indirectly by modifying the configuration of the core en-

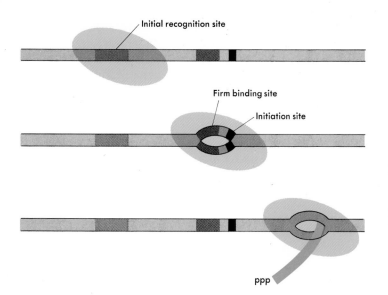

Figure 11–14
Schematic illustration of the initiation of RNA synthesis following binding of RNA polymerase holoenzyme to a start signal (promoter) for RNA synthesis.

zyme remains to be worked out. In either case its presence leads to a very tight binding of the holoenzyme to those DNA regions (*promoters*) which contain start signals. Most likely the holoenzyme initially recognizes paired sequences within intact double helices. Soon after binding, however, a very localized unwinding occurs, allowing the synthesis of a nascent RNA chain to commence on the appropriate single-stranded partner.

Until two years ago the nature of promoter base sequences was totally obscure. Now, however, a number of different promoter sequences are known and soon many more will be worked out. Already they reveal that the bases which recognize RNA polymerase are not themselves transcribed into RNA (Figure 11–14). Soon after this initial binding to the *RNA polymerase recognition site,* the polymerase diffuses closer to the start of transcription to a second group of AT-rich base pairs to which it binds more tightly (*RNA polymerase binding site*). Much variation is found in both the length and sequence of the initial recognition site. In contrast, the more firm polymerase binding site usually encompasses a seven base pair sequence all of which are derivatives of the sequence

5′   TATPuATG
3′   ATAPyTAC

Conceivably the individual variations (Figure 11-15) reflect differences in promoter efficiency, allowing different DNA functional units to be transcribed to lesser or greater degrees. The fact that both binding sequences are AT-rich

297

probably reflects the fact that such sequences denature much more easily than do GC-rich regions, and so are obvious targets for an enzyme like RNA polymerase, whose functioning demands local denaturation.

RNA Polymerase binding site sequences

fd	T G C T T C T G A C	T A T A A T A	G A C A G G G T A A A G A C C T G A T T T T T G A
T7 A3	A A G T A A A C A C G G	T A C G A T G	T A C C A C A T G A A A C G A C A G T G A G T C A
T7 A2	A G T A A C A T G C A G	T A A G A T A	C A A A T C G C T A G G T A A C A C T A G C A G
Lac-UV-5	G C T T C C G G C T C G	T A T A A T G	T G T G G A A T T G T G A G C G G A T A A C A A
lambda P_R	A C C T C T G G C G G T	G A T A A T G	G T T G C A T G T A C T A A G G A G G T T G
SV40	T T T A T T G C A G C T	T A T A A T G	G T T A C A A A T A A A G C A A T A G C A T C
Lambda P_L	A C C A C T G G C G G T	G A T A C T G	A G C A C A T C A G C A G G A C G C A C T G A C
E coli Tyr tRNA	C G T C A T T T G A	T A T G A T G	C G C C C C G C T T C C C G A T A A G G G A G C A G
Lac Wildtype	G C T T C C G G C T C G	T A T G T T G	T G T G G A A T T G T G A G C G G A T A A C A A

Figure 11–15
RNA polymerase binding site sequences for different viral and bacterial DNA's. The initially transcribed base is shown in color.

INITIATION OF CHAINS WITH pppA OR pppG

Actual chain initiation does not start in the more stable binding region but some 6 to 7 bases down the double helix. So the active site which carries out the actual polymerization may be located some 20 Å distant from the site involved in the more firm binding. Most likely this distance is not invariant since there are hints that, at least *in vitro*, some ambiguity exists in the exact base at which transcription starts. In all cases, however, the first base must be a purine, with most *E. coli* chains starting with pppG. However, when ϕX174 or T7 DNA is the template, more chains start with pppA.

Some apparent exceptions to obligatory A or G starts have a simple explanation. After synthesis, many RNA chains are cleaved by endonucleases to yield smaller fragments, many of which start with a pyrimidine. For example, although some tRNA chains start with U (see Chapter 12), they never contain terminal triphosphate groups, strongly suggesting their origin through *in vivo* cleavage of longer chains. As virtually all cells contain a variety of nucleases, unambiguous data about initiating nucleotide sequences is likely to come from *in vitro* experiments where, using highly purified enzymes, it is possible to exclude nuclease effects.

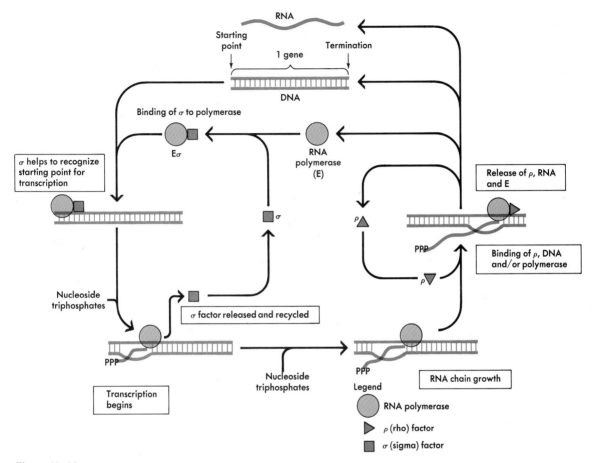

Figure 11–16
RNA synthesis—outline diagram. Here is illustrated a case where chain termination requires the ρ factor. In many other situations, RNA polymerase by itself can read the terminating signal.

DISSOCIATION OF σ FOLLOWING MAKING OF THE INITIAL INTERNUCLEOTIDE LINKAGES

After chain elongation commences, σ dissociates from the core enzyme–DNA–nascent RNA complex and becomes free to attach to another core molecule (Figure 11–16). Conceivably it is released merely because it no longer has any role to play. Alternatively, its continued presence might result in such tight binding to DNA that subsequent enzyme movement along the template would be effectively impossible. Chain elongation may best occur when the nascent chain–enzyme complex is held to the DNA template by the much weaker and sequence-unspecific interactions which occur between the core enzyme and DNA.

STOP SIGNALS PRODUCE CHAINS OF FINITE LENGTH

There also exist stop signals whose function is to terminate RNA synthesis at specific points along the DNA template. There appear to be two types. One, which seems to be read by RNA polymerase itself, produces transcripts with the terminal sequence UUUUUU(A). It thus appears that an extended Poly A sequence is a part of such terminator sequences. The other type of stop signal works only in the presence of the ρ factor, a protein of molecular weight 50,000. It functions only in a multimeric state, suggesting that it recognizes symmetrical DNA sequences like those recognized by the restriction enzymes. It is not, however, a nuclease, and so in some way may directly block the elongation process. Unfortunately, experiments which have tried to test whether ρ directly binds to DNA and/or to RNA polymerase have given ambiguous results. Further progress is likely to demand direct sequence data on ρ-mediated stops.

SUMMARY

DNA molecules are not the direct templates for protein synthesis. The genetic information of DNA is first transferred to RNA molecules. In turn, RNA molecules act as the primary templates that order amino acid sequences in proteins:

$$DNA \rightarrow RNA \rightarrow protein.$$

The covalent structure of RNA is chemically very similar to that of DNA, and its genetic information is also stored in the sequences of its four bases. In contrast to DNA, however, most RNA molecules are single-stranded and linear. The synthesis of RNA upon DNA templates shows many similarities to the DNA duplication process. Most importantly, the selection process involves formation of complementary base pairs. The enzyme RNA polymerase links together the monomeric precursors, which are the ribonucleoside triphosphates (ATP, GTP, CTP, and UTP). In a given gene, only one of the two DNA strands is copied.

RNA polymerase is constructed from five types of polypeptide chains. One of these chains, σ, easily dissociates from the other four, which collectively make up the core enzyme. The core catalyzes the formation of the phosphodiester linkages, while σ helps in the recognition of start signals along DNA molecules. After initiation occurs, σ dissociates from the DNA–RNA core complex, becoming free to attach to another core.

Synthesis of all chains starts with either adenine or guanine, depending on the specific start signal in the DNA template. Elongation proceeds in the 5' to 3' direction, so the 5' terminal nucleotides always contain a ℗~℗~℗ group. Synthesis ends upon reaching stop signals, which, like start signals, are specific nucleotide sequences. Termination sometimes depends on the presence of a specific protein, the ρ factor. The molecular mode of ρ action has yet to be solved.

REFERENCES

PRESCOTT, D. M., "Cellular Sites of RNA Synthesis," in J. N. Davidson and W. E. Cohn (eds.), *Progr. Nucleic Acid Res.*, **III,** 35–37 (1964). A very clear discussion of the techniques of cell biology which have established the nucleus as the site of most RNA synthesis.

SPIEGELMAN, S. S., "Hybrid Nucleic Acids," *Sci. Am.*, **210,** 48–56 (1964). Here are clearly explained many of the experimental techniques that have been so elegantly used to demonstrate homology in nucleotide sequences between DNA and RNA.

MARMUR, J., C. GREENSPAN, F. PALACEK, F. M. KAHAN, J. LEVINE, AND M. MANDEL, "Specificity of the Complementary RNA Formed by *B. subtilis* Infected with Bacteriophage SP8," *Cold Spring Harbor Symp. Quant. Biol.*, **XXVIII,** 191–199 (1963). A description of an experiment showing that, *in vivo*, only one of the two SP8 DNA strands is a template for RNA synthesis.

TAYLOR, K., Z. HRADECNA, AND W. SZYBALSKI, "Asymmetric Distribution of the Transcribing Regions on the Complementary Strands of Coliphage λ DNA," *Proc. Nat. Acad. Sci. U.S.*, **57,** 1618 (1967). Use of the DNA–RNA hybridization technique to establish the transcription pattern of phage λ.

BURGESS, R., A. A. TRAVERS, J. J. DUNN, AND E. K. F. BAUTZ, "Factor Stimulating Transcription by RNA Polymerase," *Nature*, **222,** 537 (1969). The classic announcement of the separation of RNA polymerase into the core and σ components.

"Transcription of Genetic Material," *Cold Spring Harbor Symp. Quant. Biol.*, **XXXV** (1970). Here are found over 90 articles revealing the state of the transcription problem as of June 1970. The most extensive collection of facts now available about the synthesis of RNA.

Control of Transcription, edited by Biswas, B. B., R. K. Mandal, A. Stevens, W. E. Cohn, Plenum, (1974). Specialized reviews arising out of a 1973 meeting on RNA synthesis.

CHAMBERLIN, M. J., "The Selectivity of Transcription," *Ann. Rev. Biochem.*, **43,** 721 (1974). A superb review emphasizing how RNA polymerase works.

PRIBNOW, D., "Nucleotide Sequence of a RNA Polymerase Binding Site at an Early T7 Promoter," *Proc. Nat. Acad. Sci.*, **72:**in press (1975). A very clear description of the working out of a promoter sequence.

Involvement of RNA in Protein Synthesis

We now begin to look at how single-stranded RNA mole-cules function during protein synthesis. When the general validity of the central dogma DNA → RNA → protein was becoming obvious in the mid-1950's, there was general belief that all RNA was template RNA. Hope also existed that, when the RNA general structure was solved, mere inspection might tell us how RNA ordered amino acid sequences. Now, however, we realize that these views were very naive and that protein synthesis is a much more complicated affair than the synthesis of nucleic acid. More-over, not all RNA molecules are templates. In addition to a template class, there exist two additional classes of RNA, each of which plays a vital role in protein synthesis.

CHAPTER **12**

AMINO ACIDS HAVE NO SPECIFIC AFFINITY FOR RNA

The fundamental reason behind this complexity has been mentioned before. There is no specific affinity between the side groups of many amino acids and the purine and pyrimidine bases found in RNA. For example, the hydro-carbon side groups of the amino acids alanine, valine, leucine, and isoleucine do not form hydrogen bonds and would be actively repelled by the amino and keto groups of the various nucleotide bases. Likewise, it is hard to imagine the existence of specific RNA surfaces with unique affinities for the aromatic amino acids phenylalanine, tyrosine, and tryptophan. It is thus impossible for these amino acids in unmodified form to line up passively in specific accurate order against an RNA template prior to peptide bond formation.

AMINO ACIDS ATTACH TO RNA TEMPLATES BY MEANS OF ADAPTORS

Before the amino acids line up against the RNA template, they are chemically modified to possess a specific surface capable of combining with a specific number of the hydrogen-bonding groups along the template. This chemical change consists of the addition of a specific adaptor molecule to each amino acid through a single covalent bond. It is this adaptor component that combines with the template; at no time does the amino acid side group itself need to interact with the template. Adding a specific adaptor residue to an amino acid is much more economical than chemically modifying the side group itself. The latter process might require many enzymes for just a single amino acid. Conceivably, a similar number would be required to change the modified side group back to its original configuration after it becomes part of a polypeptide chain. On the other hand, only a single enzyme is needed either to attach or to detach an amino acid from its specific adaptor.

SPECIFIC ENZYMES RECOGNIZE SPECIFIC AMINO ACIDS

There need not be any obvious relation between the shape of the amino acid side group and the adaptor surface. Instead, the crucial selection of an amino acid is done by a specific enzyme. The enzyme that catalyzes the attachment of the amino acid to its adaptor must be able to bind specifically to both the amino acid side group and the adaptor. Proteins are extremely suitable for this task because their active regions can be rich in either hydrophilic or hydrophobic groups. There is no difficulty in folding a suitably sequenced polypeptide chain to produce a cavity that is specific for the side group of one specific amino acid. For example, tyrosine can be distinguished from phenylalanine by an enzyme having a specific cavity containing an atom that can form a hydrogen bond to the OH group on tyrosine. Here the formation of one specific hydrogen bond yields about 4 to 5 kcal/mole of energy.

This would be lost if phenylalanine were chosen instead. Thus, with the help of a physical chemical theory, we can predict that the probability that tyrosine is found in the "tyrosine cavity" is about 1000 times greater than the probability that phenylalanine is found in the tyrosine cavity.

There is more difficulty in immediately seeing how a similar accuracy can be achieved in distinguishing between amino acids differing only by one methyl residue, a group incapable of either salt linkages or hydrogen bonding. For example, glycine must be distinguished from alanine and valine from isoleucine. There is, of course, no difficulty in understanding why the larger alanine side group cannot fit into the cavity designed for the smaller amino acid glycine. The problem arises when we ask why glycine will not sometimes fit into the alanine cavity or valine into the isoleucine hole. If this should happen, there would be loss of the van der Waals forces arising out of a snug fit around a methyl group. These forces are now thought to be about 2 to 3 kcal/mole, by themselves too small in value to account for the general accuracy with which amino acids are ordered during protein synthesis. Now we suspect that the maximum frequency at which a wrong amino acid is inserted into a growing polypeptide chain is between 1 in 1000 and 1 in 10,000. This means that the energy gained by selecting the correct amino acid must be at least 4 to 6 kcal, a value about twice that provided by the apparent van der Waals energy. So clearly the discrimination process which separates such amino acids is more complicated than initially conceived.

Resolution of this long-troublesome paradox most likely comes from the recent proposal that all such selection processes occur in multiple steps so that several opportunities exist to reject each wrongly inserted amino acid. Combination of two steps each of which utilizes 2 to 3 kcal of binding energy would produce the 4 to 6 kcal necessary to explain the 10^{-3} error rate. Later, when we discuss the chemical steps involved in attachment of amino acids to their adaptors, we will see that many supposedly one-step discriminations in fact involve several successive steps.

THE ADAPTOR MOLECULES ARE THEMSELVES RNA MOLECULES

The molecules to which the amino acids attach are a group of relatively small RNA molecules called transfer RNA (tRNA). It is really not surprising that the adaptors are also RNA molecules, since a prime requirement for a useful adaptor is the ability to attach specifically to the free keto and amino groups on the single-stranded template RNA molecules. This attachment is ideally accomplished by having the adaptor also be a single-stranded RNA molecule, since this opens the possibility of having very specific hydrogen bonds (perhaps of the base-pair variety) to hold the template and adaptor together temporarily. The tRNA adaptors for the twenty different amino acids all have different structures, each uniquely adapted for fitting a different nucleotide sequence on the template. Thus, a large number of different types of tRNA exist.

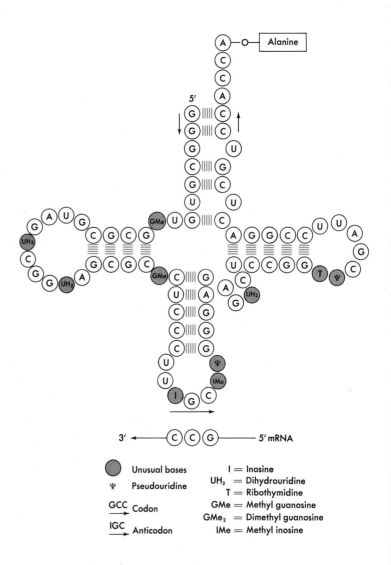

Figure 12–1
The complete nucleotide sequence of alanine tRNA showing the unusual bases and codon/anticodon position.

Each of the tRNA's contains approximately 80 nucleotides (MW ~ 25,000) linked together in a single covalently bonded chain (Figure 12–1). One end of the chain (3' end) always terminates in a CCA sequence (cytidylic acid, cytidylic acid, adenylic acid). The terminal nucleotide of the other end (5') is usually guanylic acid. Though there is only one chain, a majority of the bases are hydrogen bonded to each other. Hairpin folds often bring bases on the same chain into a DNA-like double-helical arrangement. Hydrogen bond formation can then occur because the number of adenine residues approximates the number of uracil residues and the guanine number almost equals the cytosine number. The correspondence is not exact, however, and some nucleotides that do not internally base pair are available to fit to the template.

At first, it was thought that the three-dimensional shape of different tRNA's might be radically different. The finding that a mixture of all tRNA's can form very regular three-dimensional crystals, however, tells us that the basic three-dimensional arrangement is the same for all molecules.

YEAST ALANINE tRNA CONTAINS 77 NUCLEOTIDES

During December, 1964, the first nucleotide sequence of a specific tRNA molecule (from yeast) was reported. This tRNA specifically attaches to alanine and, therefore, is called *alanine tRNA*. It contains 77 nucleotides arranged in a unique sequence (Figure 12–1). The most striking aspect of this sequence is the high content (10/77) of unusual bases (Figure 12–2). ("Unusual" means a base other than A, G, C, or U.) Many of these unusual bases differ from normal bases by the presence of one or more methyl ($—CH_3$) groups. Most, if not all, of the methyl groups are enzymatically added after the nucleotides are linked together by 3'–5' phosphodiester linkages. Very probably, the other unusual bases also arise by the enzymatic modification of a preexisting polynucleotide.

The function of the unusual bases is not yet clear. They are not limited to alanine tRNA but occur in varying proportions in all tRNA molecules. Our only hint of their role is the fact that several unusual bases cannot form conventional base pairs. Some of the unusual bases may thus have the function of disrupting double-helical hairpin regions, thereby exposing free keto and amino groups which can then form secondary bonds. Depending upon the specific bases, the free groups may form secondary bonds to template RNA, to a ribosome, or to the enzyme needed to attach a specific amino acid to its specific tRNA molecule.

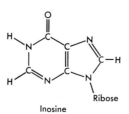

Inosine

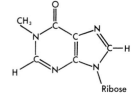

1-methylinosine (Me I)

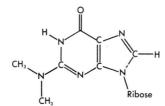

N²-dimethylguanosine (DiMeG)

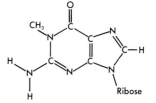

1-methylguanosine (MeG)

Figure 12–2
The structures of the rare nucleotides found in yeast alanine tRNA.

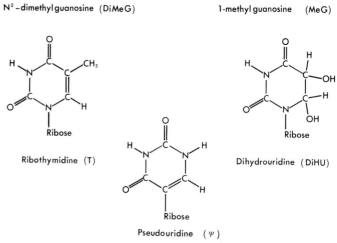

Ribothymidine (T)

Pseudouridine (ψ)

Dihydrouridine (DiHU)

CLOVERLEAF FOLDING OF tRNA MOLECULES

The exact nucleotide sequence of alanine tRNA does not provide sufficient information to guess unambiguously the way various bases hydrogen bond to each other. Given only this structure, a number of possible hairpin configurations could be imagined. But after several other yeast tRNA sequences had been determined, it was realized that certain sequences were common to all types of tRNA molecules. There was only one way (the cloverleaf) to fold the chain in two dimensions which maximized the number of base pairs and led to a common shape (Figure 12–1). Each cloverleaf contains the following nonhydrogen-bonded sections:

1. The 3' end consists of ACC plus a variable fourth nucleotide. The amino acid always becomes attached to the 3' terminal A.
2. As one moves along the backbone away from the 3' end, one encounters the first loop of the cloverleaf. It is made up of seven unpaired bases and always contains the sequence 5'-TψCG-3'. Binding to the ribosomal surface may involve this region.
3. Next appears a loop of highly variable size, often called "the lump."
4. The third loop, which consists of seven unpaired bases, contains the anticodon—the three adjacent bases which bind (by base pairing) to the successive bases comprising the codon. The anticodon is bracketed on its 3' end by a purine and on its 5' end by U. The purine is often alkylated.
5. The fourth loop contains eight to twelve unpaired bases and is relatively rich in dihydro-U (D loop). It is thought to be involved in the binding to the specific activating enzyme (amino-acyl synthetase).

Over the past few years, tRNA sequences from a variety of organisms in addition to yeast have been worked out. All show the same general features favoring the formation of the cloverleaf. The fact that some yeast tRNA molecules can be activated (see below) by enzymes from quite unrelated organisms is thus not surprising.

CRYSTALLINE tRNA

Discovery of the cloverleaf pattern does not by itself tell us how the various loops are arranged in space. This information can only come from x-ray diffraction studies of tRNA crystals. Fortunately, the gloomy impression of former years that tRNA might be impossible to crystallize is gone. In fact, very good crystals of several tRNA species can now be routinely obtained (Figure 12–3). Most detailed progress so far has come from the study of yeast phenylalanine tRNA (tRNA[phe]). Its 3-D (tertiary) structure has recently (1974) been elucidated to 3Å resolution, allowing the locating of virtually all its major groups. Most pleasantly, all the double-helical stems predicted by the cloverleaf

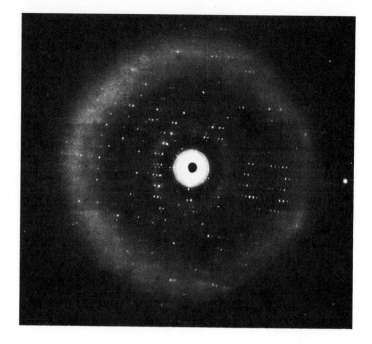

Figure 12–3
X-ray diffraction pattern of a mixed tRNA crystal. This crystal was formed by the co-crystallization of an unfractionated tRNA preparation from yeast. The ability of the unfractionated material to crystallize so regularly (the reflections extend out to about 10 Å, and in other photographs, order to the 3-Å level has been observed) indicates that the shapes of all tRNA species are very similar. [From Blake, Fresco, and Langridge, *Nature*, **225**, 32 (1970). Photograph kindly supplied by Dr. Robert Langridge, Princeton University.]

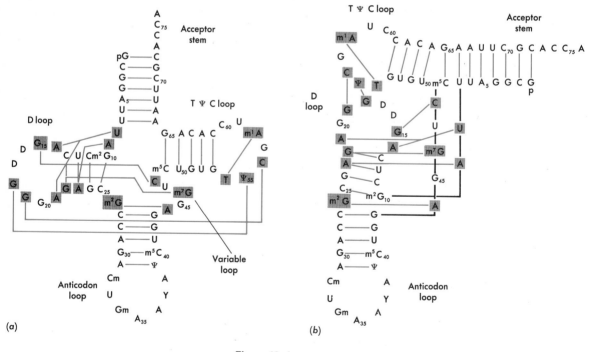

Figure 12–4
The tertiary interactions in yeast phenylalanine tRNA are indicated by solid lines connecting bases that are hydrogen bonded. (a) Conventional cloverleaf configuration. (b) Sequence rearranged to illustrate the close interaction between the TψC and D loops. [Reproduced from Kim, *et al.*, *Science* **185**, 435 (1974).]

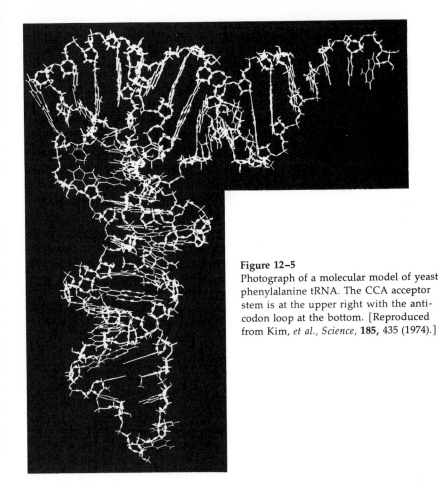

Figure 12–5
Photograph of a molecular model of yeast phenylalanine tRNA. The CCA acceptor stem is at the upper right with the anti-codon loop at the bottom. [Reproduced from Kim, *et al., Science,* **185,** 435 (1974).]

model for tRNA[phe] exist (Figure 12–4). In addition, there exist a series of additional hydrogen bonds which bend the cloverleaf into a stable *tertiary* structure that has a rough L-shaped appearance (Figures 12–5 and 12–6(a) and (b)). Examination of this structure reveals:

1. The amino-acid acceptor CCA group is located at one end of the "L" some 70Å from the anticodon which occurs at the opposite end.
2. The dihydrouracil-rich (D) and TψC loops form the corner of the "L."
3. Many tertiary hydrogen bonding interactions involve bases which are invariant in all known tRNA's (Figure 12–4(a)), strongly supporting the belief that all tRNA's have basically the same tertiary configuration.
4. Most of the hydrogen bonds which form the tertiary structure involve base pairs (triplets) dif-

Figure 12–6 (a)

A schematic diagram illustrating the folding of the yeast tRNA^{phe} molecule. The ribose-phosphate backbone is drawn as a continuous cylinder, and internal hydrogen bonding is indicated by crossbars which have gold segments in them. Positions of single bases are indicated by rods which are intentionally shortened. The variable regions, in terms of the number of nucleotides in different tRNA molecules, are shown in dotted outline; and two of the variable regions in the D loop, α and β, are labeled. The anticodon is at the bottom of the figure, while the amino acid acceptor is at the upper left. The numbering of nucleotides is the same as in Figure 12–4. The D loop and stem are shaded in gray, whereas the TψC loop is unshaded. [Redrawn from Kim *et al., Proc. Nat. Acad. Sci. USA* **71**, 4970 (1974).]

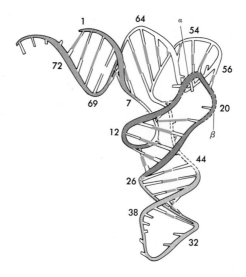

(a)

Figure 12–6(b)

A schematic diagram illustrating the hydrophobic stacking interactions between the nucleotides of yeast tRNA^{phe}. Full stacking and partial stacking are indicated. In this orientation the anticodon is at the bottom, the CCA acceptor stem is at the upper right, and the TψC loop is shown at the upper left corner. Where adjacent stacking nucleotides are connected by a ribose-phosphate chain, the linkage is noted by a thin line. The heavy solid lines attached to the nucleotide symbols represent in schematic fashion the purine or pyrimidine bases, while the hydrophobic interaction is indicated by the blocks between the bases. The connectivity of the molecule is indicated only by the numbering scheme. This accounts for the fact that many nucleotides appear to be unconnected. Bases not involved in stacking (bases 16, 17, 20, 47) are omitted from the figure. [Redrawn from Kim, *et al., Proc. Nat. Acad. Sci.,* **71**, 4970 (1974).]

(b)

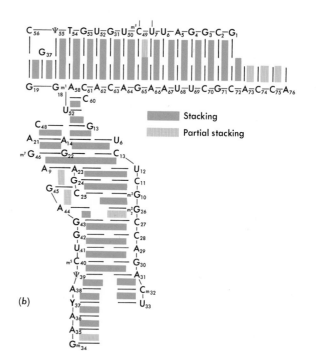

ferent from the conventional A···T and G···C pairs used in double-helical DNA (Figure 12–7).

5. Almost all the bases are so oriented that they can stack next to each other, thereby maximizing interaction between their hydrophobic flat faces (Figure 12–6). Even the apparently unstabilized anticodon region is held relatively rigid by stacking interactions. Most likely, stacking is as major a factor in stabilizing the tRNA configuration as are the tertiary hydrogen bonds.

6. Only several tertiary hydrogen bonds hold the anticodon stem to the remainder of the molecule, raising the possibility that the relative orientation of the anticodon region may change during protein synthesis.

ADDITION OF THE ADAPTOR ALSO ACTIVATES THE AMINO ACID

The link between the 3' terminal adenosine and the amino acid is a covalent bond between the amino acid carboxyl group and the terminal ribose component of the RNA (Figure 12–8). The use of the amino acid carboxyl group to attach to the adaptor has several interesting implications. In the first place, before the carboxyl group can form a peptide bond, the adaptor must be released. Thus, peptide bond formation and adaptor removal occur in a coordinated fashion. In the second place, the bond linking the tRNA to its specific amino acid is a high-energy bond, making these complexes "activated" precursors. The energy in the amino acid–tRNA bond (an amino–acyl bond) can be used in the formation of the lower-energy peptide bond.

The energy required for forming the amino–acyl bond comes from a high-energy pyrophosphate linkage ($\circledP \sim \circledP$) in ATP. Prior to formation of the AA~tRNA compounds, the amino acids are activated by enzymes (amino–acyl synthetases) to form amino acid adenylates (AA~AMP) in which the amino acid carboxyl group is attached by high-energy bonding to an adenylic acid (AMP) group (Figure 12–8).

$$\text{AA} + \text{ATP} \underset{\substack{\text{amino acyl} \\ \text{synthetase}}}{\rightleftharpoons} \text{AA} \sim \text{AMP} + \circledP \sim \circledP \qquad (12\text{–}1)$$

The AA ~ AMP intermediate normally remains tightly bound to the activating enzyme until collision with a tRNA molecule specific for the amino acid. Then the same activating enzyme transfers the amino acid to the terminal

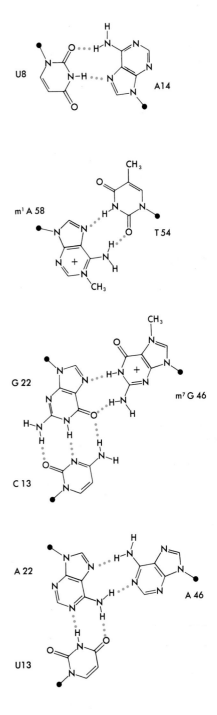

Figure 12–7
Diagrams showing various types of hydrogen bonding that are seen in yeast tRNA^{phe} and postulated for other tRNA's. The filled black circles represent the 1'-carbon of the ribose residues.

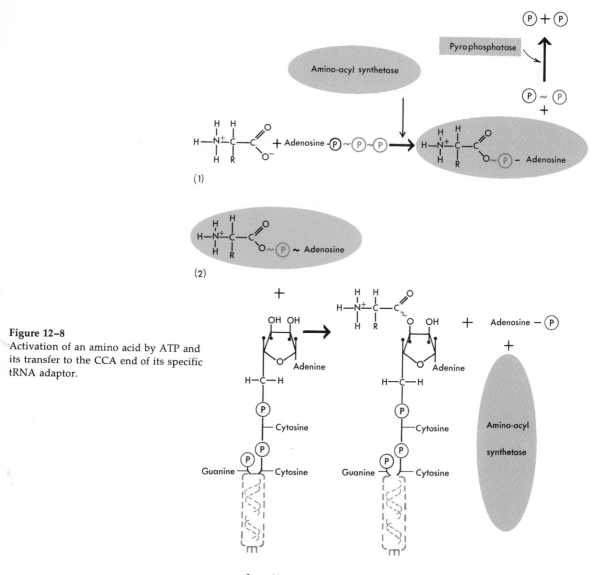

Figure 12–8
Activation of an amino acid by ATP and its transfer to the CCA end of its specific tRNA adaptor.

adenylic acid residue of the tRNA.

$$\text{AA} \sim \text{AMP} + \text{tRNA} \underset{}{\overset{\substack{\text{amino acyl} \\ \text{synthetase}}}{\rightleftharpoons}} \text{AA} \sim \text{tRNA} + \text{AMP} \qquad (12\text{–}2)$$

We therefore see that the activating enzymes are able to recognize (bind to) specifically both a given amino acid and its tRNA adaptor. For this purpose, the enzymes must have two different combining sites: one that recognizes the side group of an amino acid, and another that recognizes the tRNA specific for that amino acid. Similarly, each tRNA molecule must have two specific recognition sites: one for its activating enzyme, the other for a specific group of template nucleotides. It follows that the amino acid side group itself never has to come into contact with a template

314

molecule. It needs only to bind specifically to the correct activating enzyme.

Every cell needs at least twenty different kinds of activating enzymes and at least twenty kinds of tRNA molecules. There must be at least one of each for every amino acid. It was first thought that only twenty different tRNA molecules were used, but now it is clear that several cases exist in which at least two different types of tRNA molecules are specific for the same amino acid. This is connected with the fact that the genetic code often uses more than one nucleotide sequence (codon) for a given amino acid (degeneracy; see Chapter 13). Frequently, there is a unique tRNA molecule for each functional codon. There need not be, however, a separate activating enzyme for each of the several tRNA's corresponding to a given amino acid. Since the tRNA binds to the templates and to the activating enzymes at two distinct sites, tRNA molecules that differ in their template-binding nucleotides can possess identical nucleotide sequences in the region that combines with the enzyme. Thus, tRNA molecules that bind to different codons can bind to the same activating enzyme.

AA~tRNA FORMATION IS VERY ACCURATE

How given amino–acyl synthetases accurately select between very similar amino acids at first seemed very mysterious. So the discovery of how isoleucyl synthetase precisely discriminates isoleucine from valine was particularly pleasing. These amino acids differ by only single methyl groups, and so the difference in their binding energy to isoleucyl synthetase is only 2 to 3 kcal, an amount of energy at first sight insufficient to prevent frequent mistakes. For example, when isoleucyl synthetase is presented with a mixture of equimolar amounts of isoleucine and valine, approximately one valine~AMP activation will occur for every 100 correctly activated isoleucine~AMP complexes. If all these activated valine molecules were later to be transferred to isoleucyl tRNA molecules, an unacceptably high level of mistakes would result. Such transfers, however, are very infrequent, for when isoleucine tRNA binds to a valine~AMP–isoleucyl synthetase complex, almost all valine~AMP molecules fall apart into their valine and AMP components (Figure 12–9). This form of "proofreading" demands continued binding of the amino acid to the active site of the synthetase through both the enzymatic steps catalyzed by the synthetase. Discrimination between isoleucine and valine can thus occur twice, giving rise to an acceptable $10^{-2} \times 10^{-2} = 10^{-4}$ final error level. Now we would guess that the selection

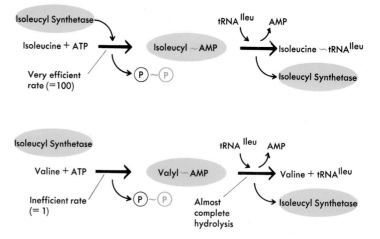

Figure 12–9
Discrimination between isoleucine and valine in the formation of isoleucine-tRNA.

between the very similar amino acids glycine and alanine occurs in the same way.

PEPTIDE BOND FORMATION OCCURS ON RIBOSOMES

Once the amino acids have acquired their adaptors, they diffuse to the ribosomes, the spherical particles on which protein synthesis occurs. Protein synthesis never takes place free in solution, but only on the surfaces of the ribosomes, which might be regarded as miniature factories for making protein. Their chief function is to orient the incoming AA~tRNA precursors and the template RNA so that the genetic code can be read accurately. Ribosomes thus contain specific surfaces that bind the template RNA, the AA~tRNA precursors, and the growing polypeptide chain in suitable stereochemical positions. There are approximately 15,000 ribosomes in a rapidly growing *E. coli* cell. Each ribosome has a molecular weight of slightly less than 3 million. Together the ribosomes account for about one-fourth of the total cellular mass, and hence a very sizable fraction of the total cellular synthesis is devoted to the task of making proteins. Only one polypeptide chain can be formed at a time on a single ribosome. Under optimal conditions, the production of a chain of MW = 40,000 requires about ten seconds. The finished polypeptide chain is then released and the free ribosome can be used immediately to make another protein.

All ribosomes are constructed from two subunits, the larger subunit approximately twice the size of the smaller one (Figure 12–10). Both subunits contain both RNA and

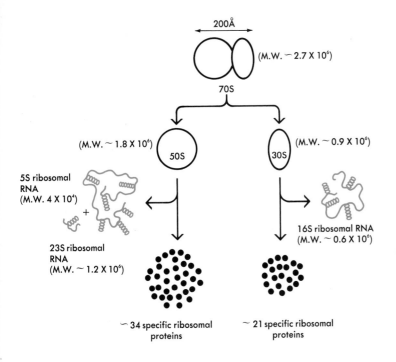

Figure 12–10

The structure of the *E. coli* ribosome. It is usually called the 70-S ribosome since 70S (Svedbergs) is a measure (the sedimentation constant) of how fast this ribosome sediments in the centrifuge. Likewise, the designations 30S and 50S are the sedimentation constants of the smaller and larger ribosomal subunits; 16S and 23S are the sedimentation constants of the smaller and larger ribosomal RNA molecules. All bacterial ribosomes have sizes similar to those of *E. coli,* possessing 30-S and 50-S subunits. In higher organisms (including yeast, etc.), ribosomes are somewhat larger (80S), with 40-S and 60-S subunits. For convenience below, we shall use 30S to designate all smaller particles and 50S for larger particles.

protein. In *E. coli* ribosomes the RNA/protein ratio is about 2:1; in many other organisms, it is about 1:1. Most of the protein serves a structural role (as opposed to an enzymatic role). Both the large and small subunits contain a large number of different proteins whose chief function is to help bring the template RNA and the AA~tRNA precursors together correctly. So far, most intensive work has been done with the ribosomal proteins from *E. coli.* All twenty-one proteins from the smaller subunit (30S) have been characterized (Figure 12–11) and show a variety of sizes with most present in only one copy per ribosome. Characterization of the 34 proteins in the larger 50-S subunit is less complete. It appears, however, that like 30-S proteins, most are present only once in a given ribosome.

317

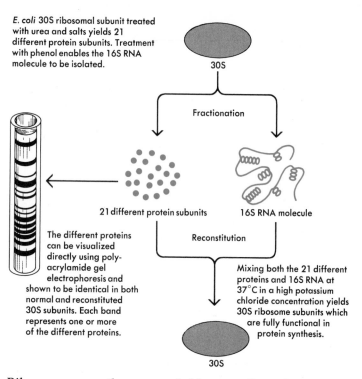

E. coli 30S ribosomal subunit treated with urea and salts yields 21 different protein subunits. Treatment with phenol enables the 16S RNA molecule to be isolated.

30S

Fractionation

21 different protein subunits

16S RNA molecule

The different proteins can be visualized directly using poly-acrylamide gel electrophoresis and shown to be identical in both normal and reconstituted 30S subunits. Each band represents one or more of the different proteins.

Reconstitution

Mixing both the 21 different proteins and 16S RNA at 37°C in a high potassium chloride concentration yields 30S ribosome subunits which are fully functional in protein synthesis.

30S

Figure 12–11
Reconstitution of the 30S ribosome from a mixture of 21 specific ribosomal proteins and a 16S rRNA molecule.

Ribosomes are thus remarkably complicated structures whose complete chemical characterization may lie decades in the future.

RIBOSOME RECONSTITUTION

The complete reassembly of the smaller *E. coli* subunit from its RNA and protein constituents was first accomplished in 1968. Such reconstituted particles are active in protein synthesis, showing identical behavior to the normal 30S subunits. Reconstitution of functional 50S *E. coli* subunits has been much harder to achieve, and solid success has been obtained only over the last year. Finding the right conditions for reconstitution is a very important achievement, and opens up the possibility of preparing particles lacking one or more specific proteins and seeing the effects of their absence on specific ribosomal functions (e.g., binding of tRNA, peptide bond formation, etc.).

RIBOSOME-ASSOCIATED RNA DOES NOT USUALLY CARRY GENETIC INFORMATION

When ribosomes became implicated in protein synthesis (1953), it seemed natural at first to suppose that their tightly bound RNA component was the template that ordered the amino acids. In fact, it was initially supposed that all cellular RNA was located in the ribosomes and that the lighter, slowly sedimenting, soluble fraction (~ 20 percent of total RNA) was a degradation product of the ribosomal RNA templates. The identification (1956) of the soluble RNA fraction as the adaptor molecules corrected this faulty guess, but it did not remove the belief that all the remaining 80 percent of cellular RNA functioned as templates. In 1960, however, RNA isolated from purified ribosomes (rRNA or ribosomal RNA) was unambiguously shown not to have a template role.

TEMPLATE RNA (mRNA) REVERSIBLY ASSOCIATES WITH RIBOSOMES

The active templates are instead an RNA fraction comprising only one to several percent of total RNA. This RNA reversibly binds to the surface of the smaller ribosome subunit and, in media of low Mg^{2+} ion concentration, can be removed without affecting ribosome integrity. Because it carries the genetic message from the gene to the ribosomal factories, it is called messenger RNA (mRNA). By moving across the ribosomal site of protein synthesis, it brings successive codons into position to select the appropriate AA~tRNA precursors. The existence of mRNA was first unambiguously established in experiments with T2-infected *E. coli* cells. After T2 DNA enters a host cell, it must turn out RNA templates for the many viral-specific proteins needed for viral reproduction (see Chapter 15 for more details). Many biochemists were, therefore, surprised by the 1959 finding that no new rRNA chains and hence no new ribosomes were synthesized following T2 infection. This result could only mean that T2-specific proteins are not synthesized on rRNA templates, and further hinted that perhaps in normal cells rRNA chains were not the templates for protein synthesis. Searches begun when the T2 result became known quickly revealed the presence of mRNA, first in viral infected cells, and soon afterward in normal *E. coli* cells.

rRNA EXISTS IN TWO MAJOR SIZE CLASSES

Two major size classes of rRNA are found in all bacterial ribosomes. They are an integral component and, unlike mRNA, cannot be removed without the complete collapse of the ribosome structure. The smaller 16-S rRNA molecule, found in the smaller ribosome subunit, has a molecular weight of about one-half million, whereas the larger 23-S molecule, a component of the larger ribosome subunit, has a molecular weight of about one million. Each larger subunit contains in addition one very short rRNA molecule which sediments at 5S. All are single-stranded and have unequal amounts of guanine and cytosine and of adenine and uracil. Nonetheless there is enough equivalence of base pairs so that many rRNA bases on the same chain are hydrogen bonded into the hairpin-type turns found in tRNA. The joint presence of single-stranded and double-helical regions gives individual rRNA molecules irregular 3-D shapes. As a result, it has not yet been possible to obtain rRNA preparations in which the molecules are regularly arranged in space. Thus, x-ray diffraction pictures of rRNA are much more distorted than the corresponding tRNA diagrams and cannot tell us precise details of rRNA structure. Similar dilemmas probably exist for elucidation of mRNA structure by x-ray analysis.

Sequence analysis thus is the only way now available to explore detailed rRNA structure. But because of their very long chain lengths (16S = 1520 nucleotides; 23S = ~3000 nucleotides), such analysis poses a much greater challenge than elucidation of a tRNA sequence. Nonetheless, a tentative sequence now exists for almost all the *E. coli* 16-S rRNA chain; and over the next few years, the 23-S rRNA sequence should be largely worked out. Inspection of such sequences reveals large numbers of potential hairpin sections, but we as yet have no clue as to how the rRNA molecule folds up in three dimensions.

THE FUNCTION OF MOST rRNA IS NOT YET KNOWN

Until recently there did not exist even a semisatisfactory hypothesis for why ribosomes contain rRNA as well as protein. But now we believe that many of the unpaired bases in rRNA are involved in the binding of tRNA and mRNA to ribosomes. As we shall see later, several unpaired bases near the 3' ends of 16-S rRNA may form temporary base pairs with the initiation sites (ribosome binding sites) of mRNA molecules. So we now guess that further searches will find still other rRNA sequences complementary to certain invariant tRNA sequences. But it is still hard to imagine why each major rRNA component has hundreds of nucleotides available for specific binding. Perhaps most such interactions are, in fact, between rRNA and the various ribosomal proteins themselves and help to hold the latter in a well defined 3-D shape. But even if this is true, we still are in doubt why RNA instead of protein itself is used to provide the structural skeleton.

ALL THREE FORMS OF RNA
ARE MADE ON DNA TEMPLATES

Since both tRNA and rRNA chains have very special roles in protein synthesis, the idea was proposed (several years ago) that perhaps these RNA forms were not made on DNA templates but were instead self-replicating, like the RNA in the single-stranded RNA viruses (see Chapter 15). This idea has been shown to be wrong by DNA–RNA hybridization experiments. In these experiments, rRNA (or tRNA) chains are mixed at a high temperature (near 100°C) with DNA isolated from the same organism. At this temperature, the double-helical regions of rRNA (tRNA) fall apart. Likewise, all hydrogen bonds in DNA break and complementary strands separate. When the temperature is then slowly dropped, stable double helices form again. In addition to reformation of many complementary DNA double helices, some single DNA strands specifically combine with rRNA (tRNA) chains to form hybrid DNA–RNA helices. These DNA–RNA complexes are very specific, for they do not form if DNA from an unrelated species is used. Hence, tRNA and rRNA are synthesized exactly as mRNA is, using DNA molecules as templates. Most interestingly, even though rRNA and tRNA together comprise over 98 percent of all RNA, less than 1 percent of the total DNA functions as their templates.

PRE-rRNA AND PRE-tRNA

Neither 16S nor 23S rRNA is transcribed as such. Instead they are derived from a larger precursor molecule which nucleases subsequently break down into mature rRNA molecules. The 16S and 23S rRNA chains arise from a single 30S pre-rRNA transcript of MW~2.1×10^6. Transcription of the 16S section begins first, with the 23S component being completed last. Breakdown of the 30S pre-rRNA results first in 17S and 25S intermediates which later are degraded to 16S and 23S chains, respectively. Most likely each pre-rRNA molecule also contains the 5S rRNA sequences, but where they are located is not yet known.

All tRNA's also are initially transcribed in the form of larger precursors (Figure 12–12). Our first precise data came from *E. coli* tRNA^{tyr}, which arises by a cleavage of 39 residues from the 5' end and 3 bases from the 3' end of pre-tRNA^{tyr} (Figure 12–13). Other pre-tRNA molecules contain the sequences of several different tRNA's, thereby assuring equal intracellular numbers of the respective tRNA products.

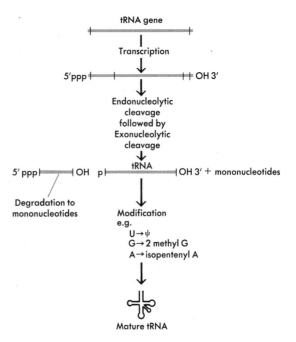

Figure 12–12
Steps in the processing of a monocistronic pre-tRNA molecule into mature tRNA.

Probably only one specific sequence of DNA nucleotides (a gene) codes for each of the 30 to 40 different *E. coli* pre-tRNA molecules. In contrast, the pre-rRNA chains appear to be coded by several *E. coli* genes located at quite separate chromosomal sites. Each of these separate pre-rRNA genes has slightly different nucleotide sequences as the result of random mutational events. In no case could they possess radically different sequences, since their rRNA products must have very similar 3-D shapes in order to fit into a ribosome.

The use of DNA to code for tRNA and for rRNA tells us that not all genes need code for specific amino acid sequences. We must thus ask the question whether other RNA forms, as yet undiscovered, may play important metabolic roles. Certainly we now cannot automatically assume that for each gene there exists a corresponding protein.

FORMATION OF RIBOSOMES IN DISCRETE STEPS

Ribosomal proteins begin to bind to pre-rRNA soon after its transcription starts. Most such proteins, however, probably add on after chain release and partial processing into smaller regions of rRNA. The first bound proteins interact directly with free rRNA, with the addition of the re-

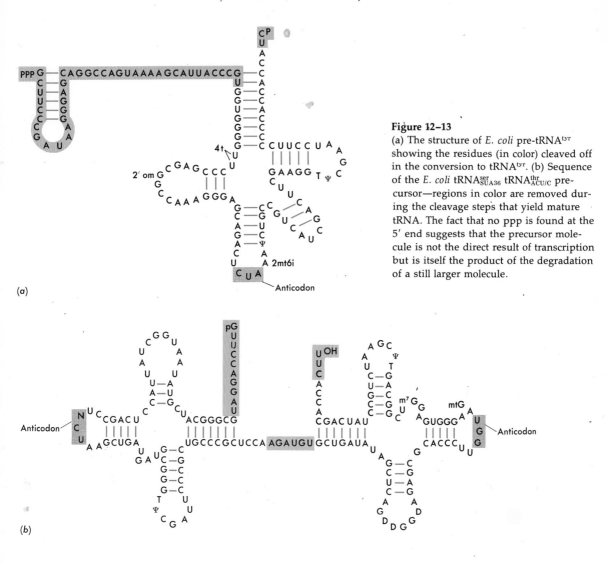

Figure 12–13
(a) The structure of *E. coli* pre-tRNA[tyr] showing the residues (in color) cleaved off in the conversion to tRNA[tyr]. (b) Sequence of the *E. coli* tRNA[ser][SUA36] tRNA[thr][ACU/C] precursor—regions in color are removed during the cleavage steps that yield mature tRNA. The fact that no ppp is found at the 5′ end suggests that the precursor molecule is not the direct result of transcription but is itself the product of the degradation of a still larger molecule.

(a)

(b)

maining proteins probably involving protein–protein secondary bonds as well as RNA–protein interactions. None of these aggregation processes occurs randomly in time. Instead, mature 30-S and 50-S particles form through precise assembly processes characterized by well defined precursor particles lacking specific ribosomal protein and containing rRNA fragments that have not yet been completely processed into mature 16-S and 23-S rRNA chains (Figure 12–14). Interestingly, the capacity to bind tRNA (mRNA) comes about late in the assembly process, and so no tRNA (mRNA) is needlessly bound to immature particles incapable of protein synthesis.

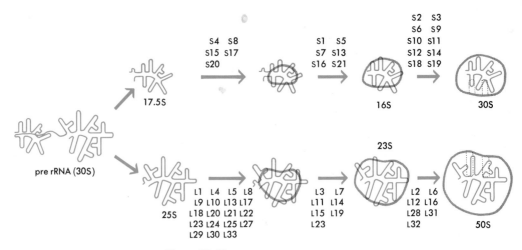

Figure 12–14
Assembly pathways in the formation of 30-S and 50-S ribosomes. The numbers refer to the current designations of the protein components.

mRNA MOLECULES EXIST IN A LARGE VARIETY OF SIZES

In contrast to tRNA molecules, which have molecular weights of about 2.5×10^4, and to rRNA molecules, which are either 5×10^5 or 10^6 in molecular weight, mRNA molecules vary greatly in chain length, and hence in molecular weight. Some of this heterogeneity reflects the large size spread in the length of polypeptide chain products. Not many polypeptide chains contain fewer than 100 amino acids, and so almost all mRNA molecules must contain at least 100×3 (because there are three nucleotides in a codon) nucleotides. Additional variations arise from unequal lengths of *leader* sequences at their 5' ends. These are sequences which do not code for any amino acids and which, at least in certain cases, function in the control of specific protein synthesis (see Chapter 14). The just recently worked out leader of *E. coli* galactose mRNA is 26 residues long (Figure 12–15), while that of *E. coli* tryptophan mRNA appears to contain at least 150 bases. As yet we have no solid data on whether appreciable untranslated nucleotides also exist at the 3' ends of mRNA, but this information should soon be available. In any case, the existence of frequent 5'-ended leaders means that most mRNA's are somewhat larger than initially predicted from their amino acid coding requirements. In *E. coli*, the mRNA's that code for the average-sized polypeptides of 300 to 500 amino acids usually contain between 1000 and 2000 nucleotides.

```
          10              20              30
ppp A U A C C A U A A G C C U A A U G G A G C G A A U U A U G A G A G U U C U G
                                                        met   arg   val   leu


   40            50            60            70
G U U A C C G G U G G U A G C G G U U A C A U U G G A A G U C A U A C C U G U
 val   thr   gly   gly   ser   gly   try   ile   gly   ser   his   thr   cys
```

Figure 12–15
The 5' end of *E. coli* galactose mRNA showing the beginning "leader" residues together with the nucleotide sequences coding for the N-terminal amino acid residues in the enzyme UDP-galactose 4-epimerase.

Still more heterogeneity in size comes from the fact that some mRNA codes for more than one polypeptide. Besides also containing leader sequences, these polygenic messengers contain intergenic regions that may be as long as the leader sequences themselves. In most, if not in all, of these polygenic messengers, the polypeptide products have related functions. For example, there exists an mRNA molecule that codes for the ten specific enzymes needed to synthesize the amino acid histidine. It has approximately 12,000 nucleotides (MW~4,000,000), or an average of 1200 nucleotides coding for each enzyme and its leader (intergenic region).

RIBOSOMES COME APART INTO SUBUNITS DURING PROTEIN SYNTHESIS

The construction of all ribosomes from easily dissociable subunits suggested a cycle during which the large and small subunits come apart at some stage in protein synthesis. This hunch was confirmed by experiments with *E. coli* and yeast which showed that most ribosomes are constantly dissociating into subunits and then reforming. Growth of cells in heavy isotopes followed by transfer to light medium was the technique used to settle this point. Soon after transfer to the light medium, hybrid (heavy 50-S–light 30-S, light 50-S–heavy 30-S) ribosomes began to appear, the rate of their appearance suggesting that the subunits exchange once during every cycle of polypeptide synthesis. Confirmatory evidence comes from *in vitro* studies of protein synthesis (see the next chapter) where the fate of "heavy" ribosomes can be followed in the presence of a great excess of light ribosomes. Within a minute or so (the time required to synthesize a complete polypeptide chain) almost all heavy ribosomes disappear, with the

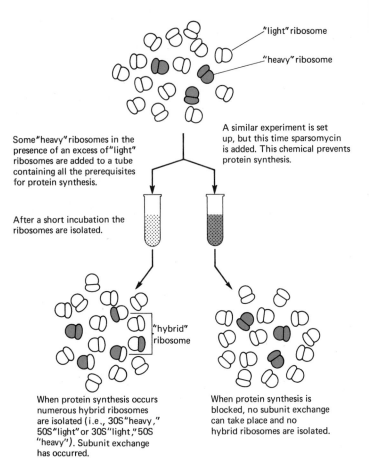

"light" ribosome

"heavy" ribosome

Some "heavy" ribosomes in the presence of an excess of "light" ribosomes are added to a tube containing all the prerequisites for protein synthesis.

A similar experiment is set up, but this time sparsomycin is added. This chemical prevents protein synthesis.

After a short incubation the ribosomes are isolated.

"hybrid" ribosome

Figure 12–16
The obligatory dependence of subunit exchange on protein synthesis.

When protein synthesis occurs numerous hybrid ribosomes are isolated (i.e., 30S "heavy," 50S "light" or 30S "light," 50S "heavy"). Subunit exchange has occurred.

When protein synthesis is blocked, no subunit exchange can take place and no hybrid ribosomes are isolated.

simultaneous appearance of hybrid ribosomes (Figure 12–16). The obligatory dependence of subunit exchange on protein synthesis is confirmed by *in vitro* experiments with the antibiotic sparsomycin. This compound, which blocks polypeptide chain elongation, also prevents any subunit exchange.

It is not yet conclusively known whether the subunits separate simultaneously with the completion of a new polypeptide, or whether they come off an mRNA molecule in the form of a 70-S ribosome which subsequently dissociates into the subunit form.

POLYPEPTIDE CHAIN GROWTH STARTS AT THE AMINO TERMINAL END

At one end of a complete polypeptide chain is an amino acid bearing a free carboxyl group, and at the other end is one bearing a free amino group. Chains always grow by

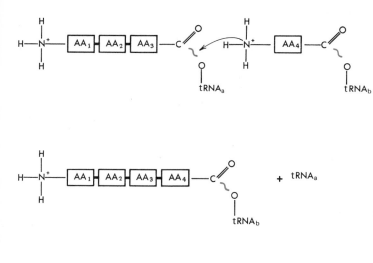

(Peptide bonds between amino acids
are indicated by a heavy line.)

Figure 12–17
Stepwise growth of a polypeptide chain.
Initiation begins at the free NH_3^+ end with
the carboxyl growing point terminated by a
tRNA molecule.

the stepwise addition of single amino acids, starting with
the amino terminal and ending with the carboxyl terminal
(Figure 12–17). This point is clearly shown by brief exposure of hemoglobin-synthesizing reticulocytes to radioactively labeled amino acids, followed by immediate isolation of newly completed hemoglobin chains. Very little
radioactivity is found in the amino acids of the amino terminal end; most is found in the amino acids of the carboxyl end (Figure 12–18). Moreover, a clear gradient of
increasing radioactivity is observed as one moves from the
amino to the carboxyl end.

STARTING OF ALL BACTERIAL POLYPEPTIDE CHAINS WITH N-FORMYL METHIONINE

The starting amino acid in the synthesis of all bacterial
polypeptides is N-formyl methionine. This is a modified
methionine which has a formyl group attached to its terminal amino group (Figure 12–19). A blocked amino acid
like N-formyl methionine can be used only to start protein
synthesis. The absence of a free amino group would prevent this amino acid from being inserted during chain
elongation. The formyl group is enzymatically added onto
methionine after the methionine has become attached to
its tRNA adaptor.

Not all methionine-tRNA molecules can be formylated. Instead, there exist two types of met–tRNA, only one
of which permits the formylation reaction. Sequence analysis of both of these met–tRNA's reveals that they have the

Figure 12–18
Experimental demonstration that hemoglobin chains grow in the $NH_3^+ \rightarrow COOH$ direction.

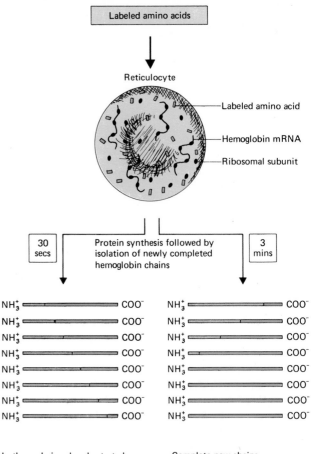

Labeled amino acids

Reticulocyte

Labeled amino acid

Hemoglobin mRNA

Ribosomal subunit

| 30 secs | Protein synthesis followed by isolation of newly completed hemoglobin chains | 3 mins |

NH_3^+ ———— COO^- NH_3^+ ———— COO^-
NH_3^+ ———— COO^- NH_3^+ ———— COO^-
NH_3^+ ———— COO^- NH_3^+ ———— COO^-
NH_3^+ ———— COO^- NH_3^+ ———— COO^-
NH_3^+ ———— COO^- NH_3^+ ———— COO^-
NH_3^+ ———— COO^- NH_3^+ ———— COO^-
NH_3^+ ———— COO^- NH_3^+ ———— COO^-
NH_3^+ ———— COO^- NH_3^+ ———— COO^-

Only those chains already started have had time to be completed. They show a gradient of radioactivity, there being more label at the COO^- end. No complete chains are found labeled at the NH_3^+ end.

Complete new chains have had time to be synthesized in the presence of the labeled amino acids. Some chains are now therefore found labeled at the NH_3^+ end.

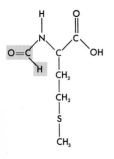

N-formyl methionine

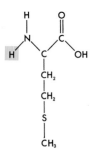

Methionine

Figure 12–19
The structure of N-formyl methionine.

same anticodon sequences (Figure 12–20), exposing the dilemma of how the same codon (AUG) can code for two different amino acids.

The discovery that synthesis of bacterial protein starts with a blocked amino acid was most unexpected because isolation of pure protein from growing bacteria revealed essentially no formylated end groups. This means an enzyme exists which removes the formyl group from the growing chain very soon after synthesis commences (Figure 12–21). In addition, another enzyme (an amino peptidase) exists which subsequently removes the terminal methionine from many proteins. It does not act on all proteins, so a large fraction of bacterial proteins commence with methionine.

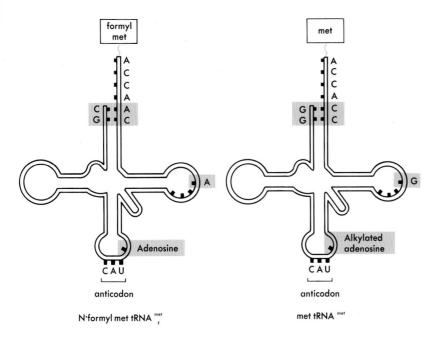

Figure 12–20
The three main points of difference between N-formyl-met-tRNA and met-tRNA.

Our picture of initiation in protein synthesis is less clear in those organisms (eucaryotic as distinct from pro-caryotic) whose cells contain discrete nuclei and have 80-S ribosomes. Though synthesis also starts with methionine, it does not seem to be blocked with a formyl group. Nonetheless, two types of methionine tRNA exist, one used exclusively in initiation and the other exclusively for elongation.

BINDING OF THE SMALLER RIBOSOMAL SUBUNIT TO SPECIFIC POINTS ALONG mRNA MOLECULES

Initiation in bacteria starts with the formation of a complex between the smaller 30-S ribosomal subunit, f-met-tRNA, and an mRNA molecule. Then a 50-S subunit attaches to form the functional 70-S ribosome. Each specific mRNA chain contains one to several ribosome-binding regions to which the free 30-S particles can stick. Within each such site are nucleotide sequences whose sole function is the correct lining up of mRNA molecules on ribosomal surfaces prior to the commencement of protein synthesis. Every mRNA molecule has one *ribosome binding site* for each of its independently synthesized polypeptide products. For example, three such sites are found on the RNA molecule of phage R17 which codes for three separate proteins.

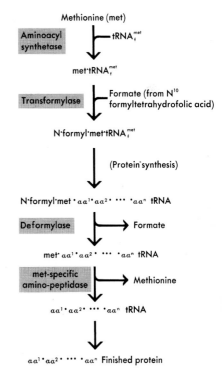

Figure 12–21
Enzymatic steps involving formyl methionine initiation of protein synthesis.

The first binding sequences were established by binding ribosomes to specific mRNA's and then adding the enzyme *ribonuclease* to break down all sequences except those protected by ribosome attachment. When nucleotide sequence methods were used upon the protected fragments, the 3' ends were found to contain the nucleotides which coded for the first 5 or 6 amino acids in the respective polypeptide products. These are not the "binding nucleotides," however, for they vary from one protein to another. Instead, this task appears to be largely played by a group of 3 or 4 purine nucleotides near the 5' end of the protected fragment. Virtually all binding sites (Figure 12–22) have an AGGA sequence (or a close derivative) centered some 8 to 13 nucleotides distant from the initial coding nucleotide. These have just recently been shown to base pair to a pyrimidine-rich section at the 3' end (tail) of 16-S rRNA chains ($\cdots$ GAUCACCUCCUUA$_{OH}$3'). Somehow this pairing brings the initiating AUG codon into position so that it can bind to the anticodon of a 30-S subunit-initiator tRNA complex.

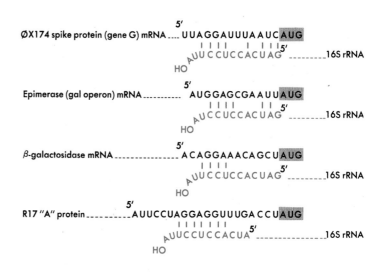

Figure 12–22
Ribosome binding sequences from viral and bacterial mRNA molecules, showing their pairing with bases at the 3' tail of 16-S rRNA.

Afterwards, when elongation commences, the binding site–16-S tail complex must somehow dissociate, leaving the mRNA free to move across the ribosomal surface. Such movement necessarily brings the ribosome into contact with mRNA regions to which it could not bind prior to synthesis. In particular, ribosome binding must temporarily disrupt the double-helical hairpin regions that characterize so many sections of free mRNA, thus creating single-stranded regions that can correctly select the anticodons of the AA~tRNA precursors.

INITIATION FACTORS

A mixture of N-formyl met-tRNA, mRNA, and the 30S and 50S subunits by itself is not sufficient for initiation. At least three separate proteins not normally part of ribosomes (IF1, IF2, and IF3) must also be present (Figure 12–23). Most likely the process starts with the attachment of IF3 to a free 30S subunit. In the second step f-met-tRNA bound to an IF2–GTP complex attaches to an IF3–30S aggregate. The resulting IF3–30S–IF2–GTP complex then binds to mRNA, a process which may also require IF1. In the last steps, the binding of a 50S subunit leads to GTP hydrolysis, as well as the release of all three initiation factors.

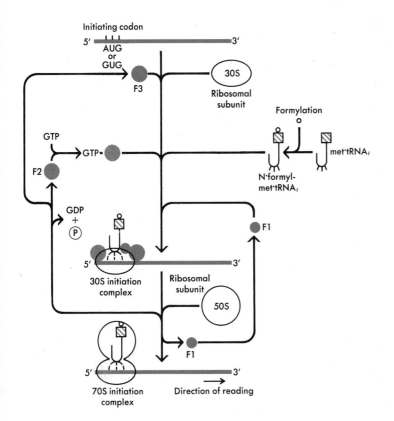

Figure 12–23
Diagrammatic view of the initiation process in protein synthesis. [Only one of the two tRNA binding sites is shown on the ribosome. It is not known which site (A or P) the f-met-tRNA enters first.]

THE DIRECTION OF mRNA READING IS 5' TO 3'

After an mRNA molecule has stuck to a ribosome, it must always move in a fixed direction during protein synthesis. It does not have the alternative of moving both to the right or to the left, which reflects the fact that RNA molecules have a direction defined by the relative orientations of the 3' and 5' ends. The end which is read first, the 5' end, is also synthesized first. This opens up the possibility that a ribosome can attach to an incomplete mRNA molecule in the process of synthesis on its DNA template. If, on the contrary, polypeptide synthesis went 3' to 5', then a length of mRNA corresponding to a complete polypeptide chain would have to be synthesized before it could stick to a ribosome. The fact that protein synthesis goes 5' to 3' most likely means that long sections of mRNA unattached to ribosomes do not normally exist in rapidly growing bacterial cells.

EACH RIBOSOME HAS TWO tRNA BINDING SITES

Each 70S ribosome contains two cavities into which tRNA molecules can be inserted (Figure 12–24). They are the "P" (peptidyl) and the "A" (amino–acyl) sites. Each hole is bounded partly by a 30S region and a 50S region and by a specific mRNA codon. Although the ribosomal bounded surface can accept any of the specific AA~tRNA's, since it binds to an unspecific region of the tRNA molecule, the codon-bounded surface is specific for a unique tRNA molecule.

We do not yet know whether the initiating f-met-tRNA first enters the "P" site or whether it first goes into the "A" site and subsequently moves into the "P" site. It is clear, however, that it must be in the "P" site before the second AA~tRNA can bind specifically to the ribosome. For all AA~tRNA's except f-met-tRNA, it has been firmly established that entry to the ribosome occurs through the "A" site. After the second AA~tRNA is correctly placed in the "A" site, a peptide bond is formed enzymatically to yield a dipeptide (two amino acids linked by a peptide bond) terminated by a tRNA molecule, the adaptor of the second amino acid. This process of amino acid addition then repeats over and over, adding one amino acid at a time to form a complete chain (Figure 12–24). In these events, the following steps should be emphasized.

1. The growing carboxyl end is always terminated by a tRNA molecule. The binding of this terminal

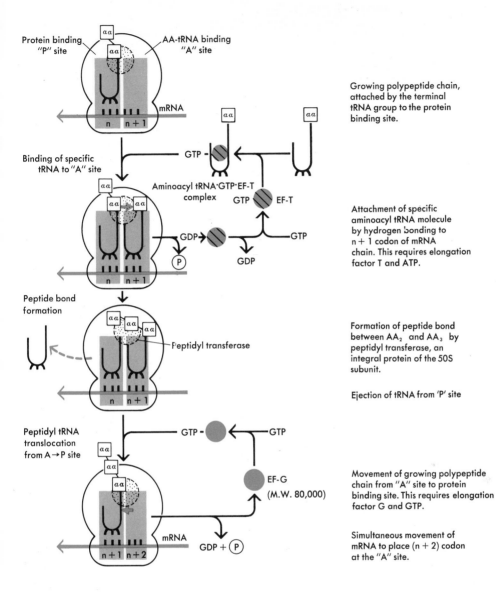

Protein binding "P" site

AA-tRNA binding "A" site

mRNA

Growing polypeptide chain, attached by the terminal tRNA group to the protein binding site.

Binding of specific tRNA to "A" site

GTP

Aminoacyl tRNA·GTP·EF-T complex

GTP   EF-T

GDP   GTP

GDP

Attachment of specific aminoacyl tRNA molecule by hydrogen bonding to n + 1 codon of mRNA chain. This requires elongation factor T and ATP.

Peptide bond formation

Peptidyl transferase

Formation of peptide bond between AA_2 and AA_3 by peptidyl transferase, an integral protein of the 50S subunit.

Ejection of tRNA from 'P' site

Peptidyl tRNA translocation from A → P site

GTP   GTP

EF-G (M.W. 80,000)

mRNA   GDP + P

Movement of growing polypeptide chain from "A" site to protein binding site. This requires elongation factor G and GTP.

Simultaneous movement of mRNA to place (n + 2) codon at the "A" site.

Figure 12–24
Diagrammatic view of peptide bond formation showing the role of the elongation factors.

tRNA to either the "P" or "A" site is the main force holding a growing polypeptide chain to the ribosome.

2. Formation of the peptide bond moves the attachment point of the growing chain from the "P" to the "A" site.

3. Soon after, if not coincident with, the formation of the peptide bond, the released tRNA molecule is ejected from the "P" site.

4. The new terminal tRNA molecule then moves (translocates) from the "A" to the "P" site. At the same time, the mRNA template bound to the smaller ribosome subunit moves to place codon $n + 1$ in the position previously occupied by codon n.

5. The now vacant "A" site becomes free to accept a new AA $\sim$ tRNA molecule whose specificity is determined by correct base pairing between its anticodon and the relevant mRNA codon.

EXISTENCE OF ELONGATION FACTORS

Because the amino acid carboxyl groups are activated by their attachment to their tRNA adaptors, the initial guess was that perhaps only one specific enzyme would be needed for making the peptide bond. Furthermore, there was no reason to suspect that more energy would be needed for the polymerization reaction. Both these hunches, however, were wrong. First, the energy-rich molecule GTP (analogous to ATP with adenine replaced by guanine) is required for the synthesis of all peptide bonds in protein. Second, two proteins (the elongation factors) have been isolated which are not normally part of ribosomes but which are also necessary. Moreover, neither of these two proteins itself precipitates the actual formation of peptide bonds.

BINDING OF AA~tRNA TO THE "A" SITE REQUIRES ELONGATION FACTOR T

The attachment of the AA~tRNA precursor to ribosomes was initially believed to be a nonenzymatic event occurring when the correct AA~tRNA randomly bumped the "A" site and its specific binding mRNA codon. Recent experiments, however, indicate that the binding reaction is far from simple. It starts when one of the elongation factors (EF-T) reacts with GTP and AA~tRNA to form an AA~tRNA–GTP–EF-T complex. This complex then transfers its AA~tRNA component to the "A" site with release of a free EF-T–GDP complex and (P).

The stay of an AA~tRNA molecule within the "A" site thus can be subdivided into two steps, the first during which it is combined with EF-T and GTP, and the second when it exists free. During both these stages, its anticodon is bound to an mRNA codon, and so if the wrong AA~tRNA has bound to the "A" site, two successive opportunities exist for it to be rejected. Conceivably, EF-T plays essentially a proofreading role, with its main function being to give ribosomes a second opportunity to reject mispaired AA~tRNA molecules.

THE PEPTIDE BOND-FORMING ENZYME IS AN INTEGRAL COMPONENT OF THE 50S PARTICLE

Enzymatic catalysis of the peptide bond itself is caused by one of the proteins of the larger ribosomal subunit. It is called peptidyl transferase, and is believed to be present in one copy per 50S subunit. All attempts so far to dissociate this enzyme from 50S particles have failed. Now that complete reconstitution of 50S particles has been achieved, it should be possible to pinpoint a specific 50S protein as the active catalyst.

PEPTIDYL-tRNA TRANSLOCATION REQUIRES ELONGATION FACTOR G

The movement of peptidyl-tRNA from the "A" to the "P" site is brought about by elongation factor G (EF-G), often called the *translocase*. In this process, an EF-G–GTP–ribosome complex first forms. Translocation, coupled with hydrolysis of GTP to GDP and $\circledP$ then occurs, with subsequent release of free reusable EF-G. The splitting of the high-energy bond is obligatory for the movement, with precise experiments showing that one GTP is split for every translocation act.

MOVEMENT OF mRNA ACROSS THE RIBOSOMAL SURFACE

Normally at each translocation step the mRNA template must be advanced precisely three nucleotides. One might thus imagine the existence of a machinery within the ribosome which obligatorily advances the mRNA chain three steps at a time. Alternatively, the mRNA may not be moved by the ribosome *per se*, but instead may be pulled in the translocation process which moves peptidyl-tRNA from the "A" site to the "P" site. Under this hypothesis, the movement of a codon triplet is a consequence of its binding to an anticodon in tRNA. Strongly favoring this latter proposal is the recent discovery of frameshift suppressor tRNA's (see Chapter 13) which have anticodons containing four nucleotides and so bind to groups of four adjacent mRNA nucleotides. When such unusual tRNA's are translocated, their respective mRNA templates are advanced by four bases, thus showing that mRNA movement is a direct consequence of tRNA movement.

335

INHIBITION OF SPECIFIC STEPS IN PROTEIN SYNTHESIS BY ANTIBIOTICS

A number of antibiotics have proved very useful in delineating the steps by which proteins are built up. For example, puromycin, a very powerful inhibitor of the growth of all cells, acts by interrupting chain elongation. Its structure (Figure 12–25) resembles the 3' end of a charged tRNA molecule and so it is able to enter the "A" ribosomal binding site very efficiently, competitively inhibiting the normal entering of AA~tRNA precursors. More importantly, peptidyl transferase will use it as a substitute, thereby transferring nascent chains to puromycin acceptors. Since the puromycin residues bind only weakly to the "A" site, the puromycin-terminated nascent chains fall off the ribosomes, thereby producing incomplete chains of varying lengths.

Another very useful antibiotic is fusidic acid, which specifically inhibits the translocation function of elongation factor G. The peptidyl transferase reaction itself is inhibited by the antibiotics sparsomycin and lincomycin, both of which specifically bind to the 50-S subunit. In contrast, streptomycin, which binds to the 30-S subunit, is a powerful inhibitor of chain initiation.

Figure 12–25

Premature peptide chain termination by puromycin.

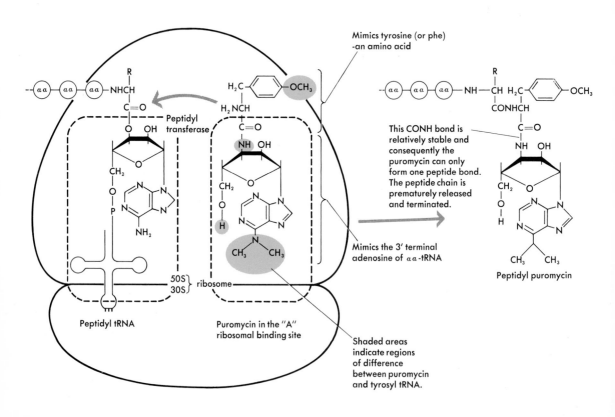

POLYPEPTIDE CHAINS FOLD UP SIMULTANEOUSLY WITH SYNTHESIS

Under optimal conditions, the time required for the synthesis of an *E. coli* polypeptide chain containing 300 to 400 amino acids is 10 to 20 seconds. During this time, the elongating chain does not remain a random coil, but quickly assumes much of its final 3-D shape through the formation of many of its secondary bonds. Thus, with many proteins, most of the final shape may be achieved before the last few amino acids are added on. As a result, trace amounts of many enzyme activities are found on ribosomes which have not yet released their polypeptide products.

CHAIN RELEASE DEPENDS UPON SPECIFIC RELEASE FACTORS WHICH READ CHAIN-TERMINATING CODONS

Two conditions are necessary for chain termination. One is the presence of a codon that specifically signifies that polypeptide elongation should stop. The other is the presence of a GTP-bound release factor which reads the chain-terminating signal. Behind all this complexity is the fact that after a polypeptide chain has reached its full length, its carboxyl end is still bound to its tRNA adaptor. Termination must thus involve the splitting off of the terminal tRNA. When this happens, the nascent chain quickly dissociates, since its binding to the ribosome occurred principally through its tRNA component.

Elucidation of the genetic code (see Chapter 13) revealed three codons specifically signifying "stop." Their existence initially led to the expectation of chain-terminating tRNA molecules; that is, tRNA's which would specifically bind to the stop codon but which had no amino acid attached to their 3'-adenosine and which in some way promoted the release of the terminal tRNA group. New experiments, however, conclusively rule out the existence of such molecules. The stop codons are instead read by specific proteins (the release factors), whose action again demands the presence of GTP. Whether the release factors (RF) are enzymes and directly catalyze the splitting off of terminal tRNA residues remains to be elucidated.

GTP MAY FUNCTION BY CAUSING CONFORMATION CHANGES

The repeated use of GTP at so many stages of the translation process (initiation, elongation, and termination) needs some simplifying explanation. No evidence exists for its use in the formation of any covalent bond and so it does not function as an ATP-like energy donor. Instead it plays a key role in the noncovalent binding of the various translational factors (IF2, EF-T, EF-G, RF) to the ribosomal surface. Such binding processes merely require the presence of GTP, while the GTP → GDP cleavage reaction is the signal for the release of the bound factor. The presence (or absence) of GTP may thus determine the conformational state of the various translational factors. Under this hypothesis, GTP attachment induces shape changes that are necessary for binding. Correspondingly, hydrolysis to GDP leads to a return of the original free configuration, resulting in an ejection of the respective factors from the ribosomal surface.

PRODUCTION OF ppGpp ON RIBOSOMES BY AN IDLING REACTION IN THE ABSENCE OF CHARGED tRNA

If by accident an uncharged tRNA molecule occupies an "A" site, not only does polypeptide growth temporarily cease, but an idling reaction occurs on the affected ribosome which uses ATP as a ℗~℗ donor to convert $pp^{5'}G^{3'}_{OH}$ (GDP) into the unusual guanine nucleotide $pp^{5'}G^{3'}pp$. While during normal bacterial growth only very low levels of ppGpp are present, very large amounts accumulate during amino acid starvation. As such, these ppGpp molecules do not play a passive role, but act as signals which specifically stop the initiation of rRNA and tRNA chains. They thus act as the control molecules that keep cells from wastefully producing more ribosomes than they can usefully employ. This specific blockage of rRNA and tRNA synthesis under conditions of amino acid starvation is called the *stringent response.* Its existence is only helpful and not obligatory, however, since there are *relaxed mutants* which continue to make rRNA and tRNA during amino acid starvation. Such relaxed cells lack a 75,000-MW enzyme, the *stringent factor,* that can bind to ribosomes and carry out the conversion of ppG to ppGpp.

BREAKS IN POLYPEPTIDE CHAINS AFTER CHAIN TERMINATION

Many examples are known where specific enzymes are modified after synthesis. We mentioned earlier the removal of formyl and methionine groups from many bacterial proteins. Very likely, cases will soon be found where more than one amino acid is removed from either the amino or carboxyl ends through the action of exopeptidases (enzymes which sequentially remove terminal amino acids). Equally important are the endopeptidases, which cut internal peptide bonds. Insulin, for many years thought to be constructed by aggregation of independently synthesized A and B chains, is synthesized first as an 82 amino acid single chain (proinsulin) whose 3-D shape is attained after formation of several disulfide (S—S) bonds. Subsequently, two internal cuts remove 31 amino acids, leaving the formerly contiguous polypeptide fragments held together only by the S—S bonds (Figure 12–26). It is thus very clear that knowledge of the amino acid sequence of a purified enzyme may not necessarily reveal the true initiating and terminating amino acids.

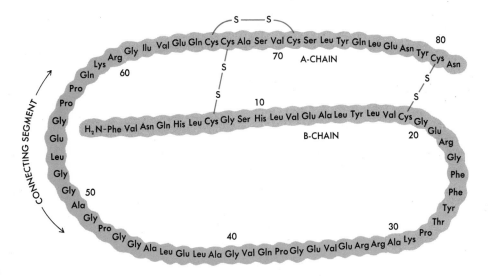

Figure 12–26
Structure of bovine proinsulin, showing amino acids removed during its conversion to insulin.

AN mRNA MOLECULE WORKS ON SEVERAL RIBOSOMES SIMULTANEOUSLY

The section of an mRNA molecule that is in contact with a single ribosome is relatively short. This allows a given mRNA molecule to work on several ribosomes at once. Single mRNA molecules can move over the surfaces of several ribosomes simultaneously (the collection of ribosomes bound to a single mRNA chain is called a *polysome* or *polyribosome*), thus functioning as a template for several polypeptide chains at once. At a given time, the lengths of chains attached to successive ribosomes in the polyribosome vary in direct proportion to the fraction of the messenger tape to which each ribosome has already been exposed (Figure 12–27). This means that at any moment the polypeptide chains being produced along the length of the mRNA are shortest at the front of the strand, and gradually lengthen toward the end. There is great variation in polysome size, which depends upon the size of the mRNA chain. At maximum utilization of an mRNA chain, there is one ribosome for every 80 mRNA nucleotides. Thus the polysomes that make hemoglobin molecules usually contain 4 to 6 ribosomes (Figure 12–28), while approximately 12 to 20 ribosomes are attached to the mRNA molecules concerned with the synthesis of proteins in the 30,000 to 50,000 MW (300 to 500 amino acids) range.

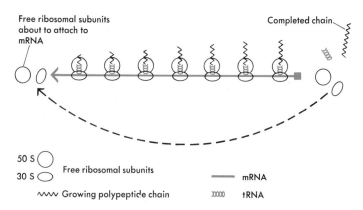

Figure 12–27
Schematic picture of a polyribosome during protein synthesis. The mRNA molecule is moving from right to left.

The ability of a single mRNA to function on several ribosomes simultaneously explains why a cell needs so relatively little mRNA. Before polysomes were discovered, and when only one ribosome was thought to be attached to a given mRNA molecule, the fact that mRNA comprised only 1 to 2 percent of the total cellular RNA seemed highly paradoxical. This followed from the fact that if the average mRNA chain were of MW about 5×10^5, then at a given instant at most only 10 percent of the ribosomes in a cell could be making protein.

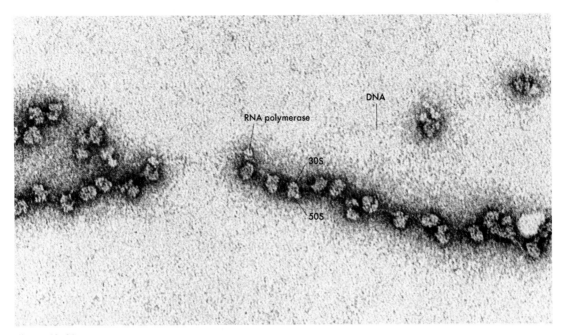

Figure 12–28
Groups of ribosomes (polysomes) moving across mRNA chains being formed on *E. coli* DNA. [Courtesy of Barbara Hamkalo, Oak Ridge National Laboratory.]

MUCH MORE MUST BE LEARNED ABOUT RIBOSOMES

It is very likely that the broad general outlines of protein synthesis are now established. Each of the key features shown in Figure 12–29 has been established by a variety of experiments. Nonetheless, much new information must be

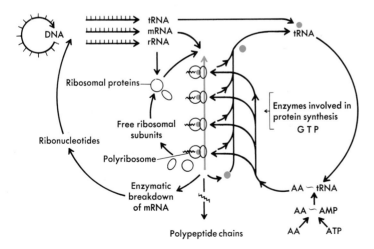

Figure 12–29
Schematic view of the role of RNA in protein synthesis.

341

obtained before we can honestly state that protein synthesis is understood at the molecular level. At the heart of our uncertainty is the role of the ribosome.

Though high-resolution electron microscope research may soon be able to tell us the relative locations of the various ribosomal proteins as well as the positions of the "P" and "A" tRNA binding sites, it cannot show details at the atomic level. This will come only when it is possible to apply x-ray diffraction analysis. For many years this approach seemed almost hopeless, for no one had been able to grow ribosome crystals, and doubts existed whether it would be possible.

Now, however, numerous examples have been found where ribosome crystals form within living cells. The cells in which they occur are those where protein synthesis has temporarily stopped. For example, excellent crystals (Figure 12–30) are obtained in chick embryo cells when incubation at high temperature ceases and the cells are allowed to cool slowly. Under such conditions, nascent chains are completed, no new initiation occurs, and mRNA dissociates from the individual ribosomes. The resulting free ribosomes can then aggregate spontaneously. The existence of this *in vivo* crystallization makes it possible that, not too far in the future, crystals can be grown in the test tube. This, however, will just be the first step in a horrendous endeavor, since x-ray methods have not yet been used to solve structures of anywhere near this complexity. We must thus be prepared for a long wait and much hard work before we understand why such a complicated structure is necessary for protein synthesis.

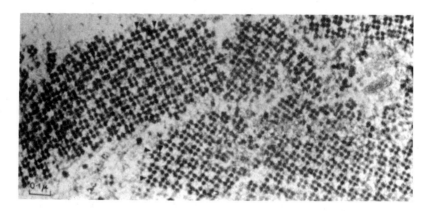

Figure 12–30
Crystals of ribosomes revealed in a thin-section electron micrograph of a chick embryo cell cooled for 6 hr at 5°C. [From Byers, *J. Mol. Biol.*, **26,** 155 (1967). Plate kindly supplied by B. Byers, The Biological Laboratories, Harvard University.]

SUMMARY

Amino acids do not attach directly to RNA templates. They first combine with specific adaptor molecules to form AA~adaptor complexes. The adaptor component has a strong chemical affinity for RNA nucleotides. All the adaptors are transfer RNA molecules (tRNA) of MW~25,000. A given tRNA molecule is specific for a given amino acid.

All tRNA chains are folded into semirigid L-shaped molecules. About half the tRNA bases are part of double-helical stems, with the remainder forming loops stabilized by interloop (stem) hydrogen bond and stacking forces. One of the loops contains the anticodon, the group of 3 bases which base pair to three successive template bases ("codon"). Amino acids attach through their carboxyl groups to the 3'-terminal adenosine of tRNA molecules by high-energy covalent bonds. Adapted amino acids are thus "activated." There is a specific activating enzyme (amino–acyl synthetase) for each amino acid; and so it is an enzyme, not tRNA, which recognizes the amino acid. The overall accuracy of protein synthesis can thus be no greater than the accuracy with which the activating enzymes can selectively recognize the various amino acids.

After activation, the AA~tRNA molecules diffuse to the ribosomes, which are spherical particles on which the peptide bonds form. Ribosomes have molecular weights of about 3×10^6 and consist of about half protein and half rRNA. They are always constructed from a large (MW ~ 2×10^6) and a small (MW ~ 10^6) subunit. About 60 different proteins are found in a single ribosome, in addition to two major RNA chains (16-S and 23-S RNA). rRNA does not contain genetic information and its function remains an important mystery. The template itself is a third form of RNA, messenger RNA (mRNA). Messenger RNA attaches to ribosomes and moves across them to bring successive codons into position to select the correct AA~tRNA precursors.

Protein chains grow stepwise, beginning at the amino terminal end. This means the growing end is always terminated by a tRNA molecule. Attachment of the nascent chain to the ribosome occurs through binding of the chain's terminal tRNA group to a hole in the ribosomal surface. Each ribosome has two holes, the "P" (peptidyl) and the "A" (amino–acyl). Precursor AA~tRNA molecules normally first enter the "A" site, allowing the subsequent formation of a peptide bond to the growing chain held in the "P" site. This transfers the nascent chain to the "A" site. Action of the enzyme translocase then moves the nascent chain back into the "P" site where another cycle can begin. Both the attachment of AA~tRNA to the "A" site and the translocation process require the splitting of GTP to GDP and Ⓟ. Movement of successive codons over the ribosomal surface may occur simultaneously with the translocation step, but nothing is known about how this happens. Initiation of synthesis requires separate enzymes from those involved in chain elongation. With bacterial ribosomes, initiation always involves N-formyl-methionine-tRNA. Chain termination results from the reading of specific stop signals by specific proteins (the release factors).

Completion of chain synthesis is followed by the dissociation of ribosomes into the large and small ribosomal subunits. Reunion of subunits occurs only after the initiating amino–acyl tRNA has first combined with an mRNA–small subunit complex. A given mRNA molecule generally works simultaneously on many ribosomes (a polysome). Thus, at a given moment, many codons of the same template are "at work."

REFERENCES

INGRAM, V. M., *Biosynthesis of Macromolecules,* W. A. Benjamin, Menlo Park, 1972. An introductory text containing experimental details of key early experiments.

The Mechanism of Protein Synthesis and its Regulation, edited by L. Bosch, North Holland-Elsevier, 1972. Contains several excellent reviews of selected aspects of protein synthesis.

Ribosomes, edited by M. Nomura, A. Tissieres, and P. Lengyel, Cold Spring Harbor Laboratory (1974). A very complete description of current knowledge on how ribosomes may function is given in this collection of advanced review articles.

CRICK, F. H. C., "On Protein Synthesis," *Symp. Soc. Exptl. Biol.,* **12,** 138–163 (1958). A presentation of the ideas that led the author to propose the adaptor hypothesis.

WATSON, J. D., "The Involvement of RNA in the Synthesis of Proteins," *Science,* **140,** 17–26 (1963). A history of work between 1953 and 1962 that established how RNA participates in protein synthesis.

HOLLEY, R. W., "The Nucleotide Sequence of a Nucleic Acid." A description of how the author worked out the structure of alanine tRNA. A 1966 *Scientific American* article reprinted in *The Molecular Basis of Life,* R. H. Haynes and P. C. Hanawalt (eds.), Freeman, San Francisco, 1968.

ROBERTUS, J. D., J. E. LADNER, J. T. FINCH, D. RHODES, R. S. BROWN, B. F. C. CLARK, AND A. KLUG, "Structure of Yeast Phenylalanine tRNA at 3Å Resolution," *Nature,* **250,** 546 (1974).

KIM, S. H., F. L. SUDDATH, F. L. QUIGLEY, A. McPHERSON, J. L. SUSSMAN, A. H. J. WANG, N. C. SEEMAN, AND A. RICH, "Three–Dimensional Tertiary Structure of Yeast Phenylalanine Transfer RNA," *Science*, **185,** 435 (1974). This and the preceding article contain independent announcements of the first 3-D structure of a tRNA molecule.

HOPFIELD, J. J., "Kinetic Proofreading: A New Mechanism for Reducing Errors in Biosynthetic Processes Requiring High Specificity," *Proc. Nat. Acad. Sci.*, **71,** 4135 (1974). An important conceptual paper that draws attention to the advantages of successive steps in a proofreading process.

BALDWIN, A. N., AND P. BERG, "Transfer Ribonucleic Acid-Induced Hydrolysis of Valyladenylate Bound to Isoleucyl Ribonucleic Acid Synthetase," *J. Biol. Chem.*, **241,** 839 (1967). A classic description of how an amino–acyl synthetase accurately discriminates between structurally very similar amino acids in the formation of AA~tRNA molecules.

NOMURA, M., "Ribosomes," *Scientific American*, December 1969. A very nice introductory survey about how ribosomes can be taken apart and then put back together.

STEITZ, J. A., "Polypeptide Chain Initiation: Nucleotide Sequences of the Three Ribosomal Binding Sites in Bacteriophage R17 RNA," *Nature*, **224,** 957 (1969). The first look at the nucleotide sequences at the beginning of genes.

SHINE, J., AND L. DALGARNO, "The 3'-Terminal Sequence of *E. coli* 16S rRNA: Complementarity to Nonsense Triplets and Ribosome Binding Sites." An imaginative structural proposal for the initial binding between mRNA and 30S ribosomal subunits.

The Genetic Code

Even when the general outline of how RNA participates in protein synthesis had been established (1960), there was little optimism that we would soon know details of the genetic code itself. At that time we believed that identification of the codons for a given amino acid would require exact knowledge of both the nucleotide sequences of a gene and the corresponding amino acid order in its protein product. As mentioned earlier, the elucidation of amino acid sequences, though a laborious objective, was already a very practical one. But on the other hand, the then current methods for determining DNA sequences were very primitive, and only today is extensive DNA base sequence information beginning to appear. Fortunately, this apparently firm roadblock did not prove a real handicap. In 1961, just one year after the discovery of mRNA, the use of artificial messenger RNA's partially cracked the genetic code with the unambiguous demonstration of a codon for the amino acid phenylalanine. To explain how this discovery was made, we must first describe some details of how biochemists study protein synthesis in cell-free systems.

ADDITION OF mRNA STIMULATES *IN VITRO* PROTEIN SYNTHESIS

The need for three RNA forms (tRNA, rRNA, and mRNA) for protein synthesis was demonstrated largely by experiments using cell-free extracts prepared from cells actively

347

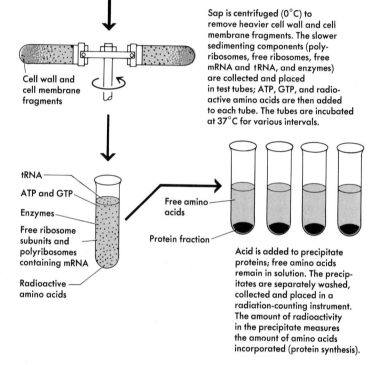

Rapidly growing *E. coli* cells are collected by centrifugation in the cold (0°C) and broken open to yield a cell sap. The enzyme deoxyribonuclease is added to break down the cellular DNA.

Sap is centrifuged (0°C) to remove heavier cell wall and cell membrane fragments. The slower sedimenting components (poly-ribosomes, free ribosomes, free mRNA and tRNA, and enzymes) are collected and placed in test tubes; ATP, GTP, and radio-active amino acids are then added to each tube. The tubes are incubated at 37°C for various intervals.

Cell wall and cell membrane fragments

tRNA
ATP and GTP
Enzymes
Free ribosome subunits and polyribosomes containing mRNA
Radioactive amino acids

Free amino acids
Protein fraction

Acid is added to precipitate proteins; free amino acids remain in solution. The precip-itates are separately washed, collected and placed in a radiation-counting instrument. The amount of radioactivity in the precipitate measures the amount of amino acids incorporated (protein synthesis).

Figure 13–1
Experimental details of *in vitro* studies of protein synthesis.

Figure 13–2
Ⓐ indicates the breakdown of en-dogenous mRNA, while Ⓑ shows the breakdown of exogenously added mRNA whose time of addition is indicated by the arrow.

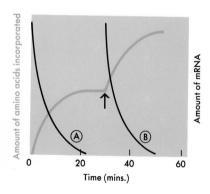

engaged in protein synthesis. In these experiments, care-fully disrupted cells were fractionated to see which cell components were necessary for incorporation of amino acids into proteins. All these experiments utilized radioac-tively labeled (H^3, C^{14}, or S^{35}) amino acids because *in vitro* protein synthesis was and still is very inefficient; that is, there is no detectable net synthesis. Only by using labeled precursors can the incorporation of precursors into pro-teins be convincingly demonstrated (Figure 13–1).

A typical time course of the *in vitro* incorporation of radioactive amino acids into proteins is illustrated by the graph in Figure 13–2. It shows that synthesis in an *E. coli* extract proceeds linearly for several minutes and then gradually stops. During this interval there is a corresponding loss of mRNA, owing to the action of degradative enzymes present in the extract. This suggests that the major cause of the inefficiency of cell-free protein synthesis is loss of the template component. The correctness of this supposition is shown by adding new mRNA to extracts that have just stopped making protein. Such an addition causes an immediate resumption of synthesis. mRNA-depleted extracts are very valuable in testing mRNA activity. Because they contain very little functional mRNA, they can be used to detect small amounts of template activity in externally added mRNA.

VIRAL RNA IS mRNA

In Chapter 15 we shall show that in many viruses the genetic component is single-stranded RNA. When these viruses infect a cell, their infecting RNA molecule must first act as a template for the synthesis of the specific proteins needed to initiate the life cycle of the virus. Among these necessary proteins is the RNA-replicating enzyme RNA synthetase. It is thus impossible for the infecting single-stranded molecule to act as a template for its complementary strand until it has *first* acted as a template for some protein synthesis. This means that the infecting viral RNA must itself be able to attach to its host's ribosomes and direct the synthesis of viral-specific proteins (i.e., that it must act as mRNA). This point is neatly shown with cell-free systems. For example, the addition of TMV RNA to an mRNA-depleted *E. coli* extract immediately stimulates the incorporation of amino acids into proteins.

SPECIFIC PROTEINS CAN BE MADE IN CELL-FREE SYSTEMS

Even with addition of excess mRNA, there is still only a small amount of amino acid incorporation into polypeptide chains in our *in vitro* systems. This means that we do not yet know the conditions that permit normal *in vivo* synthesis. Doubts were initially raised about whether, in fact, the newly made proteins had structures at all similar to those of natural proteins. Fortunately, the use of specific viral RNA showed that these doubts were unfounded and that the genetic code can be accurately read in cell-free systems. In the first successful experiments (1962), the RNA isolated from the bacterial virus F2 was added to preincubated *E. coli* extracts. It acted as a template and promoted the incorporation of amino acids into protein. Some complete polypeptide chains were made and released from the ribosomes. These newly made protein products were then compared with the F2 coat protein, which has a molecular weight of 14,000. Some of the *in vitro* products were found to have amino acid sequences identical to those of the *in vivo* synthesized coat protein, thus demonstrating that, under cell-free conditions, mRNA molecules can select the appropriate AA~tRNA precursors.

Over the ensuing decade the cell-free synthesis of a variety of other viral and bacterial proteins was achieved. One of the earliest of these successes was that of the T4 lysozyme, a 20,000 MW enzyme which helps digest bacterial cell walls. Its cell-free synthesis in enzymatically active form further showed the great fidelity of translation possible in *in vitro* systems. While for a brief period there was difficulty in making very long polypeptide chains, by 1969 the synthesis of enzymatically active β-galactosidase (1300 amino acids) was convincingly demonstrated. Moreover, the recent synthesis in cell-free conditions of a variety of important mammalian proteins (e.g., hemoglobin, light and heavy immunoglobulin chains, actin, collagen, and myosin) indicates that the conditions for *in vitro* synthesis of higher cell proteins are now as well understood as those of their bacterial counterparts.

STIMULATION OF AMINO ACID INCORPORATION BY SYNTHETIC mRNA

While the foregoing experiments with viral RNA tell us that reading specificity is preserved in cell-free systems, by themselves they say nothing about the exact form of the genetic code. They cannot tell us which three-letter words (out of the possible 64) (Table 13–1) code for any particular amino acid. To obtain such data, use was made of synthetic polyribonucleotides, whose formation involves the enzyme polynucleotide phosphorylase. This enzyme, found in all bacteria, catalyzes the reaction

$$\text{RNA} + \text{\textcircled{P}} \rightleftharpoons \text{ribonucleoside—\textcircled{P}~\textcircled{P}} \qquad (13\text{–}1)$$

Under normal cell metabolite concentrations, the equilibrium conditions favor RNA degradation to nucleoside diphosphates, so that the main cellular function of polynucleotide phosphorylase may be to control mRNA lifetime (see Chapter 14). By use of high initial nucleoside diphosphate concentrations, however, this enzyme can be made to catalyze the formation of the internucleotide 3'–5' phosphodiester bond (Figure 13–3) and thus make synthetic RNA molecules. Since it is not a biosynthetic enzyme, no template RNA is involved in the RNA synthesis; the base composition of the synthetic product depends en-

Table 13–1   The 64 Possible Three-letter Codons

First Position (5′ End)	Second Position				Third Position (3′ End)
	U	C	A	G	
U	UUU	UCU	UAU	UGU	U
	UUC	UCC	UAC	UGC	C
	UUA	UCA	UAA	UGA	A
	UUG	UCG	UAG	UGG	G
C	CUU	CCU	CAU	CGU	U
	CUC	CCC	CAC	CGC	C
	CUA	CCA	CAA	CGA	A
	CUG	CCG	CAG	CGG	G
A	AUU	ACU	AAU	AGU	U
	AUC	ACC	AAC	AGC	C
	AUA	ACA	AAA	AGA	A
	AUG	ACG	AAG	AGG	G
G	GUU	GCU	GAU	GGU	U
	GUC	GCC	GAC	GGC	C
	GUA	GCA	GAA	GGA	A
	GUG	GCG	GAG	GGG	G

tirely upon the initial concentration of the various ribonucleoside diphosphates in the reaction mixture. For example, when only adenosine diphosphate is used, the resulting RNA contains only adenylic acid, and thus is called polyadenylic acid or poly A. It is likewise possible to make poly U, poly C, and poly G. Addition of two or more different diphosphates produces mixed copolymers such as poly AU, poly AC, poly CU, and poly AGCU. In all these mixed polymers, the base sequences are approximately random, with the nearest-neighbor frequencies determined solely by the relative concentrations of the initial reactants. For example, poly AU molecules with two times as much A as U are formed in sequences like (UAAUAUAAAUAAUAAAAUAUU . . .).

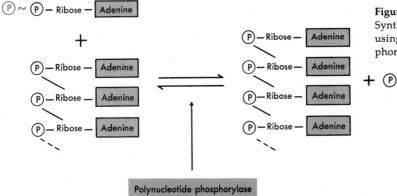

Figure 13–3
Synthesis (degradation) of RNA molecules using the enzyme polynucleotide phosphorylase.

Almost all these synthetic polymers will attach to ribosomes and function as templates. Some polymers, however, are not efficient templates. This does not necessarily mean that they lack functional base sequences (codons). Instead, they may be inactive because most of their bases are hydrogen bonded, so that they cannot attach to the ribosomes.

POLY U CODES FOR POLYPHENYLALANINE

Poly U was the first synthetic polyribonucleotide discovered to have mRNA activity. None of its bases are normally hydrogen bonded in solution, and it binds well to free ribosomes. It selects phenylalanine tRNA molecules exclusively, thereby forming a polypeptide chain containing only phenylalanine (polyphenylalanine). Thus, we know that a codon for phenylalanine is composed of a group of three uridylic acid residues (UUU) (the group number 3 initially came from the genetic experiments described in Chapter 10). Similarly, we are able tentatively to assign (CCC) as a proline codon and (AAA) as a lysine codon on the basis of analogous experiments with poly C and poly A. Unfortunately, the guanine residues in poly G firmly hydrogen bond to each other and form multistranded triple helices that do not bind to ribosomes. Thus, this type of experiment cannot tell us whether (GGG) is a functional codon.

MIXED COPOLYMERS ALLOW
ADDITIONAL CODON ASSIGNMENTS

Poly AC molecules can contain eight different codons (CCC), (CCA), (CAC), (ACC), (CAA), (ACA), (AAC), and (AAA), whose proportions depend on the copolymer A/C ratio. When CA copolymers attach to ribosomes, they cause the incorporation of asparagine, glutamine, histidine, and threonine, in addition to the proline expected from (CCC) codons and the lysine expected from (AAA) codons. The proportions of these amino acids incorporated into polypeptide products depend on the A/C ratio. Thus, since an AC copolymer containing much more A than C promotes the incorporation of many more asparagine than histidine residues, we conclude that asparagine is coded by two A's and one C and histidine is coded by two C's and one A (Table 13–2). Similar experiments with other copolymers have allowed a number of additional assignments. These experiments, however, cannot reveal the order of the different nucleotides within a codon. There is

Table 13–2 Amino Acid Incorporation into Proteins[a]

Amino Acid	Observed Amino Acid Incorporation	Tentative Codon Assignments	Calculated Triplet Frequency				Sum of Calculated Triplet Frequencies
			3A	2A1C	1A2C	3C	
Poly AC (5:1)							
Asparagine	24	2A 1C		20			20
Glutamine	24	2A 1C		20			20
Histidine	6	1A 2C			4.0		4
Lysine	100	3A	100				100
Proline	7	1A 2C, 3C			4.0	0.8	4.8
Threonine	26	2A 1C, 1A 2C		20	4.0		24
Poly AC (1:5)							
Asparagine	5	2A 1C		3.3			3.3
Glutamine	5	2A 1C		3.3			3.3
Histidine	23	1A 2C			16.7		16.7
Lysine	1	3A	0.7				0.7
Proline	100	1A 2C, 3C			16.7	83.3	100
Threonine	21	2A 1C, 1A 2C		3.3	16.7		20

[a] The amino acid incorporation into proteins was observed after adding random copolymers of A and C to a cell-free extract similar to that described in Figure 13–1. The incorporation is given as a percentage of the maximal incorporation of a single amino acid. The copolymer ratio was then used to calculate the frequency with which a given codon would appear in the polynucleotide product. The relative frequencies of the codons are a function of the probability that a particular nucleotide will occur in a given position of a codon. For example, when the A/C ratio is 5:1, the ratio of AAA/AAC = $5 \times 5 \times 5 : 5 \times 5 \times 1 = 125:25$. If we thus assign to the 3A codon a frequency of 100, then the 2A and 1C codon is assigned a frequency of 20. By correlating the relative frequencies of amino acid incorporation with the calculated frequencies with which given codons appear, tentative codon assignments can be made.

no way of knowing from random copolymers whether the histidine codon containing two C's and one A is ordered (CCA), (CAC), or (ACC). Moreover, because of the difficulty of interpreting small amounts of incorporation, a few of the assignments made in this way were wrong. For example, the experiments with AU (1:5) suggested that lysine is coded by two A's and one U, as well as by (AAA). Later experiments, however (see below), showed that U is absent from all lysine codons.

ORDERING CODONS BY tRNA BINDING

A direct way of ordering the nucleotides within some of the codons was developed in 1964. It utilizes the fact that, in the absence of protein synthesis, specific tRNA molecules bind to ribosome–mRNA complexes. For example, when poly U is mixed with ribosomes, only phenylalanine tRNA will attach. Correspondingly, the attachment of poly C to ribosomes promotes the binding of proline tRNA. Most important, this specific binding does not demand the presence of long mRNA molecules. In fact, the binding of a *trinucleotide* to a ribosome is sufficient. The addition of the trinucleotide UUU results in phenylalanine-tRNA attachment, whereas lysine tRNA specifically binds to ribosomes if AAA is added. The discovery of the trinucleotide effect opened up the possibility of relatively easily determining the order of nucleotides within many codons. Before this discovery, it seemed obvious that the order could not be determined unless organic chemists could synthesize long polynucleotides with regular repeating sequences. Now, however, the possession of trinucleotides of known sequence is sufficient to order many codons. For example, the trinucleotide $5'GUU3'$ promotes valine-tRNA attachment, $5'UGU3'$ stimulates cysteine-tRNA binding, and $5'UUG3'$ causes leucine-tRNA binding. After massive effort, all 64 possible trinucleotides were synthesized with the hope of definitely assigning the order of the majority of codons. In Table 13–3 some of the codons determined in this way are listed. It now seems likely that not all the correct combinations can be determined this way. Some of the trinucleotides bind much less efficiently than UUU or GUU, making it impossible to know whether they code for a specific amino acid.

Table 13–3 Binding of Specific tRNA Molecules to Trinucleotide–Ribosome Complexes

Trinucleotide						tRNA Bound
$5'UUU3'$	UUC					Phenylalanine
UUA	UUG	CUU	CUC	CUA	CUG	Leucine
AAU	AUC	AUA				Isoleucine
AUG						Methionine
GUU	GUC	GUA	GUG	UCU		Valine
UCU	UCC	UCA	UCG			Serine
CCU	CCC	CCA	CCG			Proline
AAA	AAG					Lysine
UGU	UGC					Cysteine
GAA	GAG					Glutamic acid

CODON ASSIGNMENT FROM REGULAR COPOLYMERS

At the same time as the trinucleotide binding technique became available, methods were developed using a combination of organic chemical and enzymatic techniques to

Table 13–4 Assignment of Codon Orders Using Regular Copolymers Building from Two Bases

Copolymer	Amino Acids Incorporated	Codon Assignments
CUC|UCU|CUC . . .	Leucine	5'CUC3'
	Serine	UCU
UGU|GUG|UGU . . .	Cysteine	UGU
	Valine	GUG
ACA|CAC|ACA . . .	Threonine	ACA
	Histidine	CAC
AGA|GAG|AGA . . .	Arginine	AGA
	Glutamine	GAG

prepare synthetic polyribonucleotides with known repeating sequences. These regular copolymers direct the incorporation of specific amino acids into polypeptides. For example, the repeating sequence CUCUCUCU . . . is the messenger for a regular polypeptide in which leucine and serine alternate. Similarly, UGUGUG . . . promotes the synthesis of a polypeptide containing two amino acids, cysteine and valine. And ACACAC . . . directs the synthesis of a polypeptide alternating threonine and histidine (Table 13–4). Use of the copolymer built up from repetition of the three-nucleotide sequence AAG (AAGAAGAAG) directs the synthesis of three types of polypeptides: polylysine, polyarginine, and polyglutamic acid. Appearance of all three chains tells us that ribosomes attach to this messenger randomly, starting equally well at its AAG, AGA, and GAA codons. Poly $(AUC)_n$ behaves the same way, being a template for polyisoleucine, polyserine, and polyhistidine. By now, a large number of such copolymers

Table 13–5 Assignment of Codon Orders Using Regular Copolymers Building from Three Bases

Copolymer	Codon Recognized	Polypeptide Made	Codon Assignment
$(AAG)_n$	AAG|AAG|AAG . . .	polylysine	5'AAG3'
	AGA|AGA|AGA . . .	polyarginine	AGA
	GAA|GAA|GAA . . .	polyglutamic acid	GAA
$(UUC)_n$	UUC|UUC|UUC . . .	polyphenylalanine	UUC
	UCU|UCU|UCU . . .	polyserine	UCU
	CUU|CUU|CUU . . .	polyleucine	CUU
$(UUG)_n$	UUG|UUG|UUG . . .	polyleucine	UUG
	UGU|UGU|UGU . . .	polycysteine	UGU
	GUU|GUU|GUU . . .	polyvaline	GUU

Table 13–6 Assignment of Codon Orders Using Regular Copolymers Building from Four Bases

Copolymer	Amino Acids Incorporated	Codon Assignments
UAU|CUA|UCU|AUC|UAU . . .	Tyrosine	$^{5'}$UAU$^{3'}$
	Leucine	CUA
	Serine	UCU
	Isoleucine	AUC
UUA|CUU|ACU|UAC|UUA . . .	Leucine	UUA
	Leucine	CUU
	Threonine	ACU
	Tyrosine	UAC

have been analyzed, giving the results shown in Table 13–5. Only a few polymers having repeating tetranucleotide sequences have been looked at so far. The codon assignments obtained from two of them are revealed in Table 13–6. The sum of all these observations permits the definite assignments of specific amino acids to 61 out of the possible 64 codons (Table 13–7). The remaining three codons, as shown in Table 13–7, code for chain termination.

Table 13–7 The Genetic Code

First Position (5' End)	Second Position				Third Position (3' End)
	U	C	A	G	
U	Phe	Ser	Tyr	Cys	U
	Phe	Ser	Tyr	Cys	C
	Leu	Ser	Term[a]	Term	A
	Leu	Ser	Term	Trp	G
C	Leu	Pro	His	Arg	U
	Leu	Pro	His	Arg	C
	Leu	Pro	GluN	Arg	A
	Leu	Pro	GluN	Arg	G
A	Ileu	Thr	AspN	Ser	U
	Ileu	Thr	AspN	Ser	C
	Ileu	Thr	Lys	Arg	A
	Met	Thr	Lys	Arg	G
G	Val	Ala	Asp	Gly	U
	Val	Ala	Asp	Gly	C
	Val	Ala	Glu	Gly	A
	Val	Ala	Glu	Gly	G

[a] Chain terminating (formerly called "nonsense").

THE CODE IS DEGENERATE

Many amino acids are selected by more than one codon (degeneracy). For example, both (UUU) and (UUC) code for phenylalanine, while serine is coded by (UCU), (UCC), (UCA), (UCG), (AGU), and (AGC). The present data suggest that when the first two nucleotides are identical, the third nucleotide can be either cytosine or uracil and the codon will still code for the same amino acid. Often adenine and guanine are similarly interchangeable. However, not all degeneracy seems to be based on equivalence of the first two nucleotides. Leucine, for example, seems to be coded by (UUA) and (UUG), as well as by (CUU), (CUC), (CUA), and (CUG) (Figure 13–4).

Codon degeneracy, especially the frequent third-place equivalence of cytosine and uracil or guanine and adenine, underlies the fact that the AT/GC ratios can show very great variations (see Chapter 9) without correspondingly large changes in the relative proportion of amino acids found in these organisms. The original explanation was that these similarities in amino acid composition were meaningless, reflecting the sequences of only those genes coding for proteins present in large quantities. But as the analysis of more individual proteins revealed no real correlation between amino acid composition and evolutionary position, this interpretation became untenable.

It was also at first guessed that a specific anticodon would exist for every codon. If so, at least 61 different tRNA's, possibly with an additional three for the chain-terminating codons, would be present. Evidence soon began to appear, however, that highly purified tRNA species of known sequence (e.g., alanyl-tRNA) could recognize several different codons. Several cases also arose where an anticodon base was not one of the four regular ones, but a fifth base, inosine. Like all the other minor tRNA bases, this arises through enzymatic modification of a base present in an otherwise completed tRNA. The base from which it is derived is adenine, whose 6-carbon is deaminated to give the 6-keto group of inosine.

THE WOBBLE IN THE ANTICODON

To explain these observations, the wobble concept was devised. It states that the base at the 5' end of the anticodon is not as spatially confined as the other two, allowing it to form hydrogen bonds with any of several bases located at the 3' end of a codon. Not all combinations are possible, with pairing restricted to those shown

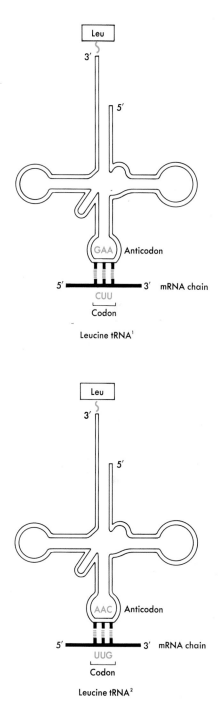

Figure 13–4

Two different tRNA molecules which accept leucine residues. Each recognizes a different code word.

Table 13-8 Pairing Combinations with the Wobble Concept

Base in Anticodon	Base in Codon
G	U or C
C	G
A	U
U	A or G
I	A, U, or C

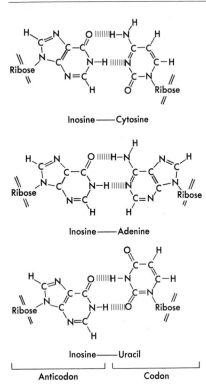

Inosine——Cytosine

Inosine——Adenine

Inosine——Uracil

| Anticodon | Codon |

"Wobble" enables this base to form hydrogen bonds with bases other than those in standard base pairs.

(a)

in Table 13-8. For example, U at the wobble position can pair to adenine and guanine, while I can pair with U, C, and A (Figure 13-5). These rules do not permit any single tRNA molecule to recognize four different codons, and three codons are recognized only when inosine occupies the third position.

Virtually all evidence now available supports the wobble concept. For example, it predicted correctly that at least three tRNA's existed for the six serine code words (UCU, UCC, UCA, UCG, AGU, and AGC). The other two amino acids (leucine and arginine) whose degenerate codons differ in the first or second position also have different tRNA's for each set.

In the recently established yeast tRNA^{phe} structure, all the anticodon bases point in roughly the same direction, with their exact conformations largely determined by stacking interactions between the flat surfaces of their respective bases. Conceivably the third anticodon base is less restricted in its movements than the other two bases—hence, third position wobble! This point, however, is still far from certain; and firm proof will only come when several more structures are worked out (and then to higher resolution than so far obtained for yeast tRNA^{phe}.)

While wobble permits a given base to recognize several bases, the binding efficiencies show considerable variation. So when the code word used does not have a tRNA which binds well to it, we might expect that the corresponding amino acid is inserted into protein at a slow rate. If this conjecture proves correct, the rate of synthesis of a given protein will be controlled in part (Chapter 14) by which codons construct its particular amino acid sequence.

Figure 13-5
Examples of wobble pairing.

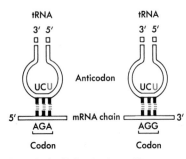

(b) U in the third anticodon position can pair with A or G.

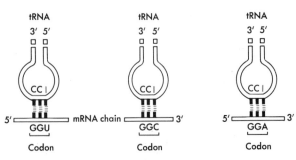

(c) I in the third position can pair with U, C, or A.

MINOR tRNA's

Sometimes several chemically distinct tRNA forms have the same anticodon. Often the only difference between two forms is the presence or absence of the 3' terminal adenylic acid. Within cells this difference is not permanent, for an "adding enzyme" exists which puts terminal adenylic acid residues on those molecules which lack them. Why some tRNA molecules lack their 3' terminal residues is still a total mystery. Perhaps the original absence was caused by an accidental exonuclease attack, in which case the adding enzyme has a repair function. But conceivably the absence is somehow involved in controlling the general rate of protein synthesis, as the deficient molecules are unable to accept their specific amino acids.

Much more intriguing are cases where internal sequence differences occur and where one species is present in very large amounts compared to the other species. There are, for example, a major and a minor tyrosine tRNA, both having the anticodon $^{3'}AUG^{5'}$. Origin of the minor component by enzymatic modification of the major component following gene transcription is ruled out by the nature of the sequence differences. They cannot arise by any known enzymatic transformations. The normal function played by these minor tRNA species remains unresolved, but as we see below, they are frequently involved in suppressor gene action.

CODON FREQUENCIES WITHIN NATURAL mRNA's

The fact that most amino acids are specified by more than one codon poses the question whether the alternative codons are used in approximately equal frequencies or whether some are preferentially employed. Clearly some bias is to be expected when the amounts of the four bases are not equal. For example, in organisms containing AT-rich DNA, we might expect that the most frequently used codons have U or A at the third position. Now our most complete group of codon assignments come from the RNA phage MS2, which has approximately equal numbers of the four bases. Here it has been possible to make definitive assessments for some 420 codons. Inspection of this data (Table 13–9) reveals no widespread preferential use of given codons. While some codons are used somewhat more often than their alternatives, in almost every case there is no reason to ascribe such bias other than to chance. When the complete MS2 RNA sequence (3,000 bases) becomes known, the answer should be less obscured by mere statistical fluctuations.

Table 13–9   Known Codons at Work in MS2 RNA—Over Half of These Are Found in the Completely Worked Out Coat Protein Sequence. The Other Half Represent Codons Used to Code for Certain "A" Protein and Replicase Sequences

	U		C		A		G		
U	Phe	10	Ser	13	Tyr	8	Cys	7	U
	Phe	13	Ser	10	Tyr	13	Cys	4	C
	Leu	11	Ser	10	Term	1	Term		A
	Leu	4	Ser	13	Term	1	Trp	14	G
C	Leu	10	Pro	7	His	4	Arg	13	U
	Leu	14	Pro	3	His	4	Arg	11	C
	Leu	13	Pro	6	GluN	10	Arg	6	A
	Leu	6	Pro	5	GluN	16	Arg	4	G
A	Ileu	8	Thr	14	AspN	11	Ser	4	U
	Ileu	16	Thr	10	AspN	23	Ser	8	C
	Ileu	7	Thr	8	Lys	12	Arg	8	A
	Met	15	Thr	5	Lys	17	Arg	6	G
G	Val	13	Ala	19	Asp	18	Gly	17	U
	Val	12	Ala	12	Asp	11	Gly	11	C
	Val	11	Ala	14	Glu	9	Gly	4	A
	Val	10	Ala	8	Glu	14	Gly	4	G

Much less data is now available for DNA, as opposed to RNA, chromosomes. Table 13–10 lists codons used in a variety of bacterial and DNA viral mRNA's. While marked departures from random use appear to exist, much more data must be obtained before the matter becomes clarified.

Table 13–10   Codons Established from Preliminary Work on *E. coli* Lactose, Galactose, and Tryptophan mRNA's, ϕX mRNA, F1 mRNA, T4 mRNA, and Hemoglobin mRNA.

	U		C		A		G		
U	Phe	2	Ser	1	Tyr		Cys		U
	Phe		Ser		Tyr	2	Cys		C
	Leu		Ser	2	Term	1	Term	1	A
	Leu		Ser		Term		Trp		G
C	Leu	1	Pro		His	1	Arg		U
	Leu		Pro		His	1	Arg	1	C
	Leu		Pro	1	GluN	2	Arg		A
	Leu	3	Pro	1	GluN	2	Arg		G
A	Ileu	4	Thr	1	AspN	1	Ser	2	U
	Ileu		Thr	2	AspN		Ser	2	C
	Ileu		Thr	1	Lys	2	Arg	1	A
	Met	5	Thr	2	Lys		Arg		G
G	Val	3	Ala		Asp	1	Gly	3	U
	Val		Ala		Asp		Gly		C
	Val		Ala		Glu	3	Gly	1	A
	Val		Ala		Glu		Gly	1	G

AUG AND GUG AS INITIATION CODONS

The discovery of a tRNA specific for N-formyl methionine (tRNA$^{f\text{-met}}$) initially suggested that it would recognize a codon different from the AUG methionine codon. But even at first, the conjecture seemed strained, since no unassigned codon existed which could specifically bind to tRNA$^{f\text{-met}}$. As shown in Figure 12–20, elucidation of the tRNA$^{f\text{-met}}$ sequence revealed that its anticodon $3'UAC^{5'}$ was identical to that of tRNA^{met}. Both f-met and met must, therefore, be directly coded by AUG. Discrimination between the two forms occurs through their differential binding to an initiation (elongation) factor. Only tRNA$^{f\text{-met}}$ can bind to the initiation factors to form the 30-S initiation complex. Correspondingly, only tRNA^{met} is able to bind to EF–T during the polypeptide extension phase.

A further complication is the suggestion, obtained from *in vitro* studies with regular copolymers, that tRNA$^{f\text{-met}}$ will bind to and initiate synthesis at GUG as well as AUG codons. Normally, GUG is a valine codon and its recognition by tRNA$^{f\text{-met}}$ suggests a different sort of wobble at the f-met anticodon. It would have ambiguity at first as opposed to the third position. A possible reason for this lies in the tRNA$^{f\text{-met}}$ sequence. The nucleotide adjacent to the 3' end of the anticodon is an unmodified adenine, not the bulky alkylated derivative found in almost all other tRNA's.

While at first there was the suspicion that GUG might be a start codon only under unnatural *in vitro* situations, recently obtained nucleotide sequence data tells us that GUG is the start codon for the "A" protein of the RNA phage MS2. Moreover, GUG has also been found to initiate proteins when the normal AUG start codons have been lost as a result of genetic deletions. Since *in vitro* GUG starts are much less efficient than AUG starts, its lower affinity for tRNA$^{f\text{-met}}$ may be a device to help control the rate at which a given gene functions. Clearly many more genes must be analyzed before the role of GUG as a start codon becomes clear.

CODONS FOR CHAIN TERMINATION

The three codons UAA, UAG, and UGA do not correspond to any amino acid. Instead, they signify chain termination. As mentioned in Chapter 12, these codons are read not by special chain-terminating tRNA's but by specific proteins, the release factors. Two release factors have so far been identified, each of which recognizes two codons. One is specific for UAA and UAG and the other for UAA and UGA. Knowledge of how two different codons can be

361

recognized by each release factor must await detailed knowledge of their 3-D structures. The use of proteins to read the stop signals emphasizes a fact ignored by many biochemists. The specific hydrogen bond-forming groups along a polynucleotide can also be recognized by those amino acids which have groups prone to hydrogen bonding.

ENDING OF POLYPEPTIDE MESSAGES WITH EITHER ONE OR TWO STOP CODONS

Why three different stop codons exist is not at all obvious, and much speculation immediately arose as to which codon would be preferentially used. So elucidation of the nucleotide sequences at the ends of actual genes was eagerly awaited. Now (spring 1975), however, with tentative assignments known for some ten such stop signals, no clear picture emerges. Genes exist where each stop codon is singly used as well as several cases where genes end with two successive stop signals (Figure 13–6). The use of two stop codons may be a safety factor to take care of the rare cases where the first codon fails, but why this device is used only in special cases is unclear. For example, in the RNA phages homologous "coat protein genes" terminate with either one (QB) or two stop codons (R17 and MS2).

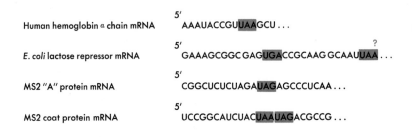

Figure 13–6
Nucleotide sequences at the 3' ends of several mRNA's, showing the stop codon(s) used to terminate the respective translational processes.

NONSENSE VERSUS MISSENSE MUTATIONS

An alteration that causes a codon specific for a given amino acid to specify another amino acid is called a *missense mutation*. On the other hand, the change to a chain-termination codon is known as a *nonsense mutation*. Given the existence of only three chain-termination codons, most mutations involving single-base replacements are likely to result in missense rather than nonsense. As new proteins arising by missense mutations contain only single amino acid replacements, they frequently possess some of the biological activity of the original proteins. The abnormal hemoglobins (see Chapter 4) are the result of missense mutations.

Table 13–11 Examples of Possible Codon Changes Underlying Some Amino Acid Replacements in the Mutant Hemoglobins

Amino Acid in Normal Hemoglobin		Amino Acid in Mutant Hemoglobin	
Lysine (AAA)	$\longrightarrow$	Glutamic Acid (GAA)	A → G
Glutamic acid (GAA)	$\longrightarrow$	Glutamine (CAA)	G → C
Glycine (GGU)	$\longrightarrow$	Aspartic acid (GAU)	G → A
Histidine (CAU)	$\longrightarrow$	Tyrosine (UAU)	C → U
Asparagine (AAU)	$\longrightarrow$	Lysine (AAA)	U → A
Glutamic acid (GAA)	$\longrightarrow$	Valine (GUA)	A → U
Glutamic acid (GAA)	$\longrightarrow$	Lysine (AAA)	G → A
Glutamic acid (GAA)	$\longrightarrow$	Glycine (GGA)	A → G

Table 13–11 shows that the amino acid replacement data obtained from these changed hemoglobin molecules strongly support the idea that these mutations result from the substitution of single nucleotides. A companion replacement series (Table 13–12) obtained from mutant TMV protein molecules points to the same conclusion. Moreover, the fact that only certain specific changes are observed (e.g., glycine to aspartic acid) also supports the hypothesis that these altered proteins arise from single nucleotide changes. If most observed mutations reflected changes in each of several adjacent nucleotides, a larger variety of amino acid switches would be observed.

Table 13–12 Amino Acid Replacements Induced by Nitrous Acid Treatment of TMV*

Proline (CCC)	$\longrightarrow$	Serine (UCC)	C → U
Proline (CCC)	$\longrightarrow$	Leucine (CUC)	C → U
Isoleucine (AUU)	$\longrightarrow$	Valine (GUU)	A → G
Isoleucine (AUA)	$\longrightarrow$	Methionine (AUG)	A → G
Leucine (CUU)	$\longrightarrow$	Phenylalanine (UUU)	C → U
Glutamic acid (GAA)	$\longrightarrow$	Glycine (GGA)	A → G
Threonine (ACA)	$\longrightarrow$	Isoleucine (AUA)	C → U
Threonine (ACG)	$\longrightarrow$	Methionine (AUG)	C → U
Serine (UCU)	$\longrightarrow$	Phenylalanine (UUU)	C → U
Serine (UCG)	$\longrightarrow$	Leucine (UUG)	C → U
Aspartic acid (GAC)	$\longrightarrow$	Glycine (GGC)	A → G

* All the observed changes can be fitted both with possible codon assignments and with the postulated mutagenic action of nitrous acid (C → U, A→ G).

NONSENSE MUTATIONS PRODUCE INCOMPLETE POLYPEPTIDE CHAINS

When a nonsense mutation occurs in the middle of a genetic message, an incomplete polypeptide is released from the ribosome due to premature chain termination. Very often these incomplete chains have no biological activity (e.g., no enzymatic action), making most nonsense mutations in vital genes easily detectable. In contrast, the majority of missense mutations have some biological activity and are usually overlooked. Thus, after treatment with a mutagen, a sizeable fraction of the *detectable* mutations is of the nonsense variety.

The size of the incomplete polypeptide chain produced depends upon the relative site of the nonsense mutations. Mutations occurring near the beginning of a gene result in very short fragments, while if the site is near the end, the fragment is of almost normal length and may have some biological activity. This fact provides a way for precisely locating a mutation within a given gene. Isolation of a series of incomplete fragments and measurements of their length unambiguously tells us the sites of the corresponding mutations.

READING MISTAKES CAN OCCUR IN CELL-FREE PROTEIN SYNTHESIS

Under certain conditions reading mistakes can be very frequent in cell-free systems. Soon after the discovery that poly U is the template for polyphenylalanine, the apparent paradox arose that in the absence of phenylalanine, poly U templates directed the synthesis of polyleucine. This meant that the (UUU) codon was selecting leucine-specific tRNA molecules. At first, the possibility was considered that a fundamental ambiguity in the (UUU) codon might exist. Now, however, it is clear that the anomalous leucine incorporation was due to the use of excessive amounts of Mg^{2+} in the incorporation experiments. When the Mg^{2+} levels are lowered, poly U-directed leucine incorporation becomes much less frequent. The result is important, as it underlines the necessity of using normal physiological conditions in experiments with cell-free systems if we want to extrapolate to events occurring within a normal cell.

SUPPRESSOR GENES UPSET THE READING OF THE GENETIC CODE

Mistakes in reading the genetic code also occur in living cells. These mistakes underlie the phenomenon of suppressor genes. Their existence was for many years very puzzling and seemingly paradoxical. Numerous examples were known where the effects of harmful mutations were reversed by a second genetic change. Some of these subsequent mutations were very easy to understand, being simple *reverse* (or back) mutations which change an altered nucleotide sequence back to its original arrangement. Much more difficult to understand were other mutations occurring at different locations on the chromosome, which suppress the change due to a mutation at site A by producing an additional genetic change at site B. Such *suppressor mutations* fall into two main categories: those due to nucleotide changes within the same gene as the original mutation but at a different site on this gene (intragenic suppression), and those occurring in another gene (intergenic suppression). Those genes which cause suppression of mutations in other genes are called *suppressor* genes.

Now we realize that these two types of suppression both work by causing the production of good (or partially good) copies of the protein made inactive by the original harmful mutation. For example, if the first mutation caused the making of inactive copies of one of the enzymes involved in making arginine, then the suppressor mutation allows the production of arginine by restoring the synthesis of some good copies of this same enzyme. However, the mechanisms by which intergenic and intragenic suppressor mutations cause the resumption of the synthesis of good proteins are completely different.

Those mutations which can be reversed through additional changes in the same gene often involve insertions or deletions of single nucleotides. These shift the reading frame (see Chapter 10) so that all the codons following the insertions (or deletions) are completely changed, thereby generating new amino acid sequences. More rarely, the shifted reading frame generates premature nonsense codons, and as a result the mutant polypeptides are correspondingly shorter (Figure 13–7).

Intragenic suppression may occur when a second mutation deletes (or inserts) a new nucleotide near the original change and thus restores the original codon arrangement beyond the second change (Figure 13–7). Even though there are still scrambled codons between the two changes, there is a good probability, because of degeneracy, that the scrambled codons all code for some amino acid. If so, full-length, often functional, proteins may be produced.

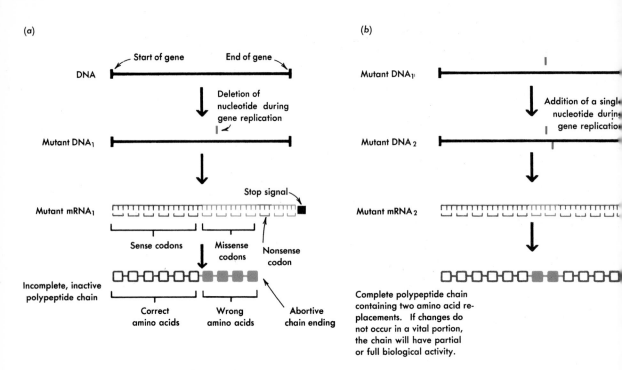

Figure 13–7

(a) The effect of a single nucleotide deletion (insertion) mutation upon the reading of the genetic message. (b) The mechanism by which a nucleotide addition (deletion) mutation can suppress the havoc caused by a previous deletion (insertion) mutation.

Intragenic suppression can also result from a second missense mutation. In these cases, the original loss of enzymatic activity is due to an altered 3-D configuration resulting from the presence of a wrong amino acid. A second missense mutation in the same gene brings back biological activity if it somehow restores the original configuration around the functional part of the molecule. An example of this type of suppression in the tryptophan synthetase system was shown in Chapter 8 (Figure 8–13).

SPECIFIC CODONS ARE MISREAD BY SPECIFIC SUPPRESSOR GENES

Suppressor genes do not act by changing the nucleotide sequences of the mutant DNA. Instead they change the way in which the mRNA templates are read. There are a number of different suppressor genes in *E. coli*. Since each causes the misreading of specific codons, they can reverse the effects of only a small fraction of the single nucleotide changes within a given gene. For example, if we collect a large number of mutations blocking the synthesis of the

enzyme β-galactosidase (see Chapter 14), only several percent of these mutations will be suppressed by a given
suppressor gene a. These few mutations would be due to
changes in codons whose reading is specifically affected
by gene a. Similarly, a completely different small fraction
of β-galactosidase mutations can be suppressed by suppressor gene b.

It is generally observed that a given suppressor gene
can suppress mutations in a number of different genes.
This fact is easily understood by the misreading concept.
For example, the ability to synthesize both arginine and
tryptophan in certain double mutants unable to make
either amino acid can be restored by a single change in a
suppressor gene. We merely need to postulate that both
these growth requirements are caused by the same specific
changes to missense or nonsense.

NONSENSE SUPPRESSION INVOLVES MUTANT tRNA's

Suppressor genes exist for each of the three chain-terminating codons. They act by reading a stop signal as if it
were a signal for a specific amino acid. There are, for example, three well-characterized genes which suppress the
UAG codon. One suppressor gene inserts serine, another
glutamine, and a third tyrosine at the nonsense position.
All three UAG suppressors act by producing anticodon
changes in a tRNA species specific for a given amino acid.
For example, the tyrosine suppressor arises by a mutation
within a $tRNA^{tyr}$ gene which changes the anticodon from
$3'AUG^{5'}$ to $3'AUC^{5'}$, thereby enabling it to recognize UAG
codons (Figure 13–8). The suppression of UAA codons also
appears to be mediated by mutant tRNA's. Because of the
general inefficiency of UAA suppression, however, we still
know much less about its detailed operation.

tRNA-mediated UAG suppression was first demonstrated with a nonsense mutation which blocked synthesis
of the coat protein of the RNA phage R17. When mutant
RNA containing this nonsense codon was used in an *in
vitro* system, no coat protein was produced unless purified
tRNA from the correct suppressor strain was added.

UGA suppression also involves mutant tRNA molecules, with the active agent being a tryptophan tRNA
($tRNA^{trp}$) which inserts tryptophan at the nonsense positions. Normally $tRNA^{trp}$ reads only the UGG codon, but as
a result of the suppressor mutation it becomes able to read
UGA as well as UGG. Much to everyone's initial surprise,
the basis for this change is not a base change at the anticodon but instead a G → A replacement at position 24.

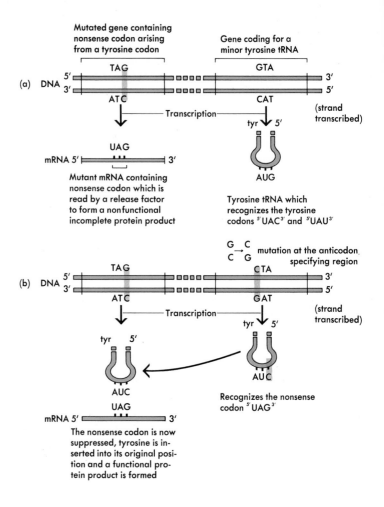

Figure 13–8
Scheme to show the action of a minor tyrosine tRNA component as a nonsense suppressor.

How this change leads to ambiguity in the anticodon–codon interaction is not clear, since x-ray analysis of tRNA^{phe} suggests that position 24 of the tRNA^{trp} lies some 15Å distant from the anticodon. Nevertheless, we must suppose the anticodon of this suppressor tRNA frequently is improperly placed within its ribosome binding site.

Discovery of tRNA-mediated suppression raised the question of how the respective normal codons could continue to be read. In the case of the tyrosine UAG suppressor, the answers come from the discovery that two genes normally code for tyrosine tRNA. One codes for the major component; the other codes for a component present in much smaller amounts and is the site of the suppressor mutation. Moreover, it was observed that the suppressor mutations most frequently occur in strains in which the minor tRNA gene is duplicated (present as two copies). Selection of those strains having the duplication occurs because loss of the minor component, while not lethal to a host cell, slows its growth. The true function of the minor component remains a tantalizing mystery. Conceivably it plays a regulatory role and thus only small numbers need be present. No such dilemma exists for UGA suppression since the suppressing tRNA^{trp} retains its capacity to read UGG (tryptophan) codons.

NONSENSE SUPPRESSORS MUST ALSO READ NORMAL TERMINATING SIGNALS

Suppression of both the UAG and UGA codons is very efficient. In the presence of the suppressor tRNA's, over half the chain-terminating signals are read as specific amino acid codons. In contrast, suppression of the UAA codon usually averages between 1 and 5 percent. At first it was believed that the efficient suppression of UAG and UGA meant that they seldom, if ever, served as normal terminator signals. If they were frequently used, the presence of their specific suppression would prevent much normal chain termination, leading to the production of aberrantly long polypeptides and cessation of cell growth. Now, however, that a double stop signal has been seen, it is easier to understand why the UAG and UGA suppression have no effect on bacterial growth. We need merely postulate that these stop codons almost never occur singly. The case of the UAA suppressor is less clear since its acquisition always slows down the growth rate. Thus many genes may terminate with UAA alone. Much more sequence data are necessary before clear answers can emerge.

MUTATIONS IN NORMAL STOP SIGNALS

Frequent occurrences of genes terminated by single UAA stop signals might mean that mutations would occasionally be detected which convert stop codons to ones which specify specific amino acids. If the 3' ends of the respective normal messages contain untranslated nucleotide sequences, translation of such mutant messages should lead to abnormally long proteins. Now we already know of one such mutation in the human gene for the α chain of the hemoglobin molecule. Normally the α chain is 141 amino acids long, with translation ceasing upon the reading of a single UAA stop codon. But when the stop codon is converted by a U → C change, glutamine becomes inserted at position 142 and translation proceeds to produce a chain containing 172 amino acids (Figure

13–9). Thus there must exist at least 93 nucleotides at the 3' end of the α chain mRNA molecule which normally are never translated.

Figure 13–9

The origin of an extra-long human hemoglobin α chain (hemoglobin "Constant Spring") through a U → C change in its UAA stop codon. An abnormally long α chain also characterizes hemoglobin "Wayne" which has a deletion of an A in the 139 lysine codon. The deletion shifts the reading frame so as to abolish the normal UAA stop codon. A new UAG stop codon is created at position 149, leading to an α Wayne chain of 149 residues.

```
                              140
               ser  asn  thr  val  lys  leu  glu  pro  arg  term
α Wayne     ...UCC  AAUACCGUUAAGCUGGAGCCUCGGUAG
                       \
                        Deletion of A

α Wild Type ...UCC AAAU ACCGUUAA
                              140
               ser  lys  tyr  arg  term


                              140                         149        172
               ser  lys  tyr  arg  gln  ala  gly  ala  ser  val  ala .. val phe  glu
α Constant Spring . UCCAAAU ACCGU C AAGC UGGAGCCUCGGUAGC
```

tRNA-MEDIATED MISSENSE SUPPRESSION

Suppression of missense mutations can also be mediated by mutant tRNA's. This was recently demonstrated for a mutation in the tryptophan synthetase A gene which replaces glycine with arginine (see Chapter 8), thereby giving rise to an inactive enzyme. Suppressor mutations exist which cause the insertion of glycine at the newly made arginine site and thus restore enzyme function. The efficiency of suppression is low, so in the presence of the suppressor gene both active and inactive forms of the enzyme are made. The nature of this suppression was investigated using an *in vitro* synthesizing system in which the mRNA was the regular copolymer polyAG (AGAG . . .). Normally it codes only for glutamic acid and arginine, but in the presence of tRNA extracted from the suppressor strain, traces of glycine were also incorporated into polypeptides. Furthermore, the level of glycine appearance corresponded well with the frequency of *in vivo* suppression.

The suppressor tRNA's cochromatograph with glycyl tRNA fractions, suggesting that the respective mutations involve altered tRNA^{gly} anticodons. Proof that this is the case comes from further analysis of the glycyl tRNA's of *E. coli*. Three different genes code for tRNA^{gly}, each specifying a different anticodon and so recognizing different subsets of the four glycine codons (Table 13–13). One, tRNA$^{gly I}_{GGG}$ specifically recognizes GGG, a codon also recognized by tRNA$^{gly II}_{GGA/G}$ since wobble gives it the possibility of recognizing both GGG and GGA. Thus tRNA$^{gly I}_{GGG}$ is not necessary for protein synthesis, and so a change in its an-

Table 13–13 The Glycyl tRNA's of *E. coli,* Showing the Origin of Missense Suppressor tRNA's which Insert Glycine at Arginine Codons

Gene	tRNA Product	Anticodon	Glycine Codons Recognized	Derived Missense Suppressor Gene	Anticodon in tRNA Product	Arginine Codons Recognized
gly U	tRNA$_{GGG}^{gly\ I}$	$^{3'}$CCC$^{5'}$	GGG	glyUsu	UCC	AGG
gly T	tRNA$_{GGA/G}^{gly\ II}$	$^{3'}$CCU$^{5'}$	GGA	glyTsu*	UCU	AGA
			GGG			AGG
gly V	tRNA$_{GGU/C}^{gly\ III}$	$^{3'}$CCA$^{5'}$	GGU			
			GGC			

* Strains with this mutation grow very badly (semilethal). The origin of the residual ability to read GGA codons is unclear.

ticodon from $^{3'}$CCC$^{5'}$ to $^{3'}$UCC$^{5'}$ allows it to be selected by the AGG arginine codon without simultaneously leading to an inability of its respective cell to read the glycine GGG codon.

FRAMESHIFT SUPPRESSION

Suppressor genes also exist which mask the effects of certain frameshift mutations created by the insertion of nucleotides. At least four of the six known frameshift suppressor genes code for mutant tRNA's. Two code for proline tRNA's, while the two others code for glycine tRNA's. Complete sequence analysis of one of these frameshift glycine tRNA's (tRNA$_{sufD}^{gly}$ = tRNA$_{GGGG}^{gly}$) reveals the presence of an extra base in its anticodon, with a CCC sequence being replaced by CCCC (Figure 13–10). The remainder of its sequence is identical with that of tRNA$_{GGG}^{gly\ I}$, indicating that it arose from an insertion mutation in gly U, the gene that codes for this tRNA (Table 13–13). Possession of this extra base not only allows it to read four bases at a given time, but also leads to the subsequent pulling of a four-nucleotide length of mRNA over the ribosomal surface, thereby restoring the reading frame to the correct position. All known frameshift suppressor tRNA's act only on insertion mutations, never on deletions; this suggests that no way exists of reading only two bases at a given time.

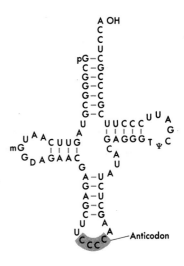

Figure 13–10
The nucleotide sequence of the frameshift suppressor tRNA$_{sufD}^{gly}$ drawn in the cloverleaf configuration.

RIBOSOMAL MUTATIONS ALSO AFFECT THE READING ACCURACY

The level of both nonsense and missense suppression is also determined by the exact ribosome structure. Specific mutations in several of the 30-S proteins affect the accuracy of reading. A number of different amino acid replacements each must distort the ribosome structure so that the disturbed template–ribosome complex is not always able to choose the correct tRNA molecule (Figure 13–11). One such mutation, ram (ribosomal ambiguity), will suppress weakly all three nonsense codons in the complete absence of any suppressing tRNA's. There are also hints that some frameshift mutations may also be suppressed by ram mutants, suggesting that specific ribosomal proteins help ensure the correct stepwise movement of mRNA across the ribosomal surface.

Figure 13–11
Schematic drawing showing how a missense mutation in a gene coding for one of the ribosomal proteins acts as a suppressor mutation.

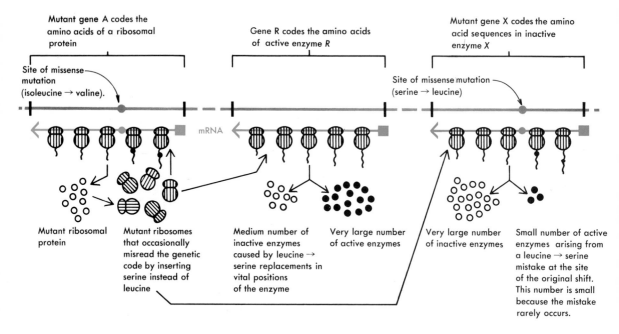

STREPTOMYCIN CAUSES MISREADING

The belief that distorted ribosomes may misread the genetic code is strongly supported by experiments showing that the addition of the antibiotic streptomycin to either *in vitro* systems or living cells promotes mistakes in the translation of the genetic code. It does this by combining with the ribosomes, thereby disturbing the normal mRNA–tRNA–ribosome interactions. The extent of the misreading depends upon whether the streptomycin is added to streptomycin-sensitive or streptomycin-resistant cells. Addition of the antibiotic to sensitive cells results in large scale misreading. The mutation to streptomycin resistance alters the ribosomes in such a way that misreadings occur much less commonly. They are, nonetheless, frequent enough to suppress a number of mutations by causing the synthesis of a small number of active enzyme molecules (Figure 13–12).

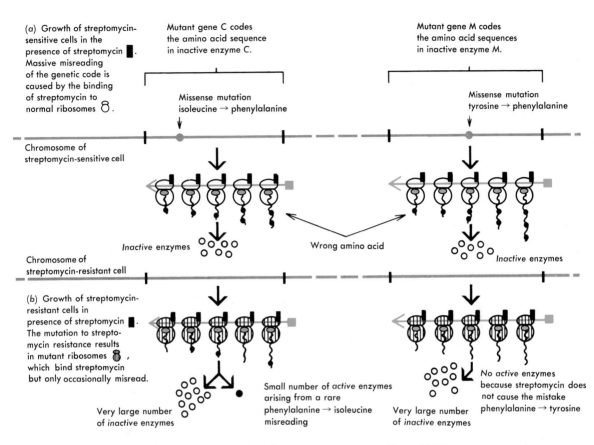

(a) Growth of streptomycin-sensitive cells in the presence of streptomycin ▪. Massive misreading of the genetic code is caused by the binding of streptomycin to normal ribosomes Ծ.

Mutant gene C codes the amino acid sequence in inactive enzyme C.

Mutant gene M codes the amino acid sequences in inactive enzyme M.

Missense mutation isoleucine → phenylalanine

Missense mutation tyrosine → phenylalanine

Chromosome of streptomycin-sensitive cell

Inactive enzymes

Wrong amino acid

Inactive enzymes

Chromosome of streptomycin-resistant cell

(b) Growth of streptomycin-resistant cells in presence of streptomycin ▪. The mutation to streptomycin resistance results in mutant ribosomes ⊛, which bind streptomycin but only occasionally misread.

Very large number of *inactive* enzymes

Small number of *active* enzymes arising from a rare phenylalanine → isoleucine misreading

Very large number of *inactive* enzymes

No *active* enzymes because streptomycin does not cause the mistake phenylalanine → tyrosine

Figure 13–12
Schematic drawing illustrating the action of streptomycin upon streptomycin-sensitive and streptomycin-resistant *E. coli* cells.

It now appears that streptomycin does not cause indiscriminate misreading. When poly U is used as a template with sensitive ribosomes, the most frequent error is the replacement of phenylalanine (UUU) by isoleucine (AUU). This hints that the presence of streptomycin normally disturbs the position of only one out of three nucleotides in the (UUU) codon (Figure 13–13).

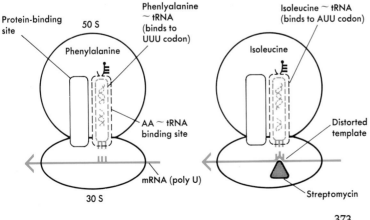

Protein-binding site

50 S

Phenylalanine

Phenlylalanine ~ tRNA (binds to UUU codon)

AA ~ tRNA binding site

mRNA (poly U)

30 S

Isoleucine ~ tRNA (binds to AUU codon)

Isoleucine

Distorted template

Streptomycin

Figure 13–13
Selection of an isoleucine~tRNA molecule by a ribosome–poly U–streptomycin complex. Here the streptomycin-induced misreading involves an isoleucine~tRNA molecule which normally attaches to the (AUU) codon.

SUPPRESSOR GENES ALSO MISREAD GOOD GENES

We thus see that suppressor genes do not specifically misread mRNA templates made on mutant genes. In fact, they affect the synthesis of essentially all proteins. Most suppressor mistakes, therefore, occur in the copying of good mRNA templates, which hinders the synthesis of sound proteins. These changes are generally not very harmful to the growing cell, since many more good copies of each protein than bad ones are produced. There is, however, no advantage for a normal cell to harbor suppressor mutations which cause it to produce even a small fraction of bad proteins. Suppressors tend to be selected against in evolution, unless a harmful mutation is present for whose effect they must compensate.

It is now possible to make a general prediction about the normal function of suppressor genes. A gene becomes a suppressor gene by mutation. Before this mutation occurs, the gene is normal and active, coding for a specific tRNA, for one of the ribosomal proteins, or for one of the enzymes involved in protein synthesis. It has evolved so that its product has the optimal configuration for accurate reading of the genetic code. If a mutation changes such a gene so that its altered product increases the misreading level, this gene becomes a suppressor gene. Only when an increased mistake level is necessary for cellular existence do its mutant products have a selective advantage over their normal counterparts.

THE CODE IS PROBABLY UNIVERSAL

Poly U stimulates phenylalanine incorporation in cell-free extracts from a variety of different organisms ranging from bacteria to higher mammals. Likewise, poly C promotes proline incorporation and poly A causes lysine incorporation in all extracts tested, regardless of their cellular source. Such indications of the universality of the code among contemporary organisms hint that the genetic code has remained constant over a long evolutionary period. But until all the codons in several organisms have been unambiguously worked out, this point will be neither rigorously proved nor disproved. Invariability in most of the code is expected. Consider what a mutation which changed the genetic code would result in. Such a mutation might, for example, alter the sequence of the serine-specific tRNA molecules of the class that corresponds to (UCU), thereby causing them to attach to (UUG) sequences instead. This would be a lethal mutation in haploid cells containing only one gene directing the production of each type of tRNA: No normal serine-specific tRNA of that class would be produced, and serine would not be inserted into many of its normal positions. Even if there were more than one gene for each tRNA type (e.g., in a diploid cell), this type of mutation would still be lethal, since it would cause the simultaneous replacement of many phenylalanine residues by serine in most cell proteins.

SUMMARY

The most direct way to study the genetic code is to examine protein synthesis in cell-free extracts. The most useful in vitro systems employ cell extracts that have been depleted of their original messenger component. Addition of new mRNA to these extracts results in the production of new proteins whose amino acid sequences are determined by the externally added mRNA. For example, the introduction of phage F2 RNA produces new proteins virtually identical to the F2 coat protein. Thus, viral genetic RNA also acts as mRNA.

The first (and probably most important) step in cracking the genetic code occurred when the synthetic polyribonucleotide poly U was found to code specifically for polyphenylalanine. A codon for phenylalanine is thus (UUU). Use of other synthetic polyribonucleotides, homogeneous (poly C, etc.) and mixed (poly AU, etc.), then produced a number of tentative codon assignments for various amino acids.

Unambiguous determination came from a study of specific trinucleotide–RNA–ribosome interactions and the use of regular copolymers as messengers. All 64 codons have been firmly established. Sixty-one signify specific amino acids; the remaining three are stop signals. The code is highly degenerate, with several codons usually corresponding to a single amino acid. A given tRNA can sometimes specifically recognize several codons. This ambiguity arises from wobble in the base at the 5' end of the anticodon.

The codon for the starting amino acid, N-formyl methionine, is usually AUG, the same codon as for methionine. GUG, which usually codes for valine, also can code for N-formyl methionine. The stop codons UAA, UAG, and UGA are read by specific proteins, not specialized tRNA molecules. Why three different stop codons need exist is unknown, though recent nucleotide sequence analysis reveals that some genes end with two successive stop codons.

Certain mutations (intergenic suppressor mutations) appear to increase the frequency of mistakes in reading the genetic code. As a result of this increase in the mistake level, a mutant gene may occasionally produce a normal product. Suppressor genes exist for both missense (amino acid replacement) and nonsense (chain-terminating) codons, as well as for frameshift mutations caused by single nucleotide insertions. tRNA's with altered anticodons are the molecular basis of many suppressor gene actions.

The genetic code appears to be essentially the same in all organisms. This is not surprising: Variations in it from organism to organism would mean that the code had evolved by mutation, and it is almost impossible to imagine a mutation which would change the letters in a codon without being lethal.

REFERENCES

NIRENBERG, M. W., AND J. H. MATTAEI, "The Dependence of Cell-Free Protein Synthesis in *E. coli* upon Naturally Occurring or Synthetic Polyribonucleotides," *Proc. Natl. Acad. Sci. U.S.*, **47,** 1588 (1961). This is the classic paper which demonstrated that poly U codes for polyphenylalanine.

CRICK, F. H. C., "The Recent Excitement in the Coding Problem," *Prog. Nucleic Acid Res.*, **I,** 164 (1963). A superb analysis of the state of the coding problem as of late 1962.

LEDER, P., AND M. W. NIRENBERG, "RNA Code Words and Protein Synthesis II: Nucleotide Sequence of a Valine RNA Code Word," *Proc. Natl. Acad. Sci. U.S.*, **52,** 420 (1964). An elegant paper that establishes the order of nucleotides within a codon for valine.

NISHIMURA, S., D. S. JONES, AND H. G. KHORANA, "The *in Vitro* Synthesis of a Copolypeptide Containing Two Amino Acids in Alternating Sequence Dependent upon a DNA-like Polymer Containing Two Nucleotides in Alternating Sequence," *J. Mol. Biol.*, **13,** 302 (1965). Another classic paper on the genetic code. Here is found an unambiguous demonstration that each codon contains three nucleotides.

GORINI, L., "Antibiotics and the Genetic Code." A 1966 *Scientific American* article reprinted in *The Molecular Basis of Life*, R. H. Haynes and P. C. Hanawalt (eds.), Freeman, San Francisco, 1968.

"The Genetic Code," *Cold Spring Harbor Symp. Quant. Biol.*, **XXXI** (1966). A most impressive collection of papers, presented in June, 1966, just after the general features of the code became clear.

FIERS, W., "Chemical Structure and Biological Activity of Bacteriophage MS2 RNA," in *RNA Phages*, N. Zinder (ed.),

Cold Spring Harbor Laboratory, 1975. A summary of the author's massive sequence analysis that aims to establish the complete sequence of a viral chromosome.

CRICK, F. H. C., "Codon–Anticodon Pairing: The Wobble Hypothesis," *J. Mol. Biol.,* **19,** 548 (1966). An important argument which led to the understanding of how a single tRNA species binds to more than one codon.

WOESE, C. R., *The Genetic Code,* Harper and Row, New York, 1967. A recent summary of current data, written from a historical point of view.

YCAS, M., *The Biological Code,* Wiley (Interscience), New York, 1969. The most complete monograph on the code, written by one of the early workers in the field.

NICHOLS, J. L., "Nucleotide Sequence from the Polypeptide Chain Termination Region of the Coat Protein Cistron in Bacteriophage R17," *Nature,* **225,** 147 (1970). The first analysis of *in vivo* stop codons.

GAREN, A., "Sense and Nonsense in the Genetic Code," *Science,* **160,** 149 (1968). An easy review about how suppressor genes act.

SMITH, J. D., "Genetics of Transfer RNA," *Ann. Rev. Genetics,* **6,** 235 (1972). A very useful summary with much data on suppressor tRNA's.

RIDDLE, D. L., AND J. CARBON, "Frameshift Suppression: A Nucleotide Addition in the Anticodon of a Glycine Transfer RNA," *Nature New Biology,* **242,** 230 (1973).

ROTH, J. R., "Frameshift Mutations," *Ann. Rev. Genetics,* **8,** 317 (1974).

Regulation of Protein Synthesis and Function

The working out of the general features of the participation of nucleic acid molecules in protein synthesis provides a solid base from which we can examine how the rate of synthesis of the various protein molecules is controlled. Within a given cell a great variation exists in the number of molecules of its different proteins; thus, devices to ensure the selective synthesis of those proteins needed in large numbers must exist. Until recently this problem was approached chiefly with ignorance, speculation, and hope. Now, however, we realize that the rate of the synthesis of a protein is itself partially under internal genetic control and partially determined by the external chemical environment. To show how these factors can interact, we shall focus attention on microbial systems, since they have been the basis of most of the important concepts up to now.

CHAPTER **14**

PROTEINS ARE NOT ALL PRODUCED IN THE SAME NUMBERS

Earlier we estimated from its length that the *E. coli* chromosome codes for between 2000 and 4000 different polypeptide chains. Exactly how many different proteins are simultaneously present in a given cell is not yet known. Based upon the probable number of enzymes needed to make the various necessary metabolites, general estimates argue for the presence of at least 600 to 800 different enzymes in a cell growing with glucose as its sole carbon source. Some of these enzymes, particularly those connected with the first steps in glucose degradation and with the reactions which make the common amino acids and nucleotides, are present in relatively large amounts. Also required in large amounts are the enzymes needed to produce the energy-rich bonds in ATP. In contrast, other enzymes, particularly those involved in making the much smaller amounts of the necessary coenzymes, are present in trace quantities. There must also be relatively large amounts of the various structural proteins used to construct the cell wall, the cell membrane, and the ribosomes.

VARIATIONS IN THE AMOUNTS OF DIFFERENT *E. COLI* PROTEINS

Precise values for the number of protein molecules normally present within a bacterial cell are known for only a few proteins. The best studied case is the *E. coli* enzyme β-galactosidase (MW $= 5.4 \times 10^5$), which splits the sugar lactose into its glucose and galactose moieties (Figure 14–1). Each active molecule has a tetrameric structure, being composed of four identical polypeptide chains of molecular weight 135,000. This is a very important enzyme because lactose cannot be used as either a carbon or energy source unless it is first broken down to the simpler sugars galactose and glucose. *E. coli* cells growing with lactose as their exclusive carbon source generally contain about 3×10^3 molecules of β-galactosidase, which represents about 3 percent of the total protein. This is the maximum quantity that can be synthesized if just one gene coding for the β-galactosidase amino acid sequence is present on each *E. coli* chromosome. If this gene is present in two copies, 6 percent of the total protein produced by the cell can be β-galactosidase. There exist, in fact, superproducing mutant strains, probably containing many copies of this gene, that can synthesize almost 15 percent of their protein as β-galactosidase. This supersynthesis, however, is achieved only at the expense of making too little of other necessary proteins. Cells making excessively large amounts of β-galactosidase grow poorly and tend to be replaced by mutants that have a balanced protein synthesis.

Good data also exist for the amounts of the structural proteins of the ribosomes. There are approximately 55 of these proteins (average MW $\sim 20,000$) that collectively comprise about 10 percent of the total protein in rapidly growing cells. Thus the average ribosomal protein repre-

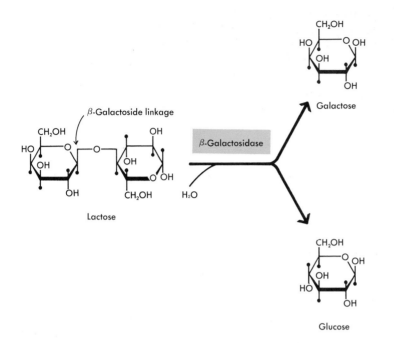

Figure 14–1
The sugar lactose can be hydrolytically cleaved to galactose and glucose by the enzyme β-galactosidase. Mutants that fail to make this protein cannot utilize lactose as a carbon source.

sents 0.2 percent of the total *E. coli* protein. No similar quantitative data have yet been obtained for the enzymes required in the biosynthesis of the coenzymes. In some cases we expect that only very few molecules will be present. This point, however, will be hard to establish, since the isolation of even one of these enzymes in the amounts necessary for a molecular weight determination will require very large amounts of cells.

RELATION BETWEEN AMOUNT OF AND NEED FOR SPECIFIC PROTEINS

Great variation can exist between the amount of a protein present when it is needed and when the environmental conditions are such that it would serve no useful function. For example, there are approximately 3000 β-galactosidase molecules in each normal *E. coli* cell growing in the presence of β-galactosides, such as lactose, and less than one one-thousandth of this number in cells growing upon other carbon sources. Substrates like lactose, whose introduction into a growth medium specifically increases the amount of an enzyme, are known as *inducers;* their corresponding enzymes are called *inducible* enzymes. An entirely different form of response is shown by many enzymes involved in cellular biosynthesis. For example, *E. coli* cells growing in a medium without any amino acids contain all the enzymes necessary for the biosynthesis of the 20 necessary amino acids. When, on the other hand, the growth medium contains these amino acids, their corresponding biosynthetic enzymes are almost entirely missing. Biosynthetic enzymes whose amount is reduced by the presence of their *end products* (e.g., histidine is the end product of the histidine biosynthetic enzymes) are called *repressible* enzymes. Those end-product metabolites whose introduction into a growth medium specifically decreases the amount of a specific enzyme are known as *corepressors.* The inductive and repressive responses are equally useful to bacteria: when enzymes are needed to transform a specific food molecule or to synthesize a necessary cell constituent, they are present; when they are unnecessary, they are effectively absent.

Adaptation is not, however, an all-or-nothing response, for under conditions of intermediate need, there may be an intermediate enzyme level (Figure 14–2). Similar variation can exist in the quantities of structural proteins. This is best shown by the variation in the number of the ribosomes themselves. When bacteria are growing at their maximum rate, ribosomes amount to 25 to 30 percent of the cell mass. If, however, their growth rate is cut down by unfavorable nutritional conditions, the bacteria need fewer ribosomes to maintain their slower rate of protein synthesis, and the ribosome content can drop to as little as 20 percent of its maximum value.

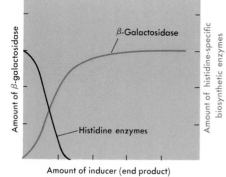

Figure 14–2
Variation in the amount of enzyme per cell as a function of the amount of inducer (end product) present in the growth medium.

VARIATION IN PROTEIN AMOUNT CAN REFLECT THE NUMBER OF SPECIFIC mRNA MOLECULES

In actively dividing bacteria, most individual protein molecules, once synthesized, are quite stable. Variation in the amount of proteins thus generally reflects rates of synthesis, not relative stability. This variation in the rate of synthesis is in turn partially related to differences in the number of available mRNA templates. The number of β-galactosidase templates in cells actively making β-galactosidase, for example, greatly exceeds the number found in cells not engaged in synthesizing this enzyme. Now our best estimate is that during maximal β-galactosidase synthesis, 35 to 50 β-galactosidase mRNA molecules are present in each cell. In contrast, when no lactose is present, the average cell contains fewer than one mRNA molecule specific for β-galactosidase synthesis.

REPRESSORS CONTROL THE RATE OF MUCH mRNA SYNTHESIS

A special group of molecules called *repressors* helps to decide when the mRNA molecules which code for many of the inducible and repressible enzymes are made. Each repressor blocks the synthesis of one or more proteins and, like all other proteins, repressors are coded by chromosomal DNA. The genes which code for them are called *regulatory genes*. A number of mutant regulatory genes, unable to code for functional repressors, have been isolated. Cells containing inactive regulatory genes produce their respective proteins independent of need (Figure 14–3). These mutants are called *constitutive mutants*, while those proteins produced in fixed amounts, independent of need, are called *constitutive proteins*.

Some mutations in regulatory genes can be suppressed by the occurrence of suppressor mutations in other genes. That is, the presence of a suppressor gene can restore the synthesis of functional repressors. This finding suggested that mistakes in the reading of the mRNA message change the structure of repressors, which in turn hinted that some, if not all, repressors are protein molecules.

REPRESSORS ARE PROTEINS

The protein nature of repressors has been confirmed by the recent isolation of five different repressors—one which controls the synthesis of E. coli β-galactosidase, a second which controls the synthesis of the E. coli enzymes involved in galactose metabolism, a third which controls the synthesis of E. coli enzymes involved in tryptophan biosynthesis, a fourth which controls the synthesis of enzymes involved in the breakdown of histidine by *Salmonella typhimurium* and a fifth which blocks the synthesis of most phage λ-specific proteins when the λ chromosome is inserted as a prophage in the E. coli chromosome (see Chapter 15). The β-galactosidase (lac) repressor is so far the best characterized. Its fundamental polypeptide chain, of molecular weight 37,000, readily ag-

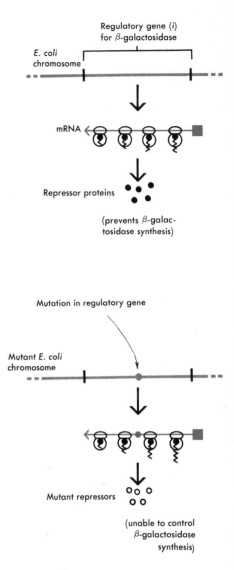

Figure 14–3
The control of repressors by normal and mutant genes.

gregates into tetramers of molecular weight 148,000. The active form appears to be the tetramer, of which only 10 to 20 copies are usually present for each *E. coli* chromosome. The smallness of this number made their original detection and isolation a remarkable feat. Now, fortunately, there exist mutations (see below) which result in very much larger amounts (>1 percent of total cellular protein), making possible isolation of sufficient quantities for amino acid sequence determination. This task was completed in 1973, revealing a sequence of 347 amino acids.

REPRESSORS ACT BY BINDING TO DNA

All known repressors bind at specific sites on their respective DNA molecules, blocking the initiation of transcription of the corresponding mRNA molecules. The specific nucleotide sequences that bind the repressors are called *operators.* In general, an operator must be at least 10 to 12 bases long in order to interact specifically with the appropriate hydrogen bond-forming groups on a repressor. A base number this large avoids the possibility that by random chance a similar sequence will exist somewhere else along the same chromosome. If a smaller number were used, too many false bindings would occur. Use of a large number of specific interactions has the added consequence that the binding can be very strong. Once a lac repressor has bound to DNA, it effectively remains attached until subsequent interaction with its inducer.

The nucleotide sequence of the lac operator region was determined by study of the DNA fragment protected from nuclease digestion by specific binding of the lac repressor to *E. coli* DNA. Twenty-one base pairs are found within this region, with some sixteen related by a twofold axis of symmetry centered at the base pair in position eleven (Figure 14–4). Such symmetry allows two subunits of the tetrameric lac repressor to simultaneously bind to the operator. That the symmetry is not perfect was originally interpreted as meaning that not all base pairs on a given side of the operator bind to the repressor. This idea led to the speculation that only the symmetrical pairs are involved in the respective protein–nucleic acid interactions. Analysis, however, of mutant operators reveals that base pair changes which directly reduce the strength of repressor–operator binding occur in the nonsymmetrical as well as symmetrical regions.

Deeper understanding will require the working out of the 3-D structure using x-ray diffraction methods. Unfortunately, this may take many years, since the crystals

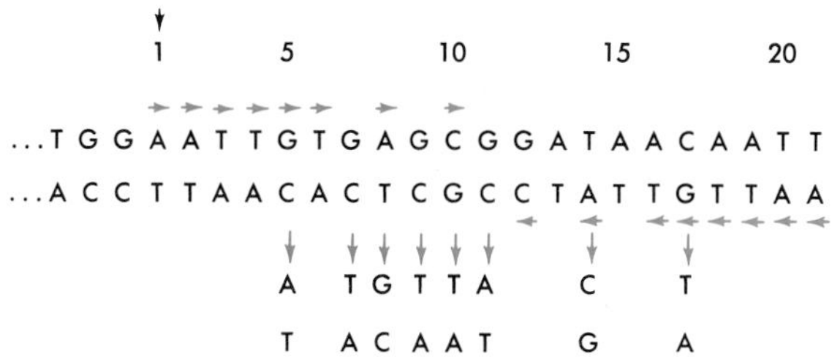

```
   1      5        10        15        20
...T G G A A T T G T G A G C G G A T A A C A A T T
...A C C T T A A C A C T C G C C T A T T G T T A A
        A   T G T T A       C       T
        T   A C A A T       G       A
```

Figure 14–4
The lac operator sequence, showing its relation to the starting point (arrow) for transcription of lac mRNA. Base pair changes which lead to poor binding of the lac repressor are shown below their wild type equivalents.

of the lac repressor so far obtained are much too small for detailed three-dimensional analysis. As the growing of large protein crystals is still essentially a matter of chance, we have no assurance as to when properly sized lac repressor crystals will become available.

COREPRESSORS AND INDUCERS DETERMINE THE FUNCTIONAL STATE OF REPRESSORS

Repressors must not always be able to prevent specific mRNA synthesis. If they could, they would permanently inhibit the synthesis of their specific proteins. Instead, all repressor molecules can exist in both an active and an inactive form, depending on whether they are combined with their appropriate *inducers* (*corepressors*). The attachment of an inducer inactivates the repressor. For example, when combined with a β-galactoside[1] (inducer), the β-galactosidase repressor cannot bind to its specific operator. Thus, the addition of β-galactosides to growing cells permits β-galactosidase synthesis by decreasing the concentration of active β-galactosidase repressors. In contrast, the binding of a corepressor changes an inactive repressor into an active form. For example, the addition of amino acids to cells activates repressors which control the synthesis of enzymes involved in amino acid biosynthesis. This quickly shuts off synthesis of their specific mRNA molecules (Figure 14–5).

---

[1] Now there are suspicions that lactose itself is not the true inducer of β-galactosidase synthesis. Instead, some lactose molecules are first transformed into a related compound, which in turn attaches to the repressor.

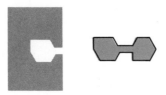
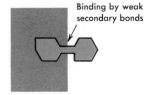

(a) *Active β-galactosidase repressor* + *β-Galactoside (inducer)* ⇌ Inactive repressor-inducer complex

(prevents β-galactosidase synthesis)

(unable to control β-galactosidase synthesis)

Binding by weak secondary bonds

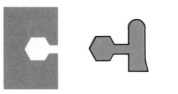

Binding by weak secondary bonds

(b) *Inactive* histidine repressor + Histidine (corepressor) ⇌ *Active* repressor-corepressor complex

(unable to control synthesis of enzymes for histidine synthesis)

(controls rate of synthesis of enzymes for histidine synthesis)

Figure 14–5
Schematic drawing illustrating the opposite effects of corepressors and inducers upon the activity of repressors. We see here that, depending upon whether the enzymes are inducible or repressible, the free repressors are either active or inactive.

No covalent bond is formed between repressors and their specific inducers or corepressors. Instead, a portion of each repressor molecule is complementary in shape to a specific portion of its inducer (corepressor), and the two are joined by weak secondary bonds (hydrogen bonds, salt linkages, or van der Waals forces). Since these bonds are weak, they are rapidly made and broken, allowing the repressor state (active or inactive) to adjust quickly to the physiological need. For example, the synthesis of β-galactosidase mRNA ceases almost immediately after the removal of lactose.

REPRESSORS CAN CONTROL MORE THAN ONE PROTEIN
In some cases, repressors may control the synthesis of only one protein. Often, however, a single repressor affects the synthesis of several enzymes. The β-galactosidase repressor of *E. coli*, for example, controls at least three en-

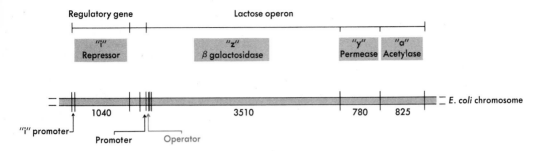

zymes: β-galactosidase itself; galactoside permease, which controls the rate of entry of β-galactosides into the bacteria; and galactoside acetylase, whose function is not yet known. When active β-galactoside permease is absent, E. coli cells are unable to concentrate β-galactosides within themselves. Since β-galactosidase and β-galactoside permease and galactoside acetylase (?) are ordinarily needed to metabolize β-galactosides, their *coordinated* synthesis is clearly desirable. Coordinated synthesis is brought about by having the respective enzymes coded by adjacent genes (Figure 14–6), thereby allowing a single mRNA molecule to carry all their genetic messages (Figure 14–7). An even larger number of genes (10) is coordinately repressed by the repressor of histidine biosynthesis. Again, this is achieved by having a single mRNA molecule carry the messages of all these genes.

Figure 14–6
The lactose operon and its associated regulatory gene drawn to scale based on known sizes of their gene products. The numbers give the number of base pairs found in the several genes.

Figure 14–7
How the interaction of repressor, inducer, and operator controls the synthesis of the E. coli proteins β-galactosidase, β-galactoside permease, and galactoside acetylase.

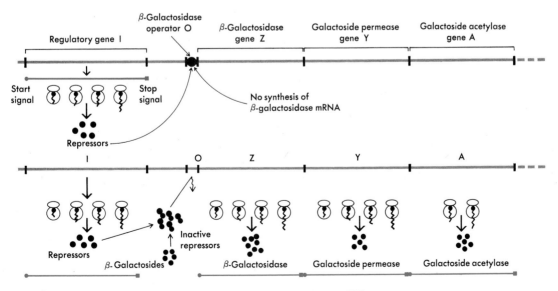

The collections of adjacent nucleotides that code for single mRNA molecules (under the control of a single promoter; see below) are called *operons*. Some operons thus contain one gene, others two, and still others, several genes. At first, it was thought that repressors were specific for single operons. Several years ago, a case was found that is most simply interpreted by assuming that a specific repressor can act on three different operons: the genes responsible for *E. coli* arginine biosynthesis have been found distributed among three unlinked operons. Nonetheless, there is evidence that one regulatory gene controls the level of enzymes belonging to all the operons. Likewise, the λ repressor can act at more than one site. It specifically binds to two sets of operators, one set to the left and one to the right of the corresponding regulatory gene (see Chapter 15).

ABSENCE OF AN OPERATOR LEADS TO CONSTITUTIVE SYNTHESIS

Operators (repressor binding sites) are all located very close to the regions where mRNA transcription starts. In fact, they are so closely linked that at first it was suspected that the operator region might overlap with the nucleotides coding for the first several amino acids in β-galactosidase. Now there is firm evidence against any overlap.

Presence of a bound repressor prevents only mRNA initiation, having no effect on chain growth once elongation has already commenced. Thus operators have essentially *negative* functions: If a functional operator is absent, the corresponding repressor cannot inhibit the synthesis of the specific mRNA and, as a result, there is constitutive synthesis of its corresponding protein product(s).

The existence of operators was first revealed by genetic analysis. The structure of the operator can mutate to an inactive form, preventing the working of repressors. When this happens, constitutive enzyme synthesis results. These mutants are, therefore, called O^c (constitutive) mutants. O^c mutations can easily be distinguished from mutations in the repressor genes by measuring enzyme synthesis in special, partially diploid cells containing two copies of the relevant chromosomal regions. Cells containing one nonfunctional and one functional repressor gene are still repressible, since good repressor molecules can act on both operators (Figure 14–8). In contrast, cells containing only one bad operator will always be constitutive, no matter what the condition of the repressor gene (Figure 14–9).

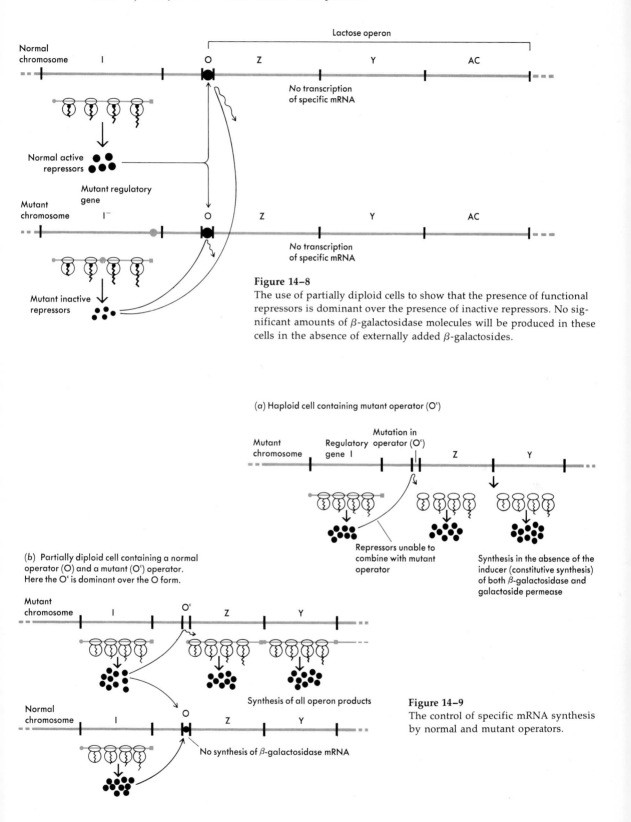

Figure 14–8
The use of partially diploid cells to show that the presence of functional repressors is dominant over the presence of inactive repressors. No significant amounts of β-galactosidase molecules will be produced in these cells in the absence of externally added β-galactosides.

(*a*) Haploid cell containing mutant operator (O^c)

(*b*) Partially diploid cell containing a normal operator (O) and a mutant (O^c) operator. Here the O^c is dominant over the O form.

Figure 14–9
The control of specific mRNA synthesis by normal and mutant operators.

POSITIVE CONTROL OF LACTOSE OPERON FUNCTIONING

The functioning of repressors in preventing mRNA synthesis is an example of negative control, for in the absence of the controlling factor, synthesis proceeds more rapidly than in its presence. While initially it was thought that the lactose operon was solely under negative control, it is now clear that, even in the presence of an inducer that neutralizes its repressor, a protein-mediated positive control signal also must be received for its normal functioning. The existence of this positive control protein was discovered through study of how the presence of glucose blocks the functioning of a number of operons (*glucose sensitive operons*), each controlling the breakdown (catabolism) of a specific sugar (e.g., lactose, galactose, arabinose, and maltose). For example, when *E. coli* is grown in the presence of both glucose and lactose, only the glucose is utilized and none of the lactose operon proteins is synthesized. Similarly, when both glucose and galactose are present, the galactose operon is inactive. The reason that glucose breakdown is preferred over that of the other sugars now can only be guessed. Perhaps, during a long evolutionary history, bacteria more often found themselves in a glucose-rich environment than in one dominated by other sugars.

In any case, glucose inhibition does not operate by influencing the rate of entry of the various sugars into the cells. Instead it acts by influencing the transcription pattern. This was first hinted by the existence of mutations within its promoter (see below), which renders the lactose operon insensitive to glucose effects. With such mutants, the lactose operon can be maximally induced even in the presence of large amounts of glucose.

GLUCOSE CATABOLISM AFFECTS THE CYCLIC AMP LEVEL

The action of glucose on transcription is not direct. Instead, one of its breakdown products (catabolites) acts by lowering the intracellular amount of *cyclic AMP (cAMP)*. This key metabolite is required for the transcription of all the operons which are inhibited by glucose catabolism. The manner in which a glucose catabolite controls the cyclic AMP level within a cell is still unknown. ATP is the direct metabolic precursor for cAMP, and the enzyme responsible for this transformation, adenylcyclase (Figure 14–10), may be directly inhibited by a specific catabolite. Alternatively, since there exists an enzyme which specifically converts cAMP to AMP, inhibition might well be controlled by the rate of cAMP breakdown. Hopefully, the correct picture will soon emerge.

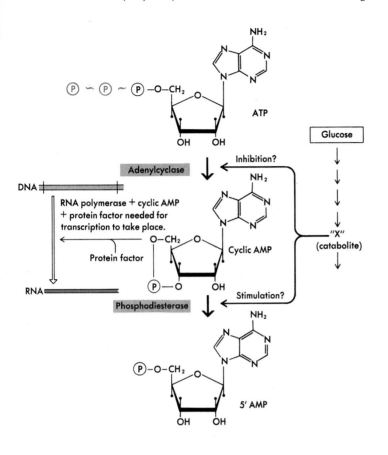

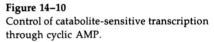

Figure 14–10
Control of catabolite-sensitive transcription through cyclic AMP.

ACTIVATION OF THE CATABOLITE ACTIVATOR PROTEIN (CAP) BY cAMP BINDING

cAMP does not directly promote lac mRNA synthesis but works instead by binding to the *catabolite gene activator protein (CAP)*, a dimeric molecule of MW ~ 44,000. CAP has no influence on transcription until cAMP has bound to it. It then requires the ability to bind to very specific sites on DNA, and by so doing increases the rate of transcription of adjacent operons. CAP is the positive control element for all the glucose sensitive operons; and so mutations which block its functioning produce cells simultaneously unable to utilize a large number of different sugars.

BOTH CAP AND SPECIFIC REPRESSORS CONTROL PROMOTER FUNCTIONING

The binding of neither CAP nor a specific repressor has any influence on the rate at which given mRNA chains grow. Instead both act, one positively, the other negatively, by controlling the rate at which RNA polymerase molecules attach to *promoters*, the regions of DNA specifi-

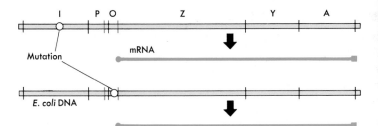

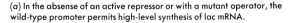

(*a*) In the absense of an active repressor or with a mutant operator, the wild-type promoter permits high-level synthesis of lac mRNA.

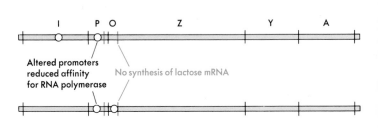

(*b*) A promoter mutation that blocks any synthesis of lac mRNA even in the absense of a functional repressor or a nonfunctional operator.

Figure 14–11
Genetic demonstration of the promoter for the lactose operon.

cally involved in the initiation of transcription. Experimental evidence for promoters first came through the isolation of several mutants, mapping next to the β-galactosidase gene, which transcribed the lac operon with highly reduced efficiency even in the total absence of the lac repressor (Figure 14–11). Now we know that not all promoter mutations lead to decreased transcription. Some, in fact, result in less restrictive conditions for the synthesis of specific mRNA (e.g., lack of requirement for CAP binding). Initially it was thought that promoters would have relatively simple structures. Today it is apparent that they often are quite complex, for they must interact with a variety of different positive and negative control elements.

REPRESSOR BINDING PREVENTS SIMULTANEOUS BINDING OF RNA POLYMERASE

The recent working out of the lac operator and promoter sequences automatically tells why the binding of the lac repressor to its respective operator prevents initiation of lac mRNA synthesis (Figure 14–12). The collection of 21 bases encompassing the operator includes the starting point for lac mRNA synthesis (Figure 14–4). Thus a mild degree of overlap exists between the lac promoter and lac operator regions; and so the presence of a bound lac repressor sterically prevents RNA polymerase from binding

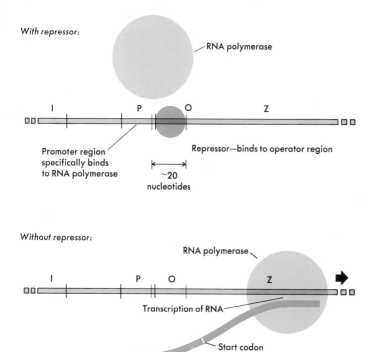

With repressor:

RNA polymerase

I   P   O   Z

Promoter region
specifically binds
to RNA polymerase

Repressor—binds to operator region

~20
nucleotides

Without repressor:

RNA polymerase

I   P   O   Z

Transcription of RNA

mRNA

Start codon

It is possible that some of the bases of
the operator region are transcribed also.

(a)

Figure 14–12

(a) The influence of the repressor upon
binding of RNA polymerase to the promoter for the lactose operon. (b) Electron
micrograph of a β-galactosidase (lactose)
repressor bound specifically to the operator for the lactose operon. [Photograph
kindly supplied by J. Griffiths of Cornell
University Physics Department.]

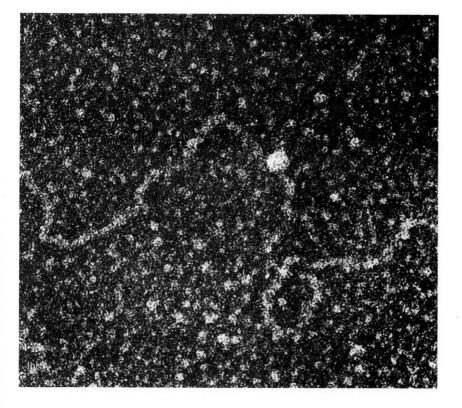

(b)

to the lac promoter. Such special overlapping, however, does not demand that the bases which directly recognize the repressor also recognize RNA polymerase. In fact, only chaos would result if a given lac repressor had the capacity to thus bind to all sites at which RNA polymerase commenced synthesis. But since many of the bases within promoter regions are not concerned with polymerase recognition *per se* (see below), there is no theoretical reason why operator and promoter regions cannot extensively intermingle. In fact, in Chapter 15, we shall see that those of phage λ show extensive overlaps.

THE LAC PROMOTER CONTAINS SOME 80 BASE PAIRS

Two functionally distinct regions have been identified within the lac promoter. One, the RNA interaction region, directly participates in the initial binding of polymerase, while the other, the CAP site, located more distal to the operator, binds the cAMP–CAP complex (Figure 14–13). Exactly which bases in the 40 base pair CAP site directly bind to CAP is not yet known, though the symmetrical sixteen base pair set found at positions −52 to −67 is a likely candidate, since CAP is a dimeric molecule. Equally unknown are the exact bases to which RNA polymerase first attaches. Though all known promoters share most base pairs within a seven base pair region(−6 to −12) near the site at which transcription starts (Figure 11–15), DNA fragments containing this region do not bind polymerase unless they also encompass bases lying some 25 base pairs farther away (−25 to −35) from the start of transcription (=1). Particularly important to the lac promoter may be a

Figure 14–13
The sequence of the *E. coli* promoter, showing its relation to the overlapping operator sequences.

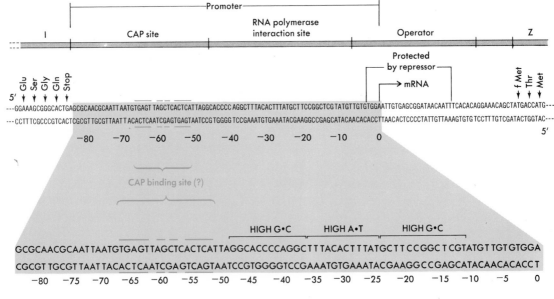

Initial recognition (entry) site   Firm binding (start) site

```
SV40   GAATGCAATTGTTGTTGTTAACTTGTTTATTGCAGCTTATAATGGTTACA
                                                         ↓
λP_R   TAACACCGTGCGTGTTGACTATTTTACCTCTGGCGGTGATAATGGTTGCA
                                                         ↓
λP_L   TATCTCTGGCGGTGTTGACATAAATACCACTGGCGGTGATACTGAGCACA
                                                         ↓
Lac    TAGGCACCCCAGGCTTTACACTTTATGCTTCCGGCTCGTATGTTGTGTGGA   Lac
                                                         ↓
```

group of AT-rich base pairs (positions −25 to −35) surrounded by two GC-rich groups. Since AT-rich regions spontaneously open up more often than GC-rich regions, it is tempting to believe that the initial entry of RNA polymerase occurs into the AT-rich section, and that the frequency with which this happens is influenced by the lengths of the surrounding GC-rich sections. As the exact location of these AT-rich regions varies within the four completely known promoter sequences (Figure 14–14), RNA polymerase may drift after entry until it becomes set in the "start" site (Figure 14–15) emcompassing bases −12 to −6. Whether any drift is in fact required is unclear since, given the great size of RNA polymerase, its binding site could easily be some 30 to 40 base pairs of length from the "active site" that catalyzes formation of phosphodiester bonds.

Also still to be worked out is how the binding of cAMP–CAP to the CAP site facilitates RNA polymerase attachment. This is far from obvious since the most likely binding site is some 30 base pairs away from the center of the AT-rich region. Most likely we shall have to wait until CAP's structure becomes known in three dimensions before we can think seriously at the molecular level about how it acts.

IN VITRO ANALYSIS OF PROMOTER FUNCTIONING

A very important step in the establishment of our concepts about how promoters and operators function has come from *in vitro* transcription experiments. For example, when lac (gal) DNA is used as the *in vitro* template, the binding of RNA polymerase and initiation of lac (gal) mRNA synthesis requires the simultaneous presence of cAMP and CAP. And when the DNA templates have been isolated from mutant cells whose lac operon is transcribed even when glucose is present, no requirement for CAP is found in the *in vitro* systems. Such experiments provide the clearest demonstrations that cAMP and CAP act at the transcriptional level and not through effects on membrane permeability.

Figure 14–14

The RNA polymerase interaction sites of the lac, λ (P_R and P_L), and SV40 promoters. Starting points of transcription are indicated by arrows, with regions of homology indicated in brackets.

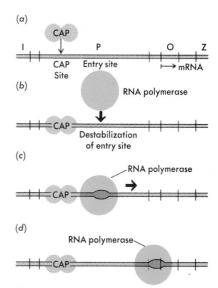

Figure 14–15

A model for the initiation of lac transcription . . . a) CAP binds to the CAP site . . . b) CAP destabilizes the entry site thereby facilitating RNA polymerase entry . . . c) RNA polymerase in the entry site "drifts" to the start site d) RNA polymerase in the start site. [Redrawn from R. C. Dickson, *et al., Science,* **187,** 27 (1975).]

Histidine

$$NH_3 \longleftarrow$$

| Histidase (H) |

Urocanate

| Urocanase (U) |

Imidazolone propionate

| Imidazolone propionate hydrolase (I) |

Formiminoglutamate

| Formiminoglutamate hydrolase (G) |

Glutamate + formate

Figure 14–16
The enzymatic pathway for the breakdown of histidine.

Figure 14–17
The two Hut operons of *Salmonella typhimurium* and how they are controlled by the interaction of positively and negatively acting protein factors.

POSITIVE CONTROL IN THE HUT OPERON MEDIATED BY THE ENZYME GLUTAMINE SYNTHETASE

All the protein factors known to influence expression of the lac, gal, trp, and λ operons have only control functions (regulatory proteins) and do not possess any enzymatic capabilities. However, there is no sound theoretical argument for this clear separation; and recently an important cellular enzyme has been found to directly affect transcription of two operons in the bacteria *Salmonella typhimurium*. The operons so controlled contain the Hut (histidine utilizing) genes that code for the four enzymes needed for the degradative breakdown of histidine, a process that generates sufficient catabolites (glutamate, formamide, and ammonia) for growth under conditions of carbon and/or nitrogen starvation (Figure 14–16). Most importantly, these operons have evolved so that when histidine is present, their respective proteins are made under conditions of either carbon or nitrogen limitation. To understand how these operons are controlled, we must thus understand both how histidine acts and through what means the signal of carbon (nitrogen) starvation is transmitted. Now it appears that histidine acts as a conven-

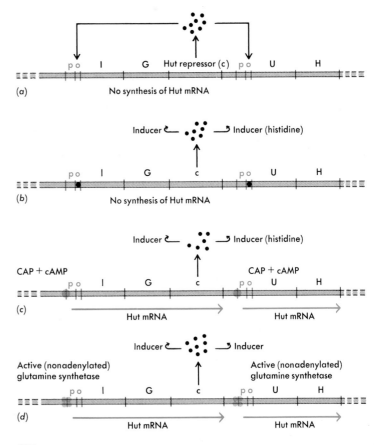

tional inducer, overcoming the negative control mediated by the Hut repressor. Thus only when the Hut repressor is prevented from binding to its respective operators can the signal of either carbon or nitrogen starvation be transmitted (Figure 14–17). This makes great sense, since existence of the Hut enzymes in the absence of appreciable histidine would be biologically very wasteful.

The signal that carbon is limiting follows the pattern seen in the lac operon, being transmitted through a rise in the cAMP level and subsequent activation of the Hut promoters by binding of cAMP–CAP complexes. When, however, cAMP levels are low, as, for example, when the glucose supply is plentiful, the signal that nitrogen is limiting must be transmitted through a positive control protein other than CAP.

Here, recent work with *in vitro* systems has shown that the activating protein for Hut mRNA synthesis is active glutamine synthetase, an enzyme of which the active nonadenylated form is present in large amounts during nitrogen limitation but which is present in an inactive adenylated form when nitrogen is plentiful. Hut operon DNA thus receives the message that nitrogen is limiting by the binding of glutamine synthetase to its respective promoter.

A PROTEIN THAT CAN MEDIATE EITHER POSITIVE OR NEGATIVE CONTROL

How the enzymes of the arabinose (ara) operon are controlled is more mysterious. In *E. coli* cells growing in the absence of arabinose, the three different enzymes involved in its breakdown are present in only very small amounts. But upon arabinose addition, all these enzymes coordinately increase in amounts. This induction process is controlled by a fourth gene "C" located adjacent to the other three, being separated only by the promoter of the ara operon (Figure 14–18). Since genetic analysis indicates that nonsense mutations occur within the "C" gene, it must be a conventional gene producing a protein product that combines with arabinose. This protein, however, is not a simple repressor since, when it is absent due to deletion of the DNA coding for it, no induction of the arabinose operon is possible. So it is tempting to speculate that after combining with arabinose it acts positively on the ara promoter in the manner of the CAP factor. Here it is important to remember that the arabinose operon is a glucose sensitive operon, and ara mRNA synthesis is also dependent upon the binding of CAP to the ara promoter. High-level functioning of the ara operon thus appears to demand the simultaneous presence of two positive control

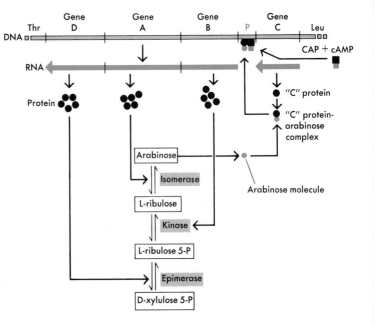

Figure 14–18
The arabinose operon of *E. coli:* an example of positive control.

signals, one saying that arabinose is present, the other that glucose is missing.

Seeming at first compatible with this hypothesis was the discovery of mutant "C" proteins (C^c) that allowed constitutive synthesis due to their activation by the binding of a normal intracellular metabolite. So it was expected that diploid cells containing one C^+ gene and one C^c gene would make the ara proteins in the absence of arabinose. But, in fact, the presence of the C^+ gene prevents its C^c allele from functioning. So it looks as if the "C" protein, in the absence of arabinose, is still able to bind to the ara promoter but now acts negatively as if it were a repressor. The "C" protein may thus have the potentiality of acting either positively or negatively, depending upon whether it is bound to an inducer.

REGULATION OF TRYPTOPHAN OPERON TRANSCRIPTION BY TWO DIFFERENT CONTROL REGIONS

Initially it seemed likely that the synthesis of five enzymes of the tryptophan (trp) operon (see Figure 8–9) also would be regulated primarily by control molecules which interacted with its promoter–operator region. Here it is important that transcription of the five contiguous trp genes (Figure 14–19) is needed only when the tryptophan is limiting. Thus tryptophan acts not as an *inducer* but as a *corepressor* that activates its specific repressor so that it can bind to the trp operator and prevent transcription of its respective operon. Correspondingly, in the absence of tryptophan the trp operator is free and the synthesis of trp mRNA commences at the adjacent promoter. Surprisingly, however, a trp mRNA molecule, once so initiated, does not automatically grow to a full length. Instead, the majority of trp mRNA molecules stop growing before the transcription of even the first trp gene (trp E) commences.

Key to the understanding of this apparently bizarre result has been the recent sequence analysis of the 5' end of trp mRNA. It reveals that a "leader" of 161 bases separates the promoter–operator region from trp E. Within this leader is an *attenuator region* (positions 123 to 150), within which most (~ 90%) of the RNA polymerase molecules stop elongation, to yield large numbers of 140-nucleotide-long fragments (Figure 14–19). Though some readthrough occurs beyond the attenuator region, it becomes frequent only when a specific control molecule acts at the attenuator. The exact nature of this molecular signal is not yet known, though it is clear that the amount of readthrough is inversely connected with the tryptophan level. In some way (either directly or indirectly) the amount of tryptophan influences both the control exercised by the tryptophan repressor and the control device employed at the attenuator region. Conceivably the coexistence of two different control regions allows a finer tuning to the level of intracellular tryptophan resulting in a two-stage response to progressively more stringent tryptophan starvation—the initial response being the cessation of repressor binding, with greater starvation leading to relaxation of attenuation.

Figure 14–19
The tryptophan operon of *E. coli*, showing the relation of the leader to the structural genes that code for the trp enzymes. Regions of dyad symmetry within the attenuator are underlined.

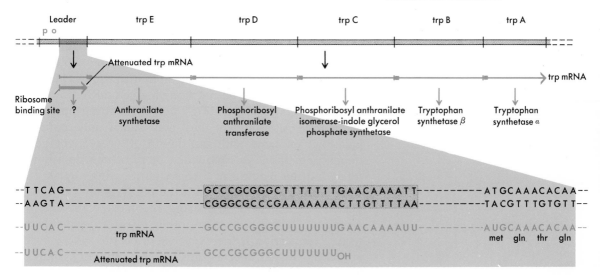

How the attenuator is normally read is still not clear. Because it is so AT-rich (Figure 14–19), it may not normally require ρ action (see page 300), though there are recent genetic hints that ρ in fact may be involved. In any case, the blockage of attenuation may be mediated by some antitermination (anti ρ?) protein specific for the attenuator. While such speculation may seem very *ad hoc*, solid evidence exists that an antitermination protein plays key roles in phage λ transcription (see page 431). So the trp operon example may soon be seen repeated many times as more operons are investigated. Already there is evidence, in both the histidine and isoleucine–valine operons, for attenuator regions located before the first structural genes.

Whether the attenuated RNA plays any functional role is not yet established. Hints that it may code for a $\sim$ 40AA-long polypeptide comes from the existence at the start of the leader of a strong ribosome binding site that contains an AUG start codon. So the large amount of leader mRNA may reflect the need for higher levels of a leader-coded polypeptide than of the polypeptides directly involved in tryptophan biosynthesis. Conceivably it may act like the λ "N" protein (see page 431) which is coded by a leader-like sequence and works as a specific antitermination factor.

UNEQUAL PRODUCTION OF PROTEINS CODED BY A SINGLE mRNA MOLECULE

Variation in the number of molecules of different proteins arises also from the fact that the proteins coded by a single mRNA molecule need not be produced in similar numbers. This point is demonstrated by the study of the lactose operon proteins: Many more copies of β-galactosidase than of galactoside permease or galactoside acetylase are synthesized. The ratios in which they appear are $1:\frac{1}{2}:\frac{1}{5}$ respectively. This may mean that ribosomes attach to the different starting points along a given mRNA molecule at different rates, depending upon the starting nucleotide sequences. Alternatively they may attach only at the β-galactosidase gene with translation of subsequent genes depending upon the frequency of ribosome detachment following the reading of a chain terminating signal. This hypothesis fits in nicely with the observation that the β-galactosidase gene is translated the most often and the acetylase the least often. Another factor conceivably affecting translation is the existence of codons whose corresponding tRNA species are present in very limiting amounts. Hypothetically, ribosomes might jam up at such codons waiting for the limiting tRNA to diffuse to them. We must emphasize, however, that so far no evidence exists for such tRNA species. In any case, it seems reasonable that mechanisms may exist that permit differential reading rates along single mRNA molecules. Although the coordinated appearance of related enzymes is obviously of great advantage to a cell, there is no reason why equal numbers should be produced. An equal number would be useful only if the specific catalytic activity rates (turnover numbers) of related proteins were equal. In general, however, there are great variations in individual turnover numbers.

BACTERIAL mRNA IS OFTEN METABOLICALLY UNSTABLE

When corepressor (inducer) molecules are added to or removed from growing bacteria, the rate of synthesis of the respective proteins is altered rapidly. This rapid adaptation to a changing environment is possible not only because growth requires continual synthesis of new mRNA molecules, but even more significantly, because many bacterial mRNA molecules are metabolically unstable. The average lifetime of many *E. coli* mRNA molecules at 37°C is about 2 minutes, after which they are enzymatically broken down. The resulting free nucleotides are then phosphorylated to the high-energy triphosphate level and reutilized in the synthesis of new mRNA molecules.

There is thus virtually complete replacement of the templates for many proteins every several minutes. For example, within several minutes after addition of suitable β-galactosides, *E. coli* cells synthesize β-galactosidase at the maximum rate possible for that particular inducer level. If on the contrary all mRNA molecules were metabolically stable, the maximum synthetic rate would not be reached until cell growth had effectively diluted out previously made mRNA molecules. Correspondingly, the existence of unstable lactose mRNA also means that, once β-galactosides are removed, synthesis of β-galactosidase quickly halts and does not resume until it is again necessary (Figure 14–20).

It now seems as if the average lifetime of different specific mRNA molecules may vary greatly. If true, this means that the mRNA lifetime is itself genetically determined. That is, the nucleotide sequence (perhaps at one end) of an mRNA molecule determines the chance of enzymatic digestion. The enzymatic mechanism by which individual mRNA molecules are broken down has not yet been clarified, though it is clear that the direction of

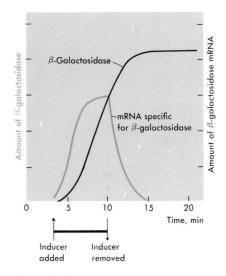

Figure 14–20
Rapid rise (fall) of β-galactosidase mRNA upon the addition (removal) of β-galactosidase inducers. The *E. coli* cells in this experiment were grown at 37°C under conditions where the cells divided every 40 minutes.

breakdown is 5' to 3' (Figure 14–21). The end which is made first is digested first. Breakdown this way will not lead to the synthesis of incomplete chains. On the contrary, if breakdown were 3' to 5', then the ends of many messengers would be destroyed before ribosomes had translated their sequences. There are suggestions that mRNA molecules are stable as long as they are bound to ribosomes. Perhaps after the 5' ends of mRNA molecules finish moving across the ribosomes, there is a choice as to whether they attach to new ribosomes or to degradative enzymes which break them down.

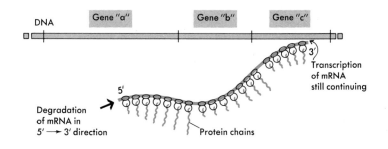

Figure 14–21
Breakdown of an mRNA in the 5'-to-3' direction. Degradation of long mRNA molecules frequently begins at the 5' end even before the 3' end has been synthesized.

PROTEINS NOT UNDER DIRECT ENVIRONMENTAL CONTROL

There are a variety of proteins within the cell whose amount does not seem to be influenced by the external environment. As an example, in *E. coli* the amounts of the enzymes controlling the degradation of glucose do not radically change when glucose is either removed from or added to the growth medium. Thus the glucose degradative enzymes seem to be *constitutive enzymes* whose rate of synthesis is controlled by neither inducers nor corepressors. We do not yet understand why this is so, since there should be a selective advantage to the cells that can change the amounts of these enzymes. We must, of course, consider that further experiments may demonstrate an inducer or corepressor. On the other hand, it now seems wiser to pose the more general question: Must all genes have repressor–operator control, or is it possible that the amounts of proteins can be controlled in other ways?

The answer is straightforward. The constitutive synthesis of any enzyme at either high or low levels is easy to imagine. High-level constitutive synthesis is what we observe when mutations cause the loss of a repressor or operator. The resulting invariant synthesis of the respective proteins occurs at the same rate as under optimal conditions of induction or repression. This tells us that repressors and operators *per se* are not required for the synthesis of mRNA molecules.

In general, the amount of constitutive synthesis is a reflection of four factors: (1) The rate at which a specific mRNA molecule can be made in the absence of repressors or operator(s)—(this is now understood to be a function of its promoter sequence); (2) the rate at which ribosomes attach to the starting point of the mRNA template; (3) the rate at which a message itself is read; and (4) the lifetime of the particular template.

Unfortunately, because we still understand very little about the factors controlling any of these rates, it is difficult to assess the absolute or even relative importance of any of them. Nonetheless, the knowledge that so many factors influence the rate of constitutive synthesis suggests that the synthesis of the many proteins needed in small amounts might be regulated without the involvement of repressors or operators.

REPRESSOR SYNTHESIS IS USUALLY UNDER PROMOTER, NOT OPERATOR, CONTROL

At a given time, there are only about 1000 mRNA molecules in a single *E. coli* cell. A guess at the minimal number of operons influenced by corepressors (inducers) is 100 to 200. It is thus hard to imagine that more than 1 or 2 specific mRNA molecules exist for each repressor. A larger number of mRNA molecules coding for repressors (regulatory mRNA) would greatly restrict the amount of mRNA coding for necessary structural and enzymatic proteins. We conclude that the synthesis of regulatory mRNA is probably carefully controlled. This cannot be done, however, by an entirely new group of repressors, since an infinite number of different repressors would be required to repress each other's synthesis. Thus, either a repressor itself can repress its own synthesis, or repressors are constitutively synthesized. The latter explanation holds for the β-galactosidase system, where many promoter mutations have already been mapped. Some of these increase the lac repressor number to more than 50 times the number found in wild type cells (Figure 14–22).

Figure 14–22

Increase in the number of lactose repressor mRNA molecules through promoter mutations. The high-level promoter is thought to arise by several additive mutational changes.

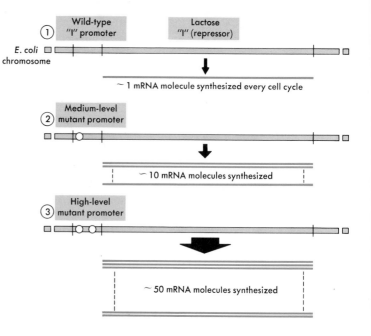

Threonine

α-Ketobutyrate

Acetohydroxybutyrate

Dihydroxyisoleucine

α-Ketoisoleucine

Isoleucine

Figure 14–23

The pathway of isoleucine biosynthesis starting from threonine. The dotted colored line shows that isoleucine inhibits the enzyme (theonine deaminase) which transforms threonine into α-ketobutyrate.

With the Hut operon, however, the repressor is coded by one of the two operons whose synthesis it specifically controls (Figure 14–17). The Hut repressor thus specifically represses its own synthesis.

REGULATION OF PROTEIN FUNCTION BY FEEDBACK INHIBITION

The catalytic activity of many proteins is affected by their binding to specific small molecules. In this way, the activity of enzymes may be blocked when they are not needed. Consider, for example, what happens when an *E. coli* cell growing on minimal glucose medium is suddenly supplied with the amino acid isoleucine. Immediately the synthesis (functioning?) of the mRNA molecules needed to code for the specific enzymes utilized in isoleucine biosynthesis ceases. Without a further control mechanism, preexisting enzymes could cause continued isoleucine production, now unnecessary because of the extracellular supply. Wasteful synthesis, however, almost never occurs, because high levels of isoleucine block the activity of the enzyme involved in the first step of its biosynthesis starting from threonine (Figure 14–23). This inhibition is due to the binding of isoleucine to the enzyme threonine deaminase. Thus bound, this enzyme is unable to convert threonine to α-ketobutyrate. Because the association between the enzyme and isoleucine is weak and reversible, relatively high isoleucine concentrations must exist before most of the enzyme molecules are inactivated. This very specific inhibition is called *feedback (end-product) inhibition,* because accumulation of a product prevents its further formation. Only the first step in a metabolic chain is blocked. With the first reaction blocked, there is no accumulation of unwanted intermediates, so inhibition of the remaining enzymes would serve no end.

The final enzymatic step in the synthesis of an end-product feedback inhibitor is often separated by several intermediate metabolic steps from the substrate (or from the product) of the enzyme involved in the first step of its biosynthesis (Figure 14–24). The structure of the inhibitor may thus only loosely resemble that of the substrate of the inhibited enzyme, so that one would not expect an end-product inhibitor to combine with the enzymatically active site (region that binds the substrate) of the enzyme it inactivates. Instead, there is the suspicion that it reversibly combines in some cases with a second site on the enzyme,

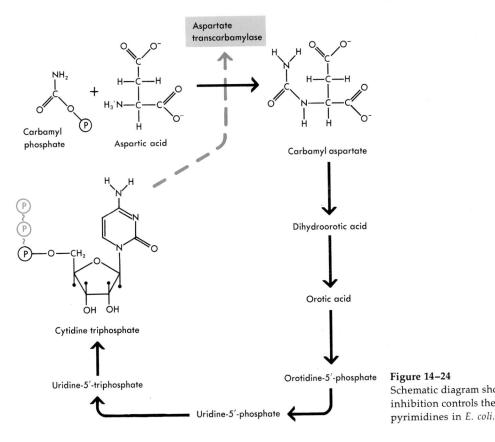

Figure 14–24
Schematic diagram showing how feedback inhibition controls the biosynthesis of pyrimidines in *E. coli.*

405

yet nevertheless causes the enzyme activity to be blocked. Perhaps by causing a change in the precise enzyme shape (allosteric transformation), the inhibitor prevents the enzyme from combining with its substrate (Figure 14–25). Such proteins, whose shapes are changed by the binding of specific small molecules at sites other than the active site, are called *allosteric proteins,* and, correspondingly, those small molecules that bring about allosteric transformations are called *allosteric effectors.* There are now only scant data on the chemical forces binding specific feedback inhibitors to proteins. As in the postulated repressor–corepressor union, the binding is believed to depend upon weak secondary forces (hydrogen bonds, salt linkages, and van der Waals forces), and not to involve any covalent bonds. Hence, feedback inhibition can be quickly reversed once the end-product concentration is again reduced to a low level.

Figure 14–25
Schematic view of how the binding of an end-product inhibitor inhibits an enzyme by causing an allosteric transformation.

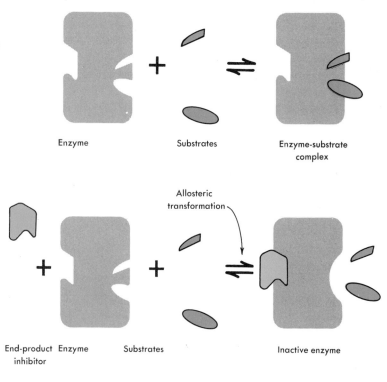

Enzyme   Substrates   Enzyme-substrate
complex

Allosteric
transformation

End-product Enzyme   Substrates   Inactive enzyme
inhibitor

SUMMARY

Cells have control mechanisms to ensure that proteins are synthesized in the required amounts. Only very recently have we begun to understand their molecular basis. Most of our knowl-

edge is limited to bacterial cells, in particular E. coli. *Bacteria contain many enzymes whose rate of synthesis depends on the availability of external food molecules. These external molecules (corepressors or inducers) control the rate of protein synthesis by controlling the synthesis of specific mRNA templates. Corepressors (inducers) often act by binding to specific molecules, the repressors. Repressors exist in an active state when they have combined with a corepressor and in an inactive state when they have combined with an inducer. Active repressors function by combining with specific regions of the DNA (operators). This binding in turn prevents specific binding of RNA polymerase to its DNA binding sites (its promoters), and so specifically stops the initiation of mRNA synthesis. The length of DNA controlled by a specific promoter, the operon, often comprises several genes with related metabolic functions (e.g., the production of successive enzymes in the synthesis of an amino acid or nucleotide). A still undiscovered mechanism brings about the differential synthesis of different proteins coded by the same mRNA molecule; some of the proteins are made much more frequently than others.*

There also exist a variety of positive control proteins which, by binding to specific promoters, encourage the simultaneous binding of RNA polymerase and the initiation of specific mRNA synthesis. Control of given operons need not be under exclusively positive or negative control, since many operons respond to both positive and negative signals. One very important positive control protein is CAP (catabolite activator protein) which binds to DNA only when it is complexed with cyclic AMP (cAMP). Receipt of this signal indicates that cAMP levels are high, and correspondingly that very little glucose is being broken down. The CAP signal thus provides a way to ensure that sugars like lactose are metabolized only after all the available glucose is used up.

For a long time it was thought that the various negative and positive control proteins were solely regulatory in nature and had no enzymatic capabilities. Recently, however, it has been found that glutamine synthetase, an important enzyme involved in the assimilation of nitrogen into organic molecules, also functions as a positive control protein. By binding to specific promoters, it transmits the signal that nitrogen is in short supply, and thus encourages the selective synthesis of enzymes which break down amino acids into utilizable nitrogen-containing molecules.

Cells that have metabolically unstable mRNA molecules can quickly shift their spectrum of protein synthesis in response to a radical change in their environment (e.g., food supply). This is true especially of bacteria, for which the average lifetime of most mRNA molecules may be as short as 2 to 3 minutes.

The rate of synthesis of many protein molecules is not controlled by repressors and inducers (constitutive synthesis). Some proteins are synthesized at fixed high rates, others at very low rates. Frequently this control is accomplished through the specific nucleotide sequence comprising their promoters. Some promoter sequences have great affinity for RNA polymerase, and others have low affinity. As repressors are usually made in very small amounts, their promoters have low affinity.

Control over cell metabolism is also quickly effected by end-product inhibition of enzyme function. An end-product metabolite can reversibly combine with the first enzyme involved in its specific biosynthetic pathway. This combination transforms the enzyme into an inactive form. Now it is suspected that the end-product inhibitor does not combine with the enzymatically active site but binds instead to a second site on the enzyme, causing a change in the enzyme shape. Proteins whose shapes and activities are changed by combination with other molecules are called allosteric proteins.

REFERENCES

WAINWRIGHT, S., *Control Mechanisms and Protein Synthesis,* Columbia University Press (1972). An advanced text looking at genetic control processes in both lower and higher cells.

The Lactose Operon, edited by J. Beckwith and D. Zipser, Cold Spring Harbor Laboratory (1970). Somewhat dated but still a very useful introduction to the best understood of operons.

MILLER, J., *Experiments in Molecular Genetics,* Cold Spring Harbor Laboratory (1972). A superior combination of a text and laboratory manual that emphasizes experiments on the lactose operon.

JACOB, F., AND J. MONOD, "Genetic Regulatory Mechanisms in the Synthesis of Proteins," *J. Mol. Biol.,* **3,** 318 (1961). A beautiful review that ties together the concept of messenger RNA with the problem of the control of protein synthesis.

GILBERT, W., AND B. MULLER-HILL, "Isolation of the Lac Repressor," *Proc. Natl. Acad. Sci., U.S.,* **56,** 1891 (1966). How the first repressor was isolated.

PTASHNE, M., "Specific Binding of the λ Phage Repressor to λ DNA," *Nature,* **214,** 232 (1967). The first proof that a repressor acts by attaching to DNA.

GILBERT, W., N. MAIZELS, AND A. MAXAM, "Sequences of Controlling Regions of the Lactose Operon," *Cold Spring Harbor Symposium* **XXXVIII**, 845 (1974). The working out of the first operator sequence.

DICKSON, R. C., J. ABELSON, W. M. BARNES, AND W. S. REZNIKOFF, "Genetic Regulation: The Lac Control Region," *Science,* **187,** 27 (1975). The sequence of the lac promoter and how it may function.

MANIATIS, T., M. PTASHNE, K. BACKMAN, D. KLEIT, S. FLASHMAN, A. JEFFREY, AND R. MAUER, "Sequences of Repressor Binding Sites in the Operators of Bacteriophage λ, *Cell,* **5,** 109 (1975). Details about the unexpectedly complex and interesting multiple operator regions of phage λ.

NAKANISHI, S., S. ADHYA, M. GOTTESMAN, AND I. PASTAN, "Studies on the Mechanism of Action of the Gal Repressor," *J. Biol. Chem.,* **248,** 5937 (1973). *In vitro* experiments showing the interaction of cAMP, CAP, and gal repressor in controlling gal promoter functioning.

ROSE, J. K., C. G. SQUIRES, L. YANOFSKY, H. YANG, AND G. ZUBAY, "Regulation of *In Vitro* Transcription of the Tryptophan Operon by Purified RNA Polymerase in the Presence of Partially Purified Repressor and Tryptophan," *Nature New Biology,* **245,** 133 (1973). *In vitro* support for the concept that the tryptophan operon is under negative control.

SMITH, G. R., AND B. MAGASANIK, "Nature and Self-Regulated Synthesis of the Repressor of the Hut Operons in *Salmonella Typhimurium*," *Proc. Natl. Acad. Sci.,* **68,** 1493 (1971). Evidence for a repressor which controls its own rate of synthesis.

TYLER, B., A. B. DeLEO, AND B. MAGASANIK, "Activation of Transcription of Hut DNA by Glutamine Synthetase," *Proc. Nat. Acad. Sci.,* **71,** 225 (1974). A very important paper showing that an important cellular enzyme directly controls the function of related operons by binding to their promoters.

MAGASANIK, B., M. J. PRIVAL, J. E. BRENCHLEY, B. M. TYLER, A. B. DeLEO, S. L. STREICHER, R. A. BENDER, AND C. G. PARIS, "Glutamine Synthetase as a Regulator of Enzyme Synthesis," *Current Topics in Cellular Regulation,* edited by B. I. Horeeker and E. Stadtman, **8,** 119 (1974), Academic Press. A review which shows the key role of glutamine synthetase in intracellular metabolism.

GREENBLATT, J., AND R. SCHLEIF, "Arabinose "C" Protein: Regulation of the Arabinose Operon *in vitro*," *Nature New Biology*, **223**, 166 (1971). Direct proof of a positive control function for the key regulatory protein of the arabinose operon.

MORSE, D. E., R. D. MOSTELLER, AND C. YANOFSKY, "Dynamics of Synthesis, Translation, and Degradation of trp Operon Messenger RNA in *E. coli*," *Cold Spring Harbor Symposium of Quant. Biol.*, **XXXIV**, 729 (1969). The most complete study so far of the kinetics of mRNA synthesis and breakdown.

BERTRAND, K., L. KORN, F. LEE, T. PLATT, C. L. SQUIRES, C. SQUIRES, C. YANOFSKY, "New Features of the Regulation of the Tryptophan Operon," *Science*, **189**, 22 (1975). A new type of regulatory site in the trp operon of *E. coli* is discussed.

MONOD, J., J. P. CHANGEUX, AND F. JACOB, "Allosteric Proteins and Cellular Control Systems," *J. Mol. Biol.*, **6**, 306–329 (1963). An important early review of the problem of allostery.

GERHART, J. C., AND A. B. PARDEE, "The Effect of the Feedback Inhibitor, CTP, on Subunit Interactions in Aspartate Transcarbamylase," *Cold Spring Harbor Symp. Quant. Biol.*, **XXVIII**, 491 (1963). An initial summary of an enzyme system, the study of which was important in developing the concept of allostery.

GERHART, J. C., "A Discussion of the Regulatory Properties of Aspartate Transcarbamylase from *E. coli*," *Current Topics in Cellular Regulation*, edited by B. Horeeker and E. Stadtman, **2**, 275 (1970), Academic Press. A more recent description of the most intensively studied allosteric enzyme.

The Replication
of Viruses

Geneticists usually have focused on viruses because they
have seemed so simple. In the beginning they were
thought of as naked genes, but gradually it became obvious that the correct analogy was the naked chromosome.
Many of the viruses first thought to be so uncomplicated
now are seen to contain several hundred genes. And when
we began to discover the ways bacteria can prevent unwanted synthesis, it seemed probable that the essence of a
virus's existence was the lack of any such regulatory devices. Being so constituted, they would be able to multiply
rapidly at the expense of their host cells' metabolism.
Again the first hunches proved wrong. The replication of
even the smallest of viruses is a very complicated affair,
achieved only with the aid of highly evolved regulatory
systems designed to see that the right molecules are synthesized at just the right time in the life cycle of a virus.
But before we look into the details of this problem, some
more general principles of viral structure and multiplication must be examined.

CHAPTER **15**

THE CORE AND COATING OF VIRUSES

Both the size and structural complexity of viruses show
great variation. Some have molecular weights as small as
several million, whereas others approach the size of very
small bacteria. However, all viruses differ fundamentally
from cells, which have both DNA and RNA, in that viruses
contain only one type of nucleic acid, which may be either

411

DNA or RNA. The genetic nucleic acid component is always present in the center of the virus particle, surrounded by a protective coat (shell). Some of the shells are quite complex; they contain several layers and are built up from a number of different proteins, as well as from lipid and carbohydrate molecules (Figures 15–1, 15–2, 15–3). In other cases, for example in tobacco mosaic virus (TMV)—a virus that multiplies in tobacco plants—or in the small

Figure 15–1
The morphology of *Herpes,* a large DNA-containing virus which multiplies in animal cells. (a) A schematic view of its general structure showing the DNA-containing core embedded in a regular shell (capsid), and an outer envelope. (b) An electron micrograph showing the 1000-Å diameter capsid surrounded by a 1500-Å envelope. (c) A high-resolution electron micrograph of the capsid. It contains 162 subunits (capsomeres) arranged about fivefold, threefold, and twofold axes of symmetry. [(b) and (c) reproduced from Wildy *et al., Virology,* **12,** 204 (1960), with permission.]

Figure 15–2
The structure of the T-even (2, 4, and 6) phage particle. (a) An electron micrograph of T2 [reproduced from R. W. Horne *et al., J. Mol. Biol.,* **1,** 281 (1959), with permission]. (b) A schematic drawing showing detailed features revealed by electron microscopy.

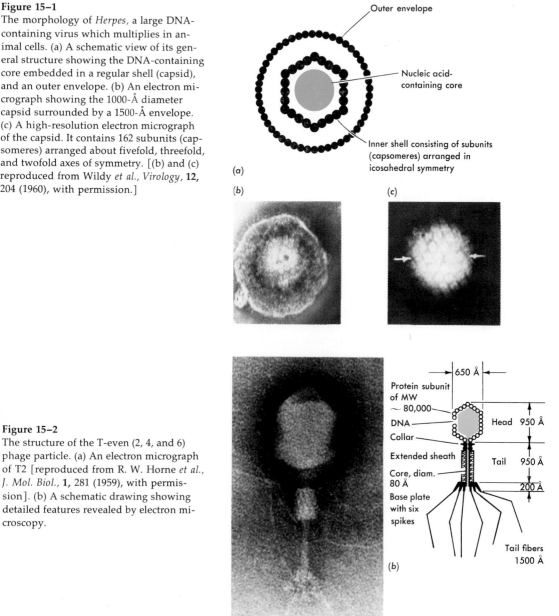

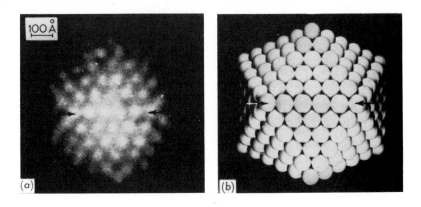

Figure 15-3
Adenovirus structure. These DNA-containing viruses, which multiply in animal cells, have very regular structures. (a) A particle at high magnification. 252 capsomeres are used to construct the outer shell. (b) A model of an icosahedron in the same orientation. [Reproduced from R. W. Horne *et al.*, *J. Mol. Biol.*, **1**, 86 (1959), with permission.]

RNA bacterial viruses F2, R17, and Qβ, the shell contains only one type of protein and no lipid or carbohydrate.

All shells contain many copies of the protein component(s), often arranged with either helical symmetry or cubical (or quasi-cubical) symmetry. Thus the TMV shell has about 2150 identical protein molecules (MW ~ 17,000) helically arranged around a central RNA molecule containing approximately 6000 nucleotides (Figure 15–4). In F2 (or R17) there are 180 identical proteins (MW ~ 14,000) cubically arranged about an inner RNA molecule with 3300 nucleotides.

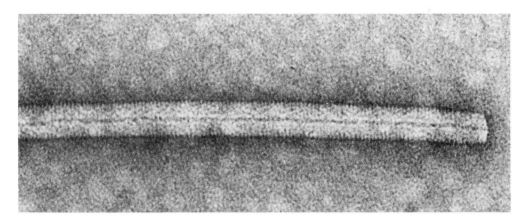

The use of a large number of identical protein molecules in the construction of the protective shell is an obligatory feature of the structures of all viruses. This follows from their limited nucleic acid content, which in turn places an absolute restriction on the maximal number of amino acids in the proteins coded by the viral chromosome. For example, the ~ 6000 nucleotides in a TMV RNA chain can code for ~ 2000 amino acids, corresponding to a

Figure 15–4
A high-resolution electron micrograph of one end of a TMV particle. The diameter is approximately 180 Å, whereas the length of a complete particle is 3000 Å. The particle is covered with a dark stain which penetrates the hollow central core. [Photograph reproduced from J. T. Finch, *J. Mol. Biol.,* **8**, 872, (1964), with permission.]

protein molecular weight of about 2.5×10^5. This is very much smaller than the molecular weight of TMV's protein shell (3.5×10^7). Thus, even if the entire TMV RNA chain coded for its coat protein (which it does not), approximately 150 identical protein molecules would be needed. This use of a large number of identical protein subunits is why the simpler viruses, which often contain only one type of protein molecule in their coat, have either helical or cubical (or quasi-cubical) symmetry. Only these two types of symmetry permit the identical protein subunits to be packed together in a regular (or quasi-regular) fashion and thus to have virtually identical (except for their contacts with the nucleic acid core) chemical environments.

NUCLEIC ACID: THE GENETIC COMPONENT OF ALL VIRUSES

Viruses afford some of the best demonstrations that genetic specificity is carried by nucleic acid molecules. Many viral nucleic acids are easily isolated from their protein shell and prepared in highly purified form. When they are added to host cells, new infective virus particles are produced; each is identical to those from which the nucleic acid was isolated (Figure 15–5). These very important experiments definitively show that the viral nucleic acid

Figure 15–5
Proof that RNA is the genetic component of TMV. The rod-shaped TMV particles can easily be separated into their protein and RNA components, which can then be separately tested for their ability to initiate virus infection. Only the RNA molecules have this ability. The virus particles produced by infecting with pure RNA are identical to those resulting from infection with intact virus.

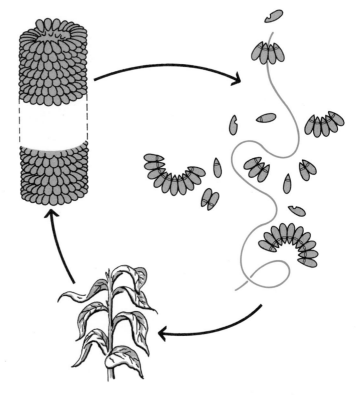

carries the genetic specificity to code both for its own replication and for the amino acid sequences in its specific coat proteins.

This is true not only for the DNA viruses, but also for viruses containing RNA. In fact, the first demonstration of infectious viral nucleic acid was made by using RNA isolated from TMV. Before this demonstration, doubts had persisted whether the RNA component of TMV was really genetic. This uncertainty existed because the large majority of TMV particles are not ordinarily infectious. Usually fewer than one in a million particles enters the tobacco leaf and serves as parent for progeny particles. It could thus be argued that perhaps this rare particle contained DNA. However, the isolation of infectious TMV RNA dispelled this uncertainty and clearly showed that sequences of nucleotide bases in RNA, like those in DNA, carry genetic messages.

VIRAL NUCLEIC ACID MAY BE EITHER SINGLE- OR DOUBLE-STRANDED

The nucleic acid of most viruses has the form of its cellular counterparts. Thus all the best known DNA viruses, such as smallpox (or its harmless relative *Vaccinia*), SV40, and the T2, T4, and T6 group of bacterial viruses, have the double-helical structure. Correspondingly, the RNA of TMV, influenza virus, poliomyelitis virus, and the bacterial viruses F2 and R17 is single-stranded. There are, however, several groups of bacterial viruses in which the DNA is single-stranded as well as at least two groups of RNA viruses (e.g., the Reoviruses) in which the RNA assumes a complementary double-helical form.

Fundamentally, it does not matter whether the genetic message is initially present as a single strand or as the double helix, for the single strand can quickly be used to form a complementary DNA chain soon after it enters a suitable host cell. The really important fact is that the genetic information is present as a sequence of nucleotide bases.

VIRAL NUCLEIC ACID AND PROTEIN SYNTHESIS OCCUR INDEPENDENTLY

Exactly what happens after a viral nucleic acid molecule enters a susceptible host cell depends on the specific viral system. Particularly important is whether the virus contains DNA or RNA. If the genetic component is DNA, then during viral replication the DNA serves as a template both for its own replication and for the viral-specific RNA necessary for the synthesis of its specific proteins. Similarly, if

415

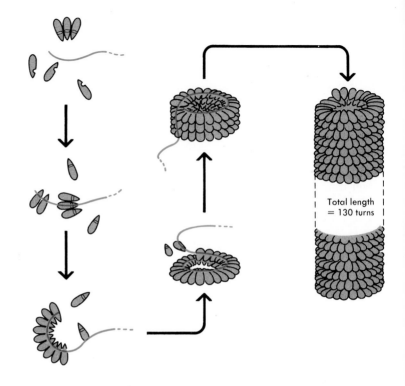

Figure 15-6
Formation of a TMV particle from its protein subunits and its RNA mole-
cule. [Redrawn from H. Fraenkel-Conrat, *Design and Function at the
Threshold of Life: The Viruses,* Figure 18, Academic, New York, 1962.]

RNA is the genetic component, the RNA molecules have
two template roles, the first to make more RNA molecules
and the second to make the viral-specific proteins. In both
cases the end result of virus infection is the same: the
production of many new copies of both the viral nucleic
acids and the coat proteins. The new progeny molecules
then spontaneously aggregate to form mature virus particles. Enzymes often play no role in these final aggregation events, since with many viruses the formation of new
covalent bonds is not needed either to build a stable protein coat or to affix it firmly to the nucleic acid core. Only
weak secondary bonds (salt linkages, van der Waals forces,
and hydrogen bonds) need be involved. This last point is
shown clearly with TMV. Here the rod-shaped particles
can be gently broken down and their free RNA and coat
protein components separated. When they are again mixed
together, new infectious particles, identical to the original
rods, quickly form (Figure 15-6). We thus see that the essential aspects of virus multiplication often are known

once we understand the principles by which viral nucleic acid and protein components are individually synthesized.

VIRAL NUCLEIC ACIDS CODE FOR BOTH ENZYMES AND COAT PROTEINS

The viral nucleic acid genetic component must code the amino acid sequences in the protein(s) that make(s) up the protective coat. These coat proteins are never found in normal uninfected cells and are completely specific to a given virus. In addition, the synthesis of one or many new enzymes usually occurs to permit successful viral multiplication.

Now there are many examples where DNA viruses carry information for the amino acid sequences of enzymes connected with the synthesis of their precursor nucleotides. One of the most striking cases involves T4 multiplication (Figure 15–7). No cytosine is present in its DNA,

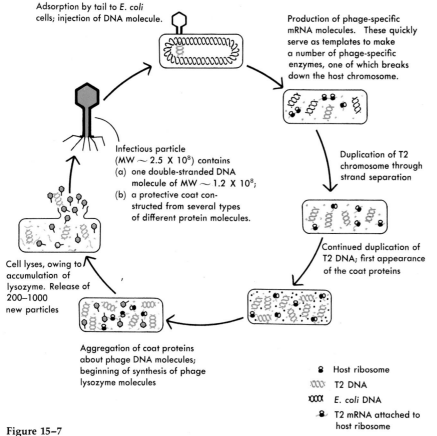

Adsorption by tail to *E. coli* cells; injection of DNA molecule.

Production of phage-specific mRNA molecules. These quickly serve as templates to make a number of phage-specific enzymes, one of which breaks down the host chromosome.

Infectious particle (MW ~ 2.5 X 10^8) contains
(a) one double-stranded DNA molecule of MW ~ 1.2 X 10^8;
(b) a protective coat constructed from several types of different protein molecules.

Duplication of T2 chromosome through strand separation

Continued duplication of T2 DNA; first appearance of the coat proteins

Cell lyses, owing to accumulation of lysozyme. Release of 200–1000 new particles

Aggregation of coat proteins about phage DNA molecules; beginning of synthesis of phage lysozyme molecules

- Host ribosome
- T2 DNA
- *E. coli* DNA
- T2 mRNA attached to host ribosome
- Phage-specific enzymes
- Phage-coat proteins

Figure 15–7
Chemical details in the life cycle of the double-stranded DNA virus T2 (T4).

a fact that initially suggested that T4 DNA might be very different from normal DNA. Instead, there is always present the closely related base 5-OH-methylcytosine, which, like cytosine, forms base pairs with guanine (Figure 15–8). The 3-D structure of T4 DNA is thus basically the same as that of normal double-helical DNA.

No 5-OH-methylcytosine is found in uninfected *E. coli* cells, and so the several new enzymes required for its biosynthesis must be coded by the T4 DNA. In addition, the rate of DNA synthesis in T4-infected cells is several times faster than in normal cells. This faster rate is achieved by having other T4 genes code for many of the enzymes involved in normal nucleotide metabolism, as well as for a DNA polymerase-like enzyme (Figure 15–9).

A new viral-specific enzyme is also frequently needed to ensure the release of progeny virus particles from their host cell. This is a vital need for those bacterial viruses multiplying in bacteria with rigid cell walls. Since these walls do not spontaneously disintegrate, they could effectively inactivate progeny particles by preventing their release and transfer to new host cells. To take care of this problem, many phages have a gene that codes for the amino acid sequence of lysozyme, a cell wall-destroying enzyme. This enzyme begins to be synthesized when the coat proteins appear, and causes the rupture of the cell wall at about the time virus maturation is complete.

MORPHOGENETIC PATHWAYS

The assembly of structurally complex viruses like λ and T4 is much more involved than the simple aggregation process needed for simple viruses like TMV. Some 40 different T4 gene products, each coded by a specific T4 gene, interact to produce the mature virus particle. Many of the genes code for the various structural proteins, while at least one other codes for an enzyme which most likely converts a precursor protein into the form found in mature T4 particles. In the assembly process the various components do not associate with each other randomly in time. Instead the assembly occurs in a definite sequence (a morphogenetic pattern), and devices not yet understood prevent the occurrence of a specific step until the preceding step has occurred. Three different branches, the first concerned with the head, the second with the tail, and the third with the tail fibers, come together as shown in Figure 15–10 to produce the final infectious particles.

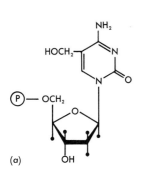

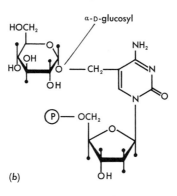

(a) (b)

Figure 15–8
Phages of the T-even group do not contain cytosine in their DNA. Instead they contain the related base 5-OH-methylcytosine (a), which base pairs exactly like cytosine. One or more glucose residues are attached to some of their 5-CH₂OH groups. (b) Shows this base with one glucose molecule attached. The biological significance of these unusual bases has not yet been clearly established. One speculative hypothesis asserts that their function is to protect T-even DNA from a phage-specific enzyme which only breaks down unmodified DNA. This hypothesis would explain how the *E. coli* DNA is selectively broken down during viral synthesis.

One of these enzymes specifically breaks down the normal DNA precursor.

A second enzyme converts.

A third enzyme adds a $P \sim P$ group to

This explains why no cytosine is found in DNA synthesized after infection.

This is a substrate for DNA polymerase and is incorporated in DNA.

A still further enzyme adds glucose to otherwise complete DNA molecules.

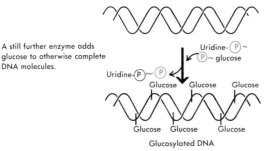

Glucosylated DNA

Figure 15–9
The biochemical mechanism that brings about the synthesis of DNA lacking cytosine and containing instead 5-OH-methylcytosine and its glucose derivatives. Immediately after infection, a number of specific enzymes are synthesized. These are coded for by the viral DNA, and each has a specific role in ensuring the successful multiplication of the virus.

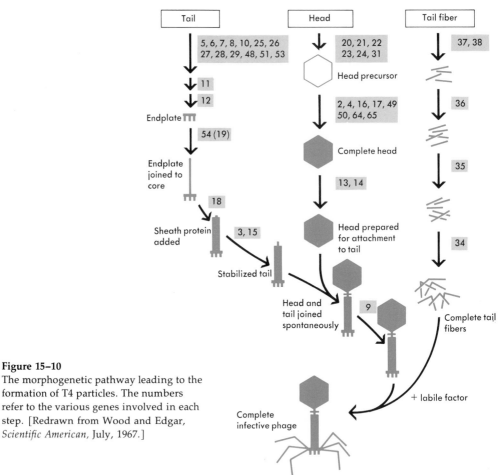

Figure 15–10
The morphogenetic pathway leading to the formation of T4 particles. The numbers refer to the various genes involved in each step. [Redrawn from Wood and Edgar, *Scientific American,* July, 1967.]

VIRAL INFECTION OFTEN RADICALLY CHANGES HOST CELL METABOLISM

Sometimes the synthesis of viral-specific nucleic acids and proteins goes hand in hand with normal cell synthesis. The chromosomes of the host cell often continue to function throughout a large fraction of the viral life cycle. In many cases, however, soon after infection, most of the cellular metabolism is directed toward the synthesis of new molecules connected exclusively with the appearance of new virus particles. In the most extreme cases, all DNA and RNA synthesis on the host chromosomes ceases, the preexisting RNA templates are degraded, and all subsequent protein synthesis occurs on new RNA templates coded by the viral nucleic acid.

The extent to which a virus is able to control its host's synthetic facilities varies greatly, depending both on the nature of the infecting virus and on the type of host cell. In general, the larger the viral nucleic acid content, the larger the number of viral genes directed toward stopping host cell functions unnecessary for the production of new viral particles. How these viral-specific genes redirect cellular metabolism is only beginning to be understood. For example, during T4 multiplication, the host *E. coli* chromosome is enzymatically broken down, most likely by an enzyme coded for by T4 DNA. In contrast, the T4 chromosome is not attacked enzymatically, most likely because it contains the unusual base 5-OH-methylcytosine. This differential destruction of the host chromosome, however, is not the sole, if even major, cause of the metabolic dominance of the viral chromosome. Long before evidence of its breakdown can be detected, the host chromosome has ceased to serve as a template for *E. coli* mRNA. Recent experiments (see below) suggest that this is due to a change in the host σ factor that prevents it from binding to the core component of the RNA polymerase.

SYNTHESIS OF VIRAL-SPECIFIC PROTEINS

Viral-specific proteins are synthesized in the same basic way as normal cellular proteins. The viral-specific messenger RNA molecules attach to host ribosomes, forming polyribosomes to which the AA $\sim$ tRNA precursors are attached. F-met-tRNA also starts all phage proteins and the terminating signals are those of their host cells. Changes may occur, however, in the structure of the host's ribosomes. After T4 infection, the *E. coli* ribosomes are so modified that they work better with T4 messenger than with host messenger. The exact nature of the modifications induced by the virus remains to be worked out. In addition, several cases are now known where, after viral infection, new tRNA species arise by enzymatic modification of preexisting tRNA molecules. Totally new tRNA molecules also appear in some virus-infected cells. And so viral, as well as host, genes can code for tRNA.

The significance of these new tRNA species is still very unclear. In some cases, they may be adaptations to the modifications in ribosome structure induced by the virus. In other cases, they may be connected with differences in base composition between the viral chromosome and the host chromosome. For example, T4 DNA has twice as many A–T base pairs as G–C base pairs, while *E. coli* DNA has roughly similar amounts of both types. As not all potential wobble pairs are equally strong, growth of an AT-rich virus using tRNA's adapted for a chromosome with equal amounts of the four bases might slow down the translation of many viral-specific proteins. Hence the possible selective advantage of new tRNA species with altered anticodons.

THE DISTINCTION BETWEEN EARLY AND LATE PROTEINS

After a viral chromosome enters a cell, its genes do not usually begin working all at the same time. Instead there is a regular time schedule by which they function. Some viral proteins appear immediately after infection, while the synthesis of others may not begin until more than halfway through a viral growth cycle. Also, some genes start working early and continue to do so throughout viral growth, while others function for only several minutes and then shut off (Figure 15–11). There are obvious advantages to such sequential appearance. Some of the early gene products are enzymes necessary to direct cellular synthesis from host to viral genes (they clearly must be turned on early), while others, such as the viral coat proteins or the lysozyme needed for cell lysis, must only appear late. For example, if lysozyme were an early enzyme, cell lysis would occur prematurely, aborting the construction of progeny virus particles. Hence, even the most cursory knowledge of a viral life cycle suggests that quite intricate control systems must exist, in order to guarantee that the right proteins appear just when they are needed. As far as we know, these control systems usually operate at the transcription level. Thus, some specific mRNA molecules appear early, while others are found only late in the growth cycle. At first it was convenient to subdivide the various mRNA's into just two categories—early and late mRNA's—and likewise to classify a particular protein as either early or late. But often the situation is more complex, and the literature is filled with terms like immediate early, delayed early, middle, quasi-late, and so on.

CONTROLLING GENE TIMING THROUGH GENE ORDER

Under optimal growth conditions, the time needed to synthesize an mRNA molecule of molecular weight ~2 million is of the order of 3 to 4 minutes. Molecules of this size (6000 nucleotides) can code for 4 to 5 average-sized proteins. But synthesis of these proteins will not begin at the same time—the proteins coded at the 5' end can be completely translated before the transcription of the gene at the 3' end has even started. Genes which need to function first thus tend to occur at the start of operons. Likewise, when the optimal time for the beginning of a gene's function is 3 to 4 minutes after viral infection, it sometimes makes sense to place it toward the 3' end of a fairly large operon (Figure 15–12).

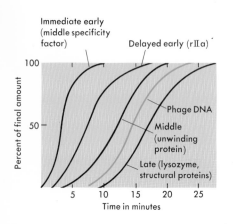

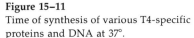

Figure 15–11
Time of synthesis of various T4-specific proteins and DNA at 37°.

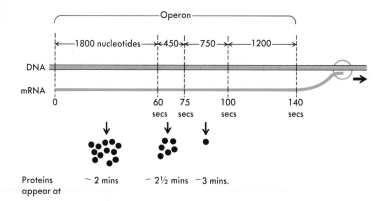

Figure 15–12
Sequential functioning of contiguous genes along an operon. At 37° *in vivo E. coli* mRNA chains grow at about 30 nucleotides per second. Complete transcription of a DNA segment containing 4200 base pairs will take about 2 minutes and 20 seconds. Attachment of the first ribosomes begins seconds after transcription starts. Since ribosome movement during translation occurs at roughly the same rate as transcription, functional gene products begin to appear in less than a minute after transcription of the corresponding gene has been completed.

The delayed appearance of the "late" proteins, however, is not completely explained by the existence of very long operons. For example, in T4 most operons are relatively short, and few have more than several thousand base pairs. Yet perhaps one-third of its 200 or so genes do not begin to function until about 10 minutes after its chromosome has entered the host cell.

THE SEARCH FOR THE ABSENT T4 REPRESSORS

To explain why these late operons were not transcribed, the hypothesis was put forward that the promoters for the "late" operons were blocked by highly specific repressors, viral-coded and synthesized immediately after penetration of the viral chromosomes into their host cell. By binding to the operator for late operons, they could specifically prevent appreciable amounts of late mRNA from being made until they were neutralized by an inducer, itself the product of a viral-specific early enzyme. Proof for such a scheme would come with the finding of mutants which upset the normal time schedule by preventing repressor synthesis and so allow all the late proteins to be made early. But, despite the isolation of a very large number of T4 nonsense mutations, not the slightest hint of a T4 repressor has been found. Instead, increasing evidence indicates that positive control, not negative control, is the key to T4 development.

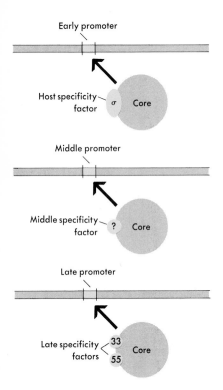

Figure 15–13
Changes in the specificity factors utilized by *E. coli* RNA polymerase to read the various classes of T4 promoters.

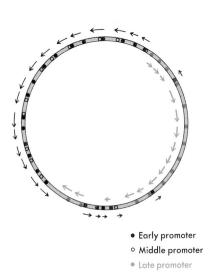

- ● Early promoter
- ○ Middle promoter
- ● Late promoter

VIRAL-SPECIFIC RNA POLYMERASE SPECIFICITY FACTORS

The heart of positive control for T4 lies in the existence of virally-coded specificity factors that change the set of promoters (early → middle → late) read by RNA polymerase as phage infection progresses. Immediately after infection, the specificity of transcription is determined by host σ factors (see page 296), which permit the recognition of some 25 different early T4 promoters (Figure 15–13). All the resulting mRNA's are copied from the same template strand, the "ℓ" strand, counterclockwise around the traditional T4 genetic map (Figure 15–14). Within most early operons, only those genes close to the promoter are quickly expressed (the immediate earlies). In contrast, the more distal genes are expressed more slowly (the delayed early genes), partly because of the greater time needed for the moving RNA polymerase molecules to reach them, but also because the growth of many early RNA chains is aborted before the more distal genes can be transcribed. Here we are not sure what is happening, but the best guess is that the host termination factor ρ often recognizes stop signals at the ends of the immediate early genes. So the more distal delayed early genes can be transcribed at high level only when ρ action becomes inhibited by a virally-coded antitermination protein similar to that thought to be coded by the "N" gene of phage λ (see Figure 15–23).

Initiation at all early promoters comes to a halt some five minutes (at 37°) after infection, when an early gene product blocks the host σ molecules from further combining with RNA polymerase cores. As this happens, the now free cores become activated, through the binding of new specificity factors, to transcribe first the middle and then the late T4 promoters. At first we thought that such specificity factors would closely resemble σ in size (MW ~ 70,000), but now it appears that relatively small proteins also can confer transcription specificity.

The middle promoters (initially referred to as quasi-late promoters) come into action before T4 DNA replication begins and continue to function throughout the re-

◀ **Figure 15–14**
Schematic diagram of the transcription pattern of T4. Genes under the control of the early and middle promoters are transcribed in a counterclockwise fashion while the late promoters control genes transcribed in the clockwise direction.

mainder of viral mutiplication. Many middle promoters control the reading of genes that are also controlled by the early promoters. Such promoters reflect the need to continue the synthesis of many (but not all) early enzymes long after the host σ ceases functioning (Figure 15–15). Other middle promoters govern the synthesis of proteins (enzymes) made in the middle to later steps of T4 development. For example, gene 32, which codes for a DNA unwinding protein, is totally controlled by a middle promoter. As yet we do not know the molecular nature of the middle specificity factor, but it may soon be known, since a T4 gene mutation has been found that specifically blocks the appearance of the mRNA that is totally under control of middle promoters.

The late promoters largely control genes that code for the T4 structural proteins as well as morphogenetic factors. Their reading, which leads to transcription exclusively from the "r" strand, is dependent upon the functioning of two early T4 genes, 33 and 55. These code for two "specificity" polypeptides of MW's 12,000 and 22,000 respectively. Both these polypeptides bind tightly to the core and so most likely act by directly allowing RNA polymerase binding to the late promoters.

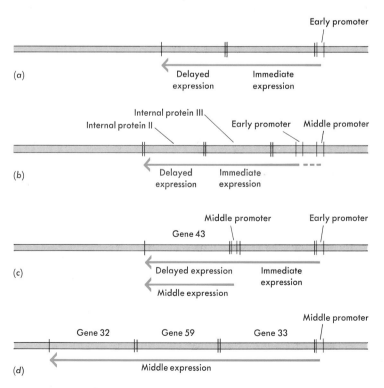

Figure 15–15
Interrelationships between genes controlled by the early and the middle T4 promoters a) A few T4 operons are totally under early promoter control. Depending upon how close the respective genes are to the promoter, they may have "immediate" or "delayed early" expression. b) Many more operons are controlled by both early and middle promoters. c) With other operons the middle promoter controls only the more distal genes. d) Still other operons are exclusively under middle promoter control.

AN ENTIRELY NEW RNA POLYMERASE CODED BY T7 DNA

Recently much attention has been given to understanding the transcription of T7-like phages (Figure 15–16). Because their DNA molecules (MW 25×10^6) are only $\frac{1}{5}$ the length of their T-even equivalents (MW 120×10^6), their genetic characterization is a much simpler task. Already some 25 new viral-specific polypeptides have been identified after T7 infection, with 24 of them related to specific genes along the linear T7 genetic map. The combined size of these T7-specific polypeptides accounts for over 90% of the T7 DNA coding capacity, and so we may expect the genetic characterization of T7 to be virtually complete within the next several years. The various T7 genes are clustered according to their function, with the left-hand 15% of its DNA coding for early genes primarily involved in the regulation of transcription. The remaining late genes fall into two quite different sets, one coding for enzymes needed in T7 DNA replication, the second coding for the structural proteins involved in T7 morphogenesis (Figure 15–17).

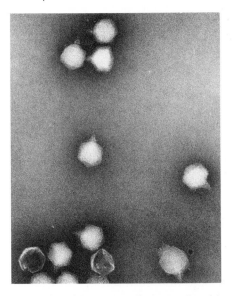

Figure 15–16
Electron micrograph of bacteriophage T7. [Courtesy of the Virus Laboratory of the University of California, Berkeley.]

The changeover from early to late T7 mRNA synthesis is not brought about by new specificity factors that interact with host polymerase cores. Instead, the late promoters are read by an entirely new type of RNA polymerase that is the product of the T7 gene 1. This gene, together with all the other early T7 genes, form a single operon at the extreme left end of the T7 genetic map and are controlled by a complex promoter which contains three initiating sites. This promoter is read exclusively by host

Figure 15–17
Genetic map of phage T7. The usual left-to-right orientation of the map is shown here from top to bottom. Indicated are the gene numbers, their estimated sizes and functions (sizes of gene 2 and 18 are still tentative). The location of the late promoters is still very tentative.

RNA polymerase molecules, with the resulting transcripts being a single class of early RNA molecules of MW $\sim 2.5 \times 10^6$. Subsequently these primary transcripts are cleaved by the enzyme ribonuclease III into a set of shorter chains (Figure 15–18). The largest of these cleavage products (MW $\sim 10^6$) codes for the T7 RNA polymerase, a polypeptide chain of MW $\sim 110,000$. Two other early genes appear to code for proteins that stop the functioning of the host RNA polymerase as the T7 RNA polymerase begins to accumulate. Thus, late in infection, only the late T7 genes are transcribed.

Several late T7 promoters exist, but it is not known how they are read or what their nucleotide sequences are. Nor do we know why the T7 RNA polymerase has only one polypeptide chain and so has a much simpler structure than the *E. coli* enzyme ($\beta'\beta\alpha_2\omega\sigma$). We can only guess that its relative simplicity is related to the need for fewer controls over its functioning. But up until now neither it nor the *E. coli* polymerase has been crystallized; as a result, many years may pass before we know exactly how they function.

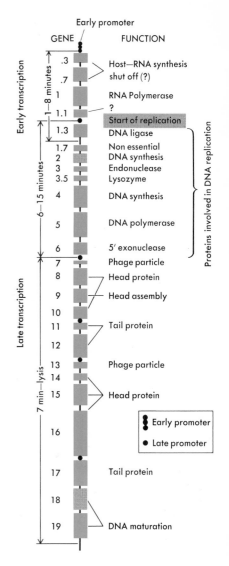

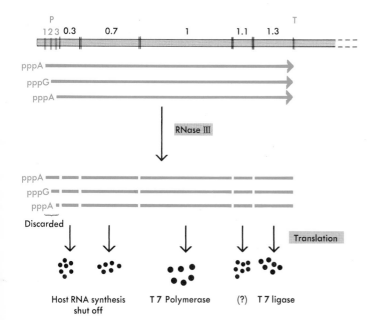

Figure 15–18
Cleavage of the primary early T7 transcripts by RNase III. Within the early promoter region three different initiation sites exist, separated from each other by 200 to 300 bases.

THE λ REPRESSOR MAINTAINS THE PROPHAGE STATE

The control of the transcription of phage λ DNA (MW 32×10^6) is even more complex (Figure 15–19). Not only does λ have a lytic life cycle characterized by early and late mRNA species, but, because it is a lysogenic phage, its genome must be able to recombine into and off host bacterial chromosomes. And while it is inserted as a prophage in a host chromosome, the transcription of the vast majority of its genes must be blocked. In this later process, the key element is the λ repressor, a protein normally found as a dimeric molecule of MW ~ 52,000. In its presence, virtually no λ-specific mRNA is made, the only exception being that made off C_1, the specific λ gene that codes for the λ repressor itself. In λ-carrying bacteria, the λ repressor normally is present in only 10 to 20 copies, most

Figure 15–19
The genetic map of λ, showing the function of many of its genes.

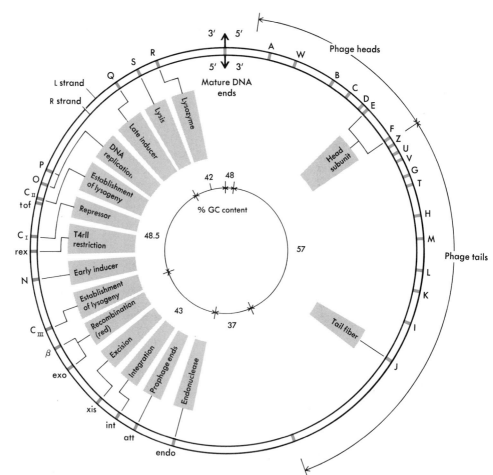

λ map

of which are bound to two specific operators, each of which controls a specific early mRNA molecule (Figure 15–20). Such binding occurs exclusively to double-helical DNA; and so the presence of a bound λ repressor does not unwind the respective operator but, instead, further stabilizes that section of duplex DNA.

Figure 15–20
During lysogeny (*a*), only λ repressor mRNA is made. Inactivation (*b*) of the repressor also allows transcription of early (left) and early (right) classes of λ mRNA.

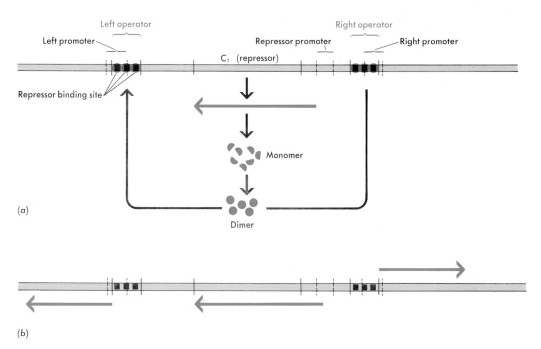

(a)

(b)

Both the left and right operators contain some 75 base pairs, with their sequences each containing three related but not identical symmetrical binding sites for repressor dimers (Figures 15–21 and 15–22). Approximately 17 base pairs are found in each binding site, with each site being separated from its neighboring site by extremely AT-rich regions. Extensively overlapping these λ operator

Figure 15–21
Sequence of the right operator of λ, showing its three repressor binding sites. Each site has sequences related by twofold symmetry and so is able to bind to repressor dimers.

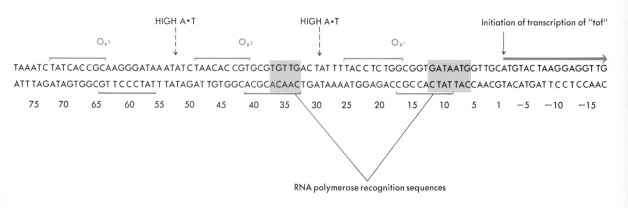

Figure 15–22

Comparison of the half sites of known λ repressor binding sites. The sequences written in the 5′ → 3′ direction are (a) of the transcribed strand and (b) of the complementary strand. Also shown, the base changes caused by three mutations (V_2, V_{101}, V_N) which lead to less effective repressor binding. All involve base pairs which are invariant in all known half sites. In the boxed insert are shown the twofold rotational symmetries in the repressor binding sites. An axis was drawn through the 9th base of each 17 base pair site, and the symmetry relations of the surrounding 16 bases were tabulated: S, perfect symmetry; P, purine-purine or pyrimidine-pyrimidine symmetry; H, hyphen, i.e., no symmetry. In a random sequence the expected symmetry would be $S_2P_4H_2$. [Data from T. Maniatis, *et al.*, 1975.]

Wild-Type Sites — Mutant Sites

$O_L{}^1$	a TATCACCGC ——— V_2 →	TATCAACGC
	b TACCACTGG ——— V_{101} →	TGCCACTGG
$O_R{}^1$	a TATCACCGC	
	b TACCTCTGG	
$O_L{}^2$	a CAACACCGC	
	b TATCTCTGG	
$O_R{}^2$	a CAACACGCA ——— V_N →	CAACAAGCA
	b TAACACCGT	
$O_R{}^3$	a TATCCCTTG	
	b TATCACCGC	
$O_L{}^3$	a TATCACCGC	

$T_9 A_{11} T_6 C_{11} A_8 C_{11} C_6 G_9 C_5$

$C_2 \quad A_3 \quad T_2 \quad T_4 C_1 G_4$

$C_2 \quad C_1 \quad G_1 T_1 A_1$

T_1

	s	p	h
$O_L{}^1$	6	2	0
$O_R{}^1$	5	2	1
$O_L{}^2$	4	2	2
$O_R{}^2$	5	1	2
$O_R{}^3$	5	1	2

sequences are the corresponding promoter sequences involved in the initial recognition of RNA polymerase and subsequent initiation of RNA synthesis. Most importantly, the sites with the highest affinity for the λ repressor are immediately adjacent to the starting points for RNA synthesis, while the more distal sites bind the repressor less well. Why each λ operator needs to have several repressor binding sites is not yet clear, beyond the obvious conjecture that sequential repressor binding makes successively more unlikely the unwinding of the promoter region that is the necessary first step in the binding of the RNA polymerase.

Inactivation of the λ repressor frequently occurs under conditions where DNA synthesis is inhibited (e.g., after treatment with uv light). This allows mRNA transcription of the two early operons to begin (Figure 15–20). One of these operons is largely occupied by genes coding for proteins necessary for the recombination events that free the λ chromosome (see Chapter 7) from the *E. coli* chromosome. The other early operon contains, among others, the genes specifically needed for the replication of λ DNA.

One of the first early genes to be read somehow stops further synthesis of the C_1 (repressor) mRNA. The exact gene involved, "tof," lies just to the right of C_1. Shutting off repressor mRNA synthesis clearly makes sense, since no new repressors are needed or wanted throughout the remaining part of the life cycle.

POSITIVE CONTROL DIRECTED BY THE "N" GENE ANTITERMINATION FACTOR

Release of repression, however, does not by itself lead to the complete transcription of the early operons. For this to occur, the key λ gene "N" must work. In the absence of the "N" gene product, mRNA synthesis starts at the two respective promoter sites, but elongation proceeds for only relatively short distances, as nearby stop signals are read by an *E. coli* RNA ρ (termination) factor. The "N" protein somehow specifically antagonizes the action of the termination factor, thereby allowing much, much larger portions of both operons to be read (Figure 15–23). The signal which terminates the synthesis of these much longer chains is not yet known, though now we would guess that it operates independently of ρ.

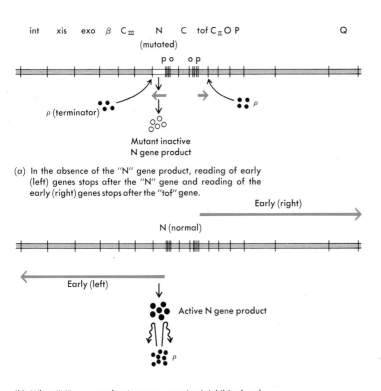

Figure 15–23
Current scheme for the action of the λ "N" gene as an antitermination factor.

(a) In the absence of the "N" gene product, reading of early (left) genes stops after the "N" gene and reading of the early (right) genes stops after the "tof" gene.

(b) When "N" gene product is present, ρ action is inhibited and both the early operons are transcribed.

A SINGLE PROMOTER FOR ALL LATE λ GENES

Almost half of λ's genetic material (~ 20 genes) codes for late proteins, most of them involved in the synthesis of λ's head and tail components. All these late genes belong to a single operon, which begins before the "S" gene and ends after gene "J" (Figure 15–24). This operon functions only in the presence of a protein made by the early gene "Q." Conceivably, "Q" product is a σ-like factor necessary for reading of the late promoter. Alternatively, it may act in a manner similar to that of the "N" protein, and lead to the extension of synthesis of a very short RNA chain whose promoter lies just before gene "S" and which, it is thought, is normally terminated by ρ action.

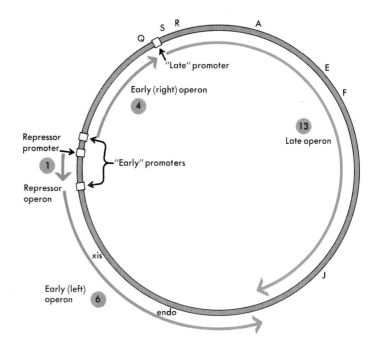

Figure 15–24
The four operons of the phage λ. The numbers give the approximate times needed at 37° for complete transcription of the respective operons.

The "Q" gene is located some 4 minutes transcriptional time away from its promoter. Thus, its first protein product only begins to appear 5 minutes after the start of infection; and it is not until 10 minutes after infection that enough "Q" products are present to allow appreciable transcription of late mRNA. Further control over the λ life cycle is achieved by the putting together of all the late genes in a single operon. For example, some 10,000 base pairs separate the genes for the main head protein from the ones coding for the tail fibers. This means that the final tail assembly can begin only 5 minutes after that of the heads.

With so much potential timing control possible through use of very long operons, the question should be posed why, in fact, a separate operon for the late proteins has come into existence. The answer lies in the fact that much larger amounts of late mRNA than of early are required. Most, if not all, of the early gene products are enzymes, while many of the late proteins are structural proteins needed in very large quantities. A device is thus required to turn on large-scale late mRNA synthesis some 8 to 10 minutes after the early operons start functioning. Evolution of a gene product permitting a new promoter to operate accomplishes this objective in a simple fashion.

SEVERAL PROMOTERS FOR THE VERY SMALL DNA PHAGES

A much, much simpler transcriptional picture holds for the very small single-stranded polyhedral (ϕX174, S13) and filamentous (F1, fd, M13) DNA phages. Both varieties have circular chromosomes; the genome of the former contains about 5500 nucleotides, and that of the latter 6000. *A priori* we would guess them to code for six to ten proteins of average chain length between 200 and 300. In fact, nine genes have already been characterized within ϕX174; and the pleasant possibility exists that almost all its genes have already been found (Figure 15–25).

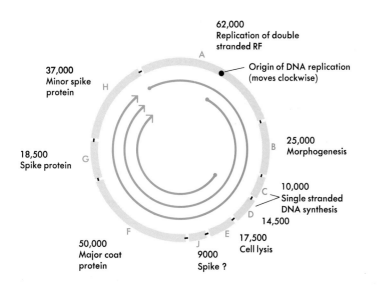

Figure 15–25
Genetic map of ϕX174 showing sizes and function of its nine genes together with possible map assignments for three of its transcripts.

Four of the ϕX174 genes are used to code for its structural proteins. Originally, because of its very small size, the protein coat was thought to be built up from the regular aggregation of only one protein. But now four distinct proteins—of molecular weights 50,000 36,000 19,000 and 9,000—have been identified (Figure 15–26). Their coding requires 3000 nucleotides, or some 50 percent of the total chromosome. One of the remaining five genes is known to be involved in the replication of the double-helical DNA intermediate (Figure 9–10); another two function in the production of the progeny single-stranded circles; and still another brings about cell lysis. Hopefully, within the next decade, the exact task of all its genes should be known.

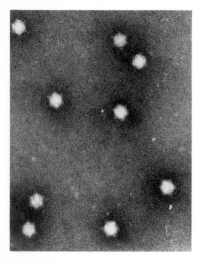

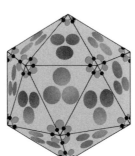

●	F = Major capsid protein	60 copies/particle
●	G = Spike protein	60 copies/particle
○	H = Spike protein	12 copies/particle
•	J = Spike (?) protein	60 copies/particle

Figure 15–26
Electron micrograph of phage φX174 [courtesy of the Virus Laboratory, University of California, Berkeley] together with a schematic drawing of its polyhedral shape showing a tentative arrangement for its coat (capsid) and spike proteins.

Synthesis of viral-specific mRNA begins as soon as the double-helical intermediates are made. All this mRNA is complementary to the viral "−" strand, with some transcribed as a single linear unit resulting from one complete reading of the genome, starting from a promoter near the start of the "A" gene (Figure 15–25). Two other promoters control the production of slightly smaller mRNA products, one of which may code for all the viral structural proteins. The relative location of the various genes is unlikely to be an important factor controlling the exact time at which the various virally-coded proteins are made—transcription of the entire genome takes only 2 minutes at 37°, while the corresponding virus life cycle occupies some 20 minutes. Further intensive research on these viruses is in order since, given their small size, the working out of all the essential details of their replication is an achievable objective.

SPECIFIC INITIATION FACTORS FOR VIRAL DNA REPLICATION

Synthesis of viral DNA molecules always starts at a fixed point along the viral chromosome. How these sites are recognized is not yet known, though we presume the existence of sequences that say "start synthesis" as well as of components of the DNA replication machinery that specifically recognize these start signals. Now we guess that both the starting sequences and the recognition factor(s) are specific for a given virus, with every DNA virus so far studied having genes that code for proteins needed to start its respective DNA synthesis. Most likely many such proteins are involved in the synthesis of the short RNA chains now thought to prime the synthesis of most (all?) DNA chains. For example, the bidirectional replication of λ DNA appears to be primed by two short RNA chains controlled by its O and P genes. Recognition of the start signals for DNA replication may thus be mediated either by completely new RNA polymerase molecules or by host molecules modified by the addition of virally-coded specificity factors.

REPEATED INITIATION OF DNA REPLICATION DURING VIRAL REPLICATION

During cellular DNA replication, a given start signal is read only once during each cell cycle. Repeated initiation cannot occur, since this would lead to an ever-increasing number of chromosomes. So it may be necessary to postulate some self-destruct mechanism that inactivates cellular DNA initiation proteins after they have functioned only once. But during viral DNA replication, there is no restriction to the number of times a given start signal can be read; and, for example, replicating T7 chromosomes with multiple forks are frequently observed (Figure 15–27). So either the initiation proteins for viral DNA synthesis can function more than once, or large numbers of such proteins are made during the early stages of viral DNA synthesis.

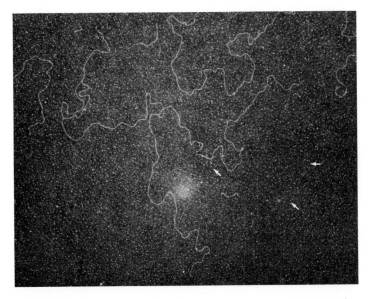

Figure 15–27
T7 DNA molecule containing multiple replication forks. [Courtesy of Dr. David Dressler, Harvard University.]

VIRAL RNA SELF-REPLICATION: REQUIREMENTS FOR A NEW VIRAL-SPECIFIC ENZYME

Cellular RNA molecules never serve as templates for the formation of new RNA strands. The replication of most RNA viruses demands the participation of a completely new enzyme capable of forming new RNA strands upon parental RNA templates. This enzyme, often called RNA replicase (synthetase), is usually formed just after the viral RNA enters the cell and attaches to host ribosomes. Like both DNA polymerase and RNA polymerase, RNA replicase catalyzes the formation of a complementary strand upon a single-stranded template. The fundamental mechanism for the copying of all nucleic acid base sequences is thus the same. Pairing of complementary bases is always used to achieve accurate replication of specific nucleotide sequences.

After their formation, the complementary RNA strands in turn serve as templates for new rounds of RNA synthesis. In this process, some of the progeny strands become part of new virus particles, while others are used as templates for specific protein formation. Exactly how this happens depends on whether the mature virus contains single-stranded or double-stranded RNA. In both cases, however, only one of the two complementary strands is used to make the protein template (i.e., mRNA) for a given gene. A close parallel thus exists between the formation of mRNA in both DNA and RNA viral systems.

During the reproduction of most single-stranded RNA viruses (e.g., polio, the various RNA phages), the strand having the template function is the same strand found in the mature particles (i. e., "+"). This allows the infectious strand to code for the RNA replicase molecules necessary to initiate viral RNA replication. Behaving quite differently, however, is the vesicular stomatitis virus, a single-stranded RNA animal virus. With this virus, the template for the viral-specific proteins is the "−" strand. This raised the question of the origin of the RNA replicase molecules necessary to make the "−" strands upon the infecting "+" strands. Just recently their origin has been found to be the infecting virus particles themselves. Each vesicular stomatitis particle has packaged within it a replicase molecule made in the previous cycle of viral infection.

RNA PHAGES ARE VERY, VERY SIMPLE

All the RNA phages (R17, F2, MS2, Qβ, etc.) so far characterized are very small (~250 Å diameter). They are the simplest viruses known (Figure 15–28), containing only

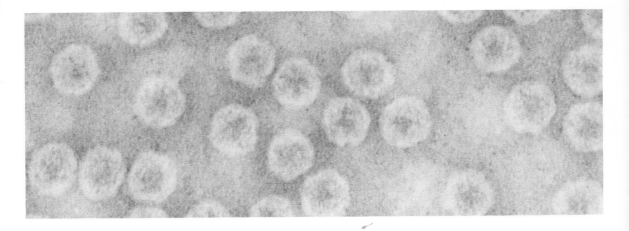

three genes on their ~ 3500 nucleotide-long RNA chromosomes and having a life cycle that produces some 5000 to 10,000 progeny phages within 30 to 60 minutes at 37° (Figure 15–29). Two out of their three genes are used to code for their two types of structural polypeptide chains. One, the coat protein (CP), has a molecular weight of 14,700 and

Figure 15–28 Electron micrograph of a collection of F2 particles negatively stained so that they appear light against a dark background. [Courtesy of Dr. Norton Zinder of Rockefeller University.]

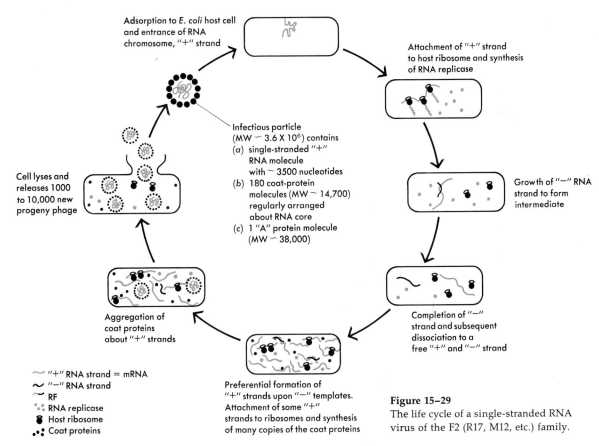

Adsorption to *E. coli* host cell and entrance of RNA chromosome, "+" strand

Attachment of "+" strand to host ribosome and synthesis of RNA replicase

Growth of "−" RNA strand to form intermediate

Infectious particle (MW $\sim 3.6 \times 10^6$) contains
(a) single-stranded "+" RNA molecule with ~ 3500 nucleotides
(b) 180 coat-protein molecules (MW $\sim 14,700$) regularly arranged about RNA core
(c) 1 "A" protein molecule (MW $\sim 38,000$)

Cell lyses and releases 1000 to 10,000 new progeny phage

Completion of "−" strand and subsequent dissociation to a free "+" and "−" strand

Aggregation of coat proteins about "+" strands

Preferential formation of "+" strands upon "−" templates. Attachment of some "+" strands to ribosomes and synthesis of many copies of the coat proteins

$\sim$ "+" RNA strand = mRNA
$\sim$ "−" RNA strand
$\sim$ RF
RNA replicase
Host ribosome
Coat proteins

Figure 15–29
The life cycle of a single-stranded RNA virus of the F2 (R17, M12, etc.) family.

is present 180 times in each virus particle. Its sequence of 129 amino acids is now completely worked out. The other structural protein is the "A" (attachment) protein needed for adsorption of these phages to their host bacteria, as well as for the subsequent penetration of the viral RNA through the bacterial cell wall and membrane. One copy is found in each virus particle. It has a molecular weight of about 38,000 and is made from about 330 amino acids. Thus, almost half of RNA phage genetic material is used to code for its structural proteins. The remaining virally specified protein is a component of the RNA replicase involved in the self-replication of its RNA chain. About 580 amino acids are found in its polypeptide chain, leaving only about 300 nucleotides that do not have a coding role.

The existence of three gene products receives direct support from genetic analysis. Use of mutagens produces nonsense mutations which fall into three complementation groups, which have been assigned to the CP, "A," and replicase (REP) genes. Their gene order, however, cannot be obtained directly by genetic tricks, since no genetic recombination occurs during RNA-phage replication. Recently, the order ($^5{}'$A–CP–REP$^3{}'$) has been established through nucleotide sequence studies (Figure 15–30).

Figure 15–30
The genes of phage R17 (MS2) drawn to scale with their sizes and relative positions as established through recent nucleotide sequence analysis.

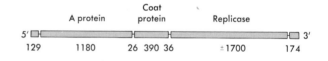

INITIAL BINDING OF RIBOSOMES TO ONE SITE ON THE PHAGE RNA

After entry of an RNA phage RNA molecule into a host cell, ribosomes attach only to the ribosomal binding site at the beginning of the CP gene (Figure 15–31). No attachment occurs at the beginning of either the "A" gene or the replicase gene, since their ribosome binding sites are blocked by the secondary structure taken up by the viral RNA. The first of these blocked sites to become accessible is the REP site, which opens up when reading of the CP gene temporarily disrupts the inhibitory hydrogen bonded hairpin loops. Functioning of the replicase gene thus always depends on prior ribosome attachment at the CP binding site. This restriction was first suggested by the existence of polar nonsense mutations in the CP gene, which prevented functioning of the replicase gene. In contrast, nonsense mutations in the replicase gene never affected the CP gene. Further experiments showed that nonsense mutations in the "A" gene never had any effect on reading of either the CP or REP genes. At first, this was interpreted to mean that the "A" protein lay at the distal (3') end. Now, however, we realize that when more than one ribosome binding site exists, polarity experiments do not necessarily give clear conclusions about gene locations.

POLARITY GRADIENTS

Not all nonsense mutations are polar. In general, only those at the 5' ends of a given gene are extreme polars, with a gradient of polarity extending toward the 3' end. When a ribosome reads a chain-terminating nonsense codon, releasing an incomplete polypeptide chain, it generally detaches from its mRNA template. It or another ribosome will not have a high probability of working on a distal gene unless that gene's binding site is exposed; and the probability that it will be exposed depends on how much of its 5' neighboring gene has been read. Thus, nonsense mutations at the 3' end of a gene will, in general, not affect the reading of distal genes. By the time a ribosome reads such nonsense mutations, the hairpin loops encompassing the next ribosome binding site have been opened up.

COAT PROTEIN CHAINS CAN REPRESS THE TRANSLATION OF THE REPLICASE GENE

The RNA phage coat protein polypeptide chains are made in much greater numbers than the replicase polypeptide chains. As far as we can tell, not more than 10 percent of the ribosomes that have moved across the CP gene move onto the REP gene. Equally, if not more, important as an *in vivo* control device, newly made CP subunits can specifically attach to the REP binding sites and prevent translation of the REP gene (Figure 15–32). Within 10 minutes following infection, sufficient CP has accumulated to block all further replicase synthesis. Thus, CP chains should be regarded as specific repressors working at the translational level.

Specifically stopping replicase synthesis less than halfway through the life cycle makes obvious sense. By this time, enough enzyme has accumulated for viral RNA replication, and the primary process needed in the latter phases of the life cycle is the production of the structural proteins of the virus.

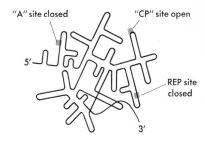

(a) In the absence of protein synthesis the "A" and REP sites are blocked.

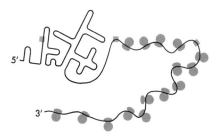

(b) Passage of ribosomes over the CP gene opens up the secondary structure normally blocking the REP site thereby making the REP gene available for translation.

Figure 15–31
Schematic diagram showing how secondary structure affects the accessibility of ribosome binding sites.

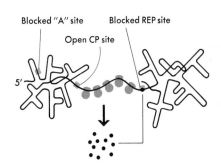

Figure 15–32
Binding of a coat protein molecule to the initiation site of the REP gene.

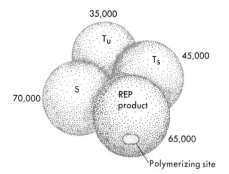

Figure 15–33
Schematic representation of Qβ replicase showing its construction from one virally-coded polypeptide chain and three host components.

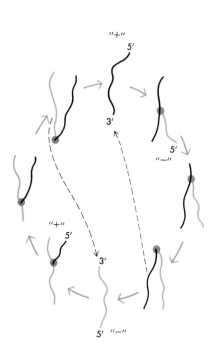

Figure 15–34
Schematic model for the replication of Qβ (MS2, R17 . . .) RNA directed by the phage replicase. Here for simplicity of visualization the single-stranded regions are shown in an extended rather than tightly collapsed form.

FUNCTIONAL COMPLEXES OF VIRALLY-CODED REPLICASE CHAINS AND HOST PROTEINS

The 65,000-MW polypeptide product of the replicase gene does not function by itself but only after it has combined with three different host polypeptide chains (Figure 15–33). Two of the host polypeptide chains (MW 35,000 and 45,000) are the Tu and the Ts components of the "T" factor, the *E. coli* protein complex involved in the placement of incoming AA ~ tRNA molecules into the "A" site of ribosomes. The remaining host factor is S1, a 70,000 MW protein of still unknown function, which is usually found tightly bound to the 30-S ribosomal subunit. Now we are certain only of the role of the virally-coded component. It contains the catalytic site responsible for the making of the internucleotide 3'-5' phosphodiester bonds. Most likely the "T" factor functions in chain initiation, helping to recognize the 3' ends of the RNA templates. Favoring this hypothesis is the presence of tRNA-like sequences (e.g., . . .CCA$_{OH}^{3'}$) at the 3' ends of both the viral "+" and "−" strands. As "T" factors are adapted specifically to recognize the tRNA molecules, their presence within the complete replicase complex may have evolved to help solve the problem of how replicase molecules preferentially attach to 3' ends of viral as opposed to cellular RNA molecules. How the S1 component functions is even less clear, though here again it seems necessary only for the initiation stage and not for chain elongation.

RNA PHAGE RNA SELF-REPLICATION DOES NOT INVOLVE DOUBLE-HELICAL INTERMEDIATES

Replication of viral RNA starts with the attachment of the replicase to the 3' end of the parental "+" strand and subsequent 5' → 3' synthesis of a complementary "−" strand, using ribonucleoside triphosphates as the precursors. Though base pairing determines the "−" strand sequences, no double-helical *replicative intermediate* is generated, and a free "−" strand is the product of the first round of synthesis (Figure 15–34). Only at the site of replicase binding is a short double-helical stretch present within the replicative intermediate. But if the replicase is removed by treatments with protein-denaturing agents, all the nascent strands base pair with their parental complements to form double-helical sections (Figure 15–35). So the replicase itself somehow is responsible for separating the progeny "−" strands from the parental "+" strand templates. Both the complete progeny "−" strand and the original "+" strand in turn serve as templates to make more "+" and "−" strands. In this process, very much

440

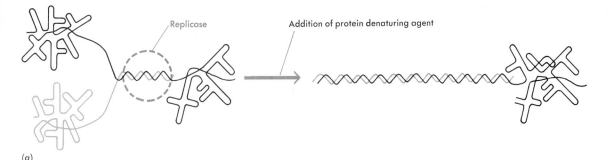

(a)

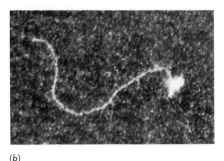

(b)

Figure 15–35
a) Schematic diagram of the conversion of the compacted replicating intermediate of an RNA phage into an extended, largely double-helical form following removal of the replicating enzyme. b) Electron micrograph of the extended form. The bright object at the end of the double-helical section is a balled up region of single stranded R17 RNA. [Courtesy of Sigmund and Robert Thach of the Washington University School of Medicine.]

larger numbers of progeny "−" strands are made than of progeny "+" strands, indicating that replicase molecules preferentially bind to the 3′ end of "−" strands, as opposed to corresponding 3′ ends of "+" strands.

Test-tube-made "+" strands sediment at the same rate as *in vivo*-made "+" chains and have equal infectivity when added to growing bacteria. Moreover, the frequency of mutants is similar to that in nature. This means that the copying achieved *in vitro* is just as accurate as that occurring within cells.

ONLY NASCENT "+" STRETCHES SERVE AS TEMPLATES FOR THE "A" PROTEIN

Reading of the "A" gene is limited to the periods of viral RNA replication that are characterized by incomplete stretches of nascent RNA. Just after a viral "+" strand has started to grow, the newly formed "A" binding site is accessible for ribosome attachment (Figure 15–36). But it quickly becomes closed as extensive patches of secondary structure form upon further chain elongation. Open "A" sites bind ribosomes even better than open "CP" sites, and so we see that their inherent affinity for ribosomes by itself does not determine how often they will be translated. Even more crucial is the degree to which they are not blocked by secondary structure.

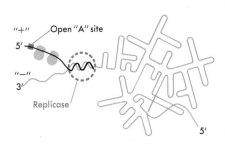

Figure 15–36
Formation of "A" protein upon nascent "+" strands.

441

SELF-ASSEMBLY OF PROGENY PARTICLES AND THE FORMATION OF INTRACELLULAR VIRAL CRYSTALS

Assembly of progeny phages spontaneously starts as soon as the "A" protein, coat protein, and progeny "+" strands begin to accumulate. Only progeny "+" strands are ever encapsulated; thus the newly made "−" strands are left free to produce still more progeny "+" strands. In the aggregation process, the "A" protein is thought to bind early to the phage RNA, forming a tight complex which persists through the next cycle of phage growth when the "A" protein enters the bacteria along with its respective RNA molecule. Conceivably, binding of the "A" protein converts progeny "+" strands into forms which hasten the aggregation of coat protein around them.

As infection progresses, the progeny protein often forms such a substantial fraction of the internal bacterial contents that they begin to form tiny crystals that are easily visualized by the EM (Figure 15–37). So far such crystals have been found only *in vivo*. Despite the fact that grams of pure RNA phage can easily be obtained, no one has yet been able to make them *in vitro*; and so it has not yet been possible to initiate x-ray diffraction analysis of RNA phage structure.

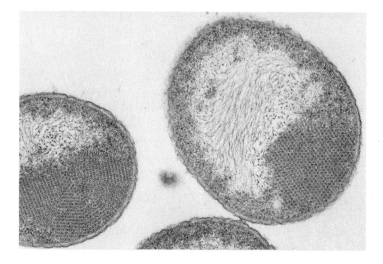

Figure 15–37
Electron micrograph of a thin section of an *E. coli* cell infected with the RNA phage F2. Most of the progeny particles lie next to each other in crystalline array. [Courtesy of Dr. Norton Zinder of Rockefeller University.]

THE COMPLETE NUCLEOTIDE SEQUENCE OF PHAGE MS2 HAS JUST BEEN ESTABLISHED

For many years, almost no one thought that the complete nucleotide sequence of a virus could be worked out. The nucleotide numbers for even the smallest viruses seemed just too large to be tackled. But over the last decade, the development of new methods for determining nucleotide sequences has changed everyone's attitude. Particularly important has been the discovery of new enzymes that cut at specific sequences, as well as the realization that the absence of any enveloping secondary structure makes certain sequences much more susceptible to nuclease cutting than equivalent sequences folded in tight hairpin turns. For example, when R17 RNA is briefly exposed to the enzyme RNase IV, the first cut creates a 40 percent piece bearing the 5' terminal fragment and a 60 percent piece containing the 3' end. Another important trick lets RNA replicase act for variable times, thereby producing incomplete chains of increasing length. This allows easy ordering of many of the polynucleotide fragments produced by limited enzymatic digestion. Even more helpful, especially when very long sequences are in question, is the fact that many of the sequences so obtained can be checked by the amino acid sequences they predict.

The first RNA phage sequences to be worked out were those of R17 and Qβ, where sizable regions of the 5' and 3' ends were worked out, as well as large parts of the CP gene. In the latter case, the data could readily be checked by comparison with the already determined CP amino acid sequences. Today the most complete data comes from phage MS2, a very close relative of R17, which possesses almost identical sequences to those found in R17. Its complete sequence has just been established (spring 1975) and models for over half of its secondary structure have been worked out. Figure 15–38 shows such a model for its CP gene, together with the intercistronic sequences that separate it from the A and REP genes. A model for the entire genome should become available over the next year.

Most importantly, this sequence data tells us that translation does not start at the 5' end, with some 129 nucleotides preceding the GUG codon that specifies the initiating formyl-methionyl group. And correspondingly there is no termination codon at the 3' end: the UAG termination signal for the REP gene is followed by an untranslated sequence of 174 nucleotides (Figure 15–38). Now we guess that many of these untranslated nucleotides (or their complements) function specifically to recognize the viral-specific RNA replicase. Since the same enzyme replicates both the "+" and "−" strand, the 3' ends of the "+" and "−" strands must look roughly similar. Hence the 5' end of the "+" strand (on which the 3' end of the "−" strand is made) must have some sequences complementary to sections found at its 3' end.

Figure 15–38

The genetic map of MS2 at the nucleotide level. a) Shows the sequence of the 5′ end up to the initiation codon (GUG) for the "A" protein. b) The complete sequence of the coat protein gene together with its adjacent intercistronic regions. c) The untranslated sequence of the 3′ end, showing the UAG codon that terminates the replicase gene (all the sequences have been worked out at the University of Ghent by Dr. Walter Fiers and his colleagues). In the inset is shown an electron micrograph of such an RNA chain in its naturally compacted 3-D form. [Courtesy of Sigmund and Robert Thach of the Washington University School of Medicine.]

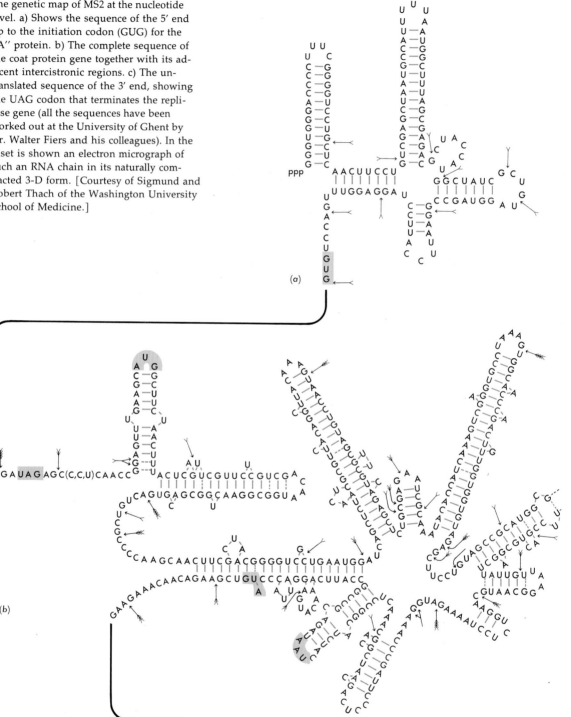

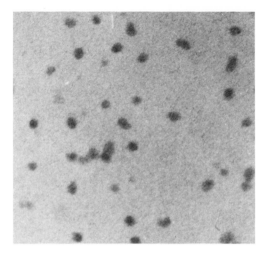

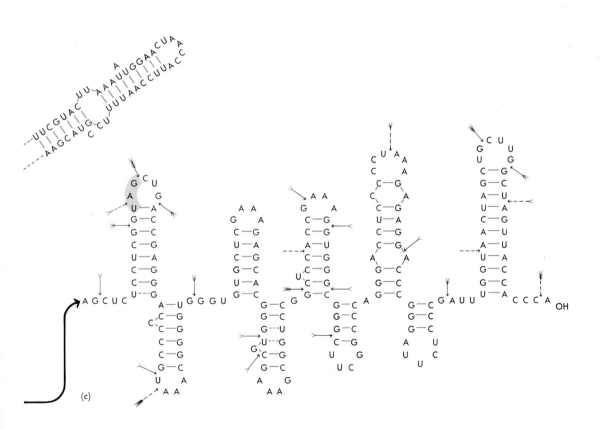

(c)

SATELLITE RNA CODES ONLY FOR COAT PROTEIN MOLECULES

Even smaller RNA-containing virus particles are found in tobacco cells infected with a tobacco necrosis virus. Their single-stranded nucleic acid chains contain only about 1000 nucleotides, just enough to code for the protein component that makes up their structural coat. These tiny particles (often called "satellites") are never found alone, but are made only when susceptible cells are simultaneously infected with a different, larger (6000-nucleotide) RNA virus. This situation suggests that the RNA of the very small virus is replicated by using viral-specific enzymes (in particular, RNA synthetase) coded for by the RNA of the larger particle. So, whether these particles should be considered the smallest known viruses is a somewhat arbitrary and perhaps not important question. Certainly relevant is the fact that their nucleic acid chains make very favorable objects for future sequence studies.

THE SMALLEST KNOWN VIRUSES ARE ALMOST AT THE LOWER LIMITS FOR A VIRUS

The need ordinarily to code for a viral-specific RNA replicase as well as their protein coats effectively places a lower limit on how small an RNA-containing virus can be. First, it would be surprising if a synthetase molecule much smaller than 30,000 could carry out all the tasks involved in RNA self-replication. Second, it is hard to imagine stable coat protein subunits made up of less than 50 to 75 amino acids. Thus the smallest complete RNA virus likely now to exist most probably contains at least 1500 nucleotides (half the R17 number). The expectation of anything much smaller seems highly unrealistic.

Analogous reasoning for the minimal size of a DNA virus is not yet possible, since it is possible that a circular viral DNA molecule might be replicated using only host enzymes. If so, then someday we might find a DNA virus whose only gene codes for its coat protein. Conceivably, such a DNA chain might contain as few as 300 to 500 base pairs.

Making intelligent guesses on the upper size for a virus is much more difficult. The largest viruses now known code for some 200 to 300 genes, but there seems to be no theoretical reason why a much larger number could not be coded for. Perhaps, above a certain size, there is no substantial need for new genes to manipulate the host cell in favor of viral products.

Always very clear, however, is the absolute difference between a virus and a cell. As few as ten years ago, however, much confusion existed. Then a key criterion for a virus was still its ability to pass through membrane filters that hold back bacteria. Under this definition, the very small disease-causing bodies known as *Rickettsiae* were generally treated as if they were viruses. But now they have been found to contain DNA and RNA, as well as a protein-synthesizing system, and so are realized to be very small bacteria. The essence of viruses is thus the absence of protein-synthesizing systems. Hence their multiplication must always involve the breakdown of their surrounding protective coats, thereby letting their chromosomes come into contact with cellular enzymes.

REPLICATING RNA MOLECULES THAT LACK PROTEIN COATS

Over the past several years evidence has been presented for the existence of disease-causing agents even smaller than any known virus, which appear to consist only of free RNA. Called *viroids*, this new class of infectious agent was first made known through the study of the spindle tuber disease of potatoes and tomatoes. A viroid sediments very slowly; and its infectivity does not depend upon the presence of any protein or lipid components. In contrast, infectivity is quickly destroyed by ribonucleases which break down single-stranded RNA chains.

Recently it has been possible to obtain the infectious material which sediments at 5.5S in highly purified form, and then to make direct measurements of its molecular weight. Now the best value is a MW $\sim 8 \times 10^5$ (~ 225 nucleotides). This value has been confirmed by electron micrographs of viroid RNA (Figure 15–39), which reveal short linear (hairpin configuration?) molecules some 400 to 500 Å in length. How viroids replicate is a complete mystery since at best they can code for only 70 to 80 amino acids, and so most likely cannot code for a viral-specific replicase. Their replication may involve a cellular enzyme that can form chains of complementary sequences. Though no such enzyme has yet been found in normal cells, the discovery of the virally-coded RNA $\rightarrow$ DNA enzyme *reverse transcriptase* (see page 677) suggests that we should have an open mind about how viroid replication may occur.

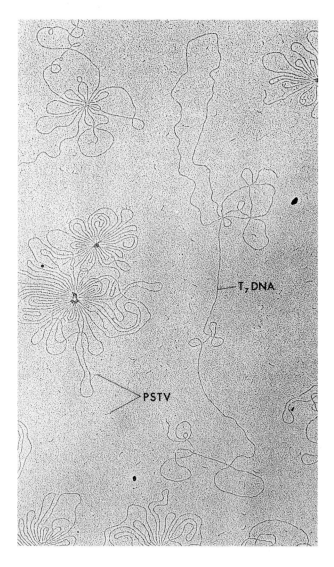

Figure 15–39
Electron micrograph of PSTV (potato
spindle tuber disease) viroids mixed
together with double-stranded T7 DNA.
[Courtesy of T. Koller and J. M. Soco, Swiss
Federal Institute of Technology, Zurich.]

A LOWER SIZE LIMIT EXISTS FOR DIVIDING CELLS

A lower size limit must exist for a cell even if it is growing in an environment containing essentially all its
required small molecules. It must be large enough to contain a functional semipermeable membrane, a protein-
synthesizing apparatus, and sufficient genetic material to
code for the various enzymes required to make its necessary proteins and nucleic acid components. In addition, its
chromosomes must carry the information for the enzymes
needed to make the small molecules that cannot be transported across the semipermeable cell membrane. For ex-

ample, even though all the amino acids and purine and pyrimidine bases can be supplied to most cells as food, phosphorylated compounds such as ATP usually cannot be supplied. Biosynthesis of such compounds must normally take place in cells; and hence genes must exist for the enzymes needed to build them up from their smaller purine and pyrimidine building blocks. It is not yet possible to say exactly what the lower size limit for a cell is. The amount of DNA in *Rickettsiae* is still not accurately known, but a conservative guess is that one *Rickettsia* chromosome contains 15 to 20 percent as much DNA as an *E. coli* chromosome. Its gene content is thus probably between 750 and 1000. There are reports of still smaller cells, but such statements must be viewed with disbelief because, if taken at face value, they imply that as few as 100 different proteins can maintain the living state.

SUMMARY

All known viruses contain a nucleic acid core (either DNA or RNA) surrounded by a protective shell that always contains protein and, in some of the more complicated particles, also lipids or carbohydrates (or both). The viral DNA may be either single-stranded or double-stranded. Likewise, both single- and double-stranded RNA viruses exist. The presence of nucleic acid is not surprising, since a genetic component must be present. What is surprising at first glance is that RNA is the genetic component of many viruses (e.g., tobacco mosaic virus and influenza). This means that viral RNA molecules serve as templates for their own formation.

After the viral nucleic acid enters a host cell, synthesis of both viral-specific nucleic acid and the specific components found in the viral protective coat occurs. These components then specifically aggregate to form new infectious viruses identical to the infecting parental particle. The selective synthesis of virus components in a host cell is often aided by prior synthesis of viral-specific enzymes involved in nucleic acid metabolism. For example, the duplication of single-stranded RNA needs the presence of RNA replicase. This enzyme uses the infecting single-stranded parental ("+") RNA molecules to make "−" strand complements. Then, in turn, it uses the "−" strands as templates to make many copies of the "+" strand. Thus the basic rules involved in the duplication of RNA molecules are the same as the rules for DNA. In all cases, new polynucleotide strands are made by the formation of complementary copies using DNA-like base pairs.

No evidence exists for the existence of any repressor–operator systems during the lytic phase of viral growth. Instead, the control mechanisms that ensure the appropriately timed appearance of virally-coded proteins all appear to act in a positive fashion. In contrast, maintenance of the lysogenic state is directly controlled by specific repressors that bind to operators and prevent the transcription of most of the viral genome.

Lytic multiplication cycles are usually separated into several phases (e.g., early, middle, and late), each of which is characterized by the functioning of specific sets of genes. The early genes, many of which code for enzymes needed for viral-specific RNA and DNA synthesis, are controlled by promoters that are read by unmodified host RNA polymerase molecules. In contrast, middle and late genes are often (e.g., phage T4) controlled by promoters whose reading depends on the existence of virally-coded specificity factors. Such specificity factors direct new forms of RNA synthesis by binding to the "core" component of host RNA polymerase molecules. With other viruses (e.g., T7), the late genes are read by a completely new RNA polymerase coded by one of the early genes.

Within given operons, the exact timing of expression is effected by the gene order, with those genes farthest from the promoters being expressed last. Expression of these more distal genes is also frequently inhibited by the action of the terminator factor ρ, which reads stop signals existing between the promoter proximal and promoter distal genes. Expressions of such distal genes depends upon the accumulation of antitermination factors, themselves coded by viral genes (e.g., gene "N" of phage λ).

The simplest viruses now known are the RNA phages. Along their 3500 nucleotide-long chromosomes are three main genes—two coding for structural proteins, the other for RNA replicase. The rates at which these genes function are controlled in part by the availability of their ribosome-binding (initiation) sites. Also, an important control device is the specific binding of newly synthesized coat protein molecules to the initiation site of the replicase gene, thereby preventing functioning of the replicase gene in the second half of the virus life cycle. New methods for nucleotide sequence analysis now permit the working out of the complete nucleotide sequence of RNA phage chromosomes.

REFERENCES

CRICK, F. H. C., AND J. D. WATSON, "Virus Structure: General Principles," *CIBA Found. Symp. Nature Viruses,*

1957, 5–13. An early statement explaining the structural consequences of the limited nucleic acid content of viruses.

Klug, A., and D. L. D. Caspar, "The Structure of Small Viruses," *Advan. Virus Res.*, **7**, 225 (1960). An elegant exposition of the principles of virus construction, with special emphasis on TMV.

Stent, G. S., *Molecular Biology of Bacterial Viruses*, Freeman, San Francisco, 1963. Now very out-of-date but still very useful for an appreciation of how bacteriophages came to play such a key role in the development of molecular biology.

Luria, S. E., and J. E. Darnell, *General Virology*, 2nd ed., Wiley, New York, 1968. An intelligent treatment of all aspects of virus multiplication.

Cohen, S. S., *Virus-Induced Enzymes*, Columbia University Press, New York, 1968. An individualistic summary of how viruses employ specific enzymes to direct cellular synthesis toward viral products.

Wood, W. B., and R. S. Edgar, "Building a Bacteria Virus." A beautifully illustrated *Scientific American* article reprinted in *The Chemical Basis of Life*, P. C. Hanawalt and R. H. Haynes (eds.), Freeman, San Francisco, 1973.

Milanesi, G., E. N. Brody, O. Grau, and E. P. Geiduschek, "Transcription of the Bacteriophage T4 Template *in vitro*: Separation of 'Delayed Early' from 'Immediate Early' Transcription," *Proc. Nat. Acad. Sci.*, **66**, 181 (1970). Evidence that early promoters control both immediate early and delayed early genes.

Travers, A. A., "Positive Control of Transcription by a Bacteriophage," *Nature*, **225**, 1009 (1970). The first experiments showing the involvement of a T4 specificity factor in the synthesis of T4 mRNA.

Schmidt, D. A., A. J. Mazaitis, R. Kasai, and E. K. F. Bautz, "Involvement of a Phage T4 σ Factor and an Antiterminator Protein in the Transcription of Early T4 Genes *in vivo*," *Nature*, **225**, 1012 (1970). The T4 life cycle imaginatively analyzed in terms of σ, ρ, and antitermination factors.

"Transcription of Genetic Material," *Cold Spring Harbor Symp. Quant. Biol.*, **XXXV** (1971). Virtually everything known by June, 1970, about the transcription of viral genes is found in this very comprehensive collection of papers.

O'FARREL, P. Z., AND L. M. GOLD, "Bacteriophage T4 Gene Expression: Evidence for Two Classes of Prereplicative Cistrons," *J. Biol. Chem.,* **248,** 5502 (1973). Experiments that help prove the distinction between early and middle (quasi-late) promoters.

MATTSON, T., J. RICHARDSON, AND D. GOODIN, "Mutant of Bacteriophage T4D Affecting Expression of Many Early Genes," *Nature,* **250,** 48 (1974). Evidence for a mutation that blocks expression of middle promoters.

STEVENS, A., "New Small Polypeptides Associated with DNA-Dependent RNA Polymerase of *E. coli* after Infection with Bacteriophage T4," *Proc. Natl. Acad. Sci.,* **69,** 603 (1972). Evidence that the T4 genes 33 and 55 control the presence of small specificity polypeptides necessary for reading of the late genes.

HORVITZ, H. R., "Polypeptide Bound to Host RNA Polymerase is Specified by T4 Control Gene 33," *Nature New Biology,* **244,** 137 (1973). Direct proof that gene 33 codes for the T4 specificity factor of MW 12,000.

CHAMBERLIN, M., J. McGRATH, AND L. WASKEIL, "New RNA Polymerase from *E. coli* Infected with Bacteriophage T7," *Nature,* **228,** 227 (1970). The unexpected discovery that an entirely new and virally-coded RNA polymerase makes late T7 RNA.

GOLOMB, M., AND M. CHAMBERLIN, "A Preliminary Map of the Major Transcription Units Read by T7 RNA Polymerase on the T7 and T3 Bacteriophage Chromosomes," *Proc. Nat. Acad. Sci.,* **71,** 760 (1974).

The Bacteriophage Lambda, edited by A. D. Hershey, Cold Spring Harbor Laboratory, 1971. A definitive book containing both general reviews as well as specialized research papers that summarize twenty years of high-level research on the nature of lysogenic phages.

ROBERTS, J. W., "Termination Factor for RNA Synthesis," *Nature,* **224,** 1168 (1969). How the discovery of termination factor ρ simultaneously led to the idea of antitermination factors.

WALZ, A., AND V. PIRROTTA, "Sequence of the P_R Promoter of Phage λ," *Nature,* **254,** 118 (1975).

MANIATIS, T., M. PTASHNE, K. BACKMAN, D. KLEID, S. FLASHMAN, A. JEFFREY, AND R. MAURER, "Recognition Sequences of Repressor and Polymerase in the Operators of Bacteriophage λ," *Cell,* **5,** 109 (1975). This and the preceding paper elegantly demonstrate the interrelationship of operator and promoter sequences within λ DNA.

Spiegelman, S., I. Haruna, I. B. Holland, G. Beaudreau, and D. Mills, "The Synthesis of a Self-Propagating and Infectious Nucleic Acid with a Purified Enzyme," *Proc. Natl. Acad. Sci.,* **54,** 919 (1965). The first report of the *in vitro* replication of infectious viral (Qβ) RNA.

RNA Phages, edited by N. Zinder, Cold Spring Harbor Laboratory, 1975. A collection of advanced review articles that brings together most of the facts currently known about this very important group of viruses.

Diener, T. O., "Viroids: The Smallest Known Agents of Infectious Disease," *Ann. Rev. Microbiol.,* **28,** 23 (1974). A review of much of the author's experimental evidence that a class of protein-free, very small RNA molecules cause certain plant diseases.

The Essence
of Being
Eucaryotic

The fact that we now know so much at the molecular level about the genetic organization of bacteria and their viruses constantly generates enthusiasm that current research will soon lead to equally firm answers about higher organisms, particularly the vertebrates, where such new knowledge might have direct applicability to the human condition. But before jumping all that far, we should realize that the fundamental biological division is not between the microorganisms and the higher plants and animals, but between those organisms that possess nuclei and those that lack them. Bacteria and their close relatives the blue-green algae have no nuclei, and collectively they are called *procaryotes.* In contrast, those organisms having nucleated cells are called *eucaryotes.* Among the latter are large groups of microorganisms like the green algae, fungi, and protozoa, as well as the larger, and more familiar, higher plants and animals. As far as we can tell, all eucaryotes are derived from a common ancestor, and so in attempting to appreciate their unique qualities we should try to understand the evolutionary factors which led to their emergence.

JUMP IN SIZE—A RESPONSE TO THE ADVANTAGES OF THE PREDATORY EXISTENCE

General agreement now exists that the first forms of life came into existence in an aqueous oceanic environment in which were suspended large masses of essential cellular building blocks like the nucleotides, the amino acids, and the fatty acids. All these organic compounds ("the primitive soup") are thought to have evolved by a prebiotic evolutionary process, in which energy from the sun transformed relatively simple molecules like CH_4 and NH_3 into a large variety of complex organic molecules. "Life" itself emerged when primitive nucleic acid and protein molecules became spontaneously encapsulated in lipid-containing membranous sacs that provided an environment where organized growth could occur. How this could have happened is still a total mystery. Particularly puzzling is how the genetic code and the respective machinery for protein synthesis could have emerged. The enormous complexity in the ways genetic information is now duplicated and used makes any attempt to reconstruct how it came into existence a seemingly insoluble problem. Though many, many speculations have been made about the nature of the first "genetic molecules," say, as to whether DNA or RNA came first, the fact that the same molecular mechanisms underlie all current forms of life means that no molecular fossils have been preserved to provide clues as to the nature of the first living organism(s).

We can assume nonetheless that the initial prebiotic soup was very rich, for many of the key chemical transformations upon which primitive life must have depended had to occur in the absence of specific protein catalysts. Even so, most of the original biological reactions may have speeded up to a rate compatible with the living state because of the presence of small catalytic molecules in the soup. Such nonenzymatic catalysts, however, were probably at best 2 to 3 orders of magnitude less efficient than today's highly evolved protein enzymes. So the first form of life probably grew much more slowly than the cells of today.

But when DNA began to code for the specific amino acid sequences of enzymes, cells could grow much faster, and much of the original organic soup was transformed into cellular material. Probably quite soon after such efficient life forms emerged, the level of prebiotic food molecules declined to the low concentrations now found in our oceans. Conditions were thus created greatly favoring not only the development of photosynthetic systems, but also the evolution of organisms that could actively seek their food material by adopting a predatory way of life. And for this to happen, a species had to first evolve to a size that would enable it to engulf smaller organisms by a phagocytic process similar to that by which amoeboid organisms now feed.

456

LARGE CELLS MUST HAVE EXTENSIVE INTERNAL MEMBRANES

The most obvious structural feature that distinguishes eucaryotes from procaryotes, other than the presence of a nucleus, is their extensive internal development of lipid-containing membranous vacuoles and organelles. While some form of internal membrane (e.g., mesosomes) is probably found in all procaryotes, only in the large, highly specialized photosynthetic bacteria and blue-green algae are internal membranes a major cellular feature. In eucaryotes, however, much more lipid is used to form internal membranes than is present as part of the exterior surface membrane (the plasma membrane). This extensive development of internal membranes is not a matter of chance, but a consequence of the relatively large size of the eucaryotic cell. Even the smallest eucaryotic cells have 5- to-10-fold larger dimensions than small unspecialized bacteria like *E. coli.* This means that, in relationship to their mass, eucaryotic cells have a much smaller relative external surface area than procaryotes. So, to the extent that vital processes have to occur on membranes, the emergence of larger and larger cells depends on their respective ability to elaborate more and more extensive internal membranous sacs.

For example, both photosynthesis and oxidative phosphorylation require separation of many key intermediates by semipermeable membranes. So not only are chloroplasts and mitochondria membrane-bounded, but each also contains extensive invaginations which greatly increase the amount of membranous surface area. Likewise, the development of the nuclear membrane may be related to the still largely not understood connection between DNA molecules and cell membranes. As cells become larger, connections of DNA to external membranes may have been replaced by binding to more closely proximate nuclear membranes.

ARRANGEMENT OF LIPIDS INTO BILAYERS

Most of the lipids found within biological membranes contain phosphate "head" groups attached to long hydrocarbon "tails" (phospholipids). The four major phospholipids found in eucaryotic membranes are phosphatidyl serine, phosphatidyl ethanolamine, phosphatidyl choline, and sphingomyelin (Figure 16–1). These are always arranged within a 70 Å-thick bilayer in which the charged phosphate "head" groups are sited outwardly. This arrangement neatly places all the hydrophobic hydrocarbon "tails" in contact with each other, leaving the

457

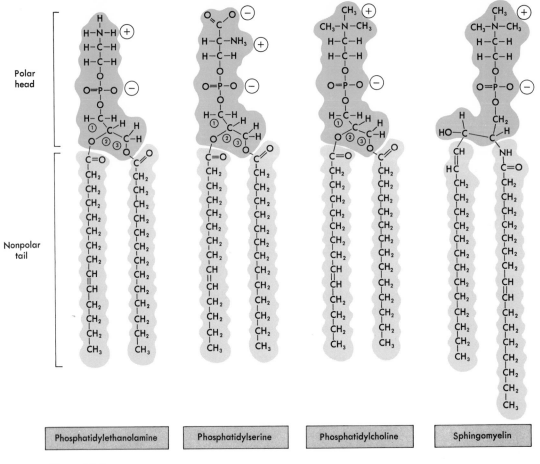

Figure 16–1
The structure of the four major phospholipids found in eucaryotic cell membranes. The exact lengths of the fatty acid nonpolar tails are variable. ①, ②, and ③ indicate the three carbon atoms of glycerol, on which three of the phospholipids are based.

highly charged phosphate "head" regions free to interact with the polar water environment. Probably from the very start of life, membranes possessed the same basic bilayer arrangement as that present today.

Most eucaryotic cell membranes also contain several types of glycolipids in which one to several water-soluble sugar groups are attached to a hydrophobic tail made up of a fatty acid and the amino alcohol sphingosine (Figure 16–2). Steroids also are found within most bilayers, with cholesterol present in large amounts in virtually all vertebrate membranes. All these latter components are also arranged so that their hydrophobic groups are in the center of the bilayer with their charged (hydrogen bonding) groups facing outwardly (Figure 16–3).

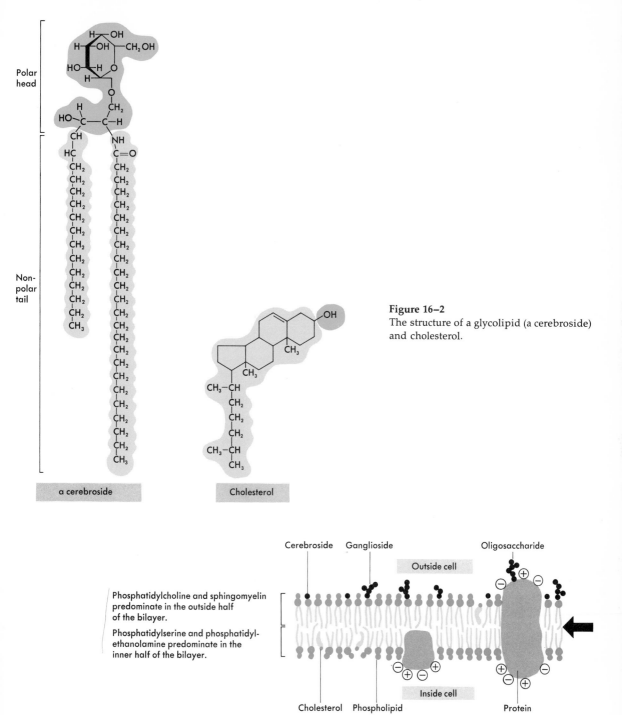

Polar head

Non-polar tail

a cerebroside

Cholesterol

Figure 16–2
The structure of a glycolipid (a cerebroside) and cholesterol.

Cerebroside Ganglioside Oligosaccharide

Outside cell

Phosphatidylcholine and sphingomyelin predominate in the outside half of the bilayer.

Phosphatidylserine and phosphatidyl-ethanolamine predominate in the inner half of the bilayer.

Inside cell

Cholesterol Phospholipid Protein

Figure 16–3
Diagram to show the probable insertion of proteins into the lipid bilayer of a plasma membrane. Some proteins penetrate right through the membrane, although most of them are concentrated in the cytoplasmic half of the membrane. Probably no protein is completely buried in the membrane. The arrow indicates the plane along which cleavage occurs during freeze-fracture.

INSERTION OF MEMBRANE PROTEINS INTO LIPID BILAYERS

Most membranes contain a variety of different protein molecules that are directly inserted into the lipid bilayer. Some run right through the bilayer, having groups that protrude from both its sides. Those amino acid side chains that lie within the bilayer are largely hydrophobic. In contrast, hydrophilic amino acids predominate in the regions that face out onto the inner (outer) surfaces (Figure 16–4). Depending upon their function, membranes have very specific protein components. Some of the most highly specialized membranes contain only one major protein component (e.g., the protein of the retinal membrane is almost exclusively composed of the visual pigment *rhodopsin*).

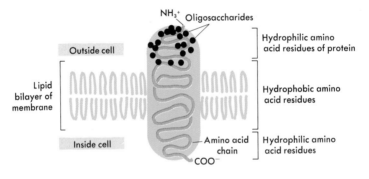

Figure 16–4
Diagrammatic representation of how glycophorin, a large glycoprotein, may be inserted in the red blood cell membrane.

Though most electron micrographs superficially suggest symmetrical membranes, there is recent compelling evidence that the two sides of a given membrane have fundamentally different chemical compositions. Phosphatidyl serine and phosphatidyl ethanolamine face toward the cytoplasmic (inner) side while phosphatidyl choline and sphingomyelin are largely found on the exterior surface. And those proteins which are inserted into membranes have specific regions which normally look exclusively inward and outward. Only those regions which face outwardly have sugar groups attached to them, suggesting that the glycosylating enzymes that transfer sugars to membrane proteins are exclusively found on the outer membrane surfaces.

When membranes fractured into their inner and outer halves are visualized in the EM, large numbers of 70 Å-diameter globular proteins appear to lie within (as opposed to going through) the lipid bilayer (Figure 16–5). Whether this arrangement is real or artifactual remains to be seen and may not be known until the functional significance of the respective proteins becomes understood.

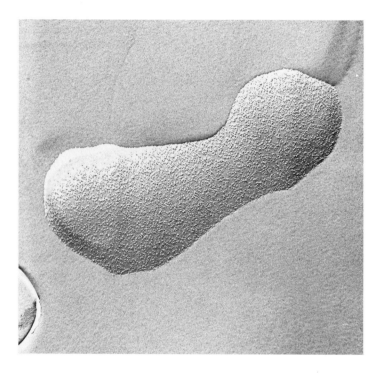

Figure 16–5
Electron micrograph (60,000×) of a red blood cell membrane prepared by the freeze-fracture method which frequently breaks membranes into their inner and outer halves. The seemingly internal 85 Å-diameter spheres may *in vivo* run through the membrane. [Courtesy of Daniel Branton, Harvard University.]

SEMIFLUIDITY OF CELL MEMBRANES

There is now very good evidence that the individual phospholipid and protein molecules do not maintain fixed orientations but are in constant movement within the two-dimensional fabric of a given membrane. Molecules which at one moment may be in touch, can move to opposite sides of a 10μ-diameter cell within an hour or so. Membranes thus can be thought of as two-dimensional liquids, virtually all of whose components are in constant diffusional motion. They have the essential quality of a self-sealing rubber tire, and holes quickly disappear due to diffusion of nearby hydrocarbon molecules into energetically unstable voids.

An equally important property is the potential ability of adjacent membranous sacs to fuse upon contact. Mem-

461

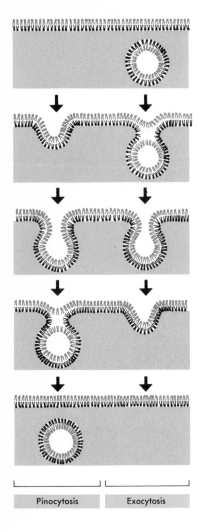

Pinocytosis · Exocytosis

Figure 16–6
The essential symmetry of pinocytosis (left) and exocytosis (right), showing the probable way in which membrane fusion is accomplished. Gold = outer half of membrane lipid bilayer; black = inner, cytoplasmic half.

branous sacs also have the capacity to divide when seemingly random motion creates unstable pinched-off regions. Probably most of these fusion (division) events are not random occurrences but responses to very well defined stimuli. Conceivably they are mediated by forces exerted on the lipid bilayers which momentarily create minute tears in the two-dimensional surface. If so, fusion (division) events may be the result of rearrangements of such torn surfaces, whereby inner surfaces fuse with other inner surfaces while outer surfaces match up with other outer surfaces (Figure 16–6).

PHAGOCYTOSIS (PINOCYTOSIS) IS REVERSIBLE

Thus, at best, we can now only speculate on the molecular details of *phagocytosis* whereby the external membranes (the plasma membranes) of large phagocytic cells tightly fold around bacteria-sized food objects and pinch off membrane-bounded internal sacs (vacuoles) containing the ingested food. This feeding process is not limited to the ingestion of cells but also occurs when the food objects are of macromolecular dimensions. Though this latter process is given the different name of *pinocytosis*, the essential molecular events are likely to be the same. Whenever phagocytosis or pinocytosis occurs, considerable amounts of the external plasma membrane are, of necessity, transferred into the interior of the cells (Figure 16–6). Such movement must be counterbalanced by the constant merging of other internal membranes with the external membrane, a process frequently called reverse pinocytosis (also called exocytosis). Similar membrane mergers occur (at least in higher cells) during cell growth, when there is a net increase in the amount of external surface. One might guess that, during evolution, pinocytosis appeared on the scene first, and only much later, when the prebiotic soup began to become significantly diluted, did phagocytic events acquire a functional biological role.

CELL MEMBRANE MOVEMENTS DIRECTED BY ACTIN–MYOSIN INTERACTIONS

Evolution of the capacity to phagocytize must have simultaneously given the respective cell(s) the capacity for some form of directed movement. Observations of amoeboid cells suggest that the membrane movements which lead to successful engulfment of food are also the basis by which such cells move over solid surfaces. Until recently, theories about amoeboid motion were very imprecise. Now, however, there is hope that its basic details will be understood within the next decade or two. The basis for this optimism is the realization that all cells capable of amoeboid motion contain many, if not most, of the structural proteins utilized in the contraction of macroscopic muscle fibers. These "muscle proteins" (e.g., myosin, actin, tropomyosin, and α-actinin) are major cellular constituents, not only of amoebae but also of many cell lines that grow in culture. For example, the most abundant protein (by mass) in chicken fibroblasts is actin. Moreover, its structure may be identical to that of the actin found in the myofibrils of striated chicken muscle. Tropomyosin, α-actinin, and myosin are also major features of all fibroblasts.

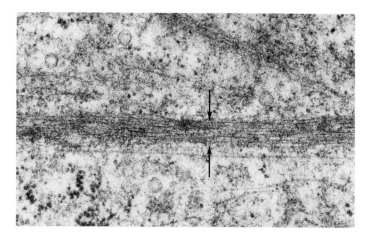

Figure 16–7
Electron micrograph of a thin section of a flattened monkey kidney cell, showing large numbers of 60 Å-diameter microfilaments (actin) within a bundle running parallel to the lower cell surface (magnification 45,000×). [From R. Goldman, Carnegie-Mellon University.]

Most actin in cultured cells is present as very long polymerized fibers of 60 Å diameter (microfilaments) which frequently are aligned parallel within bundles (initially called *stress fibers* by cytologists) just under the cell membrane (Figure 16–7). These bundles also contain myosin, tropomyosin, and α-actinin (Figure 16–8). Thus, it is initially tempting to believe that they are organized as in "striated muscles," and have as their sole function sliding filament movements like those that underlie the contraction of macroscopic striated muscle fibers (Figure 16–9). Visualization of the bundles by fluorescent antibody techniques reveals that they instead resemble the less highly organized and still poorly understood "smooth muscles" of the type found in the gizzard and heart. And while myosin, tropomyosin, and α-actinin are all periodically arranged along the intracellular bundles, the fundamental repeat is $\sim 1.6\mu$, or only two-thirds the 2.5μ repeat distance found along ordinary striated muscle fibers. Moreover, in "nonmuscle" cells there is tenfold less myosin in relation to actin than is found in striated muscle.

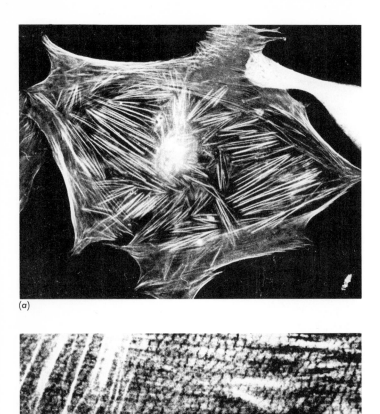

(a)

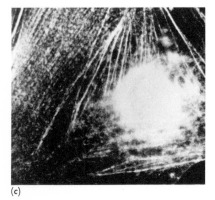

(b)

Figure 16–8
Immunofluorescent visualization of rat embryo cells using antibodies against
(a) actin, (b) tropomyosin, (c) myosin, and
(d) α-actinin. [Courtesy of E. Lazarides and K. Weber, Cold Spring Harbor Laboratory.]

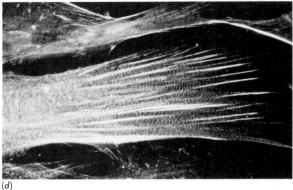

(c)

(d)

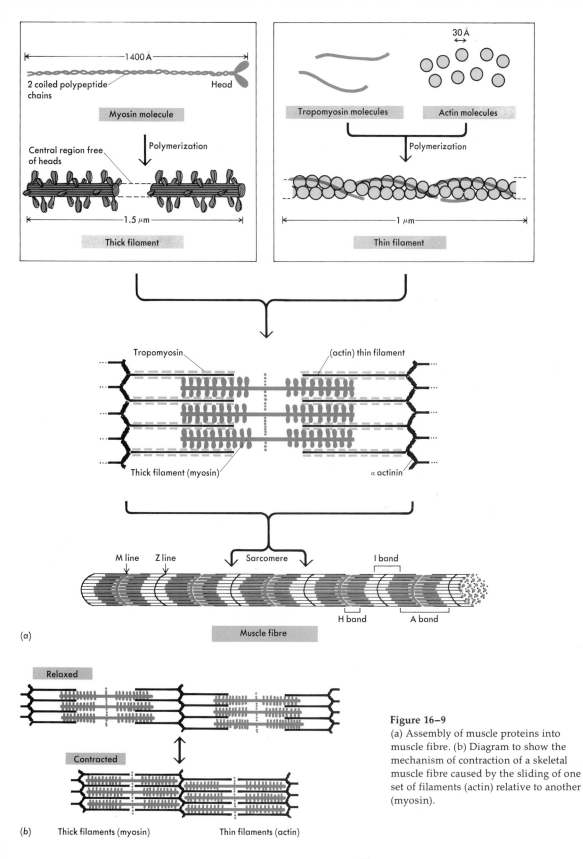

30 Å

2 coiled polypeptide chains
Head

Myosin molecule

Central region free of heads

Polymerization

1.5 μm

Thick filament

Tropomyosin molecules
Actin molecules

Polymerization

1 μm

Thin filament

Tropomyosin
(actin) thin filament

Thick filament (myosin)
α actinin

M line   Z line   Sarcomere   I band

H band   A band

Muscle fibre

(a)

Relaxed

Contracted

(b)   Thick filaments (myosin)   Thin filaments (actin)

Figure 16–9
(a) Assembly of muscle proteins into muscle fibre. (b) Diagram to show the mechanism of contraction of a skeletal muscle fibre caused by the sliding of one set of filaments (actin) relative to another (myosin).

Most strikingly, no periodicity at all can be seen in the actin arrangement, suggesting that individual actin filaments may run uninterrupted over the full length of a bundle (distances of many μ). Alternatively, the bundles (stress fibers) may represent states of complete contraction in which the opposing actin filaments are in close contact (Figure 16–9). In any case, most organized muscle-like fibers lie in relatively quiescent cellular regions and not at the sites of active membrane movement (Figure 16–10). In contrast, the actively moving pseudopodia-like projections appear to contain only completely disorganized muscle components (Figure 16–11). This apparent disorder may reflect the fact that outflowing cytoplasmic projections have very fleeting lives (periods of seconds to minutes), and so their muscular frameworks must be constantly made and broken down.

Figure 16–10
A single mouse cell moving on a glass slide stained with antibody to actin. The general direction of cell movement is from left to right. [Courtesy of R. Pollack, Cold Spring Harbor Laboratory.]

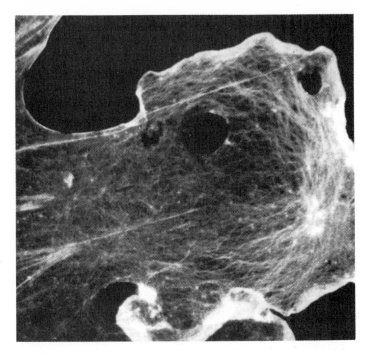

Figure 16–11
A highly magnified view of a moving pseudopodial projection of a rat embryo fibroblast visualized with anti-actin antibody. Within the pseudopodium most thin actin filaments appear disorganized compared to the thicker cable-like aggregates found in more quiescent regions of the cell. [Courtesy of R. Pollack, Cold Spring Harbor Laboratory.]

Under any scheme for contraction, the muscle-like fibers must somehow rightly interact with the exterior cell membrane; here the key protein appears to be α-actinin, the muscle protein which in striated muscles is located within the "Z" bands, where it forms the attachment sites for actin filaments (Figure 16–9). Within "nonmuscle cells," α-actinin not only runs between bundles (Figure 16–8), but also attaches their tips to the cell membrane. Here the technique of fluorescent antibody visualization has been particularly important, since it reveals a complicated polygonal array of points of attachment to the various cell membrane surfaces (Figure 16–12). Such arrangements may give considerable stability to the plasma membrane itself, and at given moments many membrane proteins may be effectively frozen into fixed patterns. But as the pseudopodial projections form and reform, most plasma-membrane proteins must change their relative positions. Hence the "semifluid" nature of the external cell membrane.

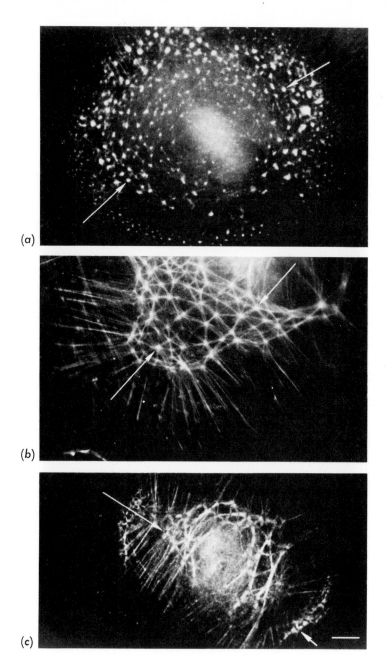

(a)

(b)

(c)

Figure 16–12
Polygonal arrays of muscle proteins attached to membranes of a rat fibroblast in the process of attaching itself to a flat surface. In (a) anti-α-actinin reveals discrete foci of α-actinin—they are linked together by (b) actin, and (c) tropomyosin—note that both actin and α-actinin are found at the central foci while tropomyosin is found only in the connecting fibers. [Courtesy of E. Lazarides, Cold Spring Harbor Laboratory.]

While these muscle proteins have been seen in virtually all forms of eucaryotic life—even rigid plant cells surrounded by thick cellulose walls have abundant actin which is thought to be involved in the generation of cytoplasmic streaming—no traces of them have been found in any procaryotes. So actin- and myosin-like proteins represent one of the main biochemical markers of the eucaryotic state.

MICROVILLI (MICROSPIKES) MAY BE SENSORY ORGANS FOR THE MOVING CELL

Once we begin to understand the basic organization of the muscle proteins of the moving cell, we can seriously study how a cell decides to move (grow) in a given direction. Given the capacity of limited regions of the membrane to respond to specific stimuli (e.g., attachment of food particles), it is simplest to imagine that "sensory bodies" exist all over the plasma membrane. One obvious group of candidates are the *microvilli (microspikes)*; these are long, narrow ($\sim 0.2\mu$-diameter) projections that are constantly popping out from and back into the cell surface (Figure 16–13). Such elongation (retraction) processes can take place in from only a few seconds to minutes, and it is very tempting to believe that their chief function is to sense the outside world. Supporting this conjecture are microscopic observations on living cells which show that the adhesion of the tip of a microspike to an appropriate solid surface acts as the signal for a cell to send out the much more extensive lamellar projections, the *ruffled edges (lamellipodia)* that mark the forward edges of moving cells.

Figure 16–13
A scanning electron micrograph of the forward leading edge of a moving mouse cell, showing several ruffled edges and numerous microvilli (MV) projecting from the upper surface (the arrows point to regions where the ruffles are attached to the glass surface). [Courtesy of J. P. Revel, California Institute of Technology, and reproduced with permission from the *Symp. Soc. Exp. Biol.*, **39**, 447 (1974).]

The insides of microvilli are largely occupied by muscle proteins; fluorescent antibody methods reveal high concentrations of actin, tropomyosin, and α-actinin. Electron-microscopic examination reveals long thin actin fibers traversing the entire length of the microvilli. They are connected both to each other and to the surrounding cell membranes by spike-like projections which may be molecules of α-actinin (Figure 16–14). Most importantly,

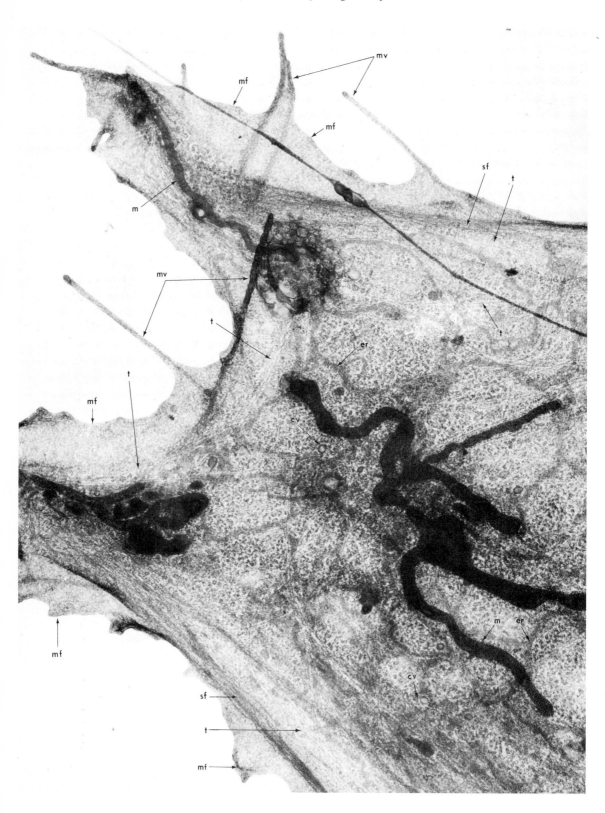

these fiber bundles do not stop at the cell surface, but extend back into the cytoplasm where they seem to interact with other muscle bundles. So the expansion and retraction of microvilli (as well as ruffles) must arise from the interaction of the muscle proteins both among themselves and with neighboring sections of the plasma membrane.

◀ Figure 16–14

The edge of a cultured fibroblast as viewed by a high-voltage electron microscope. Mitochondria (m) are surrounded by endoplasmic reticulum (er), polyribosomes (r), coated vesicles (cv), and microtubules (t) which run in many different directions. Some microtubules radiate along with peripheral bundles of filaments which go to make up the stress fibers (sf), otherwise known as muscle bundles. Others criss-cross the intervening cytoplasm and, while doing so, many undergo changes in direction. More peripherally, beyond the marginal stress fibers (muscle bundles), filament-rich subplasmalemmal cytoplasm occupies the microvilli (mv) and motile marginal flaps (mf). [Reproduced with permission from I. K. Buckley, *Tissue and Cell,* **7,** 51 (1975).]

THE RELEASE OF Ca^{++} IONS COMMENCES CYCLES OF CONTRACTION AND RELAXATION

The universal trigger to set off contraction of the muscle proteins is a sudden increase of Ca^{++} in the surrounding cytoplasm. So we suspect that the functioning of both microvilli and ruffles is very much dependent upon their Ca^{++} levels. Contraction generally requires mM Ca^{++} levels, while the free Ca^{++} level in most cytoplasmic regions is usually maintained in the μM range. One possible source of the required Ca^{++} is the extracellular milieu where Ca^{++} is invariably found in the mM range. So the external signals that lead to localized muscle contraction may act by directly causing a temporary inward flow of Ca^{++} through the plasma membrane to the adjacent muscle fibers. Alternatively, a localized change in the membrane potential may cause the release of Ca^{++} from Ca^{++}-rich internal vesicles. In either case we see that, once we ask how a cell moves, we must think of the single cell as a functioning nervous system that can integrate a constant stream of extracellular perception into the organized movements of large numbers of semiautonomous muscle fibers.

RESTRICTION OF MICROTUBULES TO EUCARYOTES

Also unique to and ubiquitous in the eucaryotic world are microtubules; hollow, cylindrical structures approximately 240 Å in diameter and usually many μ in length (Figure

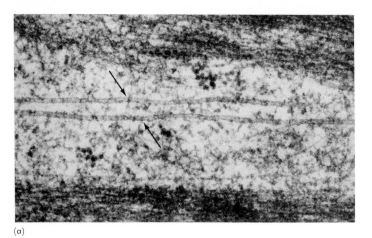

(a)

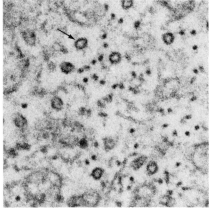

(b)

Figure 16–15 ▲
(a) Electron micrograph of a thin section of a flattened mouse fibroblast showing microtubules viewed in (a) longitudinal section, and (b) cross section (magnification 64,500×). [Courtesy of R. Goldman, Carnegie-Mellon University.]

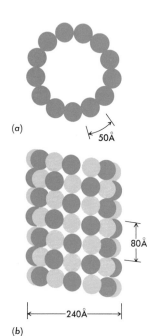

(a)

50Å

80Å

240Å

(b)

16–15). They occur in many different arrangements and can assume quite different functional roles. All microtubules have a 140 Å-diameter hollow core surrounded by a 50 Å-thick shell formed by the regular packing of a globular protein subunit called *tubulin*. At first it was believed that the fundamental tubulin subunit had a MW of 60,000, but now we realize that the functional unit is approximately twice this size and is composed of two nonidentical polypeptide chains of 55,000 and 57,000 MW.

Figure 16–16
A diagram to show the probable structure of a microtubule. (a) shows the appearance in cross section and the thirteen protofilaments from which the wall of the tubule is made. (b) The microtubule from the side, showing the helical arrangement of the globular tubulin molecules. A possible arrangement of the two types of tubulin within the tubule is indicated.

How these basic building blocks exactly fit together is not known, but cross-sectional views of microtubules frequently reveal 13 tubulin molecules at any given level. Most likely the basic arrangement is helical (as in Tobacco Mosaic Virus), with 13 tubulin molecules per helical turn (Figure 16–16).

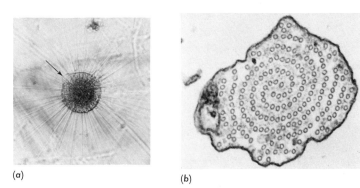

(a)

(b)

Figure 16–17
The axopodia of the protozoan *Echinosphaerium.* (a) A view of the organism showing the radiating axopodia. (b) The axopodium viewed in cross section exhibits a complex twelvefold symmetry perhaps generated by cross bridges between the different layers of the interlocked tubules. [Courtesy of L. Tilney, University of Pennsylvania.]

Many microtubules help to control cellular shape. This can be easily demonstrated using the alkaloid poison *colchicine,* which specifically binds to microtubules, with one mole binding to each mole of tubulin. When so bound, microtubules break down, leading to the retraction of those organelles whose skeletal support is largely provided by microtubules. For instance, the rigid axopodia of the protozoan *Echinosphaerium* (Figure 16–17) quickly retreat upon addition of colchicine, indicating that they are supported in part by a microtubular skeleton. Microtubules are also essential for both elongation and maintenance of the long axonal elements of nerves.

In cells, microtubules are in equilibrium with free tubulin, but as yet there are only vague hints as to how their formation (breakdown) is biologically controlled. Tubulin subunits phosphorylated with a cAMP-dependent kinase appear to aggregate more readily than the unphosphorylated subunits. This may explain the observation that cAMP promotes the formation of microtubules within epithelial cells growing in culture. The morphology of such cells is very much a function of their nutritional environment. When grown in the presence of appreciable cAMP, they often assume an elongated "fibroblast appearance," as opposed to the more rounded shape characteristic of epithelial cells. Such shape changes strictly correlate with the presence of large numbers of microtubules that run along the major projectional axes of the respective fibroblasts (Figure 16–18).

473

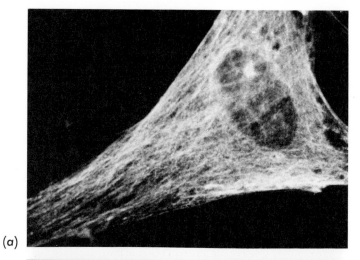

(a)

Figure 16–18
Immunofluorescent views of the microtubules of (a) a flattened chick embryo fibroblast, (b) an enucleated monkey cell in the process of flattening down on a solid surface. (c) A high-resolution view of part of a mouse (3T3) cell showing that most microtubules have irregular shapes, suggesting that they are not held together by regular cross bridges. [Courtesy of Klaus Weber and Robert Pollack, Cold Spring Harbor Laboratory.]

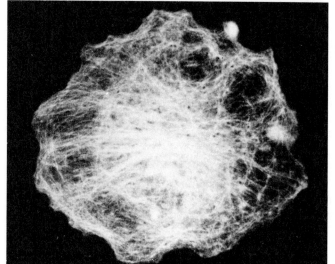

(b)

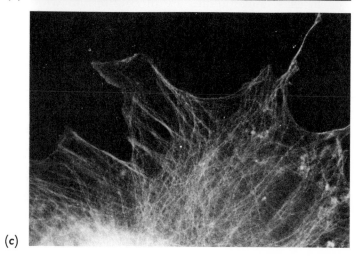

(c)

MICROTUBULES WITHIN CILIA

Another function of microtubules is to provide the main skeletal framework for cilia (flagella), locomotory organelles that each contain a precise arrangement (the axoneme) of 9 microtubular doublets surrounding 2 singlets (Figure 16–19). Each axoneme is surrounded by an outpocketing of the plasma membrane and so the microtubules are in direct contact with ATP produced by the mitochondria that lie in large numbers near the base of cilia. Ciliary motion is completely dependent upon the continued generation of ATP and many experiments have been done recently to find out how the splitting of ATP leads to their wave-like movements.

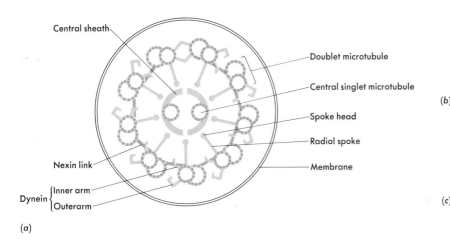

(a)

Central sheath

Doublet microtubule

Central singlet microtubule

(b)

Spoke head

Radial spoke

Nexin link

Membrane

Dynein { Inner arm / Outerarm

(c)

Figure 16–19
(a) The structure of a cilium (flagellum). The nine doublet microtubules are shown in section surrounding two singlet microtubules (solid color). Each singlet and the A tubule of each doublet contain the typical thirteen longitudinal protofilaments. The B tubule of the doublet contains 10 protofilaments. The various arms and bridges involved in the motility of the organelle are shown. (b) An electron micrograph of a section of a cilium from *Tetrahymena* showing the "9 + 2" arrangement of microtubules. (c) A section of a basal body from *Tetrahymena* showing the nine triplets and no central microtubule.

Now there is increasing evidence that this movement is the result of the periodic sliding of the doublet microtubules over each other (Figure 16–20), and that the key process in such sliding motions is the making and breaking of cross bridges between the adjacent doublets. These cross bridges are made by projecting lateral spikes called *dynein*. The two dynein bridges that connect each doublet to its neighbor spontaneously form in the absence

475

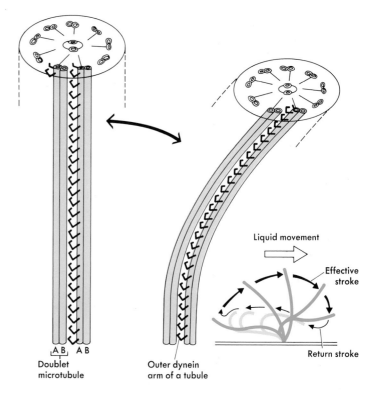

Figure 16–20
A simplified drawing to indicate how active sliding between adjacent microtubules can result in movement, in this case of a cilium or flagellum. The dynein arms are important for generating the sliding force between adjacent doublet microtubules which is converted to a bending of the cilium by the interaction of the radial spokes with the central microtubules and their sheath. Only two doublets have been shown, for clarity. The inset shows the final pattern of bending which gives rise to the beat of the cilium.

of ATP. Upon addition of ATP the cross bridges fall apart, followed by the splitting of the bound ATP into ADP and Ⓟ. Somehow, in a way yet to be worked out, the making (breaking?) of a cross bridge leads to the sliding of the respective doublet along the surface of the adjacent doublet. And given shear resistance within the remaining interlocked doublets, all of which are connected to the central microtubules by radial spikes, the resulting bending motion leads to the characteristic waves of ciliary motion. We might thus guess that the tubulin–dynein system evolved from the actin–myosin system (or vice versa). But no traces of homology in their respective polypeptide chains have yet been found. So, if they had a common molecular ancestor, it must have existed very early in the evolution of the eucaryotic cell.

At the base of each cilium is a *basal body*, a microtubule-containing organelle containing nine cylindrically arranged microtubule triplets with no microtubules in the center (Figure 16–19). They are thought to be involved in the organization (growth) of the ciliary axoneme by somehow promoting the aggregation of free intracellular tubulin. Large amounts of this protein are present in many cells. For example, following the removal of the cilia of *Chlamydomonas* by agitation, partial regeneration occurs in

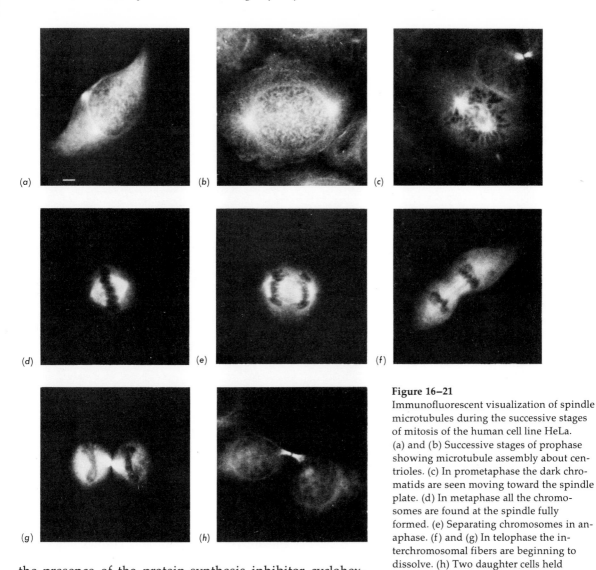

Figure 16–21
Immunofluorescent visualization of spindle microtubules during the successive stages of mitosis of the human cell line HeLa. (a) and (b) Successive stages of prophase showing microtubule assembly about centrioles. (c) In prometaphase the dark chromatids are seen moving toward the spindle plate. (d) In metaphase all the chromosomes are found at the spindle fully formed. (e) Separating chromosomes in anaphase. (f) and (g) In telophase the interchromosomal fibers are beginning to dissolve. (h) Two daughter cells held together by the microtubules in the intercellular bridge. [Reproduced with permission from Weber, Bibring, and Osborn, *Experimental Cell Research,* in press, 1975.]

the presence of the protein-synthesis inhibitor cycloheximide.

THE MITOTIC CYCLE AND THE TWO ORIGINS FOR THE SPINDLE TUBULES

Classical mitotic and meiotic behavior, during which chromosomes segregate along spindle fibers, is also an identifying feature of the eucaryotic cell. While it has been known for many years that colchicine prevents spindle formation, the origin of this specificity became clear only with the realization that microtubules are the major fibrous spindle component (Figure 16–21) and the subsequent finding of the specific binding of colchicine to tubulin. Spindle microtubules originate in two different locations: at the centriolar regions just outside the nuclear

Figure 16–22

Electron micrograph of a section through the centromere of a kangaroo rat metaphase chromosome, showing the attachment of microtubules to the kinetochores (K). Each kinetochore contains two filaments (k_1 and k_2), each of which serves as an attachment point for microtubules. [Reproduced with permission from B. R. Brinkley and E. Stubblefield in *Advances in Cell Biology*, D. M. Prescott *et al.* (eds.), Vol. I, Appleton-Century-Crofts, 1970.]

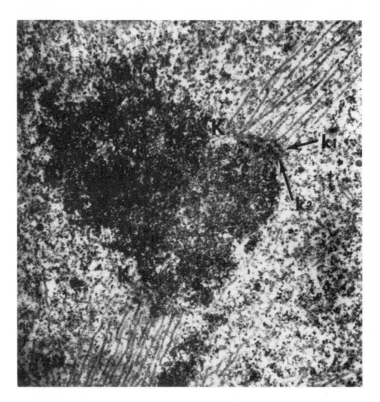

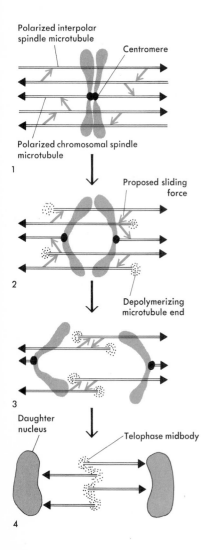

membrane (the *polar tubules*) and at the kinetochores (Figure 16–22), the attachment sites for microtubules that are fixed to the centromeric regions of each chromosome (*chromosomal tubules*). The polar tubules form first and start to grow during prophase. Later, when the nuclear membrane breaks down, the kinetochores become exposed to the cytoplasmic supply of free tubulin, allowing the chromosomal tubules to assemble.

During metaphase the polar and chromosomal tubules interpenetrate on the spindle, forming numerous cross bridges which eventually lead to the alignment of the homologous chromosomes at the center of the spindle (the *spindle plate*) (Figure 16–23). Anaphase starts when the polar and chromosomal sets begin to slide apart, leading to the segregation of homologous chromosomes to the opposite poles. In this process, the cross bridges are

Figure 16–23

A very diagrammatic possible mechanism of chromosome movement during mitosis as a result of sliding force between the polar and chromosomal spindle microtubules. Only one chromosome bivalent is shown. 1 = metaphase, 2 = early anaphase, 3 = late anaphase, 4 = telophase. Each gold arrow is the resultant productive force generated as the result of the making and breaking of cross bridges between adjacent oppositely polarized microtubules. [Redrawn from McIntosh *et al., Nature,* **224,** 659 (1969).]

frequently broken and then remade, these repetitive acts being the primary cause of the sliding motion. Here crucial evidence comes from the *in vitro* isolation of functional spindles capable of anaphase chromosomal separation. Such material reveals that as the spindle poles move apart during chromosomal separation, the individual spindle fibers do not increase in length. Instead, the lengthening movements during anaphase involve the polar and chromosomal tubules sliding over each other. The question thus becomes whether, as in ciliary motion, dynein-like cross bridges are used to convert chemical energy into mechanical movement. Now this seems likely, but since new experiments show that actin is also found within spindles, the possibility that actin–myosin interactions also play key roles is not yet ruled out.

HISTONES AND THE POSSIBILITY OF CHROMOSOMAL CONTRACTION

Equally important to the evolution of regular mitosis was the emergence of a mechanism for the regular cycles of chromosomal coiling and uncoiling that mark the mitotic cycle. Conceivably the key step was the emergence of *histones*, small basic proteins that tightly coat eucaryotic DNA. While procaryotic DNA is normally coated with basic protein, the binding is easily disrupted and, until recently, virtually everyone thought that it largely existed free. In contrast, most eucaryotic DNA is tightly complexed to proteins and forms the nucleoprotein fibers that are commonly called *chromatin*. Except for certain types of sperm, where *protamines* (short (~ 45AA) arginine-rich polypeptides) are the predominant protein component, the major protein constituents of chromatin are the five histones: H1, H2a, H2b, H3, and H4. Electron micrographs often show chromatin to have a beaded appearance built up of spherical 100 Å particles held together by thin fibers (Figure 16–24). Each of these spheres (often called *"nu"* particles or *nucleosomes*) contains some 200 base pairs held together in a still unknown configuration by their binding to discrete groups of histones. Now we guess that each "nu" particle contains two of the H2a, H2b, H3, and H4 components, and probably one of the H1 chains. Most importantly, several of the histones, in particular H3 and H4, have shown virtually no sequence variation in the course of evolution. This suggests that they play key roles in the formation of the "nu" particles, whose size and shape likewise appear invariant among all forms of eucaryotic chromatin.

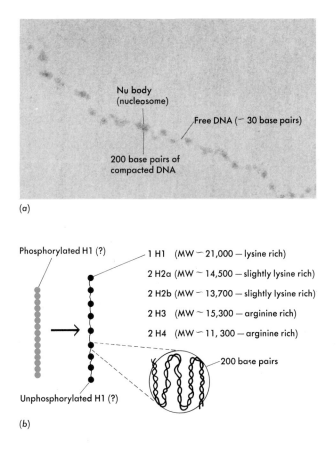

(a)

(b)

Figure 16–24
(a) Electron micrograph of slightly opened
Drosophila chromatin, showing beaded (nu
particle) appearance. [Courtesy of C. Laird,
University of Washington, Seattle.]
(b) Schematic drawing showing chromatin
in two different stages of contraction.

An almost equal mass of acidic proteins is found more or less firmly attached to the DNA–histone complexes. These nonhistone chromosomal proteins, which include the various DNA and RNA polymerases, number roughly 100. As far as we can tell, none participate in the basic chromatin structure. Instead, they function in either the replication or transcription of DNA, as well as in the respective control processes governing DNA and RNA synthesis. As such, these proteins can show great variation from one type of differentiated cell to another within the same organism.

The basic chromatin arrangement persists throughout the cell cycle, with the highly condensed mitotic chromatin having the same histone content as the much more highly extended interphase fibers. Chromosome condensation, however, may be related to cyclic chemical modification of its histones. These can be phosphorylated, acetylated, and methylated, and evidence is accumulating that their modification level may be a function of the position in the cell cycle. The clearest conclusion so far involves phosphorylation of the very lysine-rich histone H1. Phosphorylation reaches its maximum level per cell in late interphase, just before chromosome condensation becomes visible with the light microscope. Following completion of mitosis, it falls to less than 20% of its peak value. Interestingly, histone H1 itself is not involved in the basic chromatin configuration, since *in vitro*-reconstructed chromatin fibers lacking it give the same x-ray diffraction pattern as native chromatin. In contrast, *in vitro*-made chromatin, lacking any of the other four histones, fails to give the characteristic chromatin pattern.

THREE DIFFERENT RNA POLYMERASES WITHIN EUCARYOTIC CELLS

Another feature unique to eucaryotes is the presence of three quite distinct and easily separable major forms of RNA polymerase; one programmed for rRNA synthesis, a second for mRNA synthesis, and the third for tRNA synthesis. This distinction may be related to the presence within all eucaryotes of the nucleoli, small roundish nuclear bodies that contain tandemly-linked DNA copies of the gene that codes for rRNA (see page 528). The enzyme found within the nucleoli is called eucaryotic RNA polymerase 1, while the major enzymes in the nucleoplasm are called eucaryotic RNA polymerases 2 and 3. Polymerase 2 is specifically inhibited by low levels of the highly toxic mushroom poison α-amanitin, while RNA polymerase 3 is inhibited only at high concentration. In contrast, RNA polymerase 1 is resistant to α-amanitin action. Use of this inhibitor thus provides an easy way to completely determine which eucaryotic polymerase is responsible for the synthesis of a given RNA species.

Molecular characterization of the three different forms reveals that each contains a unique collection of polypeptide chains. Like their single procaryotic equivalent, each eucaryotic RNA polymerase has a MW $\sim 5.6 \times 10^5$ and is built up from two large polypeptide chains (MW $\sim 1.6 \times 10^5$ and 1.4×10^5) and 4 to 6 smaller polypeptides. The significance of this apparent diversity is still unclear, especially since the suspicion exists that certain of the smaller chains may have originated during enzyme purification by proteolytic cleavage of initially much larger chains. So, until more detailed protein chemistry is done, the various factors controlling the specificity of eucaryotic DNA transcription will be difficult to sort out.

481

BIZARRE 5' END GROUPS ON MANY EUCARYOTIC mRNA'S

Examination of the 5' ends of most eucaryotic mRNA's reveals the surprising fact that very few mRNA chains contain the expected pppNp . . . groups (where N = A or G). Instead most 5' ends are modified after transcription by the addition of a guanine nucleotide. This by itself would not be too strange were it not for the fact that it is not linked by a conventional 5'–3' phosphodiester bond. Instead, GTP reacts with the 5' end of an mRNA chain in a 5'–5' condensation, to form ends with the general structure ^{3}'G^5'ppp^5'N^3'p . . . Subsequently, methyl groups are added to the now terminal guanine residue, as well as to the 2'OH at the adjacent nucleotide (Figure 16–25). Why the 5' ends of eucaryotic mRNA need to be so blocked is still a total mystery. Though evidence is accumulating that many eucaryotic mRNA's cannot be translated unless blocked, there are several eucaryotic mRNA's (e.g., polio) which lack such ends but appear to function perfectly normally as templates for protein synthesis.

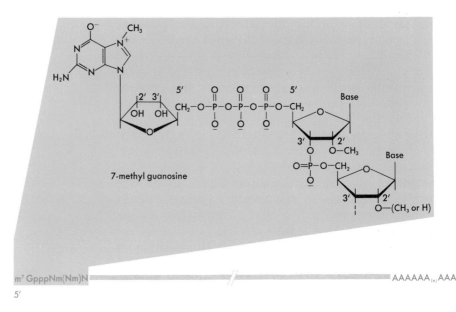

Figure 16–25
A generalized structure for eucaryotic mRNA, showing the post-transcriptional modifications at the 5' and 3' ends.

POLY A AT THE 3' END OF mRNA

Also very mysterious is the observation that most eucaryotic mRNA molecules contain relatively long stretches of Poly A ($\sim$ 200) residues at the 3' end. These poly A tracts are not coded by nuclear DNA, but are added after transcription has stopped and the newly made mRNA molecules have been released into the nuclear sap. The enzyme responsible, Poly A polymerase, specifically recognizes mRNA since no traces of Poly A are found at the 3′ end of rRNA chains. Because Poly A only very infrequently terminates procaryotic mRNA, the suggestion arises that Poly A may somehow function in the movement of mRNA through the nuclear membrane pores. Histone mRNA, however, appears to lack Poly A, and so whatever Poly A's function is, it may not be indispensable. In any case, Poly A remains attached to the 3′ ends after mRNA binds to ribosomes. There are hints, however, that as repeated polypeptides are made on the same mRNA, the lengths of the Poly A tracts shorten.

Another difference between eucaryotic and procaryotic polyribosomes is the attachment to eucaryotic mRNA molecules of two specific proteins. One (MW 22,000) binds to the 5′ end while the other (MW 52,000) is probably bound to the Poly A residues at the 3′ end. Whether these proteins first bind to their mRNA chains in the cytoplasm or are part of a complex that passes through the nuclear membrane pores is not known. Arguing for the latter possibility is the fact that virtually no free mRNA exists in the nucleus. Soon after, if not coincident with, its release from DNA, it becomes complexed with protein, some of which may remain bound after passage into the cytoplasm.

80S VS. 70S RIBOSOMES

The ribosomes of all eucaryotes are characteristically larger than the 70S procaryotic ribosomes and always sediment around 80S. Correspondingly, they have bigger subunits that move at 60S and 40S. Each 60S subunit contains a 28S rRNA chain, while the 40S particle has an 18S rRNA chain. Why these differences exist and hence why they are so evolutionarily invariant, no one knows. The important point is that on the structural level a larger gap exists between procaryotic and eucaryotic microorganisms (e.g., yeast, *Neurospora, Aspergillus*) than between the most evolutionarily divergent eucaryotic cells.

At the molecular level, proteins are made in almost exactly the same way in both eucaryotes and procaryotes. One seemingly consistent difference is that methionine, which starts polypeptide synthesis in all cells, is not blocked with a formyl group in eucaryotes. However, eucaryotes do possess two different types of methionyl

483

tRNA, with one specifically serving as an initiator tRNA. Interestingly, the eucaryotic initiator tRNA can be specifically formylated with the bacterial transformylase that normally adds formyl groups to bacterial f–met tRNA.

MONOCISTRONIC mRNA MOLECULES

All the mRNA molecules that have so far been isolated from eucaryotic cells or their viruses code for only one primary polypeptide product. Such diverse messages as those for myosin, collagen, the individual hemoglobin chains, and the several histones, all bind ribosomes at only one site. None are like *E. coli* lac mRNA or R17 RNA, which contains multiple translational start and stop signals. Even the very long animal viral messengers, such as those which code for the polio and EMC proteins, are initially translated as single long polypeptides that later become cleaved by cellular proteases into a number of functionally distinct proteins. The initial translational product of polio RNA, some 2000 amino acids in length, is cut into 7 functionally distinct proteins (Figure 16–26). Superficially this process seems very inefficient, because some of these products are structural proteins needed in large numbers while the others are enzymes required in much more minute quantities. We must, therefore, assume that the requirement for a single start signal is generated by some unique feature of the eucaryotic polyribosome.

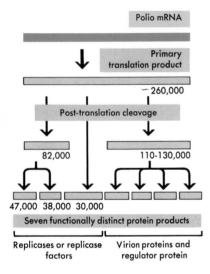

Figure 16–26
The monocistronic gene of polio virus type I. The initial primary product is proteolytically cleaved to eventually form seven functionally distinct proteins. The small figures are approximate molecular weights of the protein products.

MEMBRANE-BOUND RIBOSOMES

In procaryotes, most ribosomes exist either free or as parts of polyribosomes, and their protein products are directly released into the surrounding cytoplasm. Some procaryotic ribosomes, however, are bound to the inner surface of the outer plasma membrane. When so attached (usually in polyribosomal clusters), their respective polypeptide products grow into their bordering membranes, with some passing completely through to be released (secreted) into the cavity that separates the outer membrane from the cell wall. Attachment of ribosomes to a membrane can thus be seen as a device to ensure preferential secretion of a protein product, so there must exist some mechanism by which the right mRNA species preferentially stick to membrane-bound as opposed to free ribosomes.

Secretion in eucaryotes also appears to involve membrane-bound ribosomes. Here, however, most such ribosomes bind not to external membranes but to inner membranous sacs called the endoplasmic reticulum (ER). Only parts of the ER are covered with ribosomes. Sections rich in ribosomes are called *rough ER,* while ribosome-free regions are called *smooth ER* (Figure 16–27). The extent of rough endoplasmic reticulum varies with the quantities of protein that need to be secreted. Cells like the reticulocyte, which synthesize mainly internally needed proteins such as hemoglobin, have very poorly developed rough ER, while cells whose primary function is secretion (e.g., liver cells) devote almost half their internal space to the rough ER. Development of extensive ER's is probably a necessary consequence of the size of eucaryotic cells. The surface area-to-volume ratio falls as cells become bigger, and the surface area occupied by the plasma membrane of the larger eucaryotic cells may not be sufficient to house the ribosome number needed for the required secretory rate.

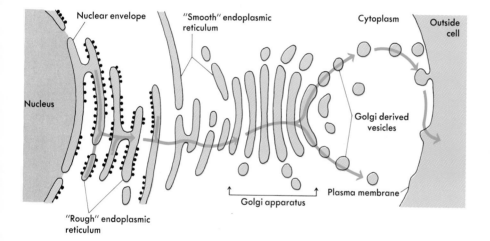

Figure 16–27
Schematic representation of the various membrane-bound cavities in the cell and their possible role in synthesis and excretion. The "rough" endoplasmic reticulum bears ribosomes on its outer surfaces. The Golgi apparatus, which secretes vesicles from one face, is formed on the other face by elements of the "smooth" endoplasmic reticulum. The solid-colored arrow indicates both the general membrane flow within the cell and also the probable route of proteins that are to be secreted. Areas cut off from the cytoplasm by membranes are shaded in color.

Why only the products of membrane-bound ribosomes seem to pass through most membranes is not yet understood. Perhaps ribosomes are so bound in order that the nascent polypeptides may become inserted into neighboring cell membranes before they have begun to fold into their final 3-D configuration. Related to this dilemma is the question of the site of synthesis of the protein products destined to function as intrinsic membrane components. Must they also be synthesized only on membrane-bound ribosomes, or do some specific proteins have the capacity to spontaneously insert themselves into a preexisting lipid bilayer?

MOVEMENT OF NEWLY MADE PROTEIN THROUGH THE SMOOTH ER AND GOLGI APPARATUS

Proteins destined for eventual secretion or positioning on other membranous components reach their final destination only after they have moved from the rough ER through the interconnecting membranous sacs of the smooth endoplasmic reticulum and the Golgi apparatus (Figure 16–27). These last two bodies appear to be the major sites for the attachment of sugar residues to nascent proteins to form glycoproteins. They are also the sites where other sugar groups become polymerized to form complex polysaccharide products like glycogen, and where many of the fatty acids that eventually end up as membranous components are produced.

Within the Golgi apparatus, specific proteins (or carbohydrate storage products) associate with each other in specific vacuoles. These eventually pinch off to become granules destined either to be secreted externally or to serve a specialized internal function. Among such internal granules are the lysosomes, which contain primarily digestive enzymes, and the peroxisomes, whose spectrum of specific enzymes includes catalase as well as several other enzymes involved in H_2O_2 metabolism (Figure 16–28).

Figure 16–28
The major stages thought to be involved in lysosome formation and function. Ingested food vacuoles (I) and membrane-enclosed cell organelles bound for breakdown (II) combine with primary lysosomes. These are acid-hydrolase-containing organelles made via the Golgi apparatus. The secondary lysosomes (III) so formed eventually lead to residual bodies containing fully hydrolyzed material and autoxidized lipids.

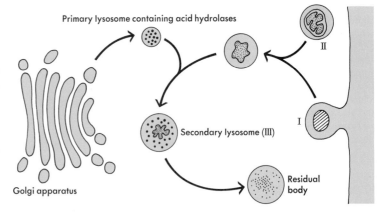

Storage within vacuoles allows cells to segregate potentially harmful hydrolytic enzymes or metabolic intermediates from most of the cellular cytoplasm. For example, cellular function would rapidly cease if the proteolytic enzymes found within lysosomes came into contact with most cellular proteins. Likewise, free H_2O_2 would quickly oxidatively destroy many cellular components if it were given access to them. By limiting its generation to within peroxisomes, where catalase almost instantly breaks it down, the cellular level of H_2O_2 need never reach dangerous levels.

DIGESTION OF INGESTED FOOD FOLLOWING FUSION OF FOOD VACUOLES WITH LYSOSOMES

Each lysosome contains a large variety of hydrolytic enzymes, such as various proteases, nucleases, phosphatases, lipases, and glycosidases. Such enzymes are essentially inert within the primary lysosome. Catalysis starts only when a primary lysosome fuses with an ingested food vacuole to form a secondary lysosome (Figure 16–28). Most food molecules are then quickly broken down, allowing their soluble hydrolysis products to pass through the lysosomal membrane into the cytoplasm. Such secondary lysosomes can then go on to fuse with more recently ingested food vacuoles. Why lysosomal membranes fuse only with phagocytic granules and not with other cellular components, such as mitochondrial membranes, is not clear. Real understanding of such fusion events will come only after there has been much detailed structural analysis of the various lysosomal membrane constituents.

NUCLEAR MEMBRANE AS OUTPOCKETING OF THE ER

In all higher eucaryotic plant and animal cells, the double-layered nuclear membrane (Figure 16–29) breaks down during spindle formation and reforms after anaphase chromosomal separation. Most likely, dissolution involves fusion of the flattened nuclear membrane sacs with preexisting cytoplasmic membrane vacuoles, in particular those of the endoplasmic reticulum. Even before its breakdown begins, the nuclear membrane might be regarded as part of the ER, since its external surface is covered with ribosomes that secrete proteins into the cavity separating the two unit membranes. Moreover, electron micrographs often show this cavity to be connected to the main channels of the ER. This leads us to speculate that nuclear membranes are formed by the coalescence of flattened outpocketings of the ER. This process, however, never goes to completion, leaving large numbers of uniformly sized (350 Å) nuclear pores scattered over the nuclear membrane surface. These pores often have eight-sided polygonal shapes, and it is tempting to ascribe the origin of their regular outlines to the shape of the plugs that fill most pores.

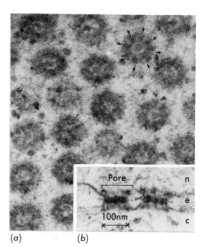

(a)　　　(b)

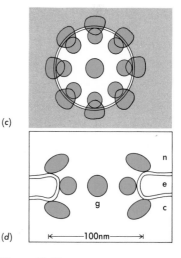

(c)

(d)

Figure 16–29

A schematic view of the structure of the nuclear membrane, together with an electron micrograph of a section containing a nuclear pore. [Courtesy of K. Roberts, John Innis Institute, Norwich, England.]

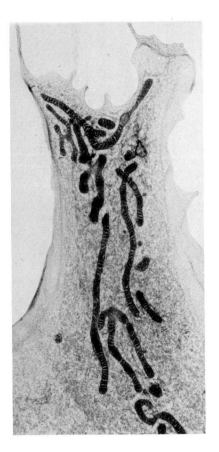

Figure 16–30
A high-voltage electron micrograph (1000 kV) of the edge of a flattened mouse cell (7000×), showing the elongated, tubular form of most mitochondria. Initially it was thought that mitochondria had regular ellipsoidal shapes, but the use of very high-voltage electron microscopes, which allow the viewing of the complete mitochondria, reveals highly irregular shapes. [Courtesy of J. Wolosenick, University of Colorado.]

Conceivably, these plugs are nucleation sites around which the ER vacuoles coalesce during formation of the nuclear membrane.

Gene products from the nucleus reach the cytoplasmic sites of protein synthesis by passing through these pores (Figure 16–29). This means either that the pores are sometimes unplugged or that channels exist within the plugs themselves. In any case, material transfer is a two-way phenomenon. Nuclear transplantation experiments in which unlabeled nuclei are inserted into radioactively labeled cytoplasm show that some newly synthesized protein quickly moves from the cytoplasm to the nucleus.

EVOLUTION OF SYMBIOTIC BACTERIA INTO MITOCHONDRIA AND CHLOROPLASTS

Until several years ago the biological origin of mitochondria and chloroplasts was very obscure. Though microscopic observations (Figure 16–30) revealed that they lengthened and divided like bacteria, the meaning of these similarities remained unknown until all mitochondria and chloroplasts were found to contain DNA molecules as well as ribosomes, tRNA, and the various enzymes needed to make protein. Even more important for our ideas about organelle biogenesis was the additional observation that many chloroplast ribosomes sedimented at the 70S value characteristic of procaryotic cells. This immediately suggested that, sometime early in eucaryotic evolution, chloroplasts, as well as mitochondria, evolved from symbiotic bacteria growing within the much larger, primitive eucaryotic cells. One obvious speculation is that they were so efficient at supplying ATP to their host that their obligatory presence as cellular organelles was selected for. Strong support for this endosymbiont hypothesis comes from amino acid sequence analysis of the chicken-mitochondrial enzyme superoxide dismutase. Its N-terminal amino acid sequence shows close homology to those of the *E. coli* dismutases (Figure 16–31), suggesting a common evolutionary origin. In contrast, it shows no significant homology with bovine erythrocyte dismutase, despite the much closer taxonomic relationship.

Chloroplasts and mitochondria now contain much less DNA than must have been present at their evolutionary origin. The DNA molecules of most chloroplasts have MW $\sim 10^8$, making them ten times smaller than any known bacterial DNA. Even less DNA is found within mitochondria. This is especially true of the mitochondria in higher animal cells, each of which invariably contains a

	1				5					10		
E. coli (Fe)	SER	PHE	GLU	LEU	PRO	ALA	LEU	PRO	TYR	ALA	LYS	ASP
E. coli (Mn)	SER	TYR	THR	LEU	PRO	SER	LEU	PRO	TYR	ALA	TYR	ASP
Chicken liver mitochondria (Mn)	LYS	HIS	THR	LEU	PRO	ASP	LEU	PRO	TYR	ASP	TYR	GLY
Bovine erythrocyte (Cu-Zn)	AcALA	THR	LYS	ALA	VAL	CYS	VAL	LEU	LYS	GLY	ASP	GLY

15						20					25					29
ALA	LEU	ALA	PRO	HIS	ILE	SER	ALA	GLU	?	ILE	GLU	TYR	HIS	TYR	GLY	LYS
ALA	LEU	GLU	PRO	HIS	PHE	ASP	LYS	GLN	THR	MET	GLU	LEU	?	HIS	?	LYS
ALA	LEU	GLU	PRO	HIS	ILE	SER	ALA	GLU	ILE	MET	GLN	LEU	HIS	?	?	LYS
PRO	VAL	GLN	GLY	THR	ILE	HIS	PHE	GLU	ALA	LYS	GLY	ASP	THR	VAL	VAL	VAL

Figure 16–31

The N-terminal amino acid sequences of superoxide dismutases from four sources. Identical residues in the sequences of the two *E. coli* and the mitochondrial dismutases are shown in dark color. Light color indicates identity between any one or more of these three dismutases and bovine erythrocyte superoxide dismutase. [Redrawn from Steinman *et al., Proc. Nat. Acad. Sci.,* **70**, 3725 (1973).]

DNA molecule of only $\sim 10^7$ MW. Many of the original functions of organelle DNA must have been taken over by host cell nuclear genes, thereby allowing the elimination of the respective organelle DNA.

NUCLEAR GENES CODING FOR ORGANELLE PROTEINS

The mitochondria of higher cells represent the most extreme example of organelle coding capability being replaced by similarly functioning host DNA. A mitochondrion's 10^7-MW DNA molecule is hardly large enough to code for all protein found in its ribosome, much less for the many different proteins involved in oxidative phosphorylation, so most of the enzymes involved in the duplication and transcription of mitochondrial DNA, as well as the subsequent translation of mitochondrial mRNA, are coded by nuclear DNA. Selection for smaller mitochondrial DNA has, in addition, led to the emergence of 60S mitochondrial ribosomes that are considerably smaller than their original 70S procaryotic progenitors. Each such ribosome contains a 45S and a 35S subunit that utilize respectively 16S (7×10^5 MW) and 12S (4.5×10^5 MW) mitochondrial rRNA. Mitochondrial DNA codes for these rRNA's as well as for 9 different tRNA molecules whose genes have now been sited relative to the rDNA genes (Figure 16–32). The fact that only 9 tRNA's are so coded means that at least half the tRNA species that function in mitochondrial protein synthesis have a nuclear origin, and so may function in both mitochondrial and cytoplasmic protein synthesis.

Still very unclear is how the nuclearly coded products reach their final organelle destinations. Both mitochondria

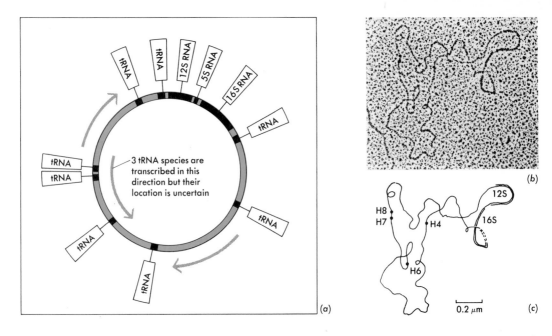

Figure 16–32
(a) A map of the relative positions of RNA genes on circular mitochondrial DNA. (b) Electron micrograph of mitochondrial "H" strand DNA hybrids with 12S RNA, 16S RNA, 4S RNA-ferritin conjugates. (c) Explanatory tracing, showing numbered 4S RNA genes and hybridized 12S and 16S RNA. [Electron micrograph and tracing reproduced with permission from Wu *et al., J. Mol. Biol.,* **71,** 81 (1972).]

and chloroplasts are surrounded by two unit membranes that electron microscopists generally assert to have no connection with the ER, yet the majority of membrane proteins, at least in mitochondria, are nuclearly coded. So either host mRNA molecules or their protein products are regularly incorporated into mitochondria. Conceivably, during growth of these organelles, their membranes temporarily fuse with outpocketings of the ER.

SUMMARY

For almost 100 years, the possession of a nucleus was the main diagnostic characteristic of the eucaryotic cell. Now, with the development of a variety of new techniques for studying cell structure, we can probe the uniqueness of eucaryotes from a variety of viewpoints. Evolution of the eucaryotic cell occurred early in evolution, conceivably in response to the transformation of most of the original primordial organic soup into living material. This process created a great selective advantage for predatory organisms, and so cells became larger and acquired the ability both to engulf other organisms by phagocytosis and to move from one region to another. Key to this evolutionary process must have been the development of muscle proteins like actin and myosin, and the emergence of microtubules to provide skeletal support as well as to direct certain types of cell movement (e.g., ciliary movement).

Equally important has been the development of a variety of membrane-bounded internal vesicles, like the endoplasmic reticulum, the Golgi bodies, and the lysosomes. Such vesicles allow the segregation away from the main mass of cytoplasmic proteins of proteins programmed either for secretion or digestion of food granules. Most such vesicles have the ability to fuse with other vesicles or with the outer cell membrane (plasma membrane). Correspondingly, the outer cell membrane can form invaginations which pinch off internally, thereby creating new membrane-bounded internal vesicles (pinocytosis).

The nuclear membrane itself represents a section of the endoplasmic reticulum which forms about the chromosomes after cell division. Why nuclear membranes exist is still unclear, though there are hints that chromosomes have to be associated with membranes at some stage in the cell cycle. So as cells become larger, it may have become much more efficient for their chromosomes to bind to an internal membrane than to the far distant outer plasma membrane.

All the DNA in eucaryotic chromosomes exists as chromatin, a complex of $\frac{1}{3}$ DNA, $\frac{1}{3}$ histone (small basic proteins), and $\frac{1}{3}$ nonbasic (acidic) chromosomal proteins. Five main histones exist, four of which (H2a, H2b, H3, and H4) fold 200-base-pair sections of double-helical DNA into distinct 125 Å spherical masses (nucleosomes). The remaining histone (H1) may control chromosomal extension and retraction during the mitotic cycle. Separation of homologous chromosomes during mitosis occurs on the spindle, a collection of microtubules of two origins (polar and chromosomal). How chromosomal movements occur on the spindle is not known. A leading possibility is the sliding of polar and chromosomal tubules past each other.

Unlike procaryotes, in which only one major form of RNA polymerase exists, all eucaryotic cells contain three forms of RNA polymerase. One functions in rRNA synthesis, another in tRNA synthesis, and the third controls mRNA synthesis. After transcription, most mRNA molecules are modified at their 5' ends by the addition of a guanine nucleotide in 5'–5' linkage and at their 3' ends by the addition of 100 to 200 adenine nucleotides (Poly A). The function of these blocking groups remains mysterious, since some eucaryotic mRNA's lack them, yet appear to function normally. Unlike procaryotic mRNA's, eucaryotic mRNA's have only one ribosome binding site and so

491

always code for only single polypeptide chains. Many of the longer translation products subsequently are broken down into smaller chains, each of which functions as a separate protein (enzyme).

Unique to all eucaryotes are the mitochondria, which are cytoplasmic ATP-generating organelles that originated by the evolution of endosymbiotic bacteria. In addition, plants have chloroplasts, which presumably had a similar evolutionary origin. Both these organelles contain mini-DNA chromosomes and show many similarities (e.g., 70S ribosomes) to procaryotic cells. While their mini-chromosomes code for both their rRNA chains and some tRNA chains, their small size precludes them from coding more than a small fraction of their respective organelle proteins. Thus, much of the protein of mitochondria (chloroplasts) is coded by cellular DNA and, in a still to be discovered way, becomes incorporated into the structural fabric of the respective organelles.

REFERENCES

NOVIKOFF, A., AND E. HOLTZMAN, *Cells and Organelles,* Holt, Rinehart, and Winston, New York, 1970. A well written and nicely illustrated introduction to the structure of eucaryotic cells.

WOLFE, S. L., *Biology of the Cell,* Wadsworth, Belmont, California, 1972. A text which emphasizes chromosomal structure and behavior during mitosis and meiosis.

ORGANIZATION AND CONTROL IN PROCARYOTIC AND EUCARYOTIC CELLS, Twentieth Symp. Soc. Gen. Microbiol., Cambridge University Press, 1970. A collection of papers emphasizing unique features of eucaryotes.

MARGULIS, L., *Origin of Eucaryotic Cells,* Yale University Press, 1970. An imaginative book which opens much of lower systematics and cell structure to an almost totally ignorant biochemical world.

MILLER, S. L., AND L. E. ORGEL, *The Origins of Life on the Earth,* Prentice-Hall, 1974. An introductory text that separates the few solid facts from much wishful speculation.

SINGER, S. J., AND G. L. NICHOLSON, "The Fluid Mosaic Model of the Structure of Cell Membranes," *Science,* **175,** 720 (1972). The first comprehensive appreciation of the semifluid nature of cell membranes.

Membrane Molecular Biology, C. F. Fox and A. D. Keith (eds.), Sinauer Associates, Stamford, Connecticut, 1972. An advanced text with review-like articles.

"Aspects of Cell Motility," *Symp. Soc. Exp. Biol.,* **XXII,** Academic Press, 1968. A collection of review-type articles about the various ways in which cell movement is generated.

TRINKHAUS, J. P., *Cells into Organs: The Forces that Shape the Embryo,* Prentice-Hall, 1969. A text that aims to understand the role of cell movements in embryological development.

WESSELS, N. K., "How Living Cells Change Shape," *Scientific American,* October, 1971. Experiments on how microfilaments and microtubules may influence cell movement.

Locomotion of Tissue Cells, CIBA Foundation Symposium XLV (new series), Elsevier-North Holland, Amsterdam, New York, 1973. An up-to-date look at cell membrane movements which generate cell motility.

POLLARD, T. D., AND R. R. WEIHING, "Actin and Myosin and Cell Movement," *CRC Critical Reviews in Biochemistry,* G. D. Fasnan (ed.), **2,** 1, Chemical Rubber Company, Cleveland, 1973. A very complete review which emphasizes the properties of the muscle proteins isolated from nonmuscle cells.

ALLISON, A. C., AND P. DAVIES, "Mechanism of Endocytosis and Exocytosis: Transport at the Cellular Level," *Symp. Soc. Exp. Biol.,* **XXVIII,** 419, Cambridge University Press, 1974. An excellent imaginative review.

LAZARIDES, E., AND K. WEBER, "Actin Antibody: The Specific Visualization of Actin Filaments in Nonmuscle Cells," *Proc. Nat. Acad. Sci.,* **71,** 2268 (1974). This and the following three papers demonstrate the great utility of fluorescent antibodies in studying the organization of muscle proteins of cells growing in culture.

GOLDMAN, R. D., E. LAZARIDES, R. POLLACK, AND K. WEBER, "The Distribution of Actin in Nonmuscle Cells: The Use of Actin Antibody in the Localization of Actin within the Microfilament Bundles of Mouse 3T3 Cells," *Exp. Cell Res.,* **90,** 333 (1975).

LAZARIDES, E., "Tropomyosin Antibody: The Specific Localization of Tropomyosin in Nonmuscle Cells," *J. Cell Biol.,* **65,** 549 (1975).

LAZARIDES, E., "Actin, α-Actinin, and Tropomyosin Interaction in the Structural Organization of Actin Filaments in Nonmuscle Cells," *J. Cell Biol.,* in press (1975).

BUCKLEY, I. K., "Three-Dimensional Fine Structure of Cultured Cells: Possible Implication for Subcellular Motility,"

Tissue and Cell, **1**, 51 (1975). An important article demonstrating the ability of million-volt electron microscopes to probe the structure of intact cells.

PORTER, K. R., "Cytoplasmic Microtubules and Their Functions," *Principles of Biomolecular Structure*, Ciba Symposium, Little, Brown and Company, Boston, 1966. A beautifully illustrated discussion of microtubules at the electron-microscopic level.

ROBERTS, K., "Cytoplasmic Microtubules and Their Functions," *Progress in Biophysics and Molecular Biology*, **28**, 371 (1974). A complete, recent overview of microtubule form and function.

SATIR, P., "How Cilia Move," *Scientific American*, October, 1974. Here we see how microtubules can be made to generate ciliary movement.

WARNER, F. D., "Macromolecular Organization of Eucaryotic Cilia and Flagella," *Adv. Cell and Mol. Biol.*, E. J. DuPraw (ed.), **2**, 193 (1972), Academic Press. A beautifully illustrated review that relates ciliary motility to microtubule movement.

BRINKLEY, B. R., AND E. STUBBLEFIELD, "Ultrastructure and Interaction of the Kinetochore and Centriole in Mitosis and Meiosis," *Advances in Cell Biology*, D. M. Prescott *et al.* (eds.), **1**, (1970), Appleton-Century-Crofts. A very comprehensive, advanced review.

McINTOSH, J. R., P. K. HEPLER, AND D. G. VAN WIE, "Model for Mitosis," *Nature*, **224**, 659 (1969). A clever attempt to understand mitosis in terms of known properties of microtubules.

CANDE, W. Z., J. SNYDER, D. SMITH, K. SUMMERS, AND J. R. McINTOSH, "A Functional Mitotic Spindle Prepared from Mammalian Cells in Culture," *Proc. Nat. Acad. Sci.*, **71**, 1559 (1974). The first successful study of mitosis at the *in vitro* level.

OLINS, A. L., AND D. E. OLINS, "Spheroid Chromatin Units (nu bodies)," *Science*, **183**, 330 (1974). The first clear demonstration that chromatin has a bead-like structure.

KORNBERG, R., "Chromatin Structure: A Repeating Unit of Histones and DNA," *Science*, **184**, 868 (1974). A proposal for how histones interact with DNA.

OUDET, P., M. GROSS-BELLARD, AND P. CHAMBON, "Electron Microscopic and Biochemical Evidence that Chromatin is a Repeating Unit," *Cell*, **4**, 281 (1975). An important EM analysis of how the addition of histone H1 to nu bodies (nucleosomes) leads to a more compacted chromatin structure.

ADAMS, J. M., AND S. CORY, "Modified Nucleosides and Bizarre 5' Termini in Mouse Myeloma Messenger RNA," *Nature,* **255,** 28 (1975).

FURUICHI, Y., M. MORGAN, S. MUTHUKRISHMAN, AND A. J. SHATKIN, "Reovirus Messenger RNA Contains a Methylated, Blocked 5'-Terminal Structure: $m^7G(5')ppp(5')$-G^mpCp," *Proc. Nat. Acad. Sci.,* **72,** 362 (1975). This and the preceding article are among the first reports of the unusual 5' blocking groups of eucaryotic mRNA.

Control of Organelle Development, Symposium of the Society for Experimental Biology, XXIV, Academic Press, 1970. A wide-ranging book covering the structure, function, and genetics of the various cytoplasmic organelles, particularly those that carry genetic information.

SAGER, R., *Cytoplasmic Genes and Organelles,* Academic Press, 1972. A text which summarizes modern ideas about the genetic nature of chloroplasts and mitochondria.

SCHATZ, G., AND T. L. MASON, "The Biosynthesis of Mitochondrial Proteins," *Ann. Rev. Biochem.* **43,** 51 (1974).

Embryology
at the
Molecular Level

The careful reader will have noticed that many statements in the preceding chapters were qualified; our understanding of how bacterial cells control the synthesis of specific proteins is less complete than our general knowledge about the mechanism of protein synthesis, and much less complete than our understanding of the structure of DNA. Although we can state unambiguously that the genetic information of cellular chromosomes resides in the nucleotide sequences of DNA molecules and are certain that all tRNA molecules are folded into a cloverleaf, we are not at all sure whether the majority of operons will follow the lactose example and be controlled by repressors, or whether positive control systems will be the rule. These uncertainties, however, do not seriously annoy us. There is every reason to believe that, within the next several years, most of our speculations about bacterial control mechanisms will harden into facts.

We must not, however, be mesmerized by our past successes into asserting uncritically that our achievements at the molecular level with bacteria can automatically be extended to the cells of higher plants and animals; we must remember that bacteria and their viruses were chosen because of their simplicity, that higher plants and animals are exceedingly complex objects, and that much wisdom must be exercised in deciding which of the genetic processes of higher organisms can be investigated profitably at the molecular level within the next ten to twenty years. In particular, we should ask if we have sufficient background at this time to attack embryology at the molecular level.

AMOUNT OF DNA PER CELL INCREASES ABOUT EIGHT HUNDREDFOLD FROM *E. COLI* TO MAMMALS

Before experts climb a high mountain, they carefully measure its height and try to anticipate how difficult the ascent will be. Likewise, it would be most useful to know how much more complex, genetically speaking, the mammalian cell is than the *E. coli* cell. One obvious approach is to determine how much DNA is present per mammalian cell; the amount (Figure 17–1) is approximately 800 times that in *E. coli*. This number gives us an upper limit of the number of different genes, since there is no reason to believe that protein size (and hence, gene size) increases from the lower to the higher forms of life. Thus, under the assumption that each gene is present in only one copy per haploid genome, a mammalian cell would be capable of synthesizing over two million different proteins. If so, the task of relating a given mutant character to a specific mutant protein will be much, much more formidable than the corresponding job with bacteria.

Figure 17–1
Haploid amounts of DNA in various cells, expressed as multiples of the amounts found in *E. coli* (4×10^{-12} mg $= 2.4 \times 10^{9}$ MW). Many of these values should be taken as approximate. [Redrawn from Holliday, *Symp. Soc. Gen. Microbiol.*, **20**, 362 (1970).]

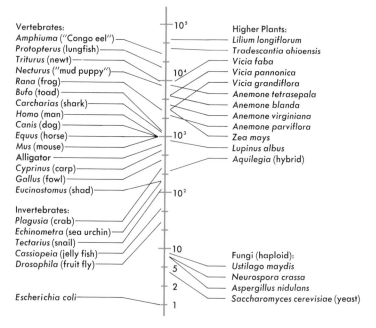

In some cases, however, the amount of DNA is certainly misleading; there are groups of amphibians that contain 25 times more DNA in their cells than is present in mammalian cells. Here, there is no obvious reason to believe that greater biological complexity is involved.

We must thus be very cautious about relating DNA content directly to the number of different proteins that may be synthesized by a given cell. Nonetheless, mere morphological examination with the electron microscope tells us that a much larger variety of subcellular structures exists in the vertebrate cell than in *E. coli*. So we must expect a correspondingly larger number of structural proteins and enzymes to be necessary for their construction and function. It would be surprising if the vertebrate cell were not at least 20 to 50 times more complex genetically than *E. coli*.

CONCENTRATION ON ORGANISMS WITH EASILY OBSERVABLE CLEAVAGE DIVISIONS

The mechanisms by which fertilized eggs develop into multicellular organisms have been a continuous source of mystery to biologists for almost a hundred years. During this time, experimental studies have been concentrated on echinoderms (especially the sea urchin) and amphibians, in particular various frogs, salamanders, and toads. The reason for such polarization lay in the ease with which echinoderm and amphibian eggs could be procured, fertilized, and subsequently observed as they passed through their regular developmental stages. In contrast, birds and mammals—where all the embryological stages take place within the thick-shelled egg (female parent)—are much more tricky to work with. Only recently has real progress been made in opening up the early stages of mouse embryogenesis to experimental manipulation.

A very important feature of both urchin and frog embryology is that all the nutrients (mostly stored in the form of yolk) required for development through the blastula and gastrula stages (Figure 17–2) must be present in the unfertilized egg. Only when the developing embryo becomes able to feed itself—the frog, for example, at the tadpole stage—does real growth in mass occur. Consequently, echinoderm and amphibian eggs are very, very large in comparison with their respective adult cells, and the cell divisions that occur after fertilization progressively produce smaller and smaller cells. Hence the name "cleavage divisions." Early cleavage divisions occur in rapid suc-

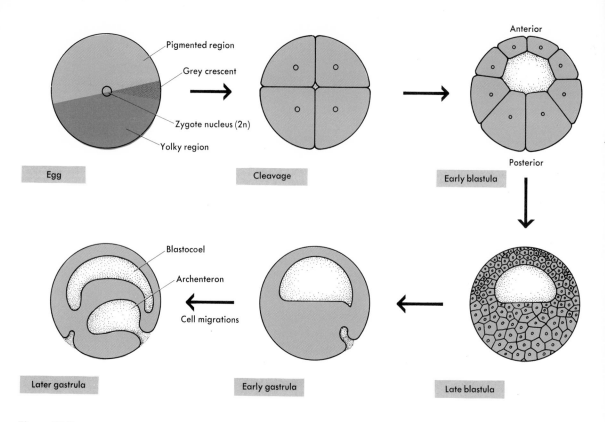

Figure 17–2
Early stages in the development of an amphibian egg.

cession, characterized by short interphase periods during which DNA synthesis occurs at a very accelerated rate: The amount of DNA in each of 1000 or more small daughter cell nuclei of a gastrula is the same as that in the nucleus of the newly fertilized egg.

THE HEART OF EMBRYOLOGY IS THE PROBLEM OF CELL DIFFERENTIATION

The cleavage divisions by themselves, of course, do not lead to embryological development. Its real essence instead is the process of cell differentiation. All higher plants and animals are constructed from a large variety of different cell types (e.g., nerve cells, muscle cells, thyroid cells, blood cells) which must arise in a regular way. In some organisms, specialization begins with the first few cell divisions after fertilization. In other organisms, a large number of divisions occur before any progeny cell is fixed in its fate. Irrespective of the exact time that differentiation occurs, however, it always results in the transformation of the parental cell into a large number of morphologically different progeny cell types.

Differentiation can be examined from three viewpoints. First, what are the external forces acting upon the original undifferentiated cell which might initiate a chain of events resulting in two progeny cells of different constitution? Sometimes the existence of asymmetrically acting external forces is easy to perceive. For example, gravity forces the yolk of a fertilized amphibian egg to the bottom. Thus, after the first few cell divisions, some of the progeny cells have more yolk than others.

The second way to analyze differentiation is to ask, What are the molecular differences between differentiated cells? Are they extreme, or do the morphological differences arise from the presence of only a few unique proteins in abnormally large numbers? Now all our evidence indicates the former conclusion: Each type of differentiated cell contains many types of molecules peculiar to that cell type. Thus, a complete description of differentiation at the molecular level would necessarily be a most formidable task.

Third, we may ask whether the various changes which bring about differentiation are irreversible, and if so, how they are hereditarily perpetuated. The answering of these questions has always been a prime goal of most biologists. But until very recently, embryology has been largely studied as an isolated subject, apart from modern genetic or biochemical ideas. Now, however, it is clear that the morphological tools of the classical embryologist cannot give satisfying answers. Instead, as in genetics, the fundamental answer must lie at the molecular level. The parallel between biochemistry and modern genetics may, in fact, be very close, since embryologists now believe that many of the basic control mechanisms that fix a cell's potential chemical reactions act at the level of the gene. Thus, the recent methodological advances which have made aspects of biochemistry and genetics indistinguishable may soon encompass developmental biology.

But here we must point out a most disquieting fact: compared to higher organisms like *Drosophila* or the mouse, the genetics of sea urchins is virtually nonexistent, and that of frogs, salamanders, and toads is almost as bad.

DIFFERENTIATION IS OFTEN IRREVERSIBLE

At present, it is possible to isolate a variety of differentiated cells and grow them (like bacteria) outside living organisms under well defined nutrient conditions. This technique of "cell culture" allows us to ask, for example, whether a nerve cell continues to look like a nerve cell when growing outside its normal cellular environment: the answer in this case is *yes.* Something has happened that has permanently destroyed this cell's capacity to synthesize proteins other than those found in nerve cells. Similar results come when many other cell types from higher animals are studied. But with higher plants, the opposite answer is more often found. A complete plant can often be regenerated, starting from either highly differentiated root or epidermal cells.

DIFFERENTIATION USUALLY IS NOT DUE TO CHROMOSOME GAIN OR LOSS

An obvious hypothesis to explain irreversible differentiation proposes that only a fraction of the genes of the fertilized egg are passed on to a nerve cell, a muscle cell, and so on. This scheme, however, appears to be completely wrong. As far as we can tell, all cells of most organisms, with the obvious exception of the haploid sex cells, contain roughly the same chromosomal complement. Virtually all cell divisions are preceded by a regular mitosis, so that daughter cells generally receive identical chromosome groups. We cannot say, however, that no permanent changes have occurred at the level of individual genes. The question remains whether it is possible to mutate specific genes selectively, thereby making them nonfunctional (or functional). Our problem is to devise methods that can test this possibility. Unfortunately, this task seems very difficult at present.

MULTICELLULAR ORGANISMS MUST HAVE DEVICES TO CONTROL WHEN GENES ACT

Irrespective of the molecular mechanism (i.e., whether a chemical change in the gene itself is involved), there is now very good evidence that in multicellular organisms, as in bacteria, not all genes in a cell function at the same time. Something must dictate that a muscle cell, for example, selectively synthesize the various proteins used to construct muscle fibers.

The understanding of embryology will thus, in one sense, be the understanding of how genes selectively function. Moreover, we must ask not only what causes two progeny cells to synthesize different proteins, but also what makes them continue to synthesize exclusively the same group of proteins. With the problem phrased in this way, it is clear that no one will ever be able to work out all the chemical details that accompany embryological development of any higher plant or animal. For even a modest approach to a comprehensive understanding, we would have to look at the behavior of hundreds of different proteins. Nonetheless, common sense tells us that, as in bacteria, there may exist some general principles governing the selective occurrence of specific proteins. For example, some differentiation might occur largely at the chromosome level by devices that control the amount of specific mRNA synthesis. In fact, there is now excellent evidence (see below) for differential rates of RNA synthesis along different regions of many chromosomes. Most important, regions which are active at one time often are inactive at another stage in development.

So now a growing number of embryologists are beginning to ask whether the fundamental control mechanisms act negatively, like the lactose repressor, or positively, like the viral-specific T4 factors. The dilemma exists, however, that most of the characters which embryologists study are hopelessly complex from the chemical viewpoint (consider the eye); only a few can be related to the occurrence of well defined chemical reactions. For example, although nerve cells are easy to identify on morphological grounds, almost nothing is known about their structure on the molecular level, and only several of the many hundreds of protein molecules in nervous tissue have been well characterized. Although we are much more familiar with the muscle proteins, again large gaps in our knowledge are likely to make the analysis of precise gene–protein relationships a most tricky endeavor.

NECESSITY OF FINDING SIMPLE MODEL SYSTEMS FOR STUDYING DIFFERENTIATION

We must ask whether any of the embryologists' systems for studying differentiation are appropriate for a serious attack at the present time. To begin with, let us emphasize the fact that merely cataloguing obvious protein and nucleic acid differences between the various differentiated cells is unlikely to yield any fundamental answer. We already know from classical morphology that there are differences. Instead, incisive answers are likely to come only from more meaningful questions. One of the most impor-

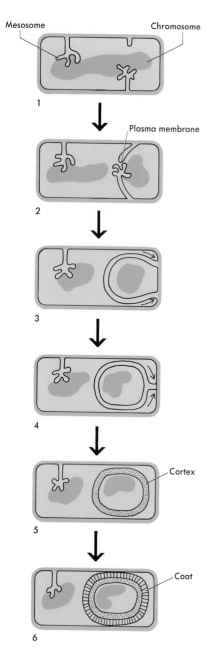

Mesosome   Chromosome

1

Plasma membrane

2

3

4

Cortex

5

Coat

6

Figure 17–3
Diagrammatic representation of the stages in sporulation. Key: (1) Preseptation—cell with axial chromosome. (2) Septation—plasma membrane infolds and cuts off an area containing intact chromosome. (3) and (4) Stages of protoplast envelopment. (5) Cortex formation—material laid down between two membranes. (6) Coat formation and maturation—eventually the spore is released.

tant goals is the identification of the external factors (embryonic inducers) which cause the directed transformations of many undifferentiated cells. For example, the differentiation of many nerve cells depends on an external factor received from nearby cells. A second important goal is to discover how the inducer changes the undifferentiated cell. In particular, we wish to understand the chain of events that relates inducers to the functioning of specific genes.

These problems are likely to be solved only when undifferentiated cells growing in cultures can be specifically transformed by the addition of their embryonic inducers. Even though there exist many claims that *in vitro* differentiation has been obtained, careful examination of the experiments reveals that they are usually over-interpreted. For example, undifferentiated embryonic chicken cells can often be transformed into nerve cells by the addition of a distinct chemical compound. At first sight, this is a most spectacular result. Unfortunately, there is no single distinct compound that induces the differentiation of a nerve cell, but rather a large variety of seemingly unrelated molecules, all of which have the same result. Under certain concentration conditions, even NaCl is an inducer. Most embryologists suspect that all currently known chemicals which induce *in vitro* differentiation act unspecifically, and that the true specific embryonic inducers have not yet been observed.

BACTERIAL SPORULATION AS THE SIMPLEST OF ALL MODEL SYSTEMS

The colossal magnitude of the task of attempting to understand the molecular basis of a complex differentiation process (e.g., the origin of a nerve cell) has led many people to look for cell systems much, much simpler than those studied by the classical embryologist. Some biochemists have gone so far as to focus their attention on the formation of bacterial spores. These are highly dehydrated cells which form inside certain bacteria (e.g., the genus *Bacillus*) as a response to a suboptimal environment. As spores possess virtually no metabolic activity, they resemble seeds of higher plants. Compared to vegetative bacteria, they are highly resistant to adverse conditions such as extreme heat or dryness. Thus, those bacteria that can produce spores have a much higher capacity to survive the extreme environmental conditions under which many forms of bacteria must live.

Some steps in the development of a spore are shown in Figures 17–3 and 17–4, which indicate that an early

504

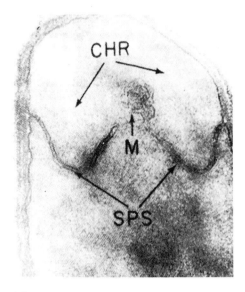

(a)

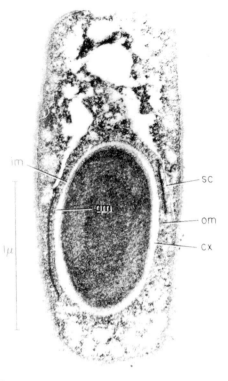

(b)

Figure 17–4
Electron micrographs of thin sections of
sporulating *Bacillus cereus* cells showing (a)
septation, (b) cortex development, and (c)
coat formation: chr, chromosome; m, mesosome; sps, transverse spore septum; cx,
cortex; sc, spore coat; ex, exosporium; im
and om, inner and outer membranes.
[Courtesy of Dr. W. G. Morrell.]

(c)

state involves invagination of a portion of the bacterial
membrane to enclose a chromosome and a small amount of
cytoplasm. Afterward, a very thick and tough surface layer
of protein is laid down on the outside. It has a completely
different composition from the surface of the corresponding vegetative forms. Parallel with these morphological transformations are important changes in enzyme
composition. The various cytochrome molecules responsible for aerobic metabolism completely disappear, and a
new electron transport system appears. Ribosome content
is also reduced, with an even greater decrease in the
number of messenger RNA molecules. Small numbers of
the enzymes necessary for protein synthesis remain—these
are necessary for the resumption of new protein synthesis
when the spore germinates under the stimulus of more
favorable nutritional conditions.

Sporulation thus involves the stopping of the synthesis of virtually all the proteins necessary for vegetative ex-

istence; conversely, germination converts spores to a condition in which they can make the vegetative proteins. So the making of a spore must entail an inhibition of most, if not all, vegetative mRNA synthesis, replacing it with the synthesis of mRNA specific for spore-specific proteins.

Until very recently the molecular basis of this changeover was a total mystery. Now very recent experiments tell us that sporulation is accompanied by the appearance of a new RNA polymerase specificity factor which reads the genes necessary for sporulation. At the same time, there is an inactivation of one (or more?) vegetative specificity factors. Thus, the key control device which regulates the sequential development of phage proteins is also the crucial factor in sporulation. Of course, many, many questions remain unanswered (e.g., elucidation of the way the stimulation for sporulation (or germination) brings about a change in the σ pattern). But after years of real frustration, there is good reason for believing that the biochemistry behind the differentiation of a vegetative cell into a spore is now amenable to successful analysis.

THERE ARE NOW MANY REASONS TO INTENSIFY WORK ON ORGANISMS LIKE YEAST

The very concentrated effort which has gone into the study of all aspects of *E. coli* is one of the major reasons why molecular biology has advanced so rapidly over the past two decades. Clearly, similar attention will soon be placed on one or more types of eucaryotic cells. For many reasons it is natural that much emphasis must go toward the study of several types of human cells. But at the same time, it may be wise to concentrate equally on the molecular biology of one or two of the simplest eucaryotes. Several reasons dictate this approach. One is that these microorganisms most certainly contain much less DNA (Figure 17–1) than human cells. Only a three-to-tenfold increase in genetic complexity is noticed in escalating to yeast or *Aspergillus* from *E. coli*. A second reason is economic: Work with higher cells is at least an order of magnitude more expensive than with microorganisms. If a choice exists between solving the same problem with human tissue culture cells or with yeast or *Neurospora*, common sense tells us to stick with the simpler system. A third, and perhaps the most important reason, is the ease with which detailed genetic analysis can be applied to many microorganisms. Despite the great advantages now brought about by the cell-fusion technique (see below), detailed genetic analysis of human cells will be extraordinarily difficult to bring about.

Thus, even if our primary interest is the human cell, this is the time for many more biologists to work with organisms like yeasts. They grow rapidly as single cells, exist in either the haploid or diploid state, and are already very well studied genetically. Mutations are as easy to obtain as with *E. coli*, with the various genes of yeast cells being located on some 17 different chromosomes. Since the total amount of DNA per haploid chromosome set is only 3 times that of the *E. coli* chromosome, the average yeast DNA molecule must be much shorter than the $\sim 1200\mu$-long *E. coli* DNA molecule. Recent EM views of yeast DNA molecules indicate MW's ranging from 10^8 to 10^9 (Figure 17–5), compatible with the idea that each yeast chromosome, like the *E. coli* one, contains only one DNA molecule. All the observed molecules appear linear, with no evidence for any circular forms. Genetic analysis so far has always yielded linear linkage maps. So if ring forms exist, they must occur only very transiently at some very specific step in the life cycle.

Unfortunately, good techniques have not yet been developed for the isolation of intact yeast nuclei. So we are still in the dark about the molecular composition of yeast chromosomes and the molecular parameters of newly made RNA. However, nuclear pores like those found in higher eucaryotes do exist; and most likely newly made RNA moves through them to reach the cytoplasm. Unlike with all eucaryotes, the nuclear membrane does not break down during mitosis. Nevertheless, spindles form in the dividing nuclei, with microtubules aligned along the direction of chromosome separation (Figure 17–6). Since large numbers of different mutations exist which block specific stages of the cell cycle, the molecular events underlying chromosome separation may soon become much more amenable to serious analysis.

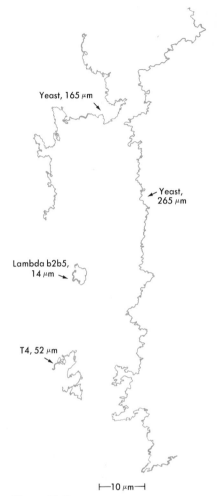

Figure 17–5
Tracing of electron micrographs of two yeast DNA molecules, together with reference λ (MW 32×10^6) and T4 (MW 1.2×10^8) DNA's. [Reproduced with permission from Petes *et al.*, *Cold Spring Harbor Symp.*, **XXXVIII**, 9 (1974).]

Figure 17–6
An electron micrograph of a thin section of a budding yeast cell, showing microtubules (mt) aligned along the spindle (sp). The dense bodies at the poles of the spindle are called the spindle plaques. They may serve as organizing points from which the polar microtubules grow out. [Reproduced with permission from Byers and Goetsch, *Cold Spring Harbor Symp.* **XXXVIII**, 128 (1974).]

REVERSIBLE STATES OF THE SLIME MOLD CELL

Clearly the most interesting microorganisms to the embryologist are those in which cell differentiation is an important aspect of the life cycle. Thus, much attention has recently been given to a cellular slime mold, *Dictyostelium discoideum,* whose life cycle is illustrated in Figure 17–7. The cycle begins with spores germinating to yield amoeboid cells called myxoamoebae. These irregularly shaped cells behave like small amoebae, living largely on bacteria and reproducing by fission. When their food supply diminishes, they begin to actively secrete cAMP into the surrounding environment. There it acts as a chemical attractant that causes the starving cells to move toward each other and form conical masses which eventually topple over on their sides. These slug-like bodies then slowly move along the surface before coming to rest. Later, they form a fruiting structure made of a basal disc, a stalk, and a mass of spores.

Figure 17–7
The life cycle of the slime mold *Dictyostelium discoideum.*

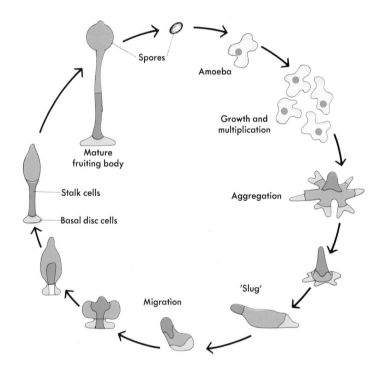

All these components arise by the direct transformation (differentiation) of amoeboid cells into more specialized cells. The type of cell to which they change depends on the order in which they become part of the initial aggregation. The first cells to come together form the lower stalk, the next group changes into upper stalk and spore cells, while the basal disc is formed from the last cells to come together. With time, all the disc and stalk cells die, leaving the spores capable of initiating a new life cycle when favorable nutritional conditions reappear. Except for the final steps of spore formation, all of the foregoing examples of differentiation are completely reversible. For example, when stalk cells are isolated from the fruiting body before cell disintegration begins, they may transform back to the amoeboid phase and subsequently divide by fission.

The synthesis of many key enzymes in the *Dictyostelium* life cycle is restricted to particular stages, with each specific enzyme synthesis requiring the prior synthesis of its respective RNA template. Only half the RNA species synthesized in the undifferentiated, dividing amoeboid cells are found in highly differentiated cells. Correspondingly, the differentiated cells possess RNA species not found earlier in development. However, as yet no one has any real clues as to how these changes in transcription products are brought about. Preliminary work with RNA polymerases I and II has not revealed any obvious modifications of their respective polypeptide subunits during development. These experiments, however, are still too incomplete to allow us to distinguish between factors that directly affect DNA and changes that occur in the polypeptides that make up the various forms of RNA polymerase.

TRANSCRIPTION AS A MEASURE OF BIOLOGICAL TIME

That eucaryotic microorganisms tend to divide at constant intervals suggests that eucaryotes as well as procaryotes have accurate clock mechanisms at the cellular level. Earlier we saw that bacterial viruses can use the distance between genes for timing, a consequence of the relatively slow rate at which RNA polymerase molecules move along DNA templates (DNA replication occurs some 100 times faster!). At 37° C the rate of movement of *E. coli* RNA polymerase is only 30 to 40 nucleotides per second, and so a complete transcription of a λ molecule by a single polymerase molecule would require over 25 minutes. Since several operons are transcribed simultaneously, this much time is not required for a single cycle of replication. Nevertheless, transcription of the very long "late" λ operon by itself counts out some 15 minutes.

Similar analysis applied to the *E. coli* chromosomes gives a total possible transcription time of 33 hours. Thus only 1 percent of its total genome would have to be used to code for periodic events occurring once every 20-minute division cycle. And even the 3 to 4 hours required for a spore to germinate could easily be directly measured out on a DNA tape. There is, of course, no reason why such relatively long intervals need be coded along one contiguous DNA section. Several operons, each coding for a unique specificity factor necessary for the reading of the subsequently transcribed operon, would also do the job (Figure 17–8).

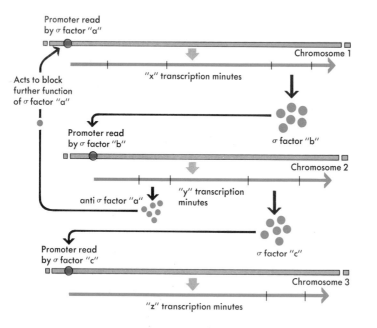

Figure 17–8

A mechanism by which sequential synthesis of different specificity factors might be used to count time.

Most likely all such biological clocks will also depend on the regular appearance of anti-specificity factors, that is, gene products that specifically prevent certain specificity factors from functioning. Such factors provide the most direct way of limiting the synthesis of specific proteins to restricted periods in a life cycle. That they exist is already known from the life cycle of the phage T4 (see page 417) where at least one viral gene codes for a product that inhibits the functioning of an *E. coli* σ factor. The obviously much more complicated bacterial life cycles must involve a number of different anti-specificity factors, each working at a different time to specifically block the functioning of certain genes.

Though the division cycles of higher cells are much longer, their DNA contents are correspondingly much greater; one polymerase molecule would require some 1000 days at 37° to completely transcribe the haploid complement of human genes. Thus, much of the timing required for a 24-hour human cell cycle is easily imagined to be at the transcription level, as may be the timing for many of the crucial steps in the embryological development of all higher organisms.

HIGHER CHROMOSOMES

Fundamental knowledge about the structure of the chromosomes of higher plants and animals is very meager. Still very confusing is the relation of individual DNA molecules to the chromosomes visible in microscopes. With the DNA viruses and bacteria, the answer is very clear: The chromosome is a pure DNA molecule, often of circular form. But with higher chromosomes the truth is much harder to reach, partly because so much more DNA is present within single chromosomes and also because the DNA is always complexed with protein to form chromatin. Only with *Drosophila melanogaster* do we have any firm answers. Here each chromosome contains a single linear double helix, with the largest molecules having MW's about 4×10^{10}. Each DNA molecule runs from one end of the chromosome to the other, independent of whether the centromere, the attachment site for the kinetochore, is located centrally (a metacentric chromosome) or whether it is terminally sited (a telocentric chromosome). The kinetochore is thus a body which attaches laterally to DNA, an inference also supported by EM observations (Figure 16–22).

Generally, chromosomes are visible in the light microscope only when they contract during mitosis and meiosis. In the EM the resulting compacted masses show enormous yet reproducible complexity, dominated by the occurrence of spaghetti-like masses of chromatin fibers. For the most part these have regular 200 to 250-Å diameters, though some sections have abruptly narrower 100 to 120-Å diameters (Figure 17–9). Most likely both these arrangements reflect bending of the double-helical DNA into at least one (100-Å fiber) or two (200-Å fiber) order(s) of superfolding.

Equally elusive until recently has been the structure of the much more extended interphase chromatin. Not only is its irregular conformation impossible to follow by EM sectioning techniques, but when it is extracted from cells it easily aggregates to form artificial arrangements unrepresentative of its true intracellular state. Now careful

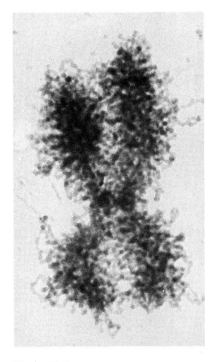

Figure 17–9
Electron micrograph of the two chromatids of human chromosome 16 isolated during metaphase prior to complete replication of the centromere. [Reproduced with permission from E. J. DuPraw, *DNA and Chromosomes*, Holt, 1970.]

511

efforts reveal a beaded appearance (see page 480) built up from 100-Å-diameter particles (nucleosomes) in which the double-helical DNA is tightly compacted. From some interphase chromatin fibers, loops of equal diameter appear to emerge. These loops are not thrown off at regular intervals but are irregularly spaced, often with 1 to 5 loops for every 10μ of fiber length (Figure 17–10). As they are an integral *in vivo* feature of both lampbrush and polytene chromosomes (see below), the belief is emerging that loops are an important feature of all chromatin.

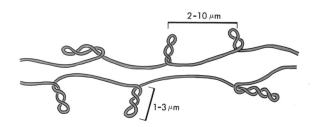

Figure 17–10
A drawing showing how the isolated and salt-cleaned chromatin from *Drosophila* egg nuclei appears in the electron microscope. Regions of comparatively straight sections (about 60 Å thick) are interspersed with "loops" every 2 to 10 μm. This does not show how the various chromatin proteins (i.e., histones and acidic proteins) are arranged in relation to the DNA.

REPLICATION OF DNA STARTS AT A NUMBER OF DIFFERENT SITES ALONG A GIVEN CHROMOSOME

In all eucaryotes, DNA replication is initiated at a large number of different sites along a given chromosome. Each initiating event creates two replicating forks which move in opposite directions to create replicating bubbles that eventually merge with adjacent replicating bubbles (Figure 17–11). The number of starting sites within a chromosome is a function of the stage in the respective life cycle. For example, DNA synthesis commences at many more starting sites during the cleavage divisions of *Drosophila* development than in growing adult cells. During each cleavage division, only some 3.5 minutes are required to replicate the *Drosophila* chromosomes, while over 200 minutes are needed for this process in *Drosophila* cells growing in culture. This difference reflects the number of sites where duplication begins, not the rate at which the replicatory

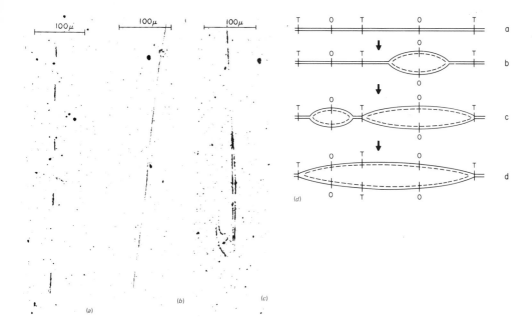

Figure 17–11
Bidirectional growth of a mammalian chromosome. (a) An autoradiograph of a Chinese hamster DNA molecule from a cell which has
briefly been labeled with tritiated thymidine. The tandem arrays of exposed grains indicate the existence of several replication points in the
portion of the DNA fiber under view. (b) Tandem arrays seen after a
pulse exposure to label followed by a chase period in nonradioactive
medium. Here the grain density declines from the center to the ends,
suggesting that the growing points move in opposite directions. (c) An
example of sister replicating molecules held together by a replicating
fork. (d) A schematic model interpreting pictures (a) through (c). [The
photographs in (a), (b), and (c) are courtesy of J. Huberman and originally appeared in Huberman and Riggs, *J. Mol. Biol.* **32,** 327 (1968).]

forks move. In both cases, at 25° they travel an average of
2600 base pairs per minute. Within the very fast replicating (once every 9.5 minutes) cleavage nuclei, DNA replication starts at some 30,000 to 50,000 different "open"
sites, each usually separated by some 3000 to 4000 base
pairs from adjacent origin sites. In adult cells, however,
fewer "open" sites exist (∼3000), and few replicating bubbles start closer than 30,000 base pairs apart.

Now hints exist that, within a given cell, replication
is initiated almost simultaneously at each "open" site. The
total time span of intracellular DNA synthesis will thus be

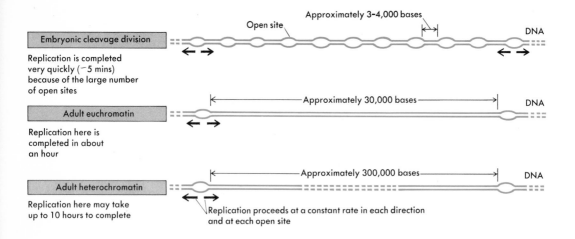

Figure 17–12
Schematic picture of DNA replication along a section of a *Drosophila* chromosome.

determined by the largest distance between adjacent "open" sites (Figure 17–12). Very few "open" sites exist within the highly compacted heterochromatic regions (see below) found near the centromere; and so the last DNA to replicate in adult cells is located adjacent to the centromere. Compacting of centromeric DNA, however, does not occur in cleavage nuclei, and so here this DNA replicates simultaneously with other DNA. However, as embryological development proceeds, the centromeric regions become compacted, and the period required for DNA synthesis grows correspondingly much longer.

Once an "open site" has been replicated, it cannot serve as an origin for further DNA replication until the next cell cycle. The molecular basis for this phenomenon is still a total and very important mystery. Conceivably a chemical change (e.g., methylation), that occurs concomitantly with initiation, blocks further replication until the next cell cycle, when a further molecular change removes the obstructing factor. Or this control mechanism may operate through changes in specific chromosomal proteins.

ACTIVE (EUCHROMATIC) VS. INACTIVE (HETEROCHROMATIC) CHROMOSOMAL REGIONS

The term *heterochromatin* is frequently used to designate highly compacted chromatin which remains visible during interphase. Early genetic experiments showed that such regions contained very few genes, giving rise to the idea that heterochromatic DNA does not have a genetic role. This inference was solidly confirmed by the finding that such compacted DNA is not transcribed into RNA. Correspondingly, the much more open chromatin (*euchromatin*) that is not normally visible in interphase, and which genetics indicates to be full of active genes, is vigorously transcribed into RNA chains. Initially it seemed possible that the nonfunctional heterochromatic regions might contain more histones, thereby being less accessible for RNA polymerase binding. But the limited measurements so far made reveal almost equal histone binding to both heterochromatin and euchromatin.

There are many instances when a gene loses activity because of a chromosomal rearrangement which inserts it into a heterochromatic region. This suggests that control devices exist which can turn on or off the functioning of a very long section of a given chromosome. This point, first demonstrated for *Drosophila,* holds equally well for mammalian systems.

The most striking findings concern the sex chromosomes. About 15 years ago, the unexpected discovery was made that in female mammals the two homologous X chromosomes look quite different. One always appears highly condensed (hinting that it does not function), whereas the other is extended. This suggestion is confirmed by biochemical analysis, which shows that only the genes on one of the two X chromosomes of a given female cell are active. Surprisingly, the inert chromosome varies from one cell to another; so female tissue is in reality a mosaic containing mixtures of two different cell types. Though the molecular basis for this bizarre condition is still unknown (it appears to be restricted to the X chromosome), it is quite important in showing that there are devices that can specifically block the functioning of an entire chromosome.

While the other eucaryotic chromosomes do not show all-or-none effects, they can also have extended sections of heterochromatin whose length depends upon when in the life cycle they are examined. Most important, many regions that are completely heterochromatic in later life appear as euchromatin early in development. Thus they most likely contain genes programmed to function in one of the early embryological stages. The function of the DNA adjacent to the centromere, which becomes permanently heterochromatic as soon as the cleavage divisions cease, is still a mystery. No RNA is made off it, either during embryological development or later in adult life. Yet it almost always amounts to some 5 to 10% of total chromosomal DNA. Conceivably it becomes compacted to slow down duplication of the DNA section immediately adjacent to the centromere, so that it replicates only just before the centromere divides during metaphase.

LAMPBRUSH CHROMOSOMES

Many fundamental insights have come from study of the seemingly very large chromosomes found in vertebrate oocytes just before the first meiotic division. At this stage (which can last hundreds of days), oocytes synthesize most of the mRNA that functions after fertilization during the early stages of development. To accomplish this selective synthesis, oocyte chromosomes are in a state of partial contraction, with most of their DNA compacted into tight masses (chromomeres) but with a small fraction extended as very long lateral loops from the main chromosomal axis (Figures 17–13 and 17–14). Hence the name "lampbrush chromosomes."

Most likely one DNA molecule runs from one end of the chromosome to the other. DNA within the chromomeric masses is inactive, while the looped-out regions of DNA are active sites of RNA synthesis. Transcription usually starts at one specific site close to the main chromosomal axis, so increasingly long RNA chains are produced as RNA polymerase molecules move along a given loop (Figure 17–15). Such loops thus have polar appearances. Soon after transcription begins, specific proteins attach to the nascent RNA chains, often to produce spherical masses of ribonucleoprotein particles attached to the respective loops. Such loops can easily be distinguished from their neighbors not only by their very reproducible lengths but also by the specific appearance of their ribonucleoprotein products.

The fact that so many loops have distinctive morphologic features suggests that each loop may in fact be a single functional unit (gene) that codes for a single primary polypeptide product. So the number of loops seen in oocyte lampbrush chromosomes (~600) may provide a minimum estimate of how many genes function during oocyte maturation. In contrast, the number of chromomeric masses along the chromosomal axis may not give us particularly meaningful information, since there is no way of telling how many genes have folded up to form a single chromomere.

Lampbrush-like chromosomes are also seen in the growing spermatocyte precursors of mature sperm. Here, however, only the male sex (Y) chromosome assumes a prominent lampbrush configuration, with the autosomes and X chromosomes forming only tiny loops. Unlike somatic cells, whose Y chromosome always looks tightly compacted (heterochromatic), in spermatocytes it throws off several very large and morphologically distinctive loops. In *Drosophila hydei,* for example, five different loops

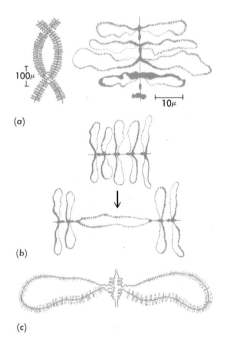

(a)

(b)

(c)

Figure 17–13
Diagrammatic view of lampbrush chromosomes. (a) The two homologous meiotic chromosomes (left) are joined together by two chiasmata. A portion of the central chromosome axis (right) shows that two loops with identical morphology emerge at a given point, evidence that each chromosome has already split into two chromatids. (b) Accidental stretching of a chromosome reveals the continuity of the loop axis with the central axis (compare with Figure 17–14). (c) A single loop pair, showing the single DNA molecules upon which RNA chains are being made. [From J. Gall, *Brookhaven Symp. in Biol,* **8,** 17 (1955).]

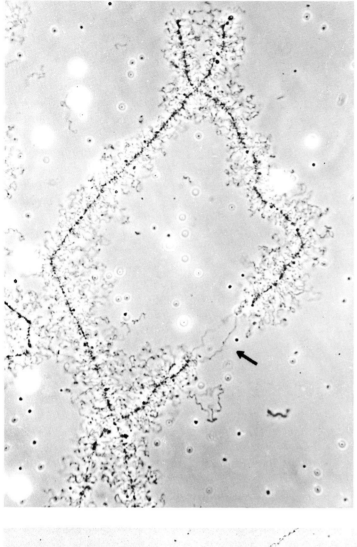

Figure 17–14
A partial view of a pair of highly extended lampbrush chromosomes from an oocyte of the newt *Triturus viridescens*. Two chiasmata are seen in this photograph, taken in the diplotene phase of meiosis. The arrow points to a section where the double-stranded character of each premeiotic chromosome is revealed. [Courtesy of J. Gall, Yale University.]

Figure 17–15
A section along a loop of a *Triturus* lampbrush chromosome, showing a gradient of increasingly long RNA chains attached to their DNA template. The black dots at the sites where the RNA chains attach are RNA polymerase molecules. [Courtesy of O. L. Miller and B. Beatty, Oak Ridge National Laboratory.] ▼

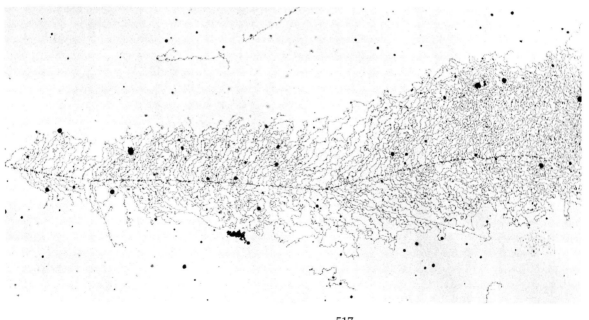

can easily be identified (Figure 17–16), each probably representing one fertility gene.

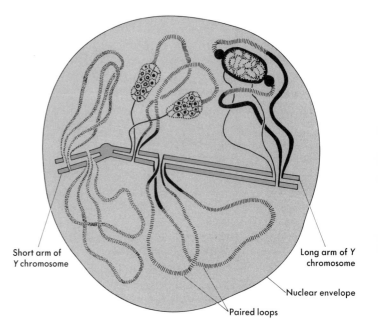

Short arm of Y chromosome

Long arm of Y chromosome

Nuclear envelope

Paired loops

Figure 17-16
The *Y* chromosome of *Drosophila hydei* during the lampbrush phase in spermatogenesis, showing the light-microscope appearance of the several loops, each morphologically distinctive. Each loop probably represents one fertility gene. [Redrawn from Hess, *Cold Spring Harbor Symp.*, **XXXVIII**, 663 (1974).]

POLYTENE CHROMOSOMES

The salivary glands of *Dipteran* flies like *Drosophila* and *Chironomus* contain giant chromosomes which are visible in interphase, having a banded appearance arising from the alternation of compacted chromomeric DNA with extended interchromomeric DNA (Figure 17–17). Band sizes vary tremendously. Some are so thin that they are hardly visible, whereas others occupy several percent of the length of a given chromosome. Each such giant chromosome consists of a large number of partially replicated chromosomes neatly stuck together in lateral array (a *polytene chromosome*). They arise by repeated cycles of DNA duplication, uncomplicated by cell division, in which most chromosomal sections are replicated some 10 times to yield 2^{10} copies. Only the DNA which has a genetic role is so duplicated, with the centromeric DNA remaining unreplicated (Figure 17–18). While some 1000 copies of most genes are produced, those coding for ribosomal RNA are duplicated less often, most likely because in ordinary chromosomes (see page 528) they are already tandemly present in large numbers.

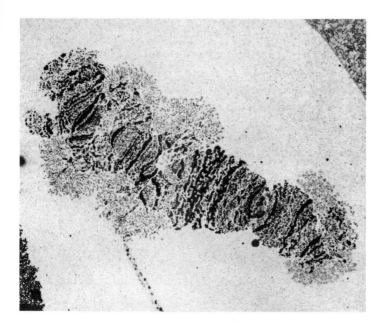

Figure 17-17
Electron micrograph of chromosome 4
from a salivary gland of *Chironomus tentans.*
[Reproduced with permission from
B. Daneholt, *Cell,* **4**, 1 (1975).]

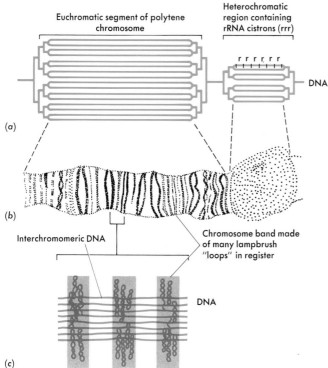

Figure 17–18
A diagrammatic view of the possible construction of a typical polytene chromosome
such as that from a *Dipteran* salivary gland
cell (b). The DNA with a genetic role is
duplicated up to one thousandfold (a),
while the heterochromatic region coding
for ribosomal RNA has fewer multiple
copies. (c) The chromomeric banding pattern seen on such chromosomes probably
arises from the parallel in register matching
of numerous lampbrush loops in each DNA
copy.

The structure of the DNA within the compacted chromomeres has proved very difficult to elucidate by electron microscopy. Now newer techniques, which remove much of the chromosomal proteins, suggest that a polytene chromosome should be regarded as a lateral array of lampbrush chromosomes in which homologous loops specifically line up together (Figure 17–18). What brings the homologous loops into such exact register is still unknown.

PUFFING

When genetic DNA is tightly compacted to form well defined bands, it is in a nonfunctional state, since autoradiographic experiments using radioactive precursors of RNA show that very little RNA is synthesized upon banded DNA. In contrast, when DNA unfolds into open loops like those of lampbrush chromosomes, it is actively transcribed into RNA (Figure 17–19). The collective groups of open loops found in polytene chromosomes are called *puffs*, and the larger and more diffuse they appear, the higher the rate of transcription. Most importantly, the locations of puffs do not remain constant during embryological development. Instead, some genes function only during very specific developmental stages.

Among the best studied puffs are those of the midge *Chironomus*, whose salivary glands synthesize and secrete the structural proteins from which the larvae spin silk-like thread. All the seven major secretory polypeptides have

Figure 17–19
Electron micrograph of a portion of a loop within the Balbiani Ring 2 of chromosome 2 of a salivary gland cell of *Chironomus tentans*. It shows spherical ribonucleoprotein granules attached by "stalks" to the thin chromatin thread (arrow). Almost simultaneous with RNA synthesis, specific proteins attach to the nascent RNA chains, leading to the formation of ribonucleoprotein granules whose shapes reflect the secondary structure of their underlying RNA chains. [Reproduced with permission from B. Daneholt, *Cell,* **4,** 1 (1975).]

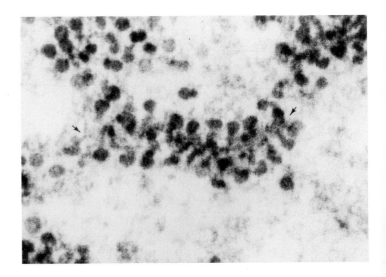

relatively high molecular weights, ranging from 1.5×10^5 to 5×10^5; and several already have been assigned to the very largest puffs that are called *Balbiani rings* (Figure 17–20). These special puffs are transcribed some 100 times more frequently than the average puff, most likely in order to produce the mRNA needed as templates for very large amounts of the silk-like proteins needed during the larval and pupal stages of insect development.

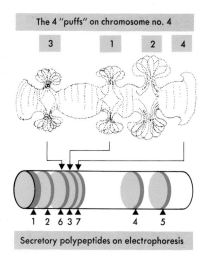

The 4 "puffs" on chromosome no. 4

Secretory polypeptides on electrophoresis

Figure 17–20
The correlation between specific gene products (salivary secretory polypeptides) and regions of active RNA transcription in Balbiani rings or chromosomal "puffs," on chromosome 4 of the salivary gland cells of *Chironomus*.

THE NUMBER OF *DROSOPHILA* GENES CORRESPONDS TO THE NUMBER OF SALIVARY CHROMOSOME BANDS

When they were first seen in the early 1930's, the bands along the salivary chromosomes were thought to be the individual genes, with the number of genes equal to the number of bands. The *Drosophila* gene number thus estimated was slightly over 5000. However, this line of reasoning became openly questioned when it became possible to measure the amount of DNA within single bands. Then it became clear that the average band contained some 30,000 base pairs, or at least 20 times more than is necessary to code for an average sized polypeptide chain. So the alternative possibility was proposed that each band might be analogous to a large bacterial operon like that for histidine, which codes for some 10 distinct yet functionally related proteins. Now, however, recent genetic data make the "band = a large operon" hypothesis most unlikely. For example, exhaustive genetic mapping of the 3A1–3C3 region (15 bands) of the X chromosome of *Drosophila melanogaster* reveals 18 complementation groups (genes) (Figure 17–21). Already, most complementation groups can be assigned to single specific bands and, considering inherent ambiguities in separating closely positioned bands from each other, it is possible that each band in this region codes for only a single functional unit. Likewise, mapping of the very small chromosome 4 of *Drosophila* indicates some 43 functional groups, while the corresponding cytological analysis indicates a minimum of 33

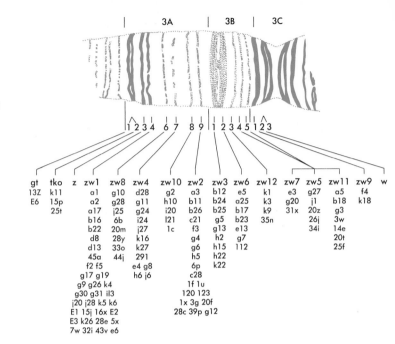

Figure 17–21
Correlation between the number of distinct bands and complementation groups in the 3A1–3C3 region of the X chromosome of *Drosophila melanogaster*. [Redrawn from Judd *et al.*, *Genetics*, **71**, 13a (1972).]

and a maximum of 50 different bands. Thus, either most *Drosophila* proteins are very, very large or most DNA within a gene does not code for amino acid sequences. A similar story most likely holds for the genes expressed in the lampbrush chromosomes of oocytes. Minimum estimates of the DNA found in loops give lengths (10μ to 100μ) much, much longer than are needed to code for the average sized protein.

THE VERY, VERY LONG TRANSCRIPTION PRODUCTS OF SINGLE CHROMOMERES (GENES)

All the DNA within a single lampbrush (polytene) loop is usually transcribed into a single RNA chain. The very, very long lengths of such products can be directly seen in electron micrographs of lampbrush loops (Figure 17–15), or indirectly measured by their rapid sedimentation in the ultracentrifuge. As far as is now known, almost all the messenger RNA of multicellular eucaryotes is initially laid down as part of such very large molecules. These primary transcripts vary greatly in size (5000 to 50,000 nucleotides), depending on the lengths of their respective chromosomal loops, and were first called heteronuclear RNA (hnRNA). Because we know them to be the precursors of messenger RNA, they should now be called pre-mRNA.

TRANSFORMATION OF PRE-mRNA INTO mRNA

Pre-mRNA exists only fleetingly in the nucleus, for it is rapidly broken down into much smaller molecules. Each pre-mRNA molecule usually gives rise to only one mRNA molecule, with most of each pre-mRNA chain being enzymatically degraded to free nucleotides. The RNA section which escapes degradation is always located at the 3' end, and so is the last part of a given pre-mRNA molecule to be synthesized (Figure 17–22).

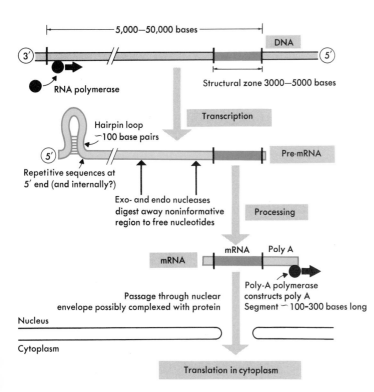

Figure 17–22
Simplified scheme outlining the steps involved in messenger RNA transcription in eucaryotic cells.

Why only this 3'-end region escapes enzymatic destruction is not known. Conceivably it is protected by combination with specific proteins. But the answer is probably not this simple, since it appears that the pre-mRNA sections destined for rapid breakdown are also complexed to protein as soon as they are synthesized. Nor does protection arise simply from the 200-nucleotide-stretches of Poly A that are post-transcriptionally added to the 3' ends of those mRNA molecules programmed for movement to the cytoplasm (see page 483). Histone mRNA lacks such Poly A termini, yet otherwise functions normally.

The fraction of the average pre-mRNA molecule that eventually reaches the cytoplasm is usually less than 10% (Figure 17–22), thereby raising the question of the function played by the great lengths of quickly discarded RNA. Most such sequences are unique and are not found within any other pre-mRNA species. Interspersed within such seemingly random arrangements are repetitive sections found in many other pre-mRNA species. Such repetitive sequences, which can comprise up to 20% of a given pre-mRNA molecule, are some 300 to 500 bases long and occur not only at the 5' end but also internally, conceivably many times, in the largest pre-mRNA molecules. The number of DNA equivalents of a given pre-mRNA repetitive sequence in a cell varies greatly: some may be present only 20 to 100 times, others are found at many thousands of chromosomal sites. It is tempting to ascribe these sequences common to many 5' (promoter) ends to some control function which modulates the coordinate expression of several functionally related genes. Why repetitive sequences need to occur internally is hard to hypothesize about. Some are made up of inverted (palindromic) sequences which spontaneously form tight (100-base-pair-long) hairpins (Figure 17–22). On the average, each pre-mRNA molecule contains one such hairpin, and so they may have a necessary role. Hints exist that they may lie adjacent to the 5' end of the mRNA component, suggesting a processing involvement.

SINGLE COPIES OF THE HEMOGLOBIN GENES PER HAPLOID COMPLEMENT

The number of DNA–RNA hybrid molecules that are formed when purified mRNA molecules are mixed with cellular DNA provides a method for measuring the number of times a given gene appears within a haploid genetic complement. Such measurements recently became possible with the isolation of specific mRNA molecules from cells differentiated for the selective synthesis of specific proteins. For example, the $\sim 9S$ mRNA molecule that codes for the several similarly sized polypeptides found in the various forms of hemoglobin has been used to show that only one copy of each of the respective hemoglobin genes is present per haploid genetic complement. This conclusion holds not only for the egg and sperm but also for the cells (*erythroblasts*) differentiated for the almost exclusive synthesis of hemoglobin. Likewise, the gene coding for the silkworm protein fibroin is present in only one copy per haploid complement, even in cells actively making silk. It seems likely that these examples will not prove exceptional, and that most genes can be transcribed at such rapid rates that usually only single copies are necessary.

MULTIPLE COPIES OF THE HISTONE GENES

Single copies of the various histone genes, however, are insufficient to produce the vast numbers of histone mRNA molecules needed during the cleavage period of early embryogenesis. Concomitantly with the very rapid rate of DNA synthesis, there is a massive requirement for newly synthesized histones. Thus, during the cleavage divisions virtually all the mRNA molecules in the 9S range code for the various histones. These ~9S histone mRNA molecules resolve under acrylamide gel electrophoresis into three size classes (Figure 17–23). The fastest migrating fraction codes for histone H4, the smallest histone; the heterogeneous middle-sized group codes for the histones H3, H2b, and H2a; while the slowest migrating histone codes for H1, the largest histone. Each of these various mRNA species is repeated some 250 to 500 times per haploid complement.

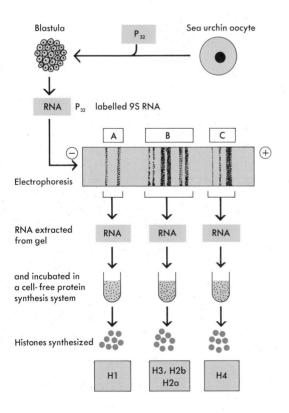

Figure 17-23
Diagram showing the demonstration that the 9S mRNA molecules in the sea urchin embryo code largely for the enormous number of histone molecules needed during cleavage. The 9S mRNA is fractionated by polyacrylamide gel electrophoresis.

HIGHLY REPETITIVE DNA SEQUENCES
NEAR THE CENTROMERE

The chromosomal region adjacent to the centromere is composed of very long blocks of *highly repetitive DNA*, in which very simple sequences are repeated hundreds if not thousands of times. In many cases these repeating simple sequences have base compositions unlike that of most DNA, and so are easily separated by centrifugation in a CsCl gradient. DNA separated by this process is referred to as *satellite DNA*. Often several different satellites of centromeric origin exist in a given species. The 3 different ones already isolated from *Drosophila virilis* have highly related sequences. One has a $^{5'}$ACAAACT$^{3'}$ basic repeat; a second contains $^{5'}$ATAAACT$^{3'}$, while the sequence repeated in the third satellite is $^{5'}$ACAAATT$^{3'}$. Why several sequences have so evolved is not clear. Since blocks of each sequence are usually located around the centromeres of more than one chromosome, none of these satellites is a completely pure repeat of the basic sequence. Other bases are occasionally present, most likely as the result of past mutational events. Often satellites present in one species appear to be absent in a closely related species. Usually, however, such differences are not meaningful, since a change in just one base of the fundamental repeating unit can sometimes convert the respective satellite density to one closely approximating that of cellular DNA.

In *Drosophila* some 25% of the total DNA is of the simple sequence variety, with most higher eucaryotes having 10 to 20% of their genomes as such DNA. It is never transcribed, most likely because its restricted nucleotide sequences lack promoter sites on which RNA polymerase can initiate RNA chains. In contrast, there must exist nucleotide sequences which recognize DNA polymerase, since centromeric DNA is replicated as rapidly as "unique sequence" DNA during the early cleavage divisions. They, however, need be present only once every 3000 base pairs, and so are undetectable using current sequencing methodologies.

VARIATION IN DNA AMOUNTS BETWEEN
CLOSELY RELATED SPECIES

After the coding significance of DNA became understood, the expectation existed that as we went from simple to more complicated organisms, the amount of DNA per haploid genome would increase in proportion to the increase in true biochemical complexity. But this is not always the case. While higher plants and animals have much more

DNA than the lower forms (Figure 17–1), no one anticipated the finding that certain fish and amphibia would have 25 times more DNA than any mammalian species. And as more and more animals and plants were examined, closely related species were sometimes found to vary in their DNA contents by a factor of five to ten (Figure 17–24).

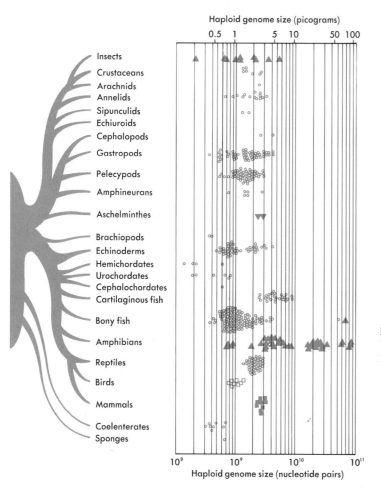

Figure 17–24
Variation in the genome sizes among related members of evolutionarily related animal groups. [Redrawn from Britten and Davidson, *Quant. Rev. Biol.*, **46**, 111 (1971).]

These variations usually do not arise by increases in chromosome number but, instead, by increases in the amount of DNA per chromosome. When this happens, the amounts of simple sequence, mildly repetitious, and unique sequence DNA all increase, sometimes in roughly proportionate amounts. As yet, however, the exact nature of the DNA within these fractions has not been worked out. For example, we do not know whether the salamander

Necturus, which has some 25 times more DNA than the toad, will also be found to contain only single copies of the various hemoglobin genes. The fact that it has proportionately larger lampbrush loops hints that this is not so; but until hard facts are available, the phenomenon of rapid shifts in total chromosome size will remain an elusive mystery.

Our only real clue may be the fact that, in closely related plants, where the DNA content can vary by a factor of ten, the amount of DNA correlates with life span. Smaller values characterize short-lived annual plants that live only 10 to 20 days, while their perennial relatives generally have the higher DNA contents. Conceivably, some loop lengths have a timing function. For example, inserting some 100,000 base pairs between the 5' and 3' ends of a pre-mRNA molecule means that some 50 minutes have to pass at 25° C before a traversing RNA polymerase can produce a required mRNA product.

THE NUCLEAR LOCATION OF rRNA SYNTHESIS

All eucaryotes have at least 50 to 1000 identical copies of the genes coding for rRNA. They are always tandemly arranged in large DNA blocks sited on one or more specific chromosomes (Figure 17–25). Such DNA is usually looped off the main chromosomal fiber masses as highly extended threads, which coalesce with specific proteins to form the nucleoli. Some organisms have only one nucleolus per haploid complement, while others have several nucleoli. In every case, however, the nucleoli are attached to specific chromosomal sites. Within each nucleolus the tandemly arranged rRNA genes, which have molecular weights of $\sim8 \times 10^6$, are transcribed separately into pre-rRNA molecules sedimenting at 45S and having molecular weights of $\sim4 \times 10^6$.

These 45S molecules are precursors for the 28S and 18S rRNA chains found in all eucaryotic ribosomes (Figure 17–26). After release from its DNA template, the 45S molecule is broken down in several stages (Figure 17–25) to yield the rRNA components present in ribosomes. Surprisingly, almost half of the initial 45S precursor does not end up as rRNA. The discarded portions are degraded to the nucleotide level by intracellular nucleases.

Within the nucleoli, the 28S chains combine with newly made ribosomal proteins to form the large (60S) ribosomal subunits. Synthesis of the ribosomal proteins does not take place in the nucleoli or, for that matter, any-

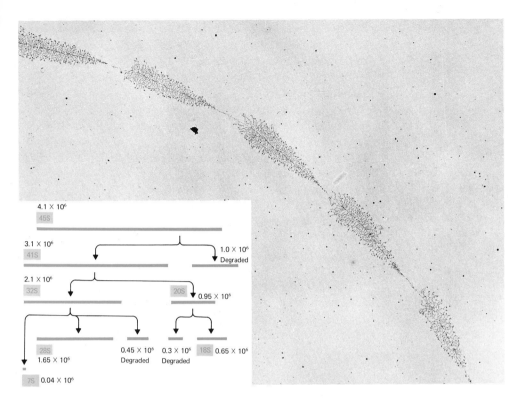

Figure 17–25
A tandemly arranged series of nucleolar rRNA genes from *Triturus viride-scens* showing growing 45S rRNA precursor chains. The insert at the lower left shows the steps in the transformation of the 45S molecule into mature mammalian molecules. [Courtesy of O. L. Miller and B. Beatty, Oak Ridge National Laboratory.]

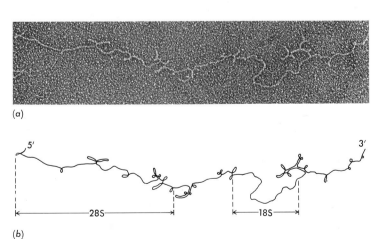

(a)

(b)

Figure 17–26
(a) Electron micrograph of a 45S pre-rRNA molecule spread in 80-percent formamide and 4M urea to preserve the stronger hairpin loops. (b) A tracing of the above which shows the regions destined to become 28S and 18S rRNA. [Reproduced with permission from Wellauer and David, *Cold Spring Harbor Symp.* **XXXVIII**, 325 (1974).]

where in the nucleus. Instead, they are made on cytoplasmic polyribosomes, afterwards somehow migrating into the nucleoli. In contrast, the smaller (40S) eucaryotic ribosomal subunits are apparently not assembled in the nucleoli. Whether they form in other parts of the nucleus or in the cytoplasm remains to be worked out. Similarly unclear is the form in which mRNA reaches the cytoplasm (as free mRNA, or attached to 40S subunits, or conceivably bound to still undiscovered proteins whose primary function is the transport of mRNA from the nucleus to the cytoplasm).

Each rRNA gene is separated from its tandem partners by well defined lengths of nontranscribable spacer DNA that may be almost as long as its adjacent genes. While a given species has spacers with very similar sequences, very little homology may exist between those of closely related organisms, indicating that there may be no selective pressure to maintain any given spacer sequence. Why all the spacers within a given species are often identical is probably related to the still unsolved mechanism which keeps the sequences of all the rRNA genes in a species effectively identical.

SELECTIVE MULTIPLICATION OF THE rRNA GENES WITHIN OOCYTES

The number of rRNA genes within vertebrate oocytes is about a hundred to a thousandfold greater than within other vertebrate cells. But this extra DNA does not result from an increase in chromosome number from the 4N characteristic of oocytes prior to the first meiotic division. Instead, almost all the increase is due to large numbers of extrachromosomal nucleoli. Each contains a single circular DNA molecule with length varying from 20 to 1000μ. Correspondingly, large numbers of tandemly arranged rRNA genes are present, each coding for the 45S rRNA precursor. Figure 17–27 shows a most elegant electron microscope picture of one of these extrachromosomal nucleoli to which are attached over 1000 growing rRNA precursor molecules.

Very large numbers of extrachromosomal nucleolar circles are needed to produce the rRNA components of the massive quantities of ribosomes found in oocytes, the largest of all cells. All the ribosomes present before gastrulation are synthesized before fertilization in the oocyte. During growth of the frog oocyte, for example, some 10^{12} ribosomes must be made over a matter of several weeks. The number of rRNA genes within the oocyte chromo-

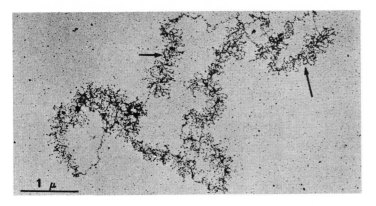

Figure 17–27
Electron microscope visualization of an extrachromosomal nucleolar core from an oocyte of the African toad *Xenopus laevis*. It contains 10 rRNA genes tandemly arranged along the circular DNA molecule. The arrows point to individual genes. [Courtesy of O. L. Miller and B. R. Beatty, Oak Ridge National Laboratory.]

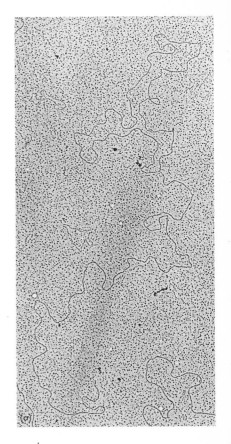

Figure 17-28
(a) An electron micrograph of a ribosomal DNA rolling circle from a *Xenopus laevis* oocyte. The circumference of the template circle corresponds to three ribosomal RNA genes and the attached tail to 3.4 ribosomal RNA genes. [Reproduced with permission from Hourcade, Dressler, and Wolfson, *Cold Spring Harbor Symp.*, **XXXVIII**, 537 (1974).]
(b) Ribosomal DNA amplification by a rolling-circle mechanism from an extrachromosomal nucleolar circle. Compare with Figure 9–20.

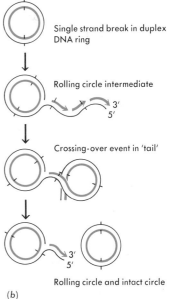

Single strand break in duplex DNA ring

Rolling circle intermediate

3′
5′

Crossing-over event in 'tail'

3′
5′

Rolling circle and intact circle

(b)

somes, however, reflects the requirements of the ordinary frog somatic cell, which is some 10^5 times smaller than its respective oocyte. Only by extensive rDNA amplification can oocyte growth proceed at a biologically acceptable rate.

Most of the nucleolar circles in mature oocytes arise by a rolling circle mechanism (see page 238) from previously made circles (Figure 17–28). Still very unclear, however, is the origin of the first extrachromosomal nucleoli. Genetic evidence obtained from hybrid frogs shows that they are not a product of nonchromosomal rRNA genes and somehow must be derived from the clustered, chromosomal rRNA genes. Conceivably, they originate by a crossing-over mechanism similar to that which releases λ prophages from the *E. coli* chromosome. But this scheme requires the later reinsertion of rDNA so that the number of chromosomal rRNA genes will remain relatively constant. Another possibility is the enzymatic creation of specific single-strand cuts in chromosomal rDNA, which lead to the displacement of long single-stranded sections. Such displaced chains could then be converted into double-stranded circles by a rolling-circle type process (Figure 17–28). This matter will most likely be settled only when such circles can be made *in vitro* by oocyte extracts.

MULTIPLE TELOCENTRIC LOCATIONS FOR THE 5S TOAD RNA GENES

Unlike the 18S and 28S rRNA components, which are coded for by one to several large blocks of genes, the genes coding for the 5S RNA components of toad ribosomes are located on almost every chromosome. Moreover, two different sets of 5S RNA genes (5S DNA) exist, one coding for the 5S RNA found in oocyte ribosomes, the other for that present in somatic cells. Both 5S RNA chains are exactly the same length, with eight nucleotide differences distinguishing their 120-nucleotide-long sequences. While only some 450 copies of the somatic 5S DNA exist, some 24,000 copies of the oocyte variety are present. Thus no extrachromosomal amplification of 5S DNA need occur during oogenesis. Instead, some control mechanism turns on the functioning of the much larger gene set needed for oocyte ribosome synthesis. Still very mysterious is the chromosomal location of 5S DNA. Both types are exclusively clustered at the ends (telomeres) of the long arms of many, if not all, toad chromosomes (Figure 17–29).

In *Drosophila melanogaster* all the 5S genes are located internally on the same chromosome that carries the rRNA genes. Why the toad and *Drosophila* have such differences may be very difficult to sort out.

CLUSTERS OF SPECIFIC tRNA GENES

The genes coding for the various toad tRNA species are also highly reiterated, with all the members of a given species tandemly clustered into long blocks of tDNA. Each haploid genome contains some 8000 tDNA genes, or an average of some 140 genes for each of 56 reported major tRNA species. For example, one methionyl tDNA class has ~330 copies, while the second met-tRNA gene has ~170 copies. Spacer DNA appears to separate each tRNA gene from its tandem partners, with the spacer lengths on the average being some 10 times longer than their respective

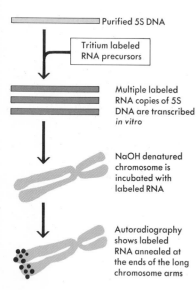

Purified 5S DNA

Tritium labeled RNA precursors

Multiple labeled RNA copies of 5S DNA are transcribed *in vitro*

NaOH denatured chromosome is incubated with labeled RNA

Autoradiography shows labeled RNA annealed at the ends of the long chromosome arms

Figure 17–29
The demonstration that the multiple copies of 5S RNA genes are located at or near the telomere region of the long arms of most of the chromosomes in the toad *Xenopus*. Labelled RNA copies of the 5S gene (shown), or *in vivo* labeled 5S RNA, anneal with denatured DNA in the chromosomes, and are revealed by autoradiography.

tRNA gene neighbors. Whether all the tRNA genes of a given species exist within a single cluster is not known. Nor do we yet know whether the cluster(s) belonging to different tRNA species have widely different chromosomal locations.

GENE AMPLIFICATION AS A MECHANISM FOR DIFFERENTIAL GENE EXPRESSION

When extrachromosomal amplification of oocyte rDNA was first discovered, much speculation arose that similar events would prove important in a variety of differentiation events. Fueling this way of thinking was the report that in *Sciarid* flies, not only did DNA of certain polytene bands increase in mass at specific stages of development, but, even more important, most of this newly synthesized genetic material also was released from its respective chromosomal locations. One then anticipated a crescendo of reports of specific gene amplification. Now, almost a decade later, we are instead struck by the finding of single copies of the hemoglobin and fibroin (silk) genes per haploid complement. These solitary numbers reflect the fact that the final level of protein synthesis is the result of two levels of amplification: the first occurring when a given gene is transcribed many times to produce large numbers of messengers, and the second occurring when these mRNA molecules in turn act as templates for many, many rounds of protein synthesis. One fibroin gene, for example, can give rise in 4 days to some 10^5 fibroin mRNA molecules, which in turn determine the amino acid sequence of some 10^{10} fibroin molecules.

When rDNA is transcribed, however, RNA chains are the final product of synthesis and no further amplification of its activity is possible. Hence the large number of rDNA copies within somatic cells and the need for their extrachromosomal amplification during oocyte growth.

POLYRIBOSOME LIFETIMES IN RAPIDLY DIVIDING CELLS

Compared to what we know about bacterial messengers, our knowledge of the lifetimes of functioning mRNA in higher cells is very limited. It seems clear that many have a limited lifetime, and average half-life estimates of 3 hours are frequently given for the mRNA in rapidly growing vertebrate cells which are dividing about once a day. If these figures are correct, then their relative lifetimes are not dissimilar to those of many bacterial mRNA molecules, which turn over some 10 times during a cell cycle.

STABLE mRNA MOLECULES EXIST IN NONDIVIDING DIFFERENTIATED CELLS

In contrast, it has begun to appear more and more likely that much of the mRNA of the highly differentiated cells of higher animals is metabolically stable. Erythroblasts (immature red blood cells) provide a good example. These cells produce virtually no RNA while they are synthesizing their principal protein, hemoglobin. If their mRNA molecules were rapidly made and broken down, it would be possible to detect incorporation of RNA precursors into RNA; none is detected. This stability has an obvious advantage, particularly since the very constant environment of red blood cells, as contrasted with the highly fluctuating growth conditions of bacteria, makes great flexibility unnecessary. Erythroblasts are designed to synthesize largely (>90 percent) hemoglobin. There is no reason to break down hemoglobin mRNA chains only to resynthesize them. Similarly, most of the mRNA found in the cytoplasm of the adult liver makes plasma protein that is to be released into the circulatory system. Correspondingly, much cytoplasmic mRNA in liver cells is relatively stable. There is, however, a slow turnover in which not only mRNA, but also the ribosomes themselves, seem to be broken down. Recent measurements of the lifetime of ribosomes give a value of about 100 hours, much shorter than the average cellular lifetime of some 100 days. Why and how these processes occur is as yet totally unexplored. Particularly puzzling is the rRNA breakdown. While it can happen in bacteria, it does so only under unfavorable nutritional conditions. Yet in the liver, as far as is known, rRNA breakdown occurs in perfectly "normal" cells.

DIFFERENTIATION IS USUALLY NOT IRREVERSIBLE AT THE NUCLEAR LEVEL

Nuclear transplantation experiments provide conclusive evidence against the notion that changes in the nucleus (e.g., generation of extrachromosomal DNA) irreversibly differentiate a cell toward the selective synthesis of a restricted set of gene products. In these experiments, diploid nuclei from differentiated cells were transplanted into unfertilized eggs whose haploid nuclei had been previously removed. The resulting genetically complete diploid eggs were then artificially induced to divide and grow, often to form adult organisms whose chromosomal makeup derived entirely from clonal reproduction of donor nuclei (Figure 17–30). So far, the frog is the most complex organism for which nuclear transplantation has allowed

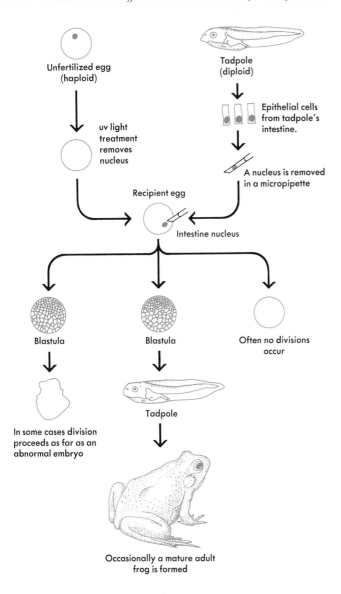

Unfertilized egg
(haploid)

Tadpole
(diploid)

uv light
treatment
removes
nucleus

Epithelial cells
from tadpole's
intestine.

A nucleus is removed
in a micropipette

Recipient egg

Intestine nucleus

Blastula

Blastula

Often no divisions
occur

In some cases division
proceeds as far as an
abnormal embryo

Tadpole

Occasionally a mature adult
frog is formed

Figure 17–30
Steps in the origin of a "clonal" frog. [Redrawn from J. R. Gurdon, *Scientific American*, December, 1968.]

"clonal" reproduction. Successes in this case owe much to the very large amphibian egg, which allows conventional microsurgical removal of the maternal nuclei. Eventually, however, we are likely to witness positive results with virtually all vertebrates for which this goal seems worth achieving.

IRREVERSIBLE CYTOPLASMIC DIFFERENTIATION CONCOMITANT WITH LOSS OF ABILITY TO DIVIDE

Many cells contain nuclei that, under normal conditions, will never divide again. Often they represent cases of extreme differentiation, where a very large fraction of the total synthesis is devoted to making just a few different proteins. Mature red blood cells (erythrocytes) are perhaps the most striking example. These cells never divide, having as their only function the production of hemoglobin molecules able to combine reversibly with oxygen. Moreover, in some species, including all mammals, the nuclei of these cells eventually disintegrate, leaving the resulting mature erythrocytes unable to make any more hemoglobin. Erythrocytes generally live only several months and must be replaced by more recently synthesized cells. Behaving somewhat similarly are the adult plasma cells, which produce specific antibodies. They likewise never divide, but produce their products for several days and die. During this time their nuclei remain fully functional, producing the mRNA molecules coding for their antibody products. In contrast, many other specialized cells (e.g., nerve cells), though never dividing, may live many, many years. The functioning of such long-lived cells always depends on the presence of a nucleus able to make mRNA. Some protein synthesis, even at a very low level, seems to be necessary for the continued functioning of all cells.

REVIVAL OF DORMANT NUCLEI THROUGH FUSION WITH MORE ACTIVE CELLS

The dormancy of many nuclei is reversed when their respective cells are fused with other cell types. Until several years ago, fusion between cells growing in tissue culture was a rare event, not at all obtainable at the command of the biologist. Then the observation was made that the adsorption of certain viruses (in particular, a flu-like virus called *Sendai*) caused many cells to fuse. Somehow, the infecting virus particles modify cellular surfaces in a way which promotes their fusion. Even more important, ultraviolet-killed viruses also promote fusion, allowing the metabolism of fused cells to be studied without the many disrupting complications which characterize viral infections. Now, as a result of the introduction of the "Sendai helpers," meaningful cell fusion experiments are very easy to carry out, and important new insights into higher cell regulation have already been obtained.

For example, when a nucleated hen erythrocyte cell is fused with the human cell line HeLa (see page 550), the erythrocyte nucleus resumes both DNA and RNA synthesis. Nuclear reactivation starts with an increase in nuclear volume, allowing the tightly compacted hen chromosomes

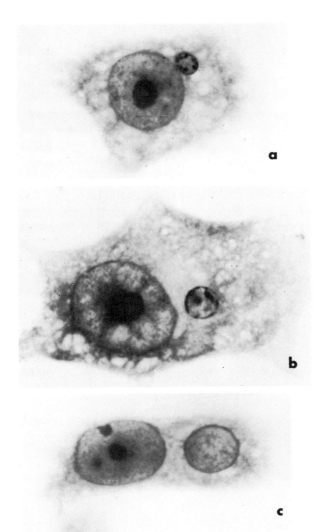

Figure 17–31
(a) A recently fused cell containing a very small hen erythrocyte nucleus and a HeLa nucleus. (b) A later stage showing enlargement of the erythrocyte nucleus. (c) A still later stage revealing still further enlargement. [Photographs by Dr. H. Harris and reproduced with permission of the *J. Cell. Science.*]

to expand and become capable of RNA synthesis (Figure 17–31). In some not yet understood way, the HeLa cytoplasm must supply components to the hen nucleus which permit it to function. Superficially at least, both DNA and RNA synthesis must be under some form of positive control. For, if the control mechanism were of the negative variety, the erythrocyte cytoplasm would prevent functioning of the HeLa nucleus. No such inhibition, however, is observed with this particular fusion pair or with any of the other cell pairs looked at so far. Further support for the hypothesis that DNA synthesis is under positive control comes from experiments in which a HeLa cell in the G_1 phase of the cell cycle (see page 560) is fused with one in S phase. Soon after fusion, DNA synthesis commences in the previously dormant G_1 nucleus.

POSITIVE CONTROL OF GENE FUNCTION

A key question thus becomes what components directly cause specific genes of higher organisms to function. Here our ignorance is still almost complete, with our only precise knowledge coming from study of the differentiation induced by specific steroid hormones. At the genetic level, the best understood steroid hormone is ecdysone, the growth hormone responsible for the successful passage of developing insects from their larval to their pupal stages. It does so by inducing the function of very specific puffs from previously dormant bands. In the early larval period, the puffing pattern of the polytene *Drosophila* salivary chromosomes is simple and stable. Then only some 10 prominent puffs exist. But as soon as ecdysone is released into the insect hemolymph, the preexisting puffs begin to regress, and a number of new puffs are quickly thrown out, with some 125 eventually appearing during the transition from the larval to pupal stages. Each band always puffs at a precise time and remains active for an again very well defined interval (Figure 17–32). The first such puffs (the "early" puffs) begin appearing within 5 minutes of hormone addition. They appear to be direct responses to the hormone, since their appearance is unaffected by inhibitors of protein synthesis. In contrast, those "late" puffs that only appear some hours after ecdysone is added do not appear when protein synthesis is blocked. It seems probable that a protein product(s) from one or more of the "early" puffs induces the formation of the "late" puffs.

How ecdysone induces the "early" puffs is still largely speculative, although the extremely rapid response suggests that this steroid directly promotes transcription from the relevant chromatin sites. This is clearly the way a number of vertebrate steroids act. For example, estrogenic steroid sex hormones stimulate RNA synthesis in the uterus before any other metabolic effects can be noticed. Such steroids, however, do not act entirely by themselves. Their action depends upon firm binding to highly specific protein *receptors*. Each target cell contains many thousands of such hormone receptors, all of which sediment at 4S and which, in the absence of their respective steroids, are exclusively present in the cytoplasm. But immediately after they complex with a steroid, the resulting receptor–steroid complexes move into the nucleus, where they bind to chromatin and unleash previously inhibited RNA synthesis. Mixing of purified receptors by themselves with DNA never results in binding. Only when their steroid partners are present can they attach to chromatin (Figure 17–33).

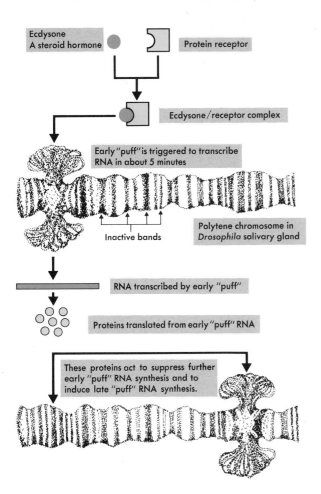

Ecdysone
A steroid hormone

Protein receptor

Ecdysone/receptor complex

Early "puff" is triggered to transcribe RNA in about 5 minutes

Inactive bands

Polytene chromosome in *Drosophila* salivary gland

RNA transcribed by early "puff"

Proteins translated from early "puff" RNA

These proteins act to suppress further early "puff" RNA synthesis and to induce late "puff" RNA synthesis.

Figure 17–32
Schematic representation of the action of ecdysone in the sequential production of late and early chromosome "puffs." The actual chromosome drawn is purely schematic, as in fact many "puffs" are formed in turn.

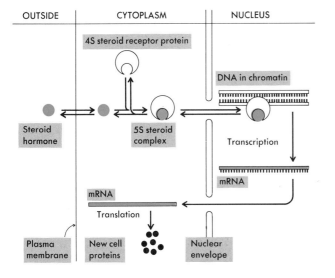

OUTSIDE	CYTOPLASM	NUCLEUS

4S steroid receptor protein

DNA in chromatin

Steroid hormone

5S steroid complex

Transcription

mRNA

mRNA

Translation

Plasma membrane

New cell proteins

Nuclear envelope

Figure 17–33
Diagrammatic representation of how steroid hormones are thought to act. The 5S steroid hormone–receptor protein complex passes through the nuclear envelope and binds to a specific chromosomal site, in turn activating specific mRNA synthesis.

Still most puzzling is the very large number (~1000) of chromosomal sites to which the steroid-receptor complexes apparently bind. Given the high specificity of most hormone responses, a limited number of binding sites might *a priori* have been predicted.

An obvious goal is the finding of *in vitro* systems where steroid hormones directly influence RNA synthesis. Unfortunately no such system has yet been worked out, though this should occur over the next few years. There do, however, exist tissue culture cells which specifically respond to steroids. One of the best studied cell lines derives from a hepatoma (liver tumor). When such cultured cells are grown in the presence of adrenal steroid hormones, there is a five to fifteen fold increase in the activity of the enzyme tyrosine amino transferase over that present in the absence of these hormones (Figure 17–34). This stimulation is highly specific, since the level of total RNA and protein synthesis is unaffected by hormone addition. Moreover, this increase in activity represents new protein synthesis, since no increase occurs if inhibitors of protein synthesis like puromycin or cycloheximide are added. Furthermore, since the RNA synthesis inhibitor actinomycin-D also blocks enzyme formation, induction most likely requires the synthesis of new mRNA molecules, thereby adding to the generalization that all steroid action works through promotion of specific RNA synthesis. Unfortunately, this system by itself can yield no molecular insights about how a steroid–receptor complex converts dormant chromatin into actively transcribing genes. Again we see the crucial need for the development of *in vitro* systems before real answers are to be established.

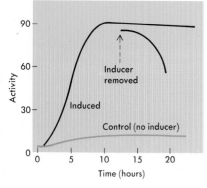

Figure 17–34
Induction of tyrosine amino transferase in rat hepatoma cells by the synthetic adrenal hormone dexamethasone phosphate.

PREFORMED mRNA ON THE WAY TO GASTRULATION

Whether translational control is an important factor in the lives of eucaryotic cells remains unclear. The obvious thought, Why waste energy making an RNA molecule which you may never use? may have greater validity for bacterial than for nucleated cells, in which a much longer time may be required for newly synthesized mRNA molecules to take up suitable positions for protein synthesis. Quick metabolic responses by higher cells may thus require the preexistence of complete yet dormant mRNA chains at suitable cytoplasmic sites. In this way, protein synthesis could begin immediately upon receipt of a molecular signal.

This is, in fact, the situation in unfertilized echinoderm eggs, which begin rapid protein synthesis immediately following fertilization. Most of the mRNA chains used up in gastrulation have been synthesized before fertilization, and so subsequent blockage of RNA synthesis with actinomycin does not at all inhibit the early cleavage divisions. So we must conclude that the unfertilized egg is filled with mRNA waiting for a signal to start serving as templates. This signal does not operate through control of the number of ribosomes, since they are also present in large numbers. Most ribosomes, however, are not combined with mRNA to form polyribosomes. Instead, fertilization releases a signal which allows specific polyribosomes to form, thereby setting into motion the rapid series of cleavage divisions which lead, first, to blastulation and, finally, gastrulation. Throughout the early cleavage divisions, some new mRNA species are synthesized, largely for function at the gastrulation stage.

Existence of preformed mRNA within unfertilized eggs immediately makes us ask whether it is also of widespread occurrence in other cells. But before this question can be tackled, the state of the preformed mRNA within eggs must be elucidated. There are claims that it is combined with specific proteins, but substantiation will depend on future experiments.

FURTHER DECIPHERING OF THE EUCARYOTIC CHROMOSOME

It is thus very clear that the molecular biologist no longer regards embryology as a hopeless problem from which only basically untestable theories can emerge. If the right questions are asked, real answers will emerge at the molecular level, admittedly after much, much work. The key to this new feeling of optimism is the discovery of the RNA polymerase specificity factors and the emergence of well defined theories for hormonal control of gene function. Now a most pressing task is to find the proper higher cells for rigorous testing of these new control concepts. At this moment these ideas still rest almost completely upon procaryotic systems. In this search, much emphasis must be placed on the choosing of experimental organisms whose chromosomes can be studied in detail at both the genetic and molecular level. Conceivably, when embryologists either converge on *Drosophila* or begin a serious assault on amphibian genetics, embryology will at last be on a solid molecular course.

SUMMARY

All too little is known about the molecular basis of the control of protein synthesis in the cells of multicellular higher organisms. In particular, little information exists as to mechanisms that bring about cell differentiation. Once a cell has become differentiated, all its descendants usually continue to produce a specific group of unique proteins. There are hints that this selective protein synthesis, like that of microbial cells, is usually based on the selective synthesis of unique types of RNA. Thus, differentiated cells must have mechanisms that control the rate at which specific DNA regions are read. A major difficulty that now hinders basic understanding is our current inability to easily study differentiation outside an intact organism. Though embryonic differentiation can be made to occur in tissue culture, it has not yet been possible to isolate the specific external factors that normally induce a cell to differentiate in a given direction.

In bacteria the differentiation process that leads to spore formation is triggered by suboptimal nutrition. This signal initiates a chain of events that causes the disappearance of one or more normal RNA polymerase specificity factors and its (their) replacement by a sporulation-specific specificity factor which reads only those genes involved in spore production. Whether similar temporal displacements of one specificity factor by another will be the basis of other differentiation processes (e.g., the slime-mold life cycle, insect metamorphosis) soon should be testable.

Still almost a complete mystery is the structure of higher chromosomes and the way in which their DNA replication can start at many different sites along a single chromosome. Also a major puzzle is the lack of correspondence between DNA content and the number of genes. While in procaryotes and in lower eucaryotes, most DNA codes for amino acid sequences, the vast majority of DNA from multicellular organisms has no apparent genetic function. In Drosophila *only ~5% of the total DNA specifies amino acid sequences, while in humans even less DNA is so employed. Much of this nongenetic DNA lies interspersed with genetic DNA, with the fundamental transcriptional products (pre-mRNA) comprising a mixture of genetic and nongenetic sequences. After transcription of pre-mRNA, the nongenetic component is processed away, leaving largely sequences that have a coding function (mRNA).*

Further complexity comes from the existence in higher organisms of three classes of DNA sequences: highly repetitive, mildly repetitive, and unique. The highly repetitive sequences, in which very simple sequences are repeated thousands of times, are found exclusively in the centromeric regions. They are never transcribed and their function is unknown. Also functionally unclear are the mildly repetitive sequences that are found throughout the genome interspersed with sections of unique DNA. Many mildly repetitive sequences are transcribed into RNA, leading to speculation that they may have some as yet undefined "control function." While many of the unique sequences code for amino acids, not all do so, and in most higher organisms the function of the vast majority of unique sequences also remains a mystery.

Most genes in higher plants and animals are present in only one copy per haploid genome. Even those genes (*e.g.*, those for the various hemoglobin chains) which code for the most common cellular proteins are present only once. Exceptions, however, to this rule exist. For example, in sea urchins the various histone genes are present in some 100 to 200 copies. Redundancy also exists for the genes coding for rRNA and tRNA. In somatic cells the large numbers of rRNA genes are found tandemly attached to each other in the one to several chromosomally attached nucleoli. Many oocytes also contain large numbers of extrachromosomal nucleoli, each containing a circular DNA molecule made up of some 10 to 1000 tandemly attached rRNA genes.

In many highly specialized cells, the nuclei normally do not divide. But this property does not reflect irreversible nuclear differentiation, as these nuclei sometimes spring back into action if they are introduced into less differentiated cytoplasm, whether by microsurgery or by cell-fusion techniques. Nuclear behavior is thus largely a function of the molecular signals received from the surrounding cytoplasm. For example, certain steroid hormones selectively influence RNA synthesis by binding to cytoplasmic steroid receptors, which then move to the nucleus where they attach to specific sites on the chromatin. While the belief exists that most control of protein synthesis in higher organisms is at the transcriptional level, in the unfertilized egg, many fully synthesized mRNA molecules somehow remain dormant until fertilization occurs. Whether "translational control" is an important feature of somatic cell existence remains to be learned.

REFERENCES

Hood, L. E., J. H. Wilson, and W. B. Wood, *Molecular Biology of Eucaryotic Cells*, W. A. Benjamin, Inc., 1975. A superior advanced problem-book which, by itself, serves as a very useful introduction to the genetic aspects of eucaryotic molecular biology.

Sussman, M., *Developmental Biology*, Prentice-Hall, 1973. A first introduction to modern embryology.

MARKERT, C. L., AND H. URSPRUNG, *Developmental Genetics*, Prentice-Hall, 1971. A detailed introduction to much of the best of modern embryology.

EBERT, J., *Interacting Systems in Development*, 2nd ed., Holt, New York, 1970. An excellent, quite extensive paperback introduction to embryology, with emphasis on the desirability of explanations on the molecular level.

GURDON, J. B., *Control of Gene Expression in Animal Development*, Harvard University Press, 1975.

Control Mechanisms of Growth Differentiation, Symposium of the Society for Experimental Biology XXV, Academic Press, 1971. An imaginative collection of articles that brings together the classical and biochemical approaches to embryology.

The Bacterial Spore, G. W. Gould and A. Hurst (eds.), Academic, New York, 1969. A now slightly dated summary of bacterial spore research.

Microbial Differentiation, Twenty-third Symposium of the Society for General Microbiology, Cambridge University Press, 1973. A look at a number of differentiating microbes.

LOSICK, R., AND A. L. SONENSHEIN, "Changes in the Template Specificity of RNA Polymerase During Sporulation," *Nature*, **224**, 35 (1969). Specificity factors and the first real wedge toward understanding the differentiation of spores.

DUPRAW, E. J., *DNA and Chromosomes*, Holt, Rinehart, and Winston, 1970. A speculative introductory text.

WHITE, M. J. D., *The Chromosomes*, 6th edition, John Wiley, New York, 1972. A simple introduction to chromosome structure and behavior.

"Chromosome Structure and Function," *Cold Spring Harbor Symposium on Quantitative Biology*, **XXXVIII**, 1974. A very comprehensive volume that contains eighty-eight articles, most of which emphasize the molecular approach.

BRITTEN, R. J., AND D. A. KOHNE, "Repeated Sequences of DNA," *Scientific American*, April, 1970. Evidence for the presence of repetitive sequences in a variety of organisms.

MILLER, O. L., "The Visualization of Genes in Action," *Scientific American*, March, 1973. A summary of the author's elegant EM investigations into gene functioning.

Developmental Studies on Giant Chromosomes, edited by W. Beermann, Springer-Verlag, 1972. Excellent reviews on salivary chromosomes.

Daneholt, B., "Transcription in Polytene Chromosomes," *Cell*, **4**, 1 (1975). A very useful summary emphasizing Balbiani rings.

Brown, D. D., "The Isolation of Genes," *Scientific American*, August, 1973. Some beautiful experiments revealing the organization of the genes coding for ribosomal RNA.

Harris, H., *Cell Fusion*, Harvard University Press, Cambridge, Mass., 1970. An early summary of the cell fusion field.

Ephrussi, B., *Hybridization of Somatic Cells*, Princeton University Press, 1972. A discussion of the genetic implications of cell fusion techniques.

Harris, H., *Cells and Cytoplasm*, 3rd ed., Clarendon Press, Oxford, 1972. An introduction to control problems in higher cells that emphasizes the cell fusion approach.

Ruddle, F. H. and R. S. Kucherlapatai, "Hybrid Cells and Human Genes," *Scientific American*, July, 1974. A recent summary of how the fusion of vertebrate cells can tell us about both the location of genes and the control of their function.

Gurdon, J. B., "Transplanted Nuclei and Cell Differentiation," *Scientific American*, December, 1968. A description of the most meaningful nuclear transplantation experiments so far accomplished.

The Control of Cell Proliferation

Biologists have always known that the control of cell division is one of the most basic aspects of multicellular existence. Throughout embryological development, as well as through all of adult life, many differentiated cells have the choice to divide or not; and only if a programmed series of correct decisions is made can the respective organisms continue to function normally. But mere desire to come to grips with the important questions does not necessarily generate clean answers; we would be less than candid if we did not admit that past research on the factors controlling cell proliferation has generated more confusion than clean facts. Today, however, we believe that some key facts have at last emerged and that with some luck the next decade of research might give the intellectual framework to understand proliferation control in a large variety of seemingly unrelated animal cell types. Key to current enthusiasm is the realization that experiments need no longer be restricted to intact organisms, but that real answers may come from growing animal cells in culture.

Always intertwined with any study of cell proliferation is the nature of cancer. This collection of horrific diseases by definition involves cells which divide when they should not, usually to produce the contiguous cellular masses called *tumors*. The different classes of cancer arise by changes in the various forms of differentiated cells, with the resulting cancer cells exhibiting many of the morphological and functional characteristics of their normal precursors. Most often they arise in cells undergoing frequent division (e.g., epithelial cells of the skin), and so their essence is not the fact that they can divide rapidly but that they lack the normal control systems to shut off unwanted cell division. So a prime scientific goal for nearly all of this century has been to find out the essential biochemical differences between normal and cancer cells.

ESTABLISHING CELLS IN CULTURE

Until recently there was much mysticism about growing cells in culture. Success apparently demanded very precise experimental conditions and, even worse, not everyone seemed to possess the necessary magic touch. For a long time only masses of excised tissue would consistently grow, and then only when implanted upon very undefined serum clots or embryonic extracts. Such "tissue cultures" occasionally multiplied indefinitely if microbial contamination could be prevented and when regular changes were made in the surrounding nutrient medium. Though repeated efforts were made to grow cultures from dissociated cells, successes were very infrequent until small amounts of the proteolytic enzyme trypsin were used to dissociate tissue masses into their component individual cells. Then the dissociated cells (*primary cells*) invariably grew well when seeded onto culture plates at high cell densities. In contrast, initial attempts to induce single cells to grow into clones did not succeed. At best the individual cells multiplied several times and then died. The main reason behind the failure of these early cloning attempts is the inherent leakiness of most vertebrate primary cells in culture. Vital nutrients diffuse out, and growth occurs best when cells are seeded at high density so that cross feeding becomes possible. Hence, the practice arose of placing single primary cells in very small drops of culture fluid. Here extensive dilution is not possible, and tiny clones result. Another way to increase cloning efficiency places dissociated single cells upon a "feeder" layer of cells sterilized by large x-ray doses. None of the irradiated feeder cells can multiply, but they remain metabolically active and release sufficient nutrients to allow most of the seeded primary cells to form visible colonies.

Cell cultures arising from multiplication of primary cells are called *secondary cells*. Such cells generally do not have infinite lives, but divide only a finite number of times before dying out. For example, secondary chicken cells multiply 20 to 40 times and then invariably die. Likewise, cultures of human skin cells at best divide some 50 to 100 times before they die. Interestingly, this number coincides with the number of divisions that such cells make during the life span of their respective vertebrate source. The temptation thus exists to wonder whether this seemingly programmed cell death is related to the normal vertebrate aging process.

Table 18–1 Some Properties of Several of the More Important Cell Lines

	Origin			Characteristics			
	Species	Tissue	Morphology	Growth in suspension	Efficiency of single cells to form clones	Form tumors	Chromosomal Makeup
3T3	Mouse 2N = 40	Endothelial	Fibroblast	No	50%	Yes (rare)	Heteroploid (differs from cell to cell) (mode = 65)
L	Mouse 2N = 40	Connective	Fibroblast	Yes	90%	No	Heteroploid (mode = 65)
CHO	Chinese hamster 2N = 22	Ovary	Epithelial	Yes	95%	?	Pseudodiploid
BHK 21	Syrian hamster 2N = 44	Kidney	Fibroblast	Yes	20%	Yes (rare)	Diploid
BSC	Monkey 2N = 42	Kidney	Epithelial	No	Low	?	Diploid
MPC	Mouse 2N = 40	Bone-marrow myeloma cell	Lymphoid	Yes	0%	?	Heteroploid
RPH	Frog 2N = 26	Egg	Epithelial	No	0%	Yes	Haploid (N = 13)
HeLa	Man 2N = 46	Cervical tumor	Epithelial	Yes	95%	?	Heteroploid (mode = 75)
KB	Man 2N = 46	Nasopharyn-geal tumor	Epithelial	Yes	95%	?	Heteroploid (mode = 75)

Fortunately not all animals yield primary cultures all of whose secondary progeny die. Though most mouse and hamster cells also die after dividing some 30 to 50 times, a few progeny pass through the "crisis period" and acquire the ability to multiply indefinitely. Progeny derived from such exceptional cells form the continuous *cell lines* (Table 18–1). Though no "normal" human cell line has yet been developed, a number of cell lines have been derived from human tumors (Figure 18–1). Somehow the possession of the cancerous phenotype often allows easier adaptation to

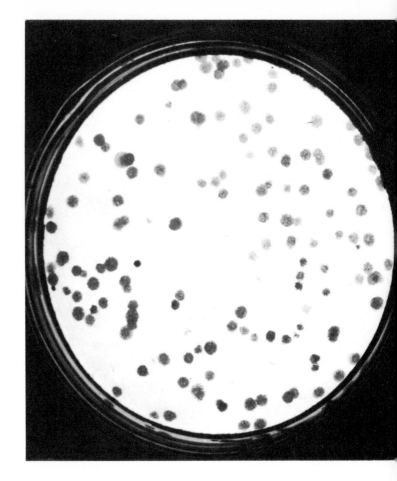

Figure 18–1
Clones of the human tissue culture strain HeLa growing on the agar-
covered surface of a petri dish. [Photo supplied by Dr. T. T. Puck, Univ.
of Colorado Medical School.]

the conditions of cell culture. This property may relate to
the fact that both cancer cells and cells adapted to permanently grow in culture have an increased chromosome
number, often nearly twice the normal diploid number.
For example, cells from the human cancer cell line "HeLa"
have some 70 to 80 chromosomes, in comparison to the
normal 46 chromosome complement (Figure 18–2). Apparently the possession of extra chromosomes (*aneuploidy*)
frequently gives cells in culture a selective advantage over
those with the normal diploid number.

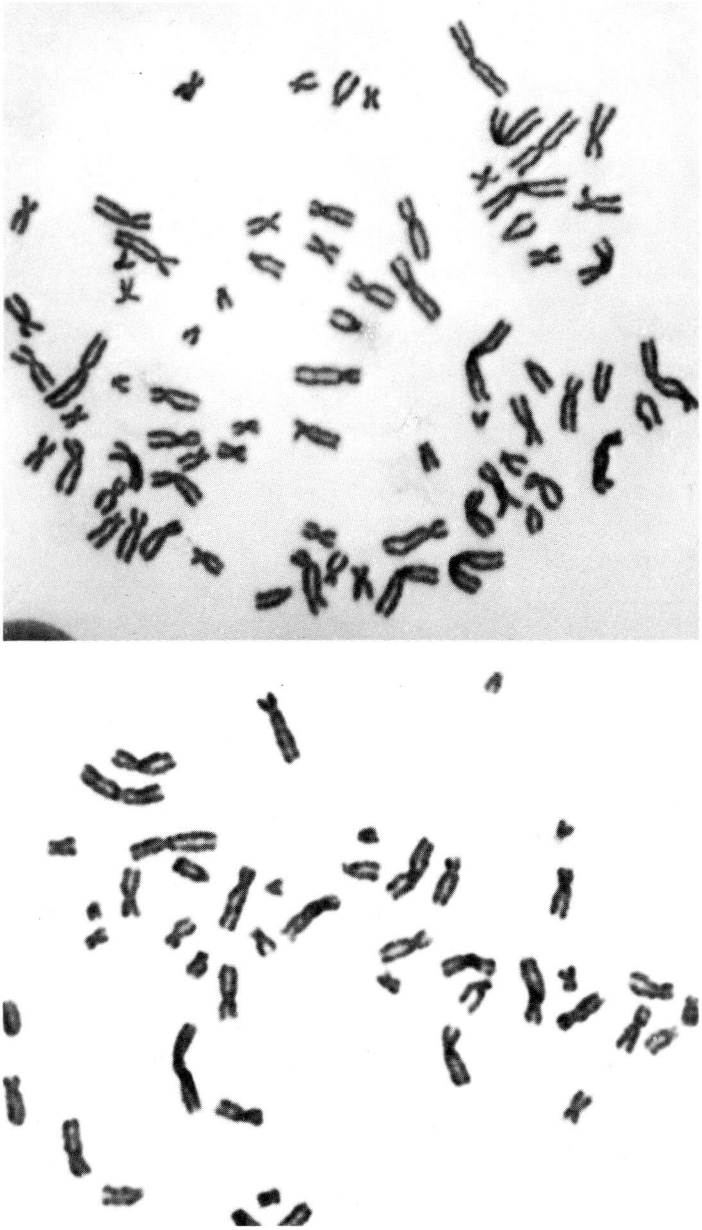

Figure 18–2
The 73 chromosomes (top) of a HeLa cell contrasted with the normal 46 complement (bottom) of a cell from a normal human male. [Courtesy of Dr. T. T. Puck, Univ. of Colorado, Denver.]

OBSCURE ORIGIN OF MANY CELL LINES

Usually primary cultures are prepared from large tissue masses, if not organs or whole embryos. Initially they comprise a large variety of differentiated cells (e.g., *fibroblasts, macrophages, epithelial cells, lymphocytes,* etc.), and so the cellular origin of a given cell line is frequently obscure. Under most conditions for growth in culture, the cells that multiply best have the spindle shape and growth rate of connective tissue cells and so are called *fibroblasts* (Figure 18–3). Just recently, however, cultural conditions have been found which selectively promote the growth of certain *epithelial cells* (Figure 18–3). Such epithelial cells can use D-valine as their sole valine source because they possess the enzyme D-amino acid oxidase that allows the transformation of D-valine into L-valine. Fibroblasts, however, do not contain this enzyme and so cannot grow when D-valine is the only source of valine in the growth medium.

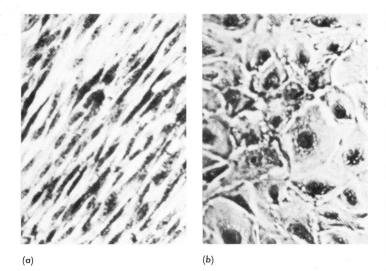

(a)   (b)

Figure 18–3
Outgrowth of secondary cells starting from primary human lung tissue. Selective proliferation of (a) fibroblasts in L-valine, and (b) epithelial cells in D-valine. [Reproduced with permission from S. Gilbert and B. Migeon, *Cell,* **5,** 11 (1975).]

The shape that cells of a given line assume in culture is not invariant. For example, some strains of HeLa cells growing in rich (high serum content) media have the stretched elongated fibroblast shape, whereas in poorer medium they pack close together and have a rounded cobblestonelike appearance (Figure 18–4). Thus, while we may refer to one cell line as "epithelial cells" and another as "fibroblasts," we must not always assume that their origin was from epithelial (connective) tissue.

Sometimes, however, the presence of specific cell products makes the origin of certain cell lines unam-

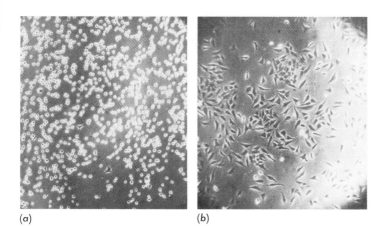

(a)   (b)

Figure 18–4
Dependence of the shape of HeLa cells upon growth conditions. In low serum (a), they have a rounded cobblestone shape, while in enriched serum (b), they have the spindle shape of fibroblasts. [Reproduced with permission from T. Puck, *The Mammalian Cell as a Microorganism,* Holden-Day, Inc., 1972.]

biguous. In particular, the origin of cells secreting specific hormones is easy to ascribe. Cells from tumors of the testes specifically secrete androgens, while ACTH and growth hormones are made by cell lines derived from pituitary tumors. Likewise, cells derived from many liver tumors preferentially secrete the major liver product albumin. Equally clear is the cellular origin of cultured neuroblastomas (tumors of nervous tissue), since their cells have excitable membranes and under certain nutritional conditions send out numerous axonal processes (Figure 18–5).

Not all the properties of differentiated cells are retained in their cultured progeny. Unless given traits are

Figure 18–5
Phase-contrast photomicrograph of mouse neuroblastoma cells growing attached to a solid surface. Many elongated processes (neurites) extend from each cell. When such cells grow in suspension they have the rounded cancerous appearance characteristic of the cells which form tumors, but when they attach to a solid surface, they become flattened and begin to send out large numbers of neurites. [Courtesy of Dr. Gunter Albrecht-Bühler, Cold Spring Harbor Laboratory.]
▼

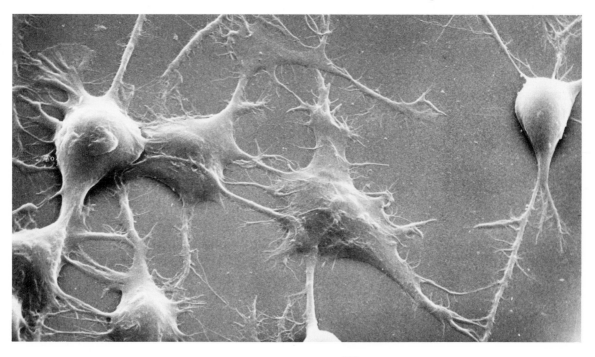

specifically selected, most cell lines seem less differentiated than their *in vivo* precursors, and often show similarities to the "undifferentiated" embryonic cells present in early development. Many so-called "normal" cell lines thus closely resemble cancer cells, many types of which also represent partially dedifferentiated states. For example, most liver-derived hepatomas fail to make many normal liver enzymes. This latter parallel becomes more plausible when we realize that continued cell growth in culture may be achieved by a loss of some of the devices which slow down, if not inhibit, cell division in the respective multicellular organism, and whose loss may result in the cancerous state. Isolation of many cell lines may thus have been dependent on the selection of cell mutants less able to respond to signals halting cell division.

ATTACHMENT TO SOLID SURFACES VS. GROWTH IN SUSPENSION

Most "fibroblasts" and "epithelial" cells grow best, if not exclusively, when attached to solid surfaces, forming thin cell lawns one to several layers thick. Under optimal conditions, primary epithelial cells form layers one cell thick (monolayers), while the lawns formed by fibroblasts are frequently 2 to 3 cells thick. Only rarely can cell lines be derived that multiply well in suspension. Such lines generally originate from tumors (transformed cells), conceivably because the surfaces of tumor cells bind to each other less well than do those of normal cells which form organized tissue masses. Cultures containing cells that grow well in suspension produce many more cells than those whose growth is restricted to surfaces. So the cell lines most commonly used to probe the molecular biology of vertebrates are those like HeLa and KB, which have suspended growth patterns.

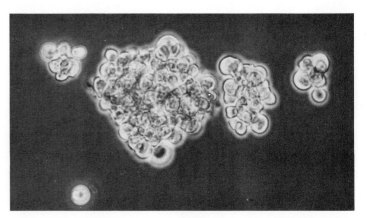

Figure 18–6
Human lymphoid cells growing in suspension. Such cells grow into tiny clumps which do not stick well to solid surfaces. [Photo courtesy of Dr. John Hlinka, Sloan-Kettering Institute (250×).]

Lymphoid cells also grow best in suspension, either as single cells or in tiny clumps (Figure 18–6). Generally, resting lymphocytes can be coaxed to divide only in the presence of certain mitogens (see page 562), or following infection with certain viruses. The resulting lymphoid lines are usually immortal, and gram amounts of such cells are easily obtainable. Though other blood cells are still quite difficult to grow in culture, the recent discovery of factors (see page 571) that promote the differentiation of erythrocytes, granulocytes, and macrophages generates hope that it will soon be possible to routinely cultivate all the major blood cells.

WORKING OUT NUTRITIONAL REQUIREMENTS

Possession of well characterized cell lines allows controlled experiments on the nutritional requirements of vertebrate cells. A semisynthetic medium consisting of 13 different amino acids, 8 vitamins and a mixture of inorganic salts, together with glucose as an energy source and dialyzed horse serum (Table 18–2), permits HeLa and "L" cells to grow optimally to high densities. Additional requirements (e.g., serine, inositol, pyruvate) sometimes emerge when cells grown at low density leak too many essential metabolites. Continuously passed cell lines tend to grow better at lower densities than newly derived lines, suggesting that cloning probably exerts a selection pressure for diminished leakiness.

Table 18–2 Basal Media for Growth of the HeLa Cell Line in Culture [After Eagle, *Science*, **122**, 501 (1955)].

L-Amino Acid (mM)		Vitamins (mM)		Salts (mM)	
Arginine	0.1	Biotin	10^{-3}	NaCl	100
Cysteine	0.05	Choline	10^{-3}	KCl	5
Glutamine	2.0	Folic acid	10^{-3}	$NaH_2PO_4 \cdot H_2O$	1
Histidine	0.05	Nicotinamide	10^{-3}	$NaHCO_3$	20
Isoleucine	0 2	Pantothenic acid	10^{-3}	$CaCl_2$	1
Leucine	0.2	Pyridoxal	10^{-3}	$MgCl_2$	0.5
Lysine	0.2	Thiamine	10^{-3}		
Methionine	0.05	Riboflavin	10^{-4}		
Phenylalanine	0.1				
Threonine	0.2			Miscellaneous	
Tryptophan	0.02				
Tyrosine	0.1			Glucose	5 mM
Valine	0.2			Penicillin	0.005%*
				Streptomycin	0.005%*
				Dialyzed human serum	5%

* To prevent microbial contamination

The requirement for serum is almost universal for all vertebrate cell lines, hinting that one or more serum factors may play key roles in regulating *in vivo* cell multiplication. *In vitro*, the final cell densities reached by specific cell lines are often a function of the serum concentration. Certain cell lines that form a one-cell-thick monolayer in 1% serum grow layers 3 to 4 cells thick in 10% serum. Also very important is the exact pH at which cells grow. The optimal pH can vary from one cell line to another, though most normal cells grow best at around pH 7.6.

"NORMAL" CELL LINES

In the early days of cell culture, the suspicion existed that cells became capable of growing in culture only when they acquired mutations that gave them the cancer cell's capability for unrestricted growth. This belief arose not only because it was generally found much easier to make cell

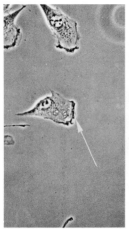

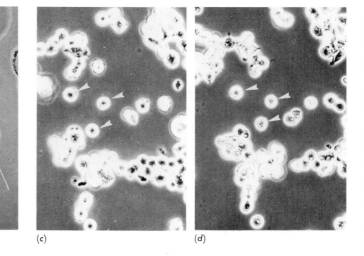

(a)   (b)   (c)   (d)

Figure 18–7

The morphologies of normal (cell line 9) vs. transformed (cell line 14b) hamster embryo cells. The phase-contrast photomicrographs (250×) in (a) and (b) show the spread-out, flat character of the normal hamster cell growing on a flat surface. The cell in the center is moving toward the right with its fan-shaped leading edge containing many ruffles (arrow). Five hours elapsed between the two photos. In (c) and (d) are seen corresponding views (300×) of cancerous hamster cells resulting from transformation by Adenovirus 2. They have a much rounder appearance and exhibit very little motility (arrow points to cells which have not moved in the 19 hours separating the two photos). Daughter cells tend to stay close together, resulting in tiny islands of cells. [Reproduced with permission from Goldman *et al.*, *Cold Spring Harbor Symp.*, **XXXIX,** 601 (1975).]

lines out of cancer cells, but also because many so-called "normal" cell lines had the roundish cancerous morphology, as well as the property of disorganized growth (Figure 18–7). Fortunately, our ideas suddenly flip-flopped when the growth properties of mouse cell lines were found to depend upon the cell density at which secondary cells are grown. If secondary cultures are always grown to high cell density, then the resulting cell lines form cultures of disorganized cells several layers thick. Such cells, though of normal origin, generally look like cancer cells, and form tumors when injected into immunologically identical mice. If, however, the secondary mouse cultures are always kept at the low densities in which cell–cell contact is infrequent, the resulting cell lines have normal fibroblast morphology, and form the thin organized monolayers characteristic of "normal" cells (Figure 18–8). Correspondingly, they only infrequently form tumors when injected into appropriate mice.

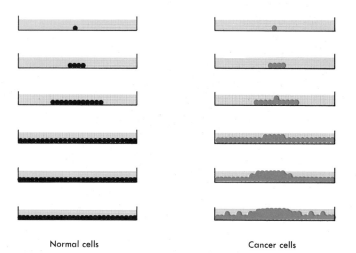

Normal cells   Cancer cells

Figure 18–8
Schematic comparison of the multiplication of a normal cell and of a cancer cell upon a solid surface. The normal cells divide until they form a solid monolayer. Cancer cells, however, often have less affinity for the solid surface and form irregular masses, several layers deep.

Realization of the powerful selective forces that exist during the "crisis period" now allows relatively easy isolation of cell lines with desired growth properties. By keeping the cell number always low, there is no selection in favor of cancerlike cells which can grow well even under crowded conditions. So there now exist a number of cell lines (e.g., 3T3) that have many of the growth properties of the "normal" cell and so can be used for meaningful comparisons with their cancerous counterparts.

CELL TRANSFORMATION

The sudden change of a cell possessing normal growth properties into one with many of the growth properties of the cancer cell is called *cell transformation.* Spontaneously, this phenomenon occurs only rarely but, as we shall see in a later chapter, transformation occurs at much higher frequencies after infection with a tumor virus, or exposure to a cancer-causing (carcinogenic) chemical or radiation source. Irrespective of its origin, however, the *transformed cell*, besides often having the potential to grow into tumors, usually has a much rounder shape reflecting less organized arrangement of cytoplasmic microtubules and microfilaments, is much less subject to contact inhibition of movement, grows to higher cell densities, is often less restricted to growth on solid surfaces, and appears nutritionally less fastidious, particularly in its serum requirements. Not all transformed cells, however, even those derived from the same cell line, are morphologically identical; and as we shall detail later, the cancerous phenotype probably arises by a number of quite distinct biochemical changes.

THE CELL CYCLE

The length of the cell cycle under optimal nutritional conditions varies from one cell line to another. At 37° a few lines can divide once every 10 to 12 hours, but the majority require some 18 to 24 hours. Now, most people divide cell cycles into the M (mitosis), G_1 (period prior to DNA synthesis), S (period of DNA synthesis), and G_2 (period between DNA synthesis and mitosis) phases (Figure 18–9). M is the shortest stage, usually lasting only an hour or so. All M phase cells, irrespective of their previous shapes, become completely spherical with accompanying depolymerization of the microtubule skeleton that runs close to the plasma membrane at other times. The length of other growth phases varies considerably from one cell type to another, with G_1 and S often of about equal 8 to 10 hour duration. But in other strains, G_1 lasts only 1 to 2 hours.

A most important tool for the analysis of the cell cycle is synchronization of growth to place virtually all cells in the same phase. One of the best methods of synchronization uses the fact that spherical M phase cells (Figure 18–10) attach much less firmly to glass surfaces than do interphase cells. So, shaking cultures allows the isolation of large numbers of uncontaminated M phase cells, and the subsequent measurement of the rates at which different

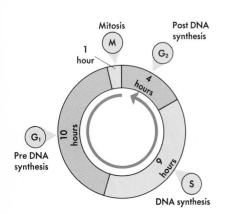

Figure 18–9
Phases in the life cycle of a mouse hepatoma cell growing in tissue culture and dividing once every 24 hours.

Figure 18–10
Scanning electron micrographs of normal hamster cells (cell line 9). (a) shows a cell (3260×) rounded up for mitosis—its surface is covered with microvilli, a characteristic feature of all "M" phase cells. (b) shows two daughter cells (1830×) in the process of separation. Note that both are covered with blebs, very characteristic protrusions of the late M–early G_1 period of normal cells. (c) shows a cell (3260×) in late G_1; no blebs are visible—instead, a number of microvilli are present, especially in the more central, less spread out, region. When the spreading is complete and S phase commences, the outer surface becomes very smooth and only the occasional microvillus can be seen. [Reproduced with permission from Goldman *et al., Cold Spring Harbor Symp.,* **XXXIX,** 601 (1975).]

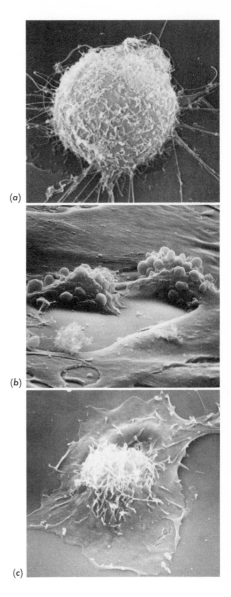

(a)

(b)

(c)

important compounds are made throughout the cell cycle. In HeLa cells, RNA and protein synthesis occur throughout interphase (G_1, S, G_2), but both show abrupt decrease, if not actual stoppage, during the mitotic phase, when the chromosomes contract and line up along the spindle fibers. While the cessation of RNA synthesis may be a consequence of shape changes in chromatin concomitant with chromosome contraction, why protein synthesis declines is not at all clear. Perhaps the supply of mRNA rapidly decreases when RNA synthesis stops; but if this is the case, then the lifetime of mRNA in higher animals must be shorter than the 3 to 4 hour period commonly suggested. In any case, virtually no polyribosomes are seen by electron microscopic examination of metaphase HeLa cells.

Mitotic cells also have much lower cAMP levels than are present during the G_1, S, and G_2 phases. During progression through G_1, S, and G_2, the cAMP level stays approximately constant but, as cells go into M, it falls to about one-half the concentration present in previous stages. Conceivably, the drastic membrane changes which lead to the spherically shaped M phase cells inhibit the functioning of adenyl cyclase, the cAMP-generating enzyme. Or perhaps it is the fall in cAMP levels which leads to the collapse of the filamentous cytoskeleton.

FUSION OF CELLS IN DIFFERENT PHASES OF THE CELL CYCLE

Fusion of a cell in interphase (G_1, S, G_2) with an M phase cell quickly results in the appearance of condensed chromosomes in the nucleus of the interphase cell (Figure 18–11). Mitotic cells thus possess a chemical factor which can promote the condensation of interphase chromosomes.

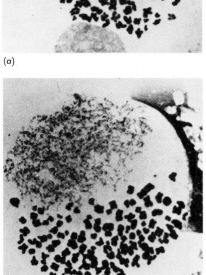

(a)

(b)

(c)

When G_1 cells are used in such experiments, their chromosomes are thinner upon condensation than their M phase partners, reflecting their single chromatid construction. In contrast, when G_2 cells are fused, their chromosomes upon condensation more closely resemble the normal double-chromatid mitotic chromosome. And when S phase cells are fused, the resulting condensed chromosomes are usually highly fragmented, conceivably because they were caught in the process of chromosome duplication and so highly susceptible to nuclease attack.

Figure 18–11
Appearance of the chromosomes in hybrid cells formed by the fusion of two HeLa cells phased in different parts of the cell cycle. (a) M and G_1; (b) M and S; (c) M and G_2. [Reproduced with permission from T. Puck, *The Mammalian Cell as a Microorganism,* Holden-Day, 1972.]

INITIATION OF DNA SYNTHESIS

Once the decision is made to reenter the proliferation cycle, a sequence of preprogrammed operations is initiated long before actual DNA synthesis begins. The levels of the deoxyribonucleoside triphosphate precursors (dATP, dGTP, dCTP, and dTTP) fall very low at the beginning of G_1 and remains so until some thirty minutes prior to initiation of DNA synthesis. Then they rise rapidly to peak amounts during S phase, remaining at relatively high levels until M ends and G_1 commences. How these levels are enzymatically controlled remains to be worked out. Likewise, the enzymology of eucaryotic DNA synthesis itself is still almost nonexistent, and so we have no real clues as to the control process which ensures that each chromosome is replicated just once during each S phase. Fusion of cells in the various phases of the cell cycle may offer an experimental route into this problem. When an S phase nucleus and a G_1 nucleus are brought together, the G_1 nucleus also begins to make DNA, indicating the existence of a positive initiating factor(s). However, fusion between an S phase cell and a G_2 cell does not lead to DNA synthesis in the G_2 nucleus or to its inhibition in the S nucleus. Conceivably the refractoriness of the G_2 chromosomes is related to the rapid coating of newly synthesized DNA by concomitantly synthesized chromosomal proteins. Cell cycle-related changes may have to occur in one or more chromosomal proteins before a new round of DNA synthesis can start.

MUTATIONS IN CELLS GROWING IN CULTURE

Increasing effort now goes toward mutational analysis of well defined cell lines. Because of their diploid (heteroploid) nature, we must anticipate that most mutations that are not sex-linked will be only infrequently expressed. So, much effort has gone toward finding subdiploid cell lines where one or more of the chromosomes of the normal diploid chromosomal complement is missing. Much work has thus been done with the Chinese hamster which not only has a normal $2N$ complement of only 24 chromosomes, but which frequently gives rise to cell lines in which one or more of the autosomes is missing. Using these lines, mutations already have been found which block the biosynthesis of several key metabolites like glycine, adenine, and inositol. Moreover, the routine employment of specific mutagens greatly increases the frequency of such mutations.

Key to such experimentation is the employment of techniques which selectively isolate desired mutants from a large background of normally appearing cells. The best such technique utilizes the fact that exposure to visible light rapidly kills cells which have incorporated the base analogue 5-BUdR into their DNA (Figure 18–12). Only actively multiplying cells incorporate 5-BUdR (or thymidine) into their DNA, and so those mutagenized cells that are unable to grow are not killed by an exposure to visible light, which kills virtually all the multiplying cells.

The BU-visible light technique has been particularly successful in the isolation of *temperature-sensitive* (*ts*) *mutants*, which are unable to multiply at high (low) temperature but which multiply normally at low (high) temperature. This class of mutants should encompass virtually all cellular functions, since their creation depends only upon induction of amino acid changes that lead to loss of protein function at either the high or low temperature. Already there exist ts mutants defective in DNA replication, RNA synthesis, protein synthesis, and ion transport, as well as others unable to traverse specific stages in the cell cycle. There are, in addition, mutant cells which can express the transformed state only at low temperature ($\sim 30°$), as well as other mutants which can express the transformed phenotype only at high temperature ($\sim 37°$)

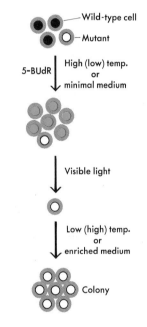

Figure 18–12
Use of the 5-BUdR-visible light technique for the isolation of mutant cells unable to multiply as the result either of a nutritional deficiency or the possession of a temperature-sensitive vital enzyme or structural protein. A mutagenized mixed population is exposed to 5-BUdR under conditions where only wild-type cells grow and incorporate 5-BUdR into DNA. Subsequent exposure to visible light kills only the wild type cells, leaving the mutant cells untouched. Subsequently, an enriched medium is added (or the temperature lowered (raised)) to allow the mutant cells to grow into colonies.

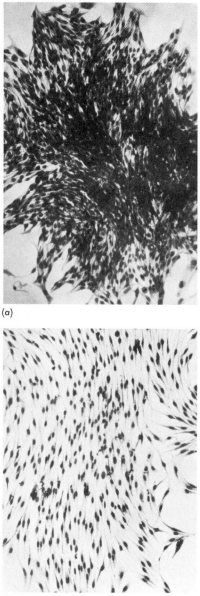

(a)

(b)

Figure 18–13
Photomicrograph of temperature-sensitive mutants of a transformed Chinese hamster lung cell line that prevent the expression of the transformed state at high temperature. (a) The transformed growth pattern of cells growing at 34°. (b) Normal growth pattern of cells growing at 39°. [Reproduced with permission from K. Miyashita and T. Kakynaga, *Cell,* **5**, 131 (1975).]

(Figure 18–13). So now it seems very likely that, as with *E. coli,* the possession of key mutants may play decisive roles in the understanding of important biological processes.

STOPPING IN EARLY G_1

Examination of fibroblasts that have stopped dividing upon formation of a confluent monolayer reveals that all the cells are blocked in early G_1. Similarly, when the growth of such cells is prematurely blocked by removal of essential growth factors like isoleucine, glutamine, phosphate, or serum, the starved cells do not stop randomly in the cell cycle but always proceed through mitosis, to yield cells with the 2N amount of DNA. In each case, they stopped growth very early in G_1, for DNA synthesis is commenced about 8 hours after reintroduction of the limiting growth factors (about the length of the G_1 stage in these cells). When so blocked in early G_1, quiescent cells remain healthy even if the needed nutrients are withheld for many generation times. This picture parallels the situation of most *in vivo* animal cells, which can remain viable and metabolically active even after months to a year in a nonproliferating state. In contrast, if poisons like colchicine or hydroxyurea are added to cells, they do not block the cells at G_1: colchicine stops cells in M, while hydroxyurea interrupts S phase DNA synthesis. Cells so inhibited do not long remain viable, and usually die within 24 hours after exposure to these respective poisons.

Many cancer cells have lost the control mechanism which sends nutritionally limited normal cells into G_1 quiescence. So when, for one reason or another, they do not divide, a spectrum of cancer cells blocked in all phases of the cell cycle is produced. As such, they are much less healthy than their normal equivalents and die off more rapidly. This concept provides a rationale for the frequent observation that the presence of antimetabolites differentially kills off cancer cells.

ACTIVATION OF G_1 QUIESCENT CELLS BY MITOGENIC STIMULI

Exposure to any of a large number of different stimuli, that collectively are called *mitogens,* triggers many quiescent G_1 blocked cells into division. Some mitogens, like the hormone *somatomedin* or the several specific "growth factors" present in normal serum, most likely are normal physiological effectors. But other strong mitogens, like some bacterial lipopolysaccharides and certain cell agglutinating proteins (*lectins*) isolated from plant seeds, normally may never provoke cell division. Yet their addition invariably sets certain quiescent cells into division. Likewise, proliferation often results when certain cell types are treated with small amounts of the proteolytic enzyme trypsin or with neuraminidase, the enzyme which removes the complex amino sugar sialic acid from surface proteins.

Between exposure to the mitogen and initiation of DNA synthesis, some 8 to 10 hours normally must pass, raising the question of which is the primary action triggered by the various mitogens. Now the best guess is that most of them act by modifying the cell surface somehow, generating signals that drastically alter intracellular metabolism.

SOMATOMEDIN AS THE INTERMEDIATE IN THE ACTION OF THE PITUITARY GROWTH HORMONE

Purification of the *serum factor(s)* that are necessary for the growth of secondary chicken and rat embryo fibroblasts reveals activity in several closely sized small polypeptides of MW $\sim$ 8,000, that differ from each other in charge. As yet, they have not been isolated in amounts sufficient for detailed chemical analysis; so whether they are chemically related is not known. Most importantly, the active substances appear similar to somatomedin, a polypeptide hormone synthesized in liver cells in the presence of pituitary growth hormone (Figure 18–14). Though the pituitary growth hormone has long been known to control animal size, how it acted at the molecular level remained totally elusive until several years ago. Then serum from hypophysectomized animals (those from whom the pituitary gland has been removed) treated with growth hormone was found to contain a factor that stimulated the incorporation of SO_4^{-2} into chondroitin (cartilage) slices, as well as the oxidation of fat and the synthesis of DNA. Since growth hormone itself generates no such activity when added to serum from pituitary-free animals, searches were made for the intermediate. It now has been characterized as an $\sim$ 8,000-MW polypeptide fraction that is given the name *somatomedin* because of its ability to regulate the growth of most forms of somatic tissue.

Work with somatomedin is still in too preliminary a state to know whether it is a homogeneous molecule or whether there are many forms of it that induce quiescent cells to multiply. Though there are claims that most sera contain a variety of differently sized growth-promoting molecules, somatomedin is very sticky and binds tightly to many different serum proteins that by themselves have no activity.

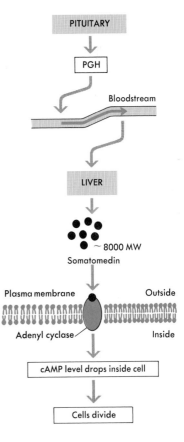

Figure 18–14

Schematic view of how the pituitary growth hormone (PGH) acts through somatomedin on the membrane-bound adenyl cyclase, reducing intracellular cAMP. Compare with the action of insulin shown in Figure 18–18.

Easily confused in its biological effects with soma-tomedin is the pancreatic hormone *insulin*. Addition of insulin to certain G_1-blocked cells (e.g., chick) suspended in serum-free medium induces a cycle of cell division that completely mimics somatomedin action, thereby raising questions whether insulin itself is an active serum factor. Now it appears not to be, since the amount of im-munologically identifiable insulin in serum is some 1000-fold less than the insulin amount necessary to pro-duce the growth-stimulatory activity seen in serum.

Some cells initially thought not to have a specific serum requirement are now realized to secrete the soma-tomedin that promotes their growth. Certain cultured liver cells, in fact, secrete so much somatomedin that media conditioned by their prior growth provide the best known material for somatomedin purification.

RECEPTORS AT THE CELL SURFACE

Neither somatomedin nor insulin enters the cells it acts upon. They work by combining with specific receptor molecules located on the plasma membrane. Each target cell contains many thousands of such receptors, which in general are specific for the respective hormone (growth factor). Somatomedin and insulin, however, can bind to the same receptors on fat and liver cells, suggesting that these two molecules may have common structural features. The insulin (somatomedin) receptor from fat cells has been extensively purified by specific adsorption (affinity chro-matography) to columns containing insulin covalently bound to an insoluble matrix. It appears to be a very large glycoprotein (MW $\sim$ 300,000) that is only soluble in hydro-carbon-rich environments like those provided by the membrane lipid bilayers.

As far as is now known, all the other small polypep-tide hormones (e.g., glucagon, ACTH, prolactin, and vaso-pressin) also bind to specific surface receptors. The response of given cells to polypeptide hormones thus de-pends upon which receptors are present on their cell sur-faces. For example, small lymphocytes (see page 610) which are insensitive to insulin action completely lack any insulin receptors.

SPECIFICITY OF THE NERVE GROWTH FACTOR FOR SYMPATHETIC NEURONS

Growth of the neurons of the sympathetic nervous system requires a specific factor called the nerve growth factor (NGF), a protein of MW $\sim$ 26,500, composed of two identical 118 amino acid-long polypeptide chains. It can be isolated both from a variety of tissue sources and from serum, but is so far found in largest amounts in certain snake venom and the salivary glands of mice. Antibodies against NGF, when injected into mice, cause the specific degeneration of the sympathetic nervous system, showing that it plays a vital role in the maintenance of the nervous system. *In vitro* studies, using sympathetic ganglion, show that addition of NGF quickly stimulates a variety of anabolic processes which lead to outgrowth of neurites from the ganglionic masses (Figure 18–15). NGF binds only to surface receptors found on sympathetic nerve cells, and shows no affinity for receptors isolated from a variety of

(a)

(b)

Figure 18–15
Outgrowth of neurites from sympathetic ganglion cells upon stimulation by the nerve growth factor. (a) Chick embryo ganglion in a control medium. (b) Ganglion in a medium supplemented with salivary NGF at a concentration of 0.01 γ/ml. [Reproduced with permission from R. Levi-Montalcini, *Science,* **143,** 105 (1964).]

other cell types (e.g., heart, kidney). Nor is its binding competitively inhibited by the presence of the pituitary growth hormone or the epidermal growth factor (see below). Its extreme tissue specificity is the result of the localization of its receptor to one very specific type of nerve cell.

Examination of its amino acid sequence shows unmistakable homologies to the sequence of the insulin precursor "proinsulin," suggesting a common evolutionary origin long in the past (Figure 18–16). NGF, however, shows no affinity for the insulin receptor, nor does insulin bind to NGF receptors.

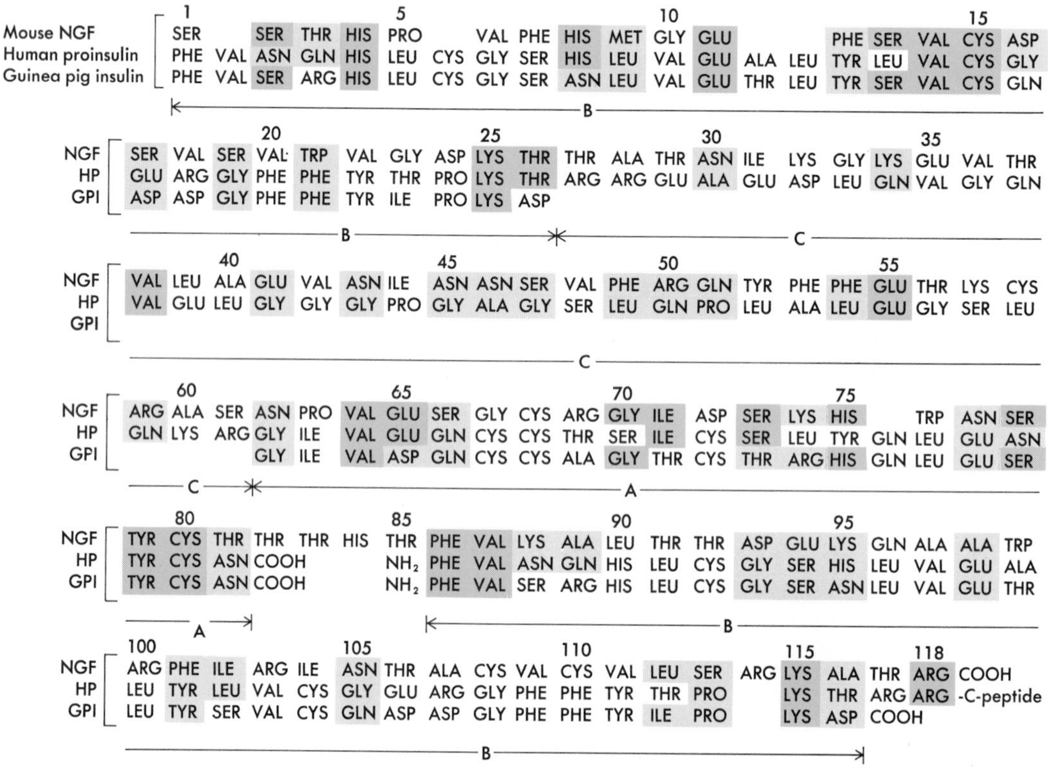

Figure 18–16

The sequence homology between nerve growth factor (NGF) and insulin. The amino acid sequence of the primary subunit of mouse nerve growth factor is aligned with that of human proinsulin and guinea pig insulin. Solid color blocks indicate sets of identical residues and shaded color blocks indicate favored amino acid replacements (1 base change in codon). The letters beneath the rows indicate the limits of each A, B, and C chain of insulin. The similarities in primary sequence also lead to a similar 3-D conformation for the two molecules. [Redrawn from Frazier et al., *Science*, **176**, 482 (1972).]

SPECIFIC RECEPTORS FOR THE EPIDERMAL GROWTH FACTOR

Salivary glands are also a rich source of the epidermal growth factor (EGF), a polypeptide of MW 6400 which specifically stimulates the growth of epidermal (skin) cells both *in vivo* and *in vitro*. The specificity of its binding has not yet been thoroughly examined, but already it is clear that it does not bind to either the insulin or NGF receptors. Work with EGF has been hindered by difficulties in making its assay reproducible, but hopefully this problem will soon be overcome, and its mode of action following binding to its receptor can then be explored.

BRAIN TISSUE AS A SOURCE FOR THE FIBROBLAST GROWTH FACTOR

Apparently unrelated to any of the above growth factors is the *fibroblast growth factor* (FGF), a newly discovered 13,000-MW polypeptide. It was initially found because of its capacity to replace the "serum requirement" of mouse embryo cells growing in the presence of glucocorticoids (a form of steroid). Now it appears, however, that FGF is not a major growth-promoting factor in serum, and *in vivo* it is largely found in brain tissue. So conceivably its normal cellular target is nervous tissue. If so, the term FGF is a complete misnomer. Supporting this conjecture are very recent experiments showing that it potentiates nerve regeneration following the severance of limbs in amphibia.

MODIFICATION OF MEMBRANE-BOUND ADENYL CYCLASE ACTIVITY BY HORMONE–RECEPTOR INTERACTIONS

Combination of insulin or somatomedin with its membrane-bound receptors leads immediately to an inhibition of the membrane-bound enzyme (adenyl cyclase) responsible for cAMP synthesis (Figure 18–17). Formation of the insulin receptor complex thus quickly leads to a lowering of the intracellular cAMP levels, which, in turn (see below) pushes the general direction of cellular metabolism in an anabolic (synthetic) manner (Figure 18–18). Not all hormone-surface receptor interactions, however, lead to adenyl cyclase inhibition. Glucagon, which has an anti-insulin effect, stimulates adenyl cyclase activity when it combines with its specific receptor. Likewise, surface binding of ACTH, epinephrine, or prostaglandin E1 leads to higher intracellular cAMP levels. As yet no one has been able to highly purify adenyl cyclase away from its normal

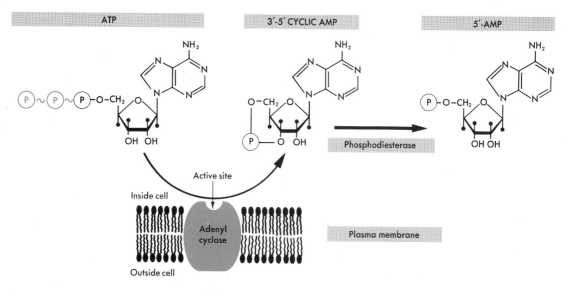

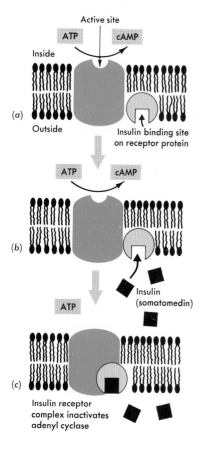

Figure 18–17

The synthesis of cyclic AMP (cAMP) by the membrane-bound enzyme adenyl cyclase. cAMP is broken down inside the cell by a phosphodiesterase. The synthesis of cAMP may be stimulated, via membrane-bound receptor proteins, by such agents outside the cell as epinephrine, prostaglandin, glucagon, and ACTH.

Figure 18–18

Schematic representation of how a hormone (insulin), by binding to a receptor protein, inactivates adenyl cyclase bound in the plasma membrane, thus leading to a drop in the intracellular cAMP level.

membrane environment, and so we are still totally in the dark as to how its activity can be modified by combination of a receptor with its respective hormone. Most likely there is direct physical contact between the receptor and adenyl cyclase. But because of the semifluid nature of the cell surface most membranes are constantly changing their exact neighbors, and so how adenyl cyclase activity is regulated may be very tricky to work out.

Mitogens like trypsin and the lectin concanavalin A (ConA) also inhibit adenyl cyclase. Trypsin enzymatically clips off portions of many membrane proteins, leading to an increase in the fluidity of the exposed membranes. So it is not clear whether it acts directly or indirectly on adenyl cyclase. ConA's mitogenic ability comes from its ability to stick tightly to the cell membrane with, in fact, a strong affinity for the insulin receptor itself.

Nerve growth factor, however, does not have any effect on cAMP levels. No changes in adenyl cyclase activity occur when NGF binds to its receptors, nor does addition of cAMP to sympathetic ganglion cells affect the rate at which neurites grow out.

PLEIOTROPIC EFFECTS OF CHANGES IN cAMP LEVELS

Whether a lowering of the cAMP level by itself triggers the appropriate quiescent cells into division is still most murky. When cAMP is present at high concentration, cells assume a quiescent state characterized by low levels of the active transport of nutrients into cells, rapid fatty acid and protein degradation, and low levels of RNA and protein synthesis. In contrast, if cAMP levels are low, sugar intake rises, synthetic reactions speed up, and degradative processes like the burning of fat slow down. Conceivably, all these diverse effects reflect the fact that high cAMP levels stimulate the activity of a variety of different phosphorylating enzymes (kinases); and whether synthesis or degradation is favored may depend upon whether the corresponding enzymes are phosphorylated.

RISE IN cGMP CONTENT AFTER MITOGENIC STIMULATION

In quiescent cells blocked in G_1, the level of cAMP is much higher (10 to 50 times) than that of the corresponding cyclic guanine nucleotide (cGMP). But a few minutes after administration of a surface-acting mitogen, the cGMP level quickly rises, often approaching that of cAMP (Figure 18–19). How the rise in cGMP occurs was at first a puzzle since, unlike adenyl cyclase, which is strictly a membrane-bound enzyme, most guanyl cyclase appeared to exist free in the cytoplasm. Now it seems that much is membrane-bound, suggesting a direct influence of a hormone-membrane receptor interaction upon such guanyl cyclase activity.

cGMP, like cAMP, appears to stimulate various protein kinases, with some kinases apparently specific for cGMP. This raises the question whether cGMP-specific kinases, not the cAMP-specific kinases, may sometimes be the key enzymes that set cell division into motion. Arguing this way is the fact that small lymphocytes can be activated by mitogens without any changes in their cAMP content. In contrast, their cGMP amount always quickly rises soon after mitogen addition.

ACTIVATION OF NUCLEAR RNA SYNTHESIS

Only some ten minutes need pass between addition of a surface acting mitogen and an increased rate of RNA synthesis. How this new RNA synthesis is activated by a fall in cAMP concentration (rise in cGMP levels?) remains to be worked out. Now, most effort goes toward studying

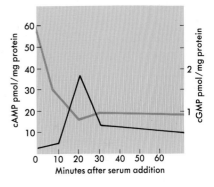

Figure 18–19
Rise in cGMP level and fall in cAMP level upon receipt of a mitogenic stimulus (serum addition) to G_1-blocked cells.

whether changes can be detected in the various chromosomal proteins as activation occurs. In particular, the level of phosphorylation of the five main histones, as well as that of the various nonhistone acidic chromosomal proteins, has been carefully studied throughout the cell cycle, with the hope of pinpointing the molecule(s) whose phosphorylation (dephosphorylation?) unlocks the specific RNA synthesis that marks the beginning of a new cell cycle. While changes can be detected in both the histone and nonhistone components during G_1, their significance as far as activation of RNA synthesis is concerned is not clear, and probably will remain unelucidated until *in vitro* systems can be worked out for the study of chromatin activation.

CELL PROLIFERATION FOLLOWING STEROID APPLICATION

The way certain steroid hormones like the estrogens lead to cell proliferation may differ fundamentally from that provoked by the surface acting polypeptides like the hormone somatomedin. They induce proliferation *per se* and do not change the differentiation state of their target cells. Estrogen, in contrast, not only triggers off the proliferative growth of the uterus (oviduct) but also radically changes its metabolic character. The first visible response of the chicken oviduct to estrogen is the very noticeable cell enlargement which accompanies the upgrading of its protein synthesizing machinery for the massive production of its major secretory product albumin. We thus do not yet know whether the proliferation events which accompany oviduct differentiation are an immediate result of estrogen action or the delayed consequence of a series of progressive biochemical changes that follow the induction of the specific protein secretory products.

In any case, the steroid hormones do not have any direct effect on cyclic nucleotide levels. Their hydrocarbon nature allows them to readily pass through the plasma membrane to interact in the cytoplasm with their respective protein receptors (Figure 17–33). The resulting steroid–receptor complexes then move to the nucleus and combine with specific chromosomal sites. The activation of specific RNA synthesis that follows is believed to be the direct result of the binding of the receptor–steroid complex to chromatin, and not to involve cyclic nucleotide-induced changes in chromosomal proteins.

ERYTHROPOIETIN INDUCTION OF RED BLOOD CELL (ERYTHROCYTE) PRODUCTION

The level of red cells in blood is under the specific control of erythropoietin, a glycoprotein of MW $\sim$ 46,000. It induces the selective proliferation and differentiation of proerythroblasts, the immediate cellular precursors of erythrocytes (Figure 18–20). The earliest detectable effect of erythropoietin is the selective stimulation of RNA synthesis. It occurs within one hour of addition and involves the synthesis of tRNA, rRNA, and mRNA. The mRNA molecules that code for hemoglobin itself appear only some 10

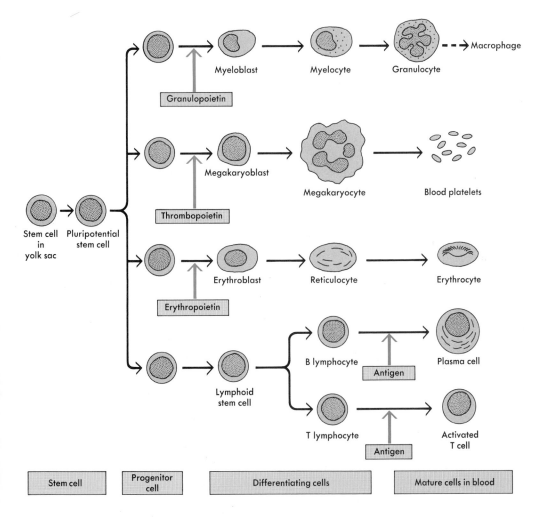

Figure 18–20
Schematic diagram showing the probable cell lineages in the differentiation of the various types of vertebrate blood cells.

hours later, indicating that hemoglobin formation is the end result of an involved differentiation process that leads not only to hemoglobin formation but also to the mitotic events which increase the number of erythropoietin-sensitive cells. Whether erythropoietin enters its target cells or binds to specific receptors on the cell surface remains unknown. Unfortunately, erythropoietin remains difficult to obtain and has not yet been prepared in pure form.

DIFFERENTIATION OF BLAST CELLS INTO GRANULOCYTES AND MACROPHAGES REQUIRES A PROTEIN INDUCER

The granulocyte and macrophage components of blood form by a differentiation–mitotic process starting from common precursor cells (blasts) found in bone marrow (Figure 18–20). These events now can be studied *in vitro* following the discovery of a specific protein inducer, *granulopoietin,* initially called the colony-stimulating factor (CSF) or the macrophage-granulocyte inducer (MGI). Most activity resides in a glycoprotein of MW $\sim$ 70,000 which is present in serum at concentrations that are physiologically effective *in vitro.* Its normal cellular source remains obscure since it is released from many types of cells (e.g., fibroblasts and macrophages themselves!). In the presence of CSF, the precursor blast cells in culture multiply to form colonies which initially contain undifferentiated cells. But as more divisions occur, granulocytes and then macrophages are found. It appears that the blast cells first differentiate into the granulocytes, which in turn can mature into macrophages. Whether both these steps are induced by the same molecule, or whether CSF is, in fact, a mixture of two different activities, remains to be clarified.

CONVERSION OF FIBROBLASTS INTO ADIPOSE CELLS

Not all differentiation processes require specific external inducing factors. The conversion of fibroblasts into *adipose (fat) cells,* for example, depends only on the continuous availability of a rich nutrient source under conditions where cell division is not possible (e.g., excessive crowding). While multiplying, 3T3 cells always maintain a fibroblast shape. But upon reaching confluence they have an increasing probability of converting into very large, round adipose cells (Figure 18–21). Not all arrested fibroblasts become so differentiated, and only certain 3T3 strains (preadipose strains) have high probabilities of undergoing

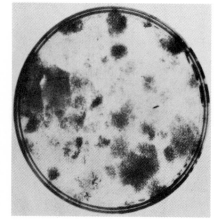

Figure 18–21
Formation of fat cells within a confluent layer of mouse fibroblasts (3T3 L1). Three weeks after the colonies became confluent, the culture was stained with Oil Red O (shows black on this photograph) that selectively stains fat deposits red. The marked variation in size and staining intensity reflects differences in the proportion of cells undergoing adipose conversion in the different colonies. [Reproduced with permission from H. Green and M. Meuth, *Cell,* **3**, 127 (1974).]

the adipose conversion. It is lack of growth, not contact between arrested cells, that leads to fat accumulation. This is shown by suspending single 3T3 cells in methyl cellulose gels. Under such conditions, individual 3T3 cells can increase in size but cannot make new DNA or divide. As they grow larger, neutral fat (triglyceride) accumulates until the center of the cell is filled with a huge fat droplet which pushes the nucleus into an eccentric position near the cell membrane (Figure 18–22).

As long as the adipose cell maintains its huge fat droplet, it is incapable of cell division. But if unfavorable nutritional conditions lead to the breakdown of its fat into energy-rich compounds, it then resumes its fibroblast appearance and is capable of further cycles of growth and cell division. So we see that differentiation *per se* is not incompatible with future cell division. Only when the functioning of a highly differentiated cell demands continuity of physical shape does mitosis become impossible. Clearly it would make no sense for the highly interconnected nerve cells to be capable of mitosis, so these cells have lost their capacity to divide. Instead, as in the case of the various blood cells, they arise by the division of partially differentiated, functionally inactive precursors that are called neuroblasts.

MAINTENANCE OF MYOBLASTS IN CONTINUOUS CELL CULTURE

Now it is also routinely possible to culture *myoblasts,* the cellular precursors of the multinucleated striated muscle cells. Though myoblasts give the appearance of essentially undifferentiated cells, in culture they usually divide only several times before they begin to aggregate and fuse into postmitotic, multinucleated muscle cells that soon acquire a cross-striated appearance and develop a contractile potentiality. By this stage, the differentiation process is irreversible, and no further DNA synthesis is possible. The resulting multinucleate cells, however, retain the capacity to make the mRNA for the many proteins whose continued synthesis is essential for normal muscle functioning.

Such irreversible differentiation, however, is not a necessary fate for all dividing myoblasts. Circumstances recently have been found that can maintain myogenic cells in continuous cultures. The key elements are conditions which inhibit the cell fusing process (e.g., low Ca^{++} or periodic additions of trypsin), together with employment of highly enriched serum of fetal origin. The transmission

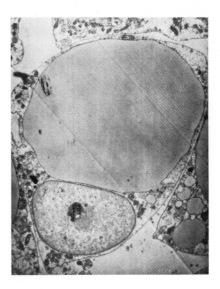

Figure 18–22
A thin section of a mature adipose cell, showing a thin cytoplasmic rim surrounding the large central fat droplet. The nucleus has the characteristic eccentric location of adipose cells. Electron micrograph by Elaine Lenk of the Massachusetts Institute of Technology. [Reproduced with permission from H. Green and M. Meuth, *Cell,* **3,** 127 (1974).]

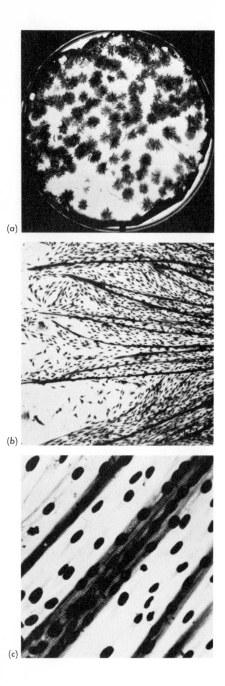

Figure 18–23
Fusion of myoblasts and the development of striated muscle fibers set within colonies of multiplying myoblasts. (a) View of a petri plate upon which 150 myoblasts had been plated 13 days previously. (b) Higher magnification of a portion of a single colony, showing large numbers of nuclei surrounding each developing fiber. (c) A still higher magnification, showing the striated nature of the individual fibers. [Reproduced with permission from D. Yaffee, *Proc. Nat. Acad. Sci.*, **61**, 477 (1968).]

of a specific differentiation potentiality to progeny cells thus is not necessarily dependent upon a visible morphological expression. So all the decendants of a single cloned myoblast can give rise to muscle-forming colonies (Figure 18–23).

SEARCH FOR CHEMICAL DIFFERENCES BETWEEN NORMAL AND CANCER CELLS

Almost as soon as scientists began to describe molecules within normal cells, they looked to see whether those same molecules were found in cancer cells. Likewise, when a new chemical reaction (or enzyme) was described in a cancer cell, a search was often made to see if the same reaction occurred in the normal cell. Often these searches initially seemed to lead somewhere, with the tumor cells containing much more (or less) of a particular compound than their normal equivalents. Further analysis, however, usually failed to justify the initial hope that the respective change was at the heart of the cancerous phenotype. A fundamental difficulty in this approach has been the inability to decide whether a given change is the primary metabolic disturbance that creates the cancer, or whether it is a minor secondary response to the altered metabolism caused by the primary change. Moreover, in many such experiments, it has been impossible to select a good control with which to compare cells from a growing tumor, since we cannot be sure in what type of normal cell the cancerous transformation has occurred. Also, comparisons have frequently been made between cells from actively growing animal tumors and their so-called normal counterparts growing in culture. But many of the so-called "normal" cell lines may have undergone a number of genetic changes during their adaptation to growth in the artificial environment of cell culture.

So now there has been an increasing tendency to compare a "normal" cell line (e.g., 3T3) with the "transformed" cell line created by exposure of cells from the "normal" cell line to a cancer-causing virus. But here again, one must guard against being misled by new sets of spontaneous mutations which occur during the further growth in cultures of the respective normal and transformed cell lines. Even using so-called parallel cultures of normal and transformed cells, far too many biochemists have confused irrelevant secondary modifications with the primary metabolic lesion that creates the cancerous state. Changes first thought to be the key to a given type of cancer are all too often not found when looked for in other cancerous cells of the same type. So the systems now most favored for probing cancer biochemistry are cell lines that have been transformed by mutated cancer viruses (see page 675). The resulting cells have a cancerous behavior when grown at a low temperature (say 32°C) but a normal behavior when grown at a higher (e.g., 38°) temperature.

WARBURG AND THE MEANING OF INCREASED GLYCOLYSIS

The first convincing biochemical difference between normal and cancer cells was discovered over fifty years ago by the German biochemist Warburg. He observed that virtually every type of cancer cell that formed a solid tumor (as opposed to cancers of the blood like leukemia) excretes much larger quantities of lactic acid than do its normal counterparts. They do this both when growing as solid tumors in animals or when multiplying as single cells in culture. Since the original observation in the early 1920s, the meaning of lactic acid overproduction (often called the Warburg effect) has been studied again and again, but its real significance remains tantalizingly elusive. The metabolic source of the excessive lactic acid, however, is known. It arises from glucose via the glycolytic pathway. But this increase in fermentation (anaerobic-metabolism) does not result from an insuffiency of any enzymes involved in the various oxidative pathways (e.g., oxidative phosphorylation proceeds normally). Nonetheless, much more glucose is consumed by these tumor cells than they need to grow and multiply.

This suggests the loss of a normal control device which regulates the rate at which glucose is taken into a cell. Unfortunately, despite years and years of countless many biochemists' careers, no one yet knows how glucose consumption is regulated. Conceivably, tumor cells lack one or more of the proteins in the cell membrane that function in the active transport of glucose across the plasma membrane. But even if this is proved true, we will still be totally in the dark about how to relate such a change to loss of normal proliferation control.

CONTACT INHIBITION OF MOVEMENT

Strong evidence that the plasma membranes of cancer cells are very different from those of normal cells comes from comparing their behavior after their ruffled edges touch each other. Such contacts between normal cells frequently result in cell adhesion, accompanied by a slowing down of the random amoeboid movements by which the respective cells move along surfaces (*contact inhibition of movement*). Early in this process (10 to 20 seconds after contact), actin-containing cables begin to form in the seemingly disorganized cytoplasmic region beneath the ruffled edges (Figure 18–24). At the same time, the ruffled edges start to lose their morphological identity, soon becoming indistinguishable from less active regions of the plasma membrane. Generation of new intracellular actin-containing cables in response to cellular contacts may thus be an essential aspect of contact inhibition of movement.

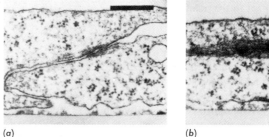

(a) (b)

Figure 18–24

(a) and (b) Electron microscopic views of the "adhesive plaques" (arrows) that form within 10 to 20 seconds of contact between ruffled edges. Organized groups of actin-containing filaments form beneath the plaques within less than a minute. Bar represents 0.5 micron. [Reproduced with permission from J. Haeysman in "Locomotion of Tissue Cells," *CIBA Foundation Symp.*, **14** (new series), 190 (1973).] (c) An electron micrograph in which the plane of thin section cuts just above the region of hamster embryo cell-surface contact. This section reveals the crystalline-like lattices of the adhesive plaques (dark patches) (42,000×). [Courtesy of R. D. Goldman, Carnegie-Mellon University.]

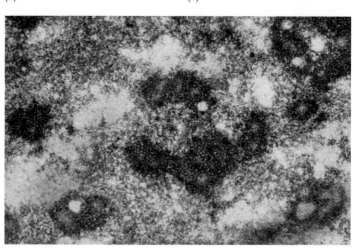

(c)

MALIGNANCY AS A LOSS OF NORMAL CELLULAR AFFINITIES

The "sticky" quality of normal cells that leads to adhesion displays considerable specificity. A given type of cell (e.g., a liver cell) prefers to stick to others of its own kind (e.g., other liver cells) and shows very little, if any, affinity for other types (e.g., kidney cells). This type of specificity has been elegantly demonstrated in experiments in which small amounts of the proteolytic enzyme trypsin are used to break apart organs such as the liver or the kidney into their single cell components. If these isolated cells are then incubated in the absence of trypsin, they reaggregate to form tissue fragments similar to those in the intact organ, that is, small fragments of liver tissue and small fragments of kidney tissue. When kidney and liver cells are mixed together, small fragments of liver and kidney are again detected. Thus, a kidney cell prefers to stick to a kidney cell, and a liver cell to a liver cell.

If this experiment is repeated with cancer cells, however, the normal cellular affinities no longer hold. For example, the mixing of cells from a malignant skin cancer with normal kidney cells results in aggregates in which the skin cancer cells are interspersed among the kidney cells. Conceivably, the inability of cancer cells to form tight adhesive junctions allows them to lie randomly adjacent to almost any cell type. This may be the reason why malignant cells invade a variety of normal organs.

The nature of the "sticky" substance(s) that normal cells use to attach to each other or to solid glass or plastic surfaces is still unclear. Electron micrographs of cells that have stuck to glass (or to each other) reveal dense adhesive regions (plaques) 60 to 80 Å thick which are located just inside the plasma membrane. What their chemical nature is has not yet been established (Figure 18–24). Over the past several years, many speculations have been put forward about how the sugar groups of opposing glycoproteins might bind cells together. But until we possess more firm facts about the molecular aspects of outer cell surfaces, there exists no possibility of testing such hypotheses.

DISORIENTATION OF THE MUSCLES OF TRANSFORMED CELLS

The fact that cancer cells show much less adhesiveness and often move to where they are not wanted, superficially suggests that they might be more organized for cell movement than are comparable normal cells. Examination, however, of many cancer cells reveals that they usually display a much less organized arrangement of the various muscle proteins than is found in their normal equivalents. Normal migratory fibroblasts, for example, contain large numbers of tiny musclelike fibers (cables) in which the various muscle proteins (e.g., actin, myosin, and tropomyosin) are regularly arranged to bring about coordinated cell movement. In contrast, transformed cells contain many fewer such fibers and those that are present are generally much thinner (Figure 18–25). Correspondingly, the touching of the cancer cells does not lead to cessation of cell movement, but instead frequently induces the uncoordinated generation of many more ruffles and blebs than are found on their normal cellular equivalents (Figure 18–26).

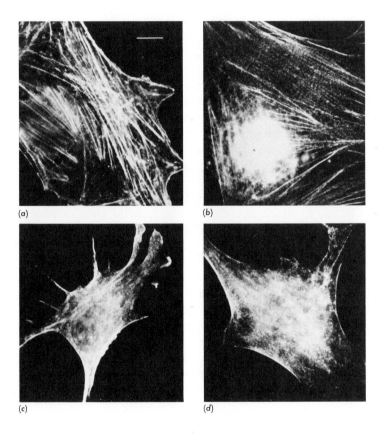

(a) (b)

(c) (d)

Figure 18–25
A comparison of the organization of "muscle" proteins within normal and transformed mouse cells. Immunofluorescent techniques are used to visualize (a) actin within a "normal" 3T3 cell, (b) myosin within a 3T3 cell, (c) actin in a 3T3 cell transformed by SV40 virus (SV3T3), and (d) the myosin of a SV3T3 cell. [Reproduced with permission from R. Pollack *et al.*, *Proc. Nat. Acad. Sci.* **72,** 994 (1975).]

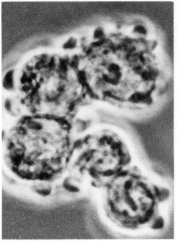
(a)

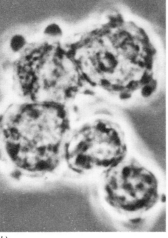
(b)

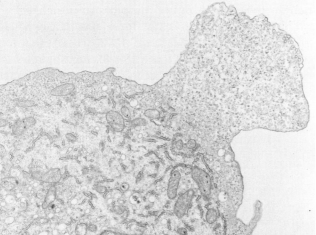
(c)

Figure 18–26

Blebs on the surface of cancer cells. (a) and (b) In these cancerous hamster cells that originated through transformation by adenovirus 5, blebs are a characteristic feature throughout the cell cycle except during M phase. As shown in these photos taken one minute apart, blebs are very unstable protrusions generally having lifetimes of less than a minute. In other types of cancer cells, blebs are less frequent. In their place are found large numbers of microvilli and ruffles which, like blebs, are constantly thrown out and then retracted. [Reproduced with permission from R. Goldman *et al., Cold Spring Harbor Symp.,* **XXXIX,** 601 (1975).] (c) An electron micrograph of a thin section of a portion of a chicken cell transformed by *Rous Sarcoma* Virus, showing a large bleb. It is largely filled with free polyribosomes. The absence of membrane-bounded bodies like mitochondria and ER suggests that they are somehow attached to the nucleus. [Courtesy of Lan Bo Chen and Elaine Lenk, Massachusetts Institute of Technology.]

Why cancer cells are less able to form large numbers of highly organized muscle fibers is not yet understood. The reason, however, is not a relative deficiency of the major muscle structural proteins. As much actin, for example, exists within tumor cells as in the corresponding normal cells. So, conceivably, the primary cancerous defect is a change in one or more components of the plasma membrane, which hinders the attachment of individual actin filaments to the inner membrane surface. More precise speculation (knowledge?) is likely to demand much more information about the detailed structure both of the various surface-membrane components and of the individual muscle proteins.

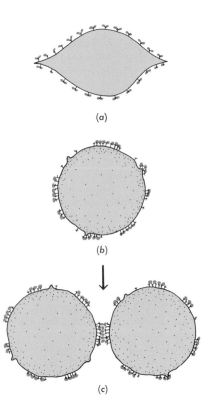

(a)

(b)

(c)

Figure 18–27
A mechanism to explain the selective agglutination of tumor cells by agglutinating
proteins (lectins). (a) In normal cells, the
agglutinin-binding sites (receptors) are
present in a dispersed random distribution
that does not favor agglutination. (b) The
agglutinin receptors of tumor cells are more
mobile and more easily form clumps cross-
linked by the multimeric agglutinins. (c)
Aggregation results when cross bridges
form between receptor clumps. [Redrawn
from Nicolson, *Nature New Biology*, **239**, 193
(1972).]

SELECTIVE PRECIPITATION OF CANCER CELLS BY LECTINS

Further evidence for the belief that the cancer cell membranes have a uniquely different quality than those of
normal cells comes from experiments which show that a
number of lectins (cell agglutinating proteins) selectively agglutinate tumor cells. For example, a glycoprotein
lectin from wheat seeds (wheat germ agglutinin) specifically
agglutinates virtually every form of cancer cell that grows
into solid tumors. At first it was thought that these agglutinating proteins attached only to tumor cells, binding to
sugar-containing protein receptors unique to the cancer
cell surface. Now, however, it appears that normal and
cancer cells possess the same number and kind of agglutinin receptors but that the receptors on the surface of cancer cells diffuse much more readily within the lipid bilayer
than do the corresponding receptors in normal cells. They
can thus much more easily form the dense patches of
cross-linked receptors (Figure 18–27) that bind together
the agglutinated cells (Figure 18–28). For some reason the
lectin receptors of normal cells do not have as much diffusional freedom as those of cancer cells.

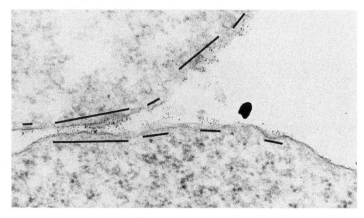

Figure 18–28
Sites of cell contact in an aggregate of SV3T3 cells agglutinated with fer-
ritin-conjugated concanavalin A. The concentration of ferritin between
the agglutinated cells is higher than on the surrounding membrane
regions (82,000×). [Reproduced with permission from G. Nicolson in
Control of Proliferation in Animal Cells, Clarkson and Baserga (eds.), Cold
Spring Harbor Laboratory, 1974.]

A vital clue to the nature of lectin-induced agglutination may come from the recent observation that normal
cells in mitosis, unlike those in G_1, S, and G_2, agglutinate
almost as readily as G_1, (or S or G_2) phase cancer cells.

This preferential agglutinability of the mitotic cell may reflect the almost total disappearance of organized muscle proteins from just beneath the surfaces of dividing cells. Attachment of intracellular muscle-like fibers to the plasma membrane may somehow restrain the diffusional freedom of the many membrane proteins.

MOLECULAR CHANGES AT THE CELL SURFACE THAT ACCOMPANY CELL TRANSFORMATION

Only recently have the chemical techniques come into existence to critically compare the plasma membranes of normal and transformed cells. So, despite intense interest in this problem, current knowledge is still very incomplete and may yet be misleading. Making this problem inherently difficult is the complexity of most plasma membranes, most of which at a minimum must easily contain hundreds of different components. Of necessity, attention has been focused either on the major structural components or on those relatively few membrane-bound enzymes that we can conveniently assay. Thus, if the essential change that defines the cancer cell membrane occurs in a minor structural component(s), or in yet-to-be-discovered membrane-bound enzyme(s), many more years may pass before this problem becomes cracked.

Moreover, it is becoming increasingly clear that normal plasma membranes are not invariant in structure. The exact groupings which protrude on their outer surfaces not only change with the various phases of the cell cycle, but are very much a function of the contacts that they make with other cells. Great caution must be shown in comparing normal cells with their transformed equivalents unless one is sure that both have been harvested in the same place in the cell cycle, and at the same cell density, under identical conditions of growth.

Now it seems fairly certain that the cancerous transformation does not change the relative amounts of the four main phospholipids that form the basic lipid bilayer. Differences, however, sometimes seemingly quite specific, have been seen in both the nature and kind of the glycolipids and glycoproteins that are inserted into the bilayer (Figure 16–3). Much study in particular has gone toward the gangliosides (Figure 18–29), the complex sialic acid-containing glycolipids. Not only are the more complex of them specifically diminished in amounts in certain mouse cell cancers, but the enzymes involved in their biosynthesis are also specifically reduced. It is not at all clear, however, that such changes are primary events in the formation of a cancer cell. Not only do such changes often lag

Figure 18-29
Ganglioside synthesis. The horizontal arrow shows the step thought to be blocked in tumor cells. Normal cells possess all four gangliosides, whereas tumor cells possess predominantly the simplest, GM3. The carbon atom of ceramide to which the successive sugar residues are added is indicated. Key: Glc = glucose, Gal = galactose, NANA = N-acetylneuraminic acid, GalNAC = N-acetylgalactosamine. Enzymes (numbered): 1 = glucosyl transferase, 2 = galactosyl transferase, 3 = sialyl transferase, 4 = N-acetylgalactosaminyl transferase, 5 = galactosyl transferase, and ▶ 6 = sialyl transferase.

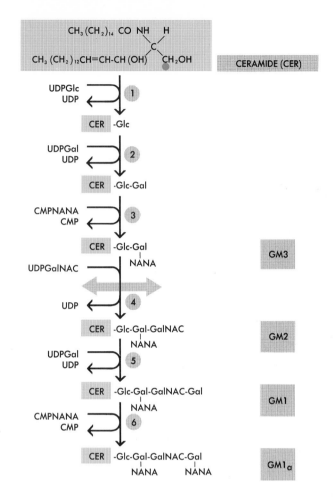

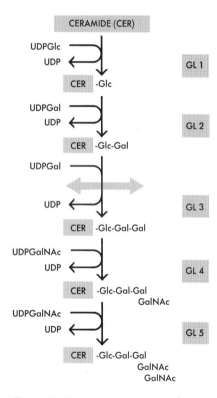

Figure 18–30
The pathway for glycolipid synthesis in hamster cells. The conversion of GL-2 to GL-3 is inhibited in hamster tumor cells, leading to lower amounts of GL-3, GL-4, and GL-5. Key: GL-1 = ceramide monohexoside, GL-2 = ceramide dihexoside, GL-3 = ceramide trihexoside, GL-4 = ceramide tetrahexoside, and GL-5 = ceramide pentahexoside = Forssman antigen. Other abbreviations as in Figure 18–29.

behind the assumption of the distinctive cancer cell morphology when viral transformation occurs but, perhaps more decisive, other transformed mouse cell lines have a perfectly normal ganglioside content. So we must postulate either that changes in a variety of quite distinct molecular components each can produce the distinctive cancer cell membrane, or that the alterations in ganglioside content are irrelevant features that are somehow selected for by a prior primary change. A similar dilemma exists in analyzing changes in those hamster cell glycolipids which lack sialic acid. Again, we see that certain transformed cell lines possess the less complex components (Figure 18–30). But such changes are again either not the earliest events in viral transformation or not present in all cell lines transformed by a given virus.

Of late, much more attention has gone toward examination of the major proteins that are inserted in the outer plasma membrane, especially of the glycoproteins whose

Table 18–3 Characteristics of the Major Chick Cell Polypeptides that Decrease in Concentration following RSV Transformation

Protein	MW†	Cellular location	Glyco-protein	Protease sensitivity	Rapidity of shut-off of appearance following transformation
Z	230,000	exterior surface	?	very sensitive to trypsin and collagenase	days
Ω	206,000 (myosin?)	attached to inner surface of cell membrane (?)	?	not sensitive	<3 hours
Δ	47,000	plasma membrane	Yes	not sensitive	3.6 hours

* Data from Robbins *et al.*, *Cold Spring Harbor Symposium* **XXXIX**, 1173, (1975).
† As determined by SDS-polyacrylamide electrophoresis, this method overestimates MW's if the protein contains large numbers of sugar groups.

exterior protruding surfaces are covered with sugar groups (Table 18–3). Here it is possible that several changes common to all types of cancer cells already have been found. One involves the disappearance from transformed cells of a surface glycoprotein of MW ~ 46,000 (Figure 18–31). This change occurs very early in the transformation process and so is an obvious candidate for a primary event in transformation. It remains to be seen whether the very diminished quantities seen in transformed cells reflects a cutback in its synthesis or its preferential excretion into the extracellular milieu. Another characteristic change is the slow disappearance of an ~ 240,000-MW protein which is present on the exterior surface. This removal occurs long after its respective cells become morphologically cancerous, and so its absence does not seem to be a fundamental feature of the cancerous state.

Conceivably more decisive to the cancerous state may be absence of an ~ 200,000-MW protein which disappears both after transformation by *Rous sarcoma* virus and by the various adenoviruses. Its molecular size is similar to that of myosin, and so we may be observing a change in the organization of the intracellular musclelike fibers.

SELECTIVE SECRETION OF PROTEASES BY TUMOR CELLS

Every form of cancer except that of the blood-forming tissues is characterized by the extracellular secretion of abnormally large amounts of proteolytic enzymes. Though normal cells generally secrete some granule- (lysosomes?)-bound proteases, the process is greatly intensified in tumor cells. The major protease so selectively secreted is called the "cell factor" and has a MW ~ 40,000. Within growing tumors, as well as in cell cultures, its major action appears to be the proteolytic activation of the inert serum protein "plasminogen" to form plasmin, a proteolytic enzyme

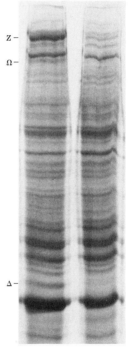

Normal   Transformed

Figure 18–31
A comparison of the major protein components of a normal chick cell with those found in chick cells transformed by *Rous Sarcoma* virus. In this photograph the proteins are visualized after electrophoresis in a polyacrylamide gel which separates proteins according to their molecular weights. Three proteins of the normal chick cell Δ, Ω, and Z are greatly reduced if not missing in the transformed chick cell. [Reproduced with permission from Robbins *et al.*, *Cold Spring Harbor Symp.*, **XXXIX**, 1173 (1975).]

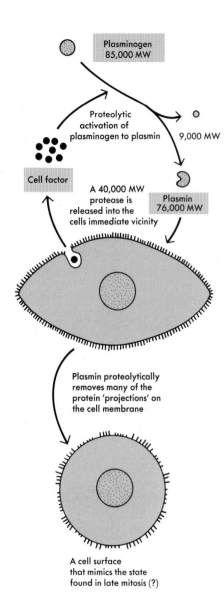

Proteolytic activation of plasminogen to plasmin

Plasminogen 85,000 MW

9,000 MW

Cell factor

A 40,000 MW protease is released into the cells immediate vicinity

Plasmin 76,000 MW

Plasmin proteolytically removes many of the protein 'projections' on the cell membrane

A cell surface that mimics the state found in late mitosis (?)

Figure 18–32
Schematic illustration of how the secretion of "cell factor" can lead to changes at the cell surface.

until now largely known for its capacity to dissolve blood clots (Figure 18–32). As a result of this cascade process, large amounts of proteolytic activity are generated in the immediate vicinity of all tumor cells. The question thus arises as to the sensitivity of cell surfaces to plasmin action. Does it in fact mimic the action of trypsin, which, in *in vitro* experiments, clips away more than half of the protein groups that protrude out from the lipid bilayers? Now there are preliminary hints that not only can plasmin digest away part of the cancer cell surface but, in so doing, it creates many of the distinctive features of certain cancer cells. The most convincing evidence comes from experiments in which cancerous mouse cells are grown in serum-containing medium from which plasminogen has been removed. Under these conditions, they still are able to grow normally, but they no longer have the distinctive cancer cell morphology and look as if they were almost normal. The possibility thus exists that the signal that sends certain cancer cells into division is not the attachment of specific growth-regulating molecules but instead the partial enzymatic digestion of their surface coats.

DECREASED SERUM REQUIREMENTS OF CANCER CELLS

Many cancer cells can grow in culture medium supplemented with much less serum than that required by the corresponding normal cells. Most probably this difference reflects a lower need for one or more of the growth-controlling factors that serum supplies. The mouse cell line 3T3, for example, multiplies faster and to higher density in growth medium containing 10% serum than in a 1%-serum-supplemented medium. But transformed 3T3 cells grow equally well in 1% and 10% serum. Such diminished serum requirements suggest that cancer cells may often multiply better than their normal equivalents because they need less outside help to push down their intracellular cAMP level low enough (raise their cGMP content high enough?) to set off a division cycle. Support for this hypothesis comes from the observation that transformed cells usually have only about half as much cAMP as the corresponding normal cell types. As with normal cells, the cAMP level is not invariant, with "M" phase cells containing less than cells in the other phases of the cell cycle.

THE AWFUL INCOMPLETENESS OF EUCARYOTIC CELL BIOCHEMISTRY

The fact that some reproducible differences are at last beginning to be seen between normal and cancer cells tells us that the study of cancer at the molecular level is no longer a premature Don Quixotic science that only exists because the responsible people never ask what their cancer research money is buying. On the other hand, we must not deceive ourselves that, because we can measure the amount of cAMP and, if we are more clever, that of cGMP, we are in any sense close to being on top of the cancer problem. Compared to *E. coli,* our knowledge of the structural organization and biochemistry of normal eucaryotic cells is still pitifully meager; and unless this situation changes, much cancer research will continue to resemble a search for a coin lost along a path only occasionally illuminated by street lights. Of necessity we look under the lighted regions for, no matter how long you look into the dark, the search will never be successful. Thus, as long as most components of the normal cell membrane remain essentially black boxes, the biochemistry of cancer may remain a mystery for some still future generation to understand.

SUMMARY

To learn how the cells of multicellular animals control their proliferation rates has long been a key objective of experimental biology. With such understanding, we will be in a much stronger position to probe both the embryological development, as well as the correct adult functioning, of higher animals. And we may at last be able to come to grips with the essence of the various forms of cancer, hereditary changes in cells which lead to uncontrolled cell proliferation.

Now almost all productive approaches to the control mechanisms governing cell proliferation utilize the in vitro *growth of cells in culture (cell culture). The cells excised from an animal to start a culture are called primary cells, while their descendants present several division cycles later are known as secondary cells. Most secondary cell cultures fail to multiply more than 20 to 50 times, and it is only the rare cell that acquires the ability to multiply indefinitely to form a cell line. Why only the exceptional cell has unlimited growth potential is very unclear, and suspicions exist that such programmed cell death may be related to the aging phenomenon in higher vertebrates. Up until now, cell lines have been obtained only from selected animal species (e.g., mouse, rat, hamster, human, but not the chicken!). And in general it proves to be much easier to obtain cell lines from tumor cells than from their normal equivalents.*

The greater potential of tumor cells to yield cell lines led for many years to the belief that perhaps all cells which grow as cell lines do so because they have acquired by mutation(s) many of the essential properties of the cancer cell. Now, how-

ever, we realize that only those secondary fibroblasts (epithelial cells) continuously passaged at high cell density later give rise to cell lines that grow to the high cell density characteristic of cancer cells. But if the secondary cells are selected at low density, the respective cell lines grow to much lower densities and generally stop dividing after they have formed organized cell layers.

Possession of well defined cell lines, like the human cancer cell line HeLa and the mouse "L" cell line, made possible the working out of the detailed nutritional requirements of cells growing in culture. In general, growth not only requires a well defined collection of amino acids and vitamins, but also depends upon protein factors present in blood serum. The nature of the "serum factors" until recently was a mystery, but now there is increasing evidence that most "serum factors" are a group of polypeptide hormones that act by binding to the surface of their target cells. In so doing, they can modify the activity of the membrane-bound enzymes, adenyl cyclase and guanyl cyclase, so that the level of cAMP in the respective cells is decreased while that of cGMP is increased. In some yet to be worked out manner, a lowering of the cAMP level (raising of the cGMP level) is a trigger which sets off cell division in certain cells. In this process, the first visible response in the nucleus is a quickening of RNA synthesis within an hour of the application of the mitogen. DNA synthesis commences only some 8 to 12 hours later.

In the absence of growth-promoting factors (mitogens), most cells enter a resting state in the G_1 phase of the life cycle, and so contain a diploid amount of DNA. Once a cell has begun to replicate its DNA, it continues through the G_2 phase and obligatorily passes into mitosis. Once into G_1, a new mitogenic stimulus must be applied before it will start another cell cycle. Otherwise the cell enters a quiescent state which often leads into a differentiation process that may create a highly specialized cell (e.g., nerve) unable to initiate a new cycle of growth and cell division. Differentiation per se, *however, does not lead to an inability to divide. For example, the commonly used fibroblasts, epithelial cells, and lymphoblasts all represent clearly defined states of differentiation. So it is only when the physiological role of a given cell makes further division unwanted (e.g., organized cell nerves, red blood cells, etc.), that the capacity to divide is lost.*

Many forms of differentiation require the presence of specific differentiation factors that appear not only to promote differentiation per se, *but also lead to selective proliferation of partially differentiated "blast" cells. In particular, the synthesis of the various types of blood cells is controlled by a variety of specific proteins (e.g., red blood cell production is controlled by the protein erythropoietin). Specific proteins also control the differentiation of specific nerve cells (nerve growth factor) as well as of certain epidermal cells (epidermal growth factor). Certain steroid hormones also can lead to differentiation and selective proliferation. Uterine growth, for example, is controlled by the steroid hormone estradiol.*

Still other steps in cell differentiation occur without the apparent need of specific inducing molecules. The formation of a fat cell from a fibroblast apparently is triggered by the presence of a rich food supply coupled with an inability to divide. And striated muscles form in culture by the spontaneous fusion of myoblasts.

Transformation is the process by which normal cells acquire many of the morphological as well as growth properties of cancer cells. In general, transformed cells have a more spherical shape than their normal counterparts and show many similarities to cells just emerging from mitosis. For example, many tumor cells show excessive blebbing throughout the cell cycle, a characteristic of normal cells only in early G_1. Likewise, many of the membrane proteins of cancer cells appear more mobile than those found in normal interphase cells, conceivably reflecting a much less organized arrangement of their muscle proteins. Transformed cells also have lost many of the normal adhesive properties, leading to an inability to be sited next to appropriate cellular neighbors. And for reasons that are totally unclear, tumor cells have lost the normal control of glucose utilization. They consume much more glucose than can be metabolized efficiently, thereby leading to massive lactic acid secretion.

Many, but not all, "transformed" cells show increased secretion of one or more proteolytic enzymes ("cell factor(s)"). When so released, these proteases can activate inactive precursors of major serum proteases (e.g., plasmin, the main enzyme that dissolves blood clots). In turn these "serum proteases" often digest away portions of tumor cell surfaces. Whether such externally acting proteases are responsible for any of the essential surface properties of tumor cells remains to be proven. Arguing for this hypothesis is the fact that the tumor-forming capacity of many transformed cells is strongly correlated with the amount of cell factor they release.

REFERENCES

Puck, T. T., *The Mammalian Cell as a Microorganism*, Holden-Day, 1972. An introduction to research on higher animal cells growing in culture.

Readings in Mammalian Cell Culture, edited by Robert Pollack, Cold Spring Harbor Laboratory, 2nd edition, 1975. A very useful collection of articles that describe many of the experiments behind the main conclusion of this chapter.

Tooze, J., *The Molecular Biology of Tumor Viruses*, Cold Spring Harbor Laboratory, 1973. An excellent summary of the animal cell culture field is found in Chapter 2.

Paul, J., *Cell and Tissue Culture*, 4th edition, Livingston, Edinburgh and London, 1970. A description of procedures used in growing cells in culture.

Mitchison, J. M., *The Biology of the Cell Cycle*, Cambridge University Press, 1971. A very useful description of both procaryote and eucaryote cell cycles.

The Cell Cycle in Development and Differentiation, M. Ballis and F. S. Billett (eds.), British Society for Developmental Biology Symposium, Cambridge University Press, 1973. A very wide-ranging look at a variety of cell cycles.

Mazia, D., "The Cell Cycle," *Scientific American*, January, 1974. A beautifully illustrated introduction to the problems faced by a cell as it grows and divides.

The Cell Cycle and Cancer, R. Baserga (ed.), Marcel Delker, New York, 1971. Reviews on aspects of vertebrate cell cycles that may lead to understanding the loss of control over cell division.

Rao, P. N. and R. T. Johnson, "Induction of Chromosome Condensation in Interphase Cells," *Advances in Cell and Molecular Biology*, **3**, 136, Academic Press (1974). A review about the premature chromosome condensation that can follow cell fusion.

Patterson, D., F. T. Kao, and T. T. Puck, "Genetics of Somatic Mammalian Cells: Biochemical Genetics of Chinese Hamster Cell Mutants with Deviant Purine Metabolism," *Proc. Nat. Acad. Sci.*, **71**, 2057 (1974). Illustrates many of the procedures now available for the genetic analysis of cells in culture.

Chasin, L. A., "Mutations Affecting Adenine Phosphoribosyl Transferase Activity in Chinese Hamster Cells," *Cell*, **2**, 37 (1974). An elegant analysis of mutation rates demonstrating that conventional gene mutations are the origin of the "variant cells" found after mutagen treatment.

Growth Control in Cell Cultures, a CIBA Foundation Symposium edited by G. E. W. Wolstenholme and J. Knight, Churchill Livingston, Edinburgh and London, 1971. An interesting collection of papers aimed at understanding the factors governing cell growth.

Control of Proliferation in Animal Cells, B. Clarkson and R. Baserga (eds.), Cold Spring Harbor Laboratory, 1974. Eighty articles from a meeting that summarized the field as of May 1973.

PASTAN, I., "Cyclic AMP," *Scientific American,* August, 1972. The multiplicity of ways this "secondary messenger" controls cell behavior.

SEIFERT, W. E., AND P. S. RUDLAND, "Possible Involvement of Cyclic GMP in Growth Control of Cultured Mouse Cells," *Nature,* **248,** 138 (1974). An example of how the cGMP level rises during mitogenesis.

GOSPODAROWICZ, D., "Localization of a Fibroblast Growth Factor and its Effect Alone and with Hydrocortisone on 3T3 Cell Growth," *Nature,* **249,** 123 (1974). The first paper on its properties.

GREEN, H., AND M. MEUTH, "An Established Pre-Adipose Cell Line and its Differentiation in Culture," *Cell,* **3,** 127 (1974). An important addition to the systems where differentiation can be studied in cell culture.

REVEL, J. P., "Contacts and Junctions Between Cells," *Transport at the Cellular Level,* Soc. Exp. Biol. Symp., **28,** 447 (1974). A very good summary of how cells attach to each other.

NICOLSON, G., "The Interaction of Lectins with Animal Cell Surfaces," *Int. Rev. Cytology,* **39,** 90 (1974). A very complete and well written discussion.

MIYASHITA, K., AND T. KAKUNAGA, "Isolation of Heat- and Cold-Sensitive Mutants of Chinese Hamster Lung Cells Affected in Their Ability to Express the Transformed State," *Cell,* **5,** 131 (1975). Illustrating the ease with which ts mutants affecting the transformed phenotype can be obtained.

Proteases and Biological Control, E. Reich, D. Rifkin, and E. Shaw (eds.), Cold Spring Harbor Laboratory, 1975. A collection of sixty-eight articles arising out of a meeting that discussed how proteolytic enzymes mediate many key control processes, including some aspects of cell proliferation.

The Problem of Antibody Synthesis

How organisms are able to generate specific immunological responses has intrigued biologists all through this century. At first, their interest had purely practical considerations: The better we understand antigens and antibodies, the easier it might be to make our immunological responses more effective against dangerous disease agents. In recent years, however, much attention has been focused upon immunology because of the realization that the induction of specific antibody synthesis involves specific irreversible cell differentiation. Particularly appealing was the fact that methods had become available for the isolation of large amounts of highly purified antibodies, thereby opening up the possibility of chemical work on their detailed structures. The belief exists that a profound breakthrough might be achieved if antibody synthesis can be understood at the molecular level, with implications not only for medicine, but also for all aspects of the biology of multicellular organisms.

ANTIGENS ARE AGENTS WHICH STIMULATE ANTIBODY SYNTHESIS

Antibody synthesis is a defense response of higher vertebrates that evolved both to combat the harmful effects of pathogenic microorganisms and to prevent the multiplication of cancer cells. For example, the introduction of a virus into the circulatory system of a higher vertebrate

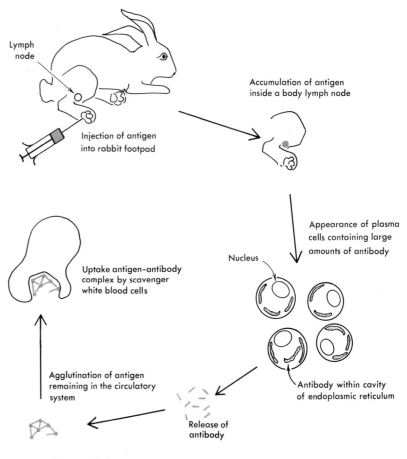

Figure 19–1
Diagrammatic view of the sequence of events between the injection of an antigen and the appearance of circulating antibodies.

stimulates specific white blood cells, the lymphocytes, to produce antibodies that combine specifically with the virus to prevent its further multiplication (Figure 19–1). An individual is *immune* to a virus as long as the corresponding antibodies are present in his circulatory system. Those objects which stimulate antibody synthesis (e.g., a virus particle) are called *antigens*. The study of antibodies and their interaction with antigens is called *immunology*.

An object is potentially antigenic when it possesses an arrangement of atoms at its surface that differs from the surface configuration of any normal host component. Thus, the immunological defense system is based on the ability of an organism to distinguish between its own molecules and foreign ones. Antibodies are produced against a virus (cancer cell) not because the system realizes that the virus (cancer cell) will produce a disease, but rather because it recognizes that the virus (cancer cell) is a foreign object and hence must be eliminated from the respective host organism.

This immediately raises the question, What are the requirements for an object to have antigenic properties? One major requirement is that an antigen must either be a macromolecule or be built up from macromolecules (e.g., a virus particle). Most proteins and some polysaccharides and nucleic acids are antigens. Small molecules by themselves can seldom induce specific circulating antibodies. This is not due to a lack of specificity, since many small molecules, nonantigenic by themselves, change the antigenic properties of a larger molecule (e.g., a protein) when covalently coupled to it. The resulting antibodies then become partially directed against the small molecule. Such potentially immunogenic small molecules are called *haptens.*

It seems unlikely that the entire surface of a large molecule is necessary for its antigenicity. Most probably, the immunological system responds to specific groups of atoms (*antigenic determinants*) located at a number of sites about a molecular surface (Figure 19–2). A given protein molecule often possesses several antigenic determinants and induces the formation of several types of antibodies, while objects the size of bacteria possess a very large number of different antigenic determinants.

Figure 19–2
Diagrammatic view of an antigen. The symbol R represents a single determinant of immunological specificity and is the actual group that combines with an antibody molecule.

At present, we are still very uncertain as to exactly how many unique antigenic determinants exist. The number is certainly large, perhaps larger than 10,000. We make this guess on the basis of experiments that test whether antibodies induced by a given protein ever accidentally combine with a completely unrelated protein. If there were only a limited number of antigenic determinants, we would expect the same determinant to be found on many proteins, leading to unexpected antigenic homologies between unrelated proteins. But, in fact, such cross-reactions virtually never occur.

CIRCULATING VS. CELL-BOUND ANTIBODIES

For a long while it was thought that antibodies functioned only after they had been secreted into the blood or other body fluids (e.g., nasal secretions, saliva) that come into contact with potentially harmful agents (*humoral antibodies*). Then it gradually became clear that many types of immunological responses were due to other antibodies

593

which remained bound to their parent lymphocytes (*cell-mediated immunity*). A large variety of seemingly unrelated antigens (e.g., the toxin of poison ivy, the tubercle bacillus, and many cancer cells) each lead to cell-mediated immunological responses. Why a given antigen predominantly leads to either a humoral or cellular response is still very unclear. Empirically, however, there is a tendency for free proteins (viruses) to stimulate humoral antibody synthesis, while antigenic determinants bound to living cells more often promote cell-mediated immunity. However, there exist so many apparent exceptions to this generalization that it may be wise to refrain from prediction until the cellular (molecular) bases of these two forms of response are better understood.

FATE OF ANTIGEN–ANTIBODY COMPLEXES

Combination of an antigen with a humoral antibody is usually only the first step in a long train of events that leads to elimination of the resulting complex from an organism. In some way, the complex has to be identified as ticketed either for phagocytosis or for specific enzymatic destruction. Such decisions are usually the result of the diverse actions of *complement,* a series of nine sequentially acting protein components (C1 to C9) of blood serum (Figure 19–3). C1, the first component, becomes activated when it binds to the antibody partner of an antigen–an-

Figure 19–3
A schematic view of how complement, a complex group of enzymes in normal blood serum, acts to cause lysis of foreign cells. The various complement components act in sequence, starting with C1 being activated by an immune complex (antibody bound to foreign antigen).

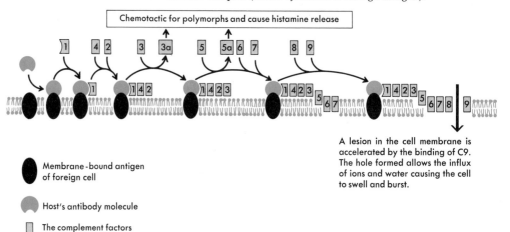

Chemotactic for polymorphs and cause histamine release

A lesion in the cell membrane is accelerated by the binding of C9. The hole formed allows the influx of ions and water causing the cell to swell and burst.

Membrane-bound antigen of foreign cell

Host's antibody molecule

The complement factors

594

tibody complex. It never binds to free antibody *per se*—it can bind only after the change in antibody shape that obligatorily follows union with antigen. Activated C1 is an esterase involved in the sequential activation of C4 and then C2. The resulting multicomponent C3-convertase always remains bound to the original antigen–antibody complex. There it enzymatically breaks down C3 into several fragments, one of which (bound C3) has a binding site for macrophages and so facilitates subsequent phagocytosis.

During several steps of this enzymatic cascade, a component activates several molecules of the next component in the sequence, thereby producing an amplified cascade effect. In this way, the thousands of (C9) molecules that result from just one initial antigen-antibody complex have the ability to create large holes in a plasma membrane. Thus, when complement binds to antibodies attached to cell-surface antigens, the resulting cascade of events very frequently can lead to cell death. It was, in fact, through this cell-killing phenomenon that the complement system was discovered. In the presence of complement, antibodies directed against sheep red blood cells quickly cause their lysis—but if the serum complement is destroyed by a half hour exposure to 56°, no lysis occurs.

The events subsequent to union of an antigen with a cell-bound antibody have yet to be worked out in such detail. Under some conditions, antigen attachment to the appropriate lymphocytes leads to release of a variety of factors (lymphokines) which mediate delayed allergenic responses (e.g., toward the poison ivy toxin) (see below). Other lymphocytes whose specificity is directed toward foreign cells have the capacity to kill such cells. Whether complement is involved in such killing events is not firmly established.

ANTIBODIES ARE ALWAYS PROTEINS

All antibodies are proteins. The selective synthesis of a specific antibody thus represents the selective synthesis of a specific protein. The term *immunoglobulin* is now used generically to refer to all antibodies. In the human, five major classes of humoral immunoglobulins exist (IgG, IgM, IgA, IgD, and IgE), each possessing a characteristic structure and most likely carrying out quite separate functional roles (Table 19–1). All are built up from regular aggregates of two types of polypeptide chains, one usually called the light chain, the other the heavy chain. Similar types of light chains are found in all the major groups. It is

The Problem of Antibody Synthesis

Table 19–1 Properties of Major Human Immunoglobulin Classes

	IgG	IgA	IgM	IgD	IgE
Sedimentation coefficient	7S	7S, 9S, 11S	19S	7S	8S
Molecular weight	150,000	160,000 320,000 480,000	900,000	185,000	200,000
Number of basic 4-peptide units	1	1, 2, 3	5	1	1
Heavy chains	γ	α	μ	δ	ϵ
Light chains	κ or λ	κ or λ	κ or λ	κ or λ	κ or λ
Molecular formula	$\gamma_2\kappa_2$ or $\gamma_2\lambda_2$	$(\alpha_2\kappa_2)_{1-3}$ or $(\alpha_2\lambda_2)_{1-3}$	$(\mu_2\kappa_2)_5$ or $(\mu_2\lambda_2)_5$	$\delta_2\kappa_2$ or $\delta_2\lambda_2$	$\epsilon_2\kappa_2$ or $\epsilon_2\lambda_2$
Valency for antigen binding	2	2, 4, 6	10	?	2
% of total serum immunoglobulins	80	13	6	1	0.002
% carbohydrate content	3	8	12	13	12

the nature of the heavy chain which generates the basic specificity of a class.

The functional differences between the several classes are only now beginning to be understood. IgG, by far the largest component in blood serum, is the major agent that combats microorganisms and ensures their phagocytosis. IgM, which also has this function, appears first in an immunological response, eventually being supplanted by IgG as the specific immunological response becomes strong. The major immunoglobulin in sero-mucous secretions is IgA, whose principal task is the defense of external body surfaces. No normal function has yet been found for IgD, while IgE may be involved in rejection of intestinal parasites. It also seems to be the principal villain in hay fever-type allergies.

IgG is still by far the best characterized class, and in subsequent sections we shall emphasize its properties.

CONSTRUCTION OF THE IgG ANTIBODY MOLECULE FROM TWO LIGHT AND TWO HEAVY CHAINS

IgG antibodies sediment in the ultracentrifuge at 7S and have molecular weights of about 150,000. Four polypeptide chains are present in each molecule. They fall into two pairs: two heavy chains (MW ~ 53,000) and two light chains (MW ~ 22,500). In a given molecule the two heavy chains are identical, as are the two light chains. Each light

596

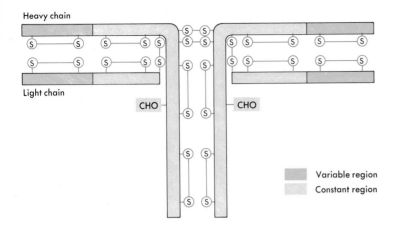

Heavy chain

Light chain

CHO   CHO

Variable region
Constant region

Figure 19–4
A model of the structure of a human IgG antibody molecule. Interchair and intrachain disulfide bonds are indicated, and CHO marks the approximate position in the heavy chain of the carbohydrate moiety.

chain is linked to a heavy chain by one covalent S—S bond, while two S—S bonds run between the two heavy chains to give the schematic picture shown in Figure 19–4. Most likely, several weak secondary bonds also hold the four chains together.

The chains are arranged so that each antibody molecule contains two identical, widely separated "active" sites, each of which can combine with an antigen, using secondary bonds to hold their complementary surfaces together (Figure 19–5).

Visualization of antibodies in the electron microscope reveals Y-shaped molecules with one active site at the end of each arm. Each arm is effectively hinged to the central

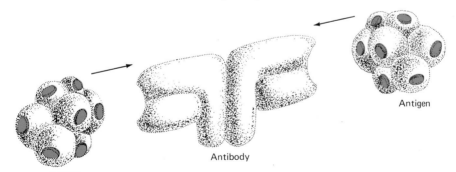

Antigen

Antibody

Antigen

Figure 19–5
Existence of two identical antigen-combining sites on each 7S antibody molecule. Both light- and heavy-chain atoms are used to form the combining sites.

597

axis, with the distance between active sites usually increasing when they are bound to antigen.

The existence of two identical binding sites permits a single antibody molecule to link together two similar antigens. This feature is of great advantage in allowing antibodies to defeat an infection by a microorganism, since all microorganisms contain a large number of identical antigenic determinants. In the presence of specific antibodies, a microorganism thus becomes linked to a large number of similar microorganisms through antibody bridges (Figure 19–6), and these aggregates then tend to be taken up and destroyed by macrophages.

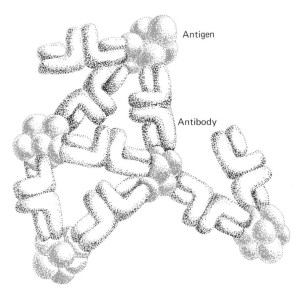

Figure 19–6
A diagrammatic view of how antigens and antibodies combine to form large aggregates.

ANTIBODY SPECIFICITY RESIDES IN AMINO ACID SEQUENCES

Until some ten years ago, two competing hypotheses existed to explain what distinguished one IgG antibody molecule from another. One theory stemmed from the fact that the gross molecular structure of all IgG antibodies is similar. This resemblance led some people to believe that the amino acid sequences of all antibody molecules are the same, and to postulate that the essential difference between different antibodies resides in their precise three-dimensional structure: the folding of the identical chains. If this were true, the antigen would determine which antibody should be formed by combining with a newly synthesized antibody chain before it had folded to the final three-dimensional form. The interaction would allow part of the antibody molecule to fold around the antigen, automatically creating a complementary shape between the two molecules. Because, according to this scheme and most variants of it, the antigen directly determines the shape of the antibody, such models are called *instructive theories* of antibody formation (Figure 19–7).

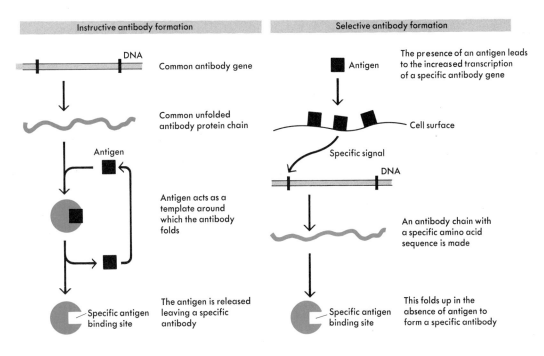

Figure 19–7
A comparison between the instructive and selective theories of making specific antigen binding sites during antibody formation.

Now, however, all instructive theories have been ruled out. Instead, everybody believes that there are not only three-dimensional differences, but also differences in primary structure (amino acid sequences) between different antibodies. One of the most compelling types of evidence comes from experiments in which the three-dimensional structure is temporarily destroyed (denatured) and allowed to reform in the absence of antigen: the antibody molecules resume their specificity! There is also a growing body of direct chemical evidence based on analysis of amino acid sequences of purified antibodies. Here definite differences exist, revealing that antibodies have two distinct regions: one common to all antibodies, which accounts for the impression that all antibodies are chemically very similar; and one whose amino acid sequence (and, hence, three-dimensional form) differs from one antibody to another.

The existence of distinct amino acid sequences for each specific antibody immediately raises the question whether there is a distinct gene for each antibody. Since any antibody-producing animal can produce a very large number of different antibodies, it is possible that within every cell a very large number of genes might code for the amino acid sequences of these antibodies. For many years this possibility has seemed repugnant to many immunologists, who were aware of the immense number of different antigenic determinants. Now, however, the existence of a distinct gene for each antibody is a possibility that can no longer be avoided. Since amino acid sequences are different, there must exist corresponding differences in their mRNA templates, and thus in the relevant DNA regions.

We are also faced with the problem that, if different genes exist, there must be a control mechanism by which the presence of an antigen tells the gene controlling a corresponding antibody to function. In some way the presence of an antigen must cause the selective synthesis of unique amino acid sequences (*the selective theory of antibody formation*).

MYELOMA PROTEINS AS MODELS FOR SINGLE ANTIBODIES

At any given time, the antibodies present in the serum of any individual are a collection of many, many different species directed against a large variety of different antigens. Even after continued immunization with a single strong antigen, no single antibody species is present in pure enough form to allow its isolation. Not only do traces remain of the many, many different antibodies previously induced in that individual's history, but even more important, a given single antigen always seems to promote synthesis of several different antibodies, each with a different amino acid sequence, yet all able to specifically bind to the same antigen.

Single antibody species can, however, be obtained from the blood of certain individuals afflicted with the bone marrow disease, *multiple myeloma*. This disease, a form of cancer, involves the uncontrolled multiplication of antibody-producing cells. Most important, a given specific tumor produces only one antibody type characterized by a specific amino acid sequence. Moreover, the antibodies produced by different myeloma tumors each have different amino acid sequences, thereby permitting the comparative study of many different antibody sequences. Until recently no one had found any antigen with a strong affinity for any myeloma protein. But now several antigens have been found that bind tightly to given proteins.

The amounts of a given myeloma protein made in a myeloma patient are very large, easily permitting amino acid sequence studies. There are, moreover, similar tumors in a number of laboratory animals. Mice, in particular, offer an excellent system for work with myeloma proteins because of the possibility of extensive genetic analysis.

BENCE-JONES PROTEINS ARE SPECIFIC LIGHT CHAINS

Many myeloma patients excrete in their urine large amounts of specific proteins named after their discoverer, a nineteenth century English physician. They were initially found because, unlike most proteins, they do not precipitate at boiling temperatures. Now we realize that a given *Bence-Jones protein* is identical to the light chain of its corresponding myeloma antibody. Its excretion into the urine is the result of overproduction of light antibody chains by the myeloma cells. Somehow the biological control device which normally ensures equal production of heavy and light antibody chains is lost, and excessive light chain synthesis results. The presence of very large amounts of Bence-Jones proteins, together with their abnormal solubility, makes their isolation extremely simple; and the first important insights into antibody amino acid sequences arose from the study of these proteins.

LIGHT AND HEAVY CHAINS HAVE CONSTANT AND VARIABLE PORTIONS

Elucidation of the amino acid sequences of the first several light chains isolated revealed that they consist of variable amino-terminal sequences (residues 1 to about 108), specific for each type of antibody, linked to constant carboxyl-terminal regions (residues 109 to about 214) common to all light chains. More recently, when data began to appear on heavy chains, the same feature emerged: variability at the amino terminus and constancy at the carboxyl end. The constant region of the heavy chain is much larger, being three times ($\sim$330 amino acids) longer than the heavy variable region ($\sim$110 amino acids). Thus, while the heavy chains are much longer than the light chains, the lengths of their variable regions are approximately the same.

Figure 19–8 shows the light chain sequence and Figure 19–9, the heavy chain sequence, of EU, the first antibody whose complete two-dimensional structure has been worked out. A total of 660 residues were ordered, a major achievement, for, so far, EU is the largest protein whose sequence has been totally elucidated.

```
1                              10                                    20
ASP—ILE—GLN—MET—THR—GLN—SER—PRO—SER—THR—LEU—SER—ALA—SER—VAL—GLY—ASP—ARG—VAL—THR—
                               30                                    40
ILE—THR—CYS—ARG—ALA—SER—GLN—SER—ILE—ASN—THR—TRP—LEU—ALA—TRP—TYR—GLN—GLN—LYS—PRO—
                               50                                    60
GLY—LYS—ALA—PRO—LYS—LEU—LEU—MET—TYR—LYS—ALA—SER—SER—LEU—GLU—SER—GLY—VAL—PRO—SER—
                               70                                    80
ARG—PHE—ILE—GLY—SER—GLY—SER—GLY—THR—GLU—PHE—THR—LEU—THR—ILE—SER—SER—LEU—GLN—PRO—
                               90                                    100
ASP—ASP—PHE—ALA—THR—TYR—TYR—CYS—GLN—GLN—TYR—ASN—SER—ASP—SER—LYS—MET—PHE—GLY—GLN—
                               110                                   120
GLY—THR—LYS—VAL—GLU—VAL—LYS—GLY—THR—VAL—ALA—ALA—PRO—SER—VAL—PHE—ILE—PHE—PRO—PRO—
                               130                                   140
SER—ASP—GLU—GLN—LEU—LYS—SER—GLY—THR—ALA—SER—VAL—VAL—CYS—LEU—LEU—ASN—ASN—PHE—TYR—
                               150                                   160
PRO—ARG—GLU—ALA—LYS—VAL—GLN—TRP—LYS—VAL—ASP—ASN—ALA—LEU—GLN—SER—GLY—ASN—SER—GLN—
                               170                                   180
GLU—SER—VAL—THR—GLU—GLN—ASP—SER—LYS—ASP—SER—THR—TYR—SER—LEU—SER—SER—THR—LEU—THR—
                               190                                   200
LEU—SER—LYS—ALA—ASP—TYR—GLU—LYS—HIS—LYS—VAL—TYR—ALA—CYS—GLU—VAL—THR—HIS—GLN—GLY—
                               210                214
LEU—SER—SER—PRO—VAL—THR—LYS—SER—PHE—ASN—ARG—GLY—GLU—CYS
```

Figure 19–8
The amino acid sequence of the light chain of the human immunoglobulin EU.

```
 1                              10                                    20
PCA—VAL—GLN—LEU—VAL—GLN—SER—GLY—ALA—GLU—VAL—LYS—LYS—PRO—GLY—SER—SER—VAL—LYS—VAL—

                                30                                    40
SER—CYS—LYS—ALA—SER—GLY—GLY—THR—PHE—SER—ARG—SER—ALA—ILE—ILE—TRP—VAL—ARG—GLN—ALA—

                                50                                    60
PRO—GLY—GLN—GLY—LEU—GLU—TRP—MET—GLY—GLY—ILE—VAL—PRO—MET—PHE—GLY—PRO—PRO—ASN—TYR—

                                70                                    80
ALA—GLN—LYS—PHE—GLN—GLY—ARG—VAL—THR—ILE—THR—ALA—ASP—GLU—SER—THR—ASN—THR—ALA—TYR—

                                90                                    100
MET—GLU—LEU—SER—SER—LEU—ARG—SER—GLU—ASP—THR—ALA—PHE—TYR—PHE—CYS—ALA—GLY—GLY—TYR—

                                110                                   120
GLY—ILE—TYR—SER—PRO—GLU—GLU—TYR—ASN—GLY—GLY—LEU—VAL—THR—VAL—SER—SER—ALA—SER—THR—

                                130                                   140
LYS—GLY—PRO—SER—VAL—PHE—PRO—LEU—ALA—PRO—SER—SER—LYS—SER—THR—SER—GLY—GLY—THR—ALA—

                                150                                   160
ALA—LEU—GLY—CYS—LEU—VAL—LYS—ASP—TYR—PHE—PRO—GLU—PRO—VAL—THR—VAL—SER—TRP—ASN—SER—

                                170                                   180
GLY—ALA—LEU—THR—SER—GLY—VAL—HIS—THR—PHE—PRO—ALA—VAL—LEU—GLN—SER—SER—GLY—LEU—TYR—

                                190                                   200
SER—LEU—SER—SER—VAL—VAL—THR—VAL—PRO—SER—SER—SER—LEU—GLY—THR—GLN—THR—TYR—ILE—CYS—

                                210                                   220
ASN—VAL—ASN—HIS—LYS—PRO—SER—ASN—THR—LYS—VAL—ASP—LYS—ARG—VAL—GLU—PRO—LYS—SER—CYS—

                                230                                   240
ASP—LYS—THR—HIS—THR—CYS—PRO—PRO—CYS—PRO—ALA—PRO—GLU—LEU—LEU—GLY—GLY—PRO—SER—VAL—

                                250                                   260
PHE—LEU—PHE—PRO—PRO—LYS—PRO—LYS—ASP—THR—LEU—MET—ILE—SER—ARG—THR—PRO—GLU—VAL—THR—

                                270                                   280
CYS—VAL—VAL—VAL—ASP—VAL—SER—HIS—GLU—ASP—PRO—GLN—VAL—LYS—PHE—ASN—TRP—TYR—VAL—ASP—

                                290                                   300
GLY—VAL—GLN—VAL—HIS—ASN—ALA—LYS—THR—LYS—PRO—ARG—GLU—GLN—GLN—TYR—ASX—SER—THR—TYR—

                                310                                   320
ARG—VAL—VAL—SER—VAL—LEU—THR—VAL—LEU—HIS—GLN—ASN—TRP—LEU—ASP—GLY—LYS—GLU—TYR—LYS—

                                330                                   340
CYS—LYS—VAL—SER—ASN—LYS—ALA—LEU—PRO—ALA—PRO—ILE—GLU—LYS—THR—ILE—SER—LYS—ALA—LYS—

                                350                                   360
GLY—GLN—PRO—ARG—GLU—PRO—GLN—VAL—TYR—THR—LEU—PRO—PRO—SER—ARG—GLU—GLU—MET—THR—LYS—

                                370                                   380
ASN—GLN—VAL—SER—LEU—THR—CYS—LEU—VAL—LYS—GLY—PHE—TYR—PRO—SER—ASP—ILE—ALA—VAL—GLU—

                                390                                   400
TRP—GLU—SER—ASN—ASP—GLY—GLU—PRO—GLU—ASN—TYR—LYS—THR—THR—PRO—PRO—VAL—LEU—ASP—SER—

                                410                                   420
ASP—GLY—SER—PHE—PHE—LEU—TYR—SER—LYS—LEU—THR—VAL—ASP—LYS—SER—ARG—TRP—GLN—GLU—GLY—

                                430                                   440
ASN—VAL—PHE—SER—CYS—SER—VAL—MET—HIS—GLU—ALA—LEU—HIS—ASN—HIS—TYR—THR—GLN—LYS—SER—
```

Figure 19–9
The amino acid sequence of the heavy
chain of the human immunoglobulin EU.

HEAVY CHAIN ORIGIN THROUGH REPETITIVE DUPLICATION OF A PRIMITIVE ANTIBODY GENE

Analysis of the foregoing sequences strongly suggests that the heavy chain evolved by successive duplication of a primitive antibody gene. To show this, its sequence is divided into four parts (Figure 19–10). The first, the variable portion, shows similarity to the variable part of the light chain. In turn, the three regions of the constant portion show strong homologies both to each other and to the constant region of the light chain. These relationships can easily be seen in Figure 19–11, which compares the sequences of C_L, C_H1, C_H2, and C_H3. Along a 100-residue length, identity between any two regions occurs at some 30 positions. Moreover, each region contains one intraregion S—S bond, with the two sulfur atoms separated by some 50 to 60 amino acids. By contrast, there are no S—S bonds joining together residues on different regions of the same chain. This suggests that each homologous region exists as a fairly separate mass, not closely bound to any of the others.

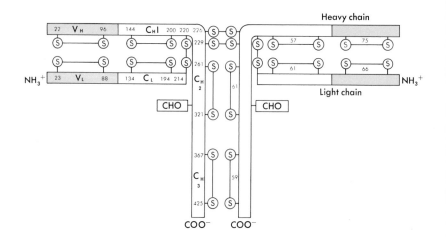

Figure 19–10
Internal homologies in the IgG antibody molecule. The variable regions V_L and V_H are homologous to each other, as are the constant regions C_L, C_H1, C_H2, and C_H3.

The fact that all variable regions also contain a single intraregion S—S bond suggests that the constant and variable portions had a common ancestor. If so, the first step in immunoglobulin evolution may have been the duplication of a primitive gene to form a compound gene with two identical nucleotide sequences. One sequence, not sig-

EU C_L (RESIDUES 109–214)
EU C_H1 (RESIDUES 119–220)
EU C_H2 (RESIDUES 234–341)
EU C_H3 (RESIDUES 342–446)

110 — 120

THR	VAL	ALA	ALA	PRO	SER	VAL	PHE	ILE	PHE	PRO	PRO	SER
SER	THR	LYS	GLY	PRO	SER	VAL	PHE	–	–	PRO	LEU	ALA
LEU	LEU	GLY	GLY	PRO	SER	VAL	PHE	LEU	PHE	PRO	PRO	LYS
GLN	PRO	ARG	GLU	PRO	GLN	VAL	TYR	THR	LEU	PRO	PRO	SER

130

ASP	GLU	GLN	–	–	LEU	LYS	SER	GLY	THR	ALA	SER	VAL	VAL	CYS	LEU	LEU	ASN	ASN	PHE
PRO	SER	SER	LYS	SER	THR	SER	GLY	GLY	THR	ALA	ALA	LEU	GLY	CYS	LEU	VAL	LYS	ASP	TYR
PRO	LYS	ASP	THR	LEU	MET	ILE	SER	ARG	THR	PRO	GLU	VAL	THR	CYS	VAL	VAL	VAL	ASP	VAL
ARG	GLU	GLU	–	–	MET	THR	LYS	ASN	GLN	VAL	SER	LEU	THR	CYS	LEU	VAL	LYS	GLY	PHE

140 — 150

TYR	PRO	ARG	GLU	ALA	LYS	VAL	–	–	GLN	TRP	LYS	VAL	ASP	ASN	ALA	LEU	GLN	SER	GLY
PHE	PRO	GLU	PRO	VAL	THR	VAL	–	–	SER	TRP	ASN	SER	–	GLY	ALA	LEU	THR	SER	GLY
SER	HIS	GLU	ASP	PRO	GLN	VAL	LYS	PHE	ASN	TRP	TYR	VAL	ASP	GLY	–	VAL	GLN	VAL	HIS
TYR	PRO	SER	ASP	ILE	ALA	VAL	–	–	GLU	TRP	GLU	SER	ASN	ASP	–	GLY	GLU	PRO	GLU

160 — 170

ASN	SER	GLN	GLU	SER	VAL	THR	GLU	GLN	ASP	SER	LYS	ASP	SER	THR	TYR	SER	LEU	SER	SER
–	VAL	HIS	THR	PHE	PRO	ALA	VAL	LEU	GLN	SER	–	SER	GLY	LEU	TYR	SER	LEU	SER	SER
ASN	ALA	LYS	THR	LYS	PRO	ARG	GLU	GLN	GLN	TYR	–	ASP	SER	THR	TYR	ARG	VAL	VAL	SER
ASN	TYR	LYS	THR	THR	PRO	PRO	VAL	LEU	ASP	SER	–	ASP	GLY	SER	PHE	PHE	LEU	TYR	SER

180 — 190

THR	LEU	THR	LEU	SER	LYS	ALA	ASP	TYR	GLU	LYS	HIS	LYS	VAL	TYR	ALA	CYS	GLU	VAL	THR
VAL	VAL	THR	VAL	PRO	SER	SER	SER	LEU	GLY	THR	GLN	–	THR	TYR	ILE	CYS	ASN	VAL	ASN
VAL	LEU	THR	VAL	LEU	HIS	GLN	ASN	TRP	LEU	ASP	GLY	LYS	GLU	TYR	LYS	CYS	LYS	VAL	SER
LYS	LEU	THR	VAL	ASP	LYS	SER	ARG	TRP	GLN	GLU	GLY	ASN	VAL	PHE	SER	CYS	SER	VAL	MET

200 — 210

HIS	GLN	GLY	LEU	SER	SER	PRO	VAL	THR	–	LYS	SER	PHE	–	–	ASN	ARG	GLY	GLU	CYS
HIS	LYS	PRO	SER	ASN	THR	LYS	VAL	–	ASP	LYS	ARG	VAL	–	–	GLU	PRO	LYS	SER	CYS
ASN	LYS	ALA	LEU	PRO	ALA	PRO	ILE	–	GLU	LYS	THR	ILE	SER	LYS	ALA	LYS	GLY		
HIS	GLU	ALA	LEU	HIS	ASN	HIS	TYR	THR	GLN	LYS	SER	LEU	SER	LEU	SER	PRO	GLY		

Figure 19–11

Sequence homology in the constant regions of the EU antibody molecule. Deletions indicated by dashes have been introduced to maximize the homology. Identical residues are darkly shaded; both dark and light shadings are used to indicate identities which occur in pairs in the same positions. [Redrawn from G. M. Edelman *et al., Proc. Nat. Acad. Sci.,* **63,** 78 (1969).]

nificantly altered further, would remain as the constant region, while the other could evolve into the variable section. Strong support for this conjecture comes from x-ray crystallographic analysis of antibody fragments created by exposure of specific myeloma proteins to papain or pepsin.

The resulting Fab (Fab') fragments (Figure 19–12) can be crystallized (Figure 19–13), and already there exist de-

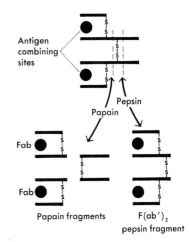

Figure 19–12

Creation of Fab and F(ab')₂ fragments by specific proteolytic digestion of intact IgG molecules.

tailed conformation data for two different fragments (Figure 19–14). As a result, we can compare the broad features of the V and C regions, as well as directly look at the amino acids that combine with the antigenic determinants.

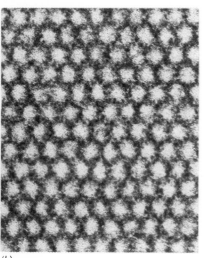

(a)

(b)

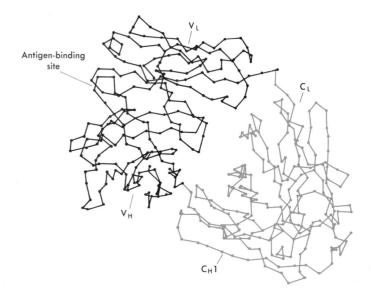

Figure 19–14
A drawing of the α carbon backbone of Fab' NEW. The C domain is shown in color and the V in black. [Redrawn from R. J. Poljak *et al.*, *Proc. Nat. Acad. Sci.*, **70**, 3305 (1973).]

Figure 19–13
(a) Optical micrograph (72×) of McPC 603 Fab' crystals. (b) Electron micrograph (400,000×) of a stained crystal section, showing the hexagonal symmetry assumed by the fragments. The Fab' fragments are stained, while the regions between them are unstained. [Reproduced with permission from Labau *et al.*, *J. Ultrastructure Res.*, **51**, 362 (1975).]

Now it appears almost certain that all regions, both V and C, have their respective polypeptide chains folded in approximately the same way (Figure 19–15). Each region contains two layers of extended β sheets (see page 139) that are roughly parallel to each other and surround an interior portion in which are packed the side groups of hydrophobic amino acids. The two V regions tend to pack together to form the variable domain, as do the two C regions to form the C_1 domain. The main difference between the V and C domains is that the V domain contains additional amino acids, many of which are used to bind to its respective antigenic determinant.

Figure 19–15

Schematic illustration of the basic immunoglobulin fold. The solid tracing shows the arrangement of the polypeptide chain of the C_L domain. The additional amino acids found in V_L are present in the loops outlined by broken lines. [Redrawn from R. J. Poljak *et al., Proc. Nat. Acad. Sci.,* **70,** 3305 (1973).]

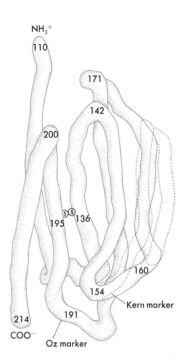

BOTH LIGHT AND HEAVY CHAINS DETERMINE THE SPECIFICITY OF ANTIBODIES

The active sites that combine with antigenic determinants include the variable parts of both light and heavy chains. At first it was suspected that most of the specificity was due to variations in the amino acid sequence of the heavy chain, but now we realize that in most cases the amino acid sequence of the light chain is also involved. This discovery helps to mitigate the dilemma posed earlier about the great number of genes needed to code for different antibodies. Since each antibody must be coded by two genes (one for the light chain, the other for the heavy chain), the number of possible antibodies may be the number of different light chains multiplied by the number of heavy chains. Thus, a million different antibodies may be formed by only two thousand genes, coding for a thousand different light chains and a thousand different heavy chains.

Not all the amino acids in the variable regions participate directly in forming active sites. About a third are involved, and it is these amino acids which show the greatest variability from one immunoglobulin to another. Their respective sites along the light and heavy chains are called the *hypervariable sites* (Figure 19–16).

Figure 19–16

Hypervariable regions of the V regions of immunoglobulin molecules.

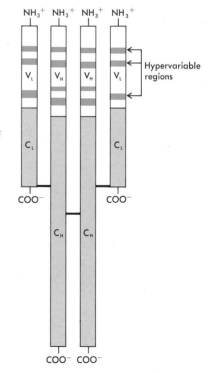

All the hypervariable amino acids occur in close spatial proximity at one end of each immunoglobulin and are completely exposed to the surrounding solvent. They prescribe a cavity (the active site) into which the respective antigenic determinant fits. The exact geometry of the active site obviously varies with the exact amino acids which line it. In the Fab' NEW fragment, the hypervariable amino acids surround a shallow cleft ($16 \times 7 \times 6$ Å) between the V_L and V_H domains into which a hydroxy derivative of vitamin K can specifically fit (Figure 19–17). A larger cavity ($15 \times 20 \times 12$ Å) is formed by the hypervariable amino acids of Fab McPC 603, a mouse myeloma

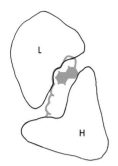

Figure 19–17
Schematic top view of the shallow cleft between the heavy and light variable chains of the human myeloma protein NEW which specifically binds the hapten, vitamin K_1OH (shown in color). [Redrawn from F. F. Richards *et al.*, in *The Immune System*, Third ICN-UCLA Symposium on Molecular Biology, Academic Press, 1974.]

protein that has phosphoryl choline-binding activity. In this case, only a small part of the active site binds the hapten, with most of the interaction being to variable amino acids of the heavy chain (Figure 19–18).

In contrast to the hypervariable amino acids involved in antigen binding, many amino acids within the variable region show very little variation among antibodies of differing specificity (Figure 19–19). This constancy, or near constancy, in the midst of variability is probably required for formation of the basic immunoglobulin fold (Figure 19–15).

Now it appears that the shape of the active site is the same, independent of whether or not an antigenic determinant is bound to it. In contrast, there appears to be more flexibility between the domains along a given L (H) chain. In particular, there exists a very flexible hinge region between C_H1 and C_H2, around which the shape of antibodies can change when antigen–antibody complexes form.

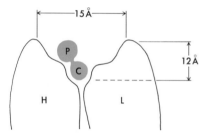

Figure 19–18
Schematic drawing of phosphoryl choline (PC, in color) in the cavity formed by the hypervariable regions of the heavy (H) and light (L) variable chains of the mouse myeloma protein McPC 603. This side view reveals that most antibody–antigen interaction involves the heavy chain. [Redrawn from E. A. Padlan *et al.*, in *The Immune System*, Third ICN-UCLA Symposium on Molecular Biology, Academic Press, 1974.]

Figure 19–19 ▶
The V_H sequences of nine human antibody chains. Identities with protein Tie are indicated with a line. Nonidentical amino acids are colored and tend to cluster in definite regions (the hypervariable sites). LAY and POM are two anti-gamma globulins of identical specificity (idiotypy) obtained from different individuals; they exhibit remarkable identity in the hypervariable regions involved in antigen binding. [From J. D. Capra and J. M. Kehoe, *Proc. Nat. Acad. Sci.*, **71**, 4032 (1974).]

```
                                    10                          20                              30
Tie (γ1)  GLU VAL GLN LEU VAL GLU SER GLY GLY GLY LEU VAL GLN PRO GLY GLY SER LEU ARG LEU SER CYS ALA ALA SER GLY PHE THR PHE SER

Was (γ1)  _____LEU_____SER_____

Jon (γ3)  ASP_____LYS_____

Zap (α1)  _____ALA_____GLY_____

Tur (α1)  _____LEU_____

Nie (γ1)  PCA_____GLN_____VAL_____ARG_____

Gal (μ)   _____ASP_____ARG_____(ASX VAL LEU

LAY       ALA_____LEU_____

POM       _____LEU_____
```

```
                                    40                          50                              60
Tie   THR SER ALA VAL TYR [ ]   TRP VAL ARG GLN ALA PRO GLY LYS GLY LEU GLU TRP VAL GLY TRP ARG TYR GLU GLY SER SER LEU THR

Was   _____ASP_____MET_____[ ]   _____ALA_____LYS_____GLN GLU ALA_____ASN SER

Jon   _____ALA TRP MET LYS [ ]   _____VAL_____VAL_____GLN VAL VAL GLU LYS

Zap   _____THR SER ARG PHE [ ]   _____GLU PHE_____VAL GLN_____ALA ILE SER

Tur   ARG VAL LEU SER SER  [ ]   _____SER GLY_____LEU ASN ALA_____ASN LEU

Nie   ARG TYR THR ILL HIS  [ ]   _____ALA VAL MET SER TYR ASX GLY ASX ASX LYS

Gal   ASX ASX PHE) MET THR [ ]   _____ALA ASN ILE  LYS GLX ASX GLY_____GLX GLX

LAY   ALA_____ MET SER [ ]   _____ALA_____LYS_____ASN GLY ASN ASP LYS

POM   SER_____ MET SER [ ]   _____ALA_____LYS_____ASX GLY ASN ASP LYS
```

```
                                    70                          80                              90
Tie   HIS TYR ALA VAL SER VAL GLN GLY ARG PHE THR ILE  SER  ARG ASN ASP SER LYS ASN THR LEU TYR LEU GLN MET LEU SER LEU GLU PRO

Was   _____PHE_____ASP THR_____ASN_____ASN ARG_____ALA

Jon   ALA PHE_____ASN_____ASN_____ILE_____VAL THR_____

Zap   _____ASP_____ALA_____ASN THR GLY_____ALA

Tur   _____PHE_____ALA_____GLN ALA

Nie   _____ASP_____ASN_____ASN_____ASN_____ARG_____

Gal   ASX_____VAL ASP_____LYS_____ASP ASN ALA_____SER_____ASN_____ARG VAL

LAY   _____ASP_____ASN_____ASN GLY_____GLN ALA

POM   _____ASP_____ASN_____LEU_____ASN_____GLN ALA
```

```
                                    100                         110                             120
Tie   GLX ASX THR ALA VAL TYR TYR CYS ALA ARG VAL THR PRO ALA ALA ALA SER LEU THR PHE SER ALA VAL TRP GLY GLN GLY THR LEU VAL

Was   _____PHE ARG GLN PRO PHE VAL GLN [     ]_____PHE ASP_____PHE_____

Jon   _____VAL VAL  SER THR [     ] SER MET ASP_____PRO_____

Zap   _____THR ARG_____GLY GLY TYR [     ] ASP_____

Tur   _____LEU SER VAL THR_____VAL [     ] ALA PHE ASP_____LYS_____

Nie   _____ILE ARG ASP THR_____MET [     ]_____PHE_____HIS_____

Gal   _____GLY TRP GLY [     ] GLY GLY ASP TYR_____

LAY   _____VAL SER_____ILE_____ASP ALA GLY PRO TYR VAL_____PRO_____PHE_____HIS_____

POM   _____LEU_____ASP ALA GLY PRO TYR VAL_____PRO_____PHE_____HIS TYR_____
```

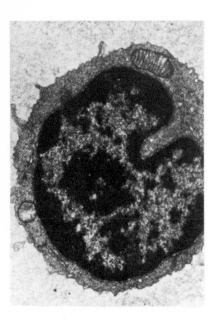

SMALL LYMPHOCYTES ARE THE PROGENITORS OF ALL IMMUNOGLOBULIN-PRODUCING CELLS

All immunological responses are dependent upon the existence of very large populations of *"small" lymphocytes,* a form of white blood cell characterized by a seemingly inert nucleus surrounded by only traces of an endoplasmic reticulum-poor cytoplasm (Figure 19–20). Under most conditions, the cells are essentially dormant, maintaining just the low metabolic level needed to keep them alive until they receive an appropriate antigenic stimulation. Then they quickly enlarge in size and commence proliferation.

Figure 19–20
Electron micrograph of a thin section of a small lymphocyte, showing the relatively small volume occupied by its cytoplasm. [Reproduced with permission from I. Roitt, *Essential Immunology,* 2nd ed., 1974, Blackwell Scientific Publications, London.]

"T" LYMPHOCYTES VS. "B" LYMPHOCYTES

Two very different populations of small lymphocytes exist. Members of one set, the *"T" lymphocytes,* originate in the *thymus,* a lymphoid gland that derives embryologically from the epithelium of the gut. Members of the second set, the *"B" lymphocytes,* are responsible for the synthesis of circulating antibodies. Their immediate origin in mammals is not clear, though in birds they are produced by a lymphoid gland called the *bursa,* which also derives from the gut in embryonic life. The cellular progenitors of both lymphocyte populations are bone marrow stem cells, some of which migrate into the thymus, where they become differentiated into "T" cells, while others move into the bursa (or its mammalian equivalent), where they differentiate into "B" cells (Figure 19–21). Thus animals from which the thymus has been removed early in life cannot mount a cell-mediated immunological attack. Correspondingly, elimination of the bursa leads to an absence of circulating antibodies.

Morphologically, "T" and "B" cells are microscopically indistinguishable, though "T" cells can be identified by a specific surface component, the θ antigen, which first appears when the "T" cells become differentiated in the thymus. Much use is now made of mutants which lack either "T" or "B" cells. Particularly useful are the so-called "nude" mice, which lack thymuses and thus are incapable of making "T" cells.

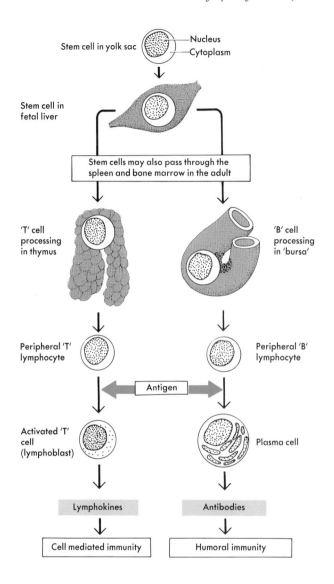

Stem cell in yolk sac — Nucleus
— Cytoplasm

Stem cell in fetal liver

Stem cells may also pass through the spleen and bone marrow in the adult

'T' cell processing in thymus

'B' cell processing in 'bursa'

Peripheral 'T' lymphocyte

Peripheral 'B' lymphocyte

Antigen

Activated 'T' cell (lymphoblast)

Plasma cell

Lymphokines

Antibodies

Cell mediated immunity

Humoral immunity

Figure 19–21
Presumed origin of B and T lymphocytes through the processing of undifferentiated stem cells.

LYMPHOCYTE TRANSFORMATION

Receipt of the appropriate antigenic stimuli changes "T" cells into *lymphoblasts* (Figure 19–22) and "B" cells into *plasma cells* (Figure 19–23). While superficially they are rather similar in appearance, they can easily be distinguished by the fact that lymphoblasts do not secrete large amounts of antibody and lack the rough endoplasmic reticulum involved in the plasma cells' export of massive amounts of circulating antibodies. Transformation of "B" cells into plasma cells does not occur directly. Instead they

Figure 19–22
Electron micrograph of a thin section of a transformed T lymphocyte (lymphoblast). [Reproduced with permission from I. Roitt, *Essential Immunology*, 2nd ed., 1974, Blackwell Scientific Publications, London.]

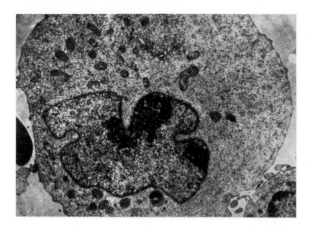

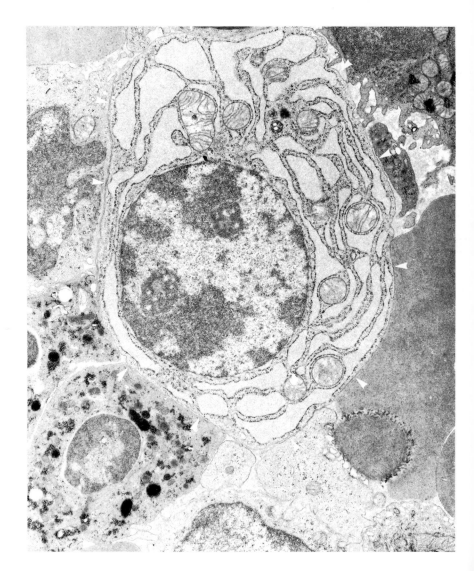

become converted into immature plasma cells called *plasmablasts*. Each plasmablast exists for only a short time before dividing to form progeny cells. These progeny are not, however, morphologically identical to their parents. Each successive division cycle results in cells having a more pronounced cytoplasm filled with an increasing number of ribosomes (Figure 19–24). By the fifth day after antigen stimulation, the cycle of successive cell divisions has produced a colony of adult plasma cells which are rapidly secreting antibodies.

Figure 19–23
Electron micrograph of a plasma cell from the spleen of a guinea pig. [Courtesy of K. R. Porter, Biological Laboratories, Harvard.] The cell margins are indicated by white arrows. The cavity of endoplasmic reticulum is greatly distended by the presence of large amounts of antibody molecules. Ribosomes are visible as black dots attached to the endoplasmic reticulum. The objects around the edges are other types of white blood cells.

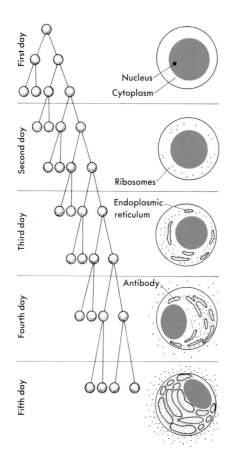

Figure 19–24
Successive stages in the development of mature plasma cells. At least 8 cell generations and 5 days of growth are required before the appearance of cells producing a great deal of antibody. The most noticeable feature of the mature cell is the extensive endoplasmic reticulum whose internal cavity is filled with antibody molecules. [Redrawn from G. J. V. Nossal, *Scientific American*, December, 1964, p. 109, with permission.]

A GIVEN PLASMA CELL PRODUCES ONLY ONE TYPE OF ANTIBODY

When a number of different antigens are injected simultaneously into an animal, the question arises as to whether a given plasma cell produces antibodies against all the foreign antigens or, instead, produces only one type of antibody. According to instructive theories, we might expect that many different types of antigen would enter a single plasma cell, and so each cell should produce a variety of antibodies. The experimental answer, however, seems to be the opposite. When the antibodies produced in a single cell are examined (this can be done by isolating single plasma cells after they have begun to produce antibodies), most cells are found to produce only one specific antibody. Thus, almost every antibody-producing plasma cell represents a most highly specialized factory, devoting much of its protein synthesis to the production of only one product.

ANTIBODY-PRODUCING CELLS NEED NOT CONTAIN ANTIGENS

At first it seemed obvious that much antigen would be present within the antibody-producing plasma cells. Now, however, there is compelling reason to believe that, at best, it is present only in traces and has no functional role. Such evidence arises from experiments in which a highly radioactive antigen is injected into an animal. Some days later, thin sections of antibody-producing regions are examined by autoradiographic techniques to find out where the antigens have gone. Much of the labelled antigen is found within the scavenger macrophages, which have no direct connection with antibody synthesis. In contrast, most plasma cells seem not to contain even a single antigen, and only a few antigens are visible on their plasmablast precursors. This observation independently rules out any instructional theory, for each antibody-producing cell is simultaneously making thousands of antibodies. Given the instructional model, we would expect that a similar number of antigens would be present, perhaps bound to the ribosomal sites of protein synthesis.

THEORY OF CLONAL SELECTION

The fact that the presence of an antigen leads to a great increase in the number of cells which produce the corresponding specific antibody supports the now generally accepted *theory of clonal selection*. (A clone is a group of identical cells all descended from a common ancestor.) This theory presumes that the sole function of the antigen is to stimulate specific cell division and so need provide no information other than the fact that it is present. It assumes that there *preexists,* prior to the appearance of the antigen, a very large variety of differentiated small lymphocytes, each species endowed with the capacity to form only one specific type of antibody. The antigen then acts to induce the amplification of the appropriate species by causing its selective proliferation (Figure 19–25).

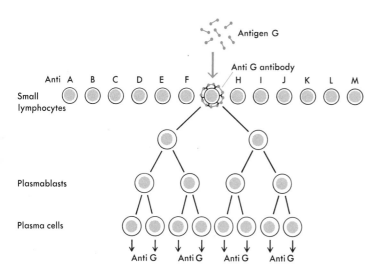

Figure 19–25
The clonal selection model of antibody formation. In this model an antigen stimulates the synthesis of complementary antibodies by causing the selective division and differentiation of those small lymphocytes which are already precommitted to the synthesis of such antibodies.

Even early in embryological development, the number of different lymphocytes differentiated to produce unique immunoglobulin species is already very large. By the time the first "B" ("T") cells can be detected leaving their bursal (thymal) sites of origin, there exist many, many thousands of different types. The belief thus exists that the differentiation process which generates so many unique specificities occurs within the bursa (thymus), and that the main, if not sole, function of these lymphoid glands is to convert unspecific bone marrow stem cells into small lymphocytes permanently endowed with a single specificity.

IMMUNOGLOBULINS ON THE SURFACE OF SMALL LYMPHOCYTES

Both "B" and "T" cells have immunoglobulin molecules imbedded in their plasma membranes in such a way that active sites face outward and are thus capable of binding specific antigens. These surface antibodies are believed to be the cellular receptors which receive the antigenic stimuli that set off the corresponding immunological response. Each lymphocyte contains only one type of surface antibody receptor; and when it becomes immunologically stimulated, its descendants continue to produce only this same unique immunoglobulin molecule. Many more immunoglobulin molecules are present on the surface of "B" cells than are bound on "T" cells, the latter of which may contain as few as several hundred receptors. Because of their paucity, we still do not know whether "T" cells carry any of the five known Ig classes.

How antigen binding triggers small lymphocytes to enlarge and commence cell division is almost a total mystery. At present, our only clue is the observation that multimeric substances containing a number of identical repeating subunits are much better antigens than molecules containing single copies of their antigenic determinants. Now it seems likely that, for a compound to be antigenic by itself, it must form cross-links between the mobile immunoglobulin surface receptors (Figure 19–26). If true, this hypothesis neatly explains why small univalent molecules like DNP (dinitrophenol) by themselves are never antigenic. Only when they become attached to large molecules can they cause aggregation of the surface immunoglobulins. How the cross-linking of the surface immunoglobulins might provide a signal to start cell proliferation is still a complete mystery.

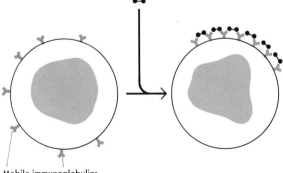

Figure 19–26
Diagram to show how a multimeric antigen can act to cluster mobile immunoglobulins in the cell membrane of a small lymphocyte.

A GIVEN ANTIGEN BINDS TO ONLY A TINY FRACTION OF THE SMALL LYMPHOCYTE POPULATION

Tagging a given antigen with a radioactive label or fluorescent dye allows a direct measurement of the percentage of small lymphocytes that contain surface immunoglobulins with complementary active sites. When the lymphocytes are taken from an animal never previously exposed to this antigen, only a very small fraction (10^{-4} to 10^{-5}) possesses radioactive label (positive fluorescence) (Figure 19–27).

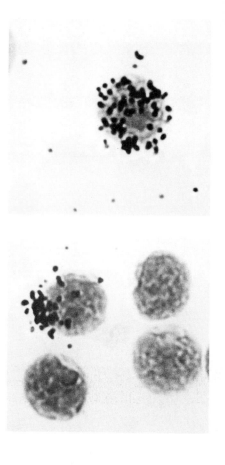

Figure 19–27
Autoradiograph of small lymphocytes from a lymph node of an unimmunized mouse following exposure *in vitro* to tritium-labeled flagellin. Approximately one cell in 10^5 shows specific binding of the label.
(a) Uniform binding following incubation at 0°C for thirty minutes.
(b) Aggregation (capping) of antigen on one cell pole after incubation at 37°C for fifteen minutes. Such capping results from the cross-aggregation of the surface immunoglobulin receptors by the polymeric antigen. [Reproduced with permission from Diener and Paetkau, *Proc. Nat. Acad. Sci.*, **69**, 2364 (1972).]

However, if the lymphocytes originate from an animal repeatedly immunized to this antigen, then a much higher percentage of the cells are labeled (brightly stained). The number of such cells always provides an upper limit for the fraction of cells containing a single specific immunoglobulin species. Not only do many antigens contain more than one strong antigenic determinant, but also a given antigenic determinant frequently can bind to more than one immunoglobulin species. Now there is evidence that, in a given individual, at least some 5 to 10 different antibodies (that is, possessing different amino acid sequences) can bind to the hapten dinitrophenol (DNP). The amount of immunological diversity is thus even greater than we guessed it to be only a few years ago.

PRIMARY VS. SECONDARY RESPONSE

While the first (*primary*) injection of an antigen into an animal provokes only a small number of small lymphocytes to become transformed into plasma cells (lymphoblasts), a second such injection coming some weeks later leads to a much more vigorous response (Figure 19–28).

Figure 19–28
A graph to show the primary and secondary response of antibody production in a rabbit. Arrows 1 and 2 represent the times of injection with the antigen. The secondary response is both more rapid and more intense than the primary response.

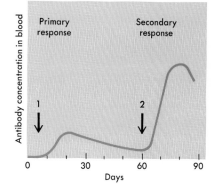

Not only do many more cells participate, but also the resulting antibodies bind much more tightly to their respective antigens. Such heightened *secondary* responses indicate the existence of *immunological memory,* the process by which an animal "remembers" a prior contact with an antigen. Its cellular mediator is the small lymphocyte itself—as shown by experiments in which their transfer from an immunized to a nonimmunized animal conveys the capacity to mount a secondary-type immunological response.

The development of a primary response thus leads not only to the large-scale production of plasma cells (lymphoblasts), but also to the selective production of many more copies of the originally stimulated small lymphocytes (*memory cells*). The origin of the new small lymphocytes is not thought to be the plasma cells (lymphoblasts) themselves, for after some days of a primary response, most of them die. Conceivably, memory cells arise either from dedifferentiation of immature plasmablasts (lymphoblasts), or by some process in which antigens provoke cell division of small lymphocytes *per se,* without concomitant differentiation into mature blast cells.

The improved binding affinities of the antibodies found in a secondary response reflect the selective multiplication of the lymphocytes which can bind the antigen the tightest. Apparently the probability that a small lymphocyte will produce a large number of memory-cell descendants is directly related to the affinity of its antibodies for the inducing antigen. Those small lymphocytes which bear immunoglobulins with relatively poor affinity for the antigen give rise to very few memory cells. A steady improvement of the binding affinity is found not only during the secondary response, but also during the progression of a primary response.

We thus see that the exact specificity of the antibodies made during the start of a secondary response has in fact been predetermined by the exact specificity of the antigen used to induce the primary response. For example, during an influenza infection the antibodies initially made are likely to be more adapted to the exact flu strain we suffered from several years ago than to the immediate troublemaker.

UNSPECIFIC TRANSFORMATION INDUCED BY SURFACE BINDING AGENTS

While a specific antigen triggers the transformation of only those lymphocytes possessing complementary surface immunoglobulins, there exist substances which induce generalized unspecific transformation. Such agents usually trigger the transformation of either "T" or "B" cells. A complex bacterial lipopolysaccharide, for example, transforms only "B" cells, while the mitogenic plant lectins concanavalin A (pokeweed) and phytohemagglutinin (jack bean) specifically provoke "T" cells to divide and differentiate (Figure 19–29). These unspecific mitogens greatly facilitate biochemical analysis of transformation, since the relevant steps can occur in a large fraction of the treated lymphocytes—not just in the minor fraction provoked by a specific antigen.

Generally, only several division cycles follow *in vitro* transformation of normal cells, thereby paralleling the situation *in vivo,* where a given antigenic stimulus leads to a self-limiting round of proliferation. Immortal cell lines, however, can be obtained when small lymphocytes are infected with certain lysogenic-like *herpes* viruses (see below).

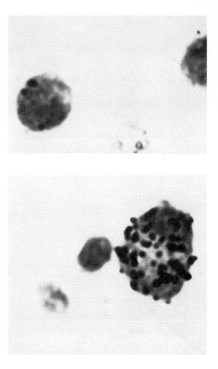

Figure 19–29
Transformation of a mouse-spleen small lymphocyte into a lymphoblast following exposure to concanavalin A (3μg/ml) for 48 hours. Tritiated thymidine was then added for two hours. (a) Unstimulated small lymphocyte incapable of making DNA. (b) Transformed blast cell actively synthesizing DNA. [Reproduced with permission from Cunningham *et al.,* in *Cellular Selection and the Immune Response,* G. M. Edelman (ed.), Raven Press, New York, 1974.]

ORIGINS OF ANTIBODY DIVERSITY

The fact that a given individual has the potential of making so many different lymphocytes ($\sim 10^6$ to 10^7), while a given small lymphocyte is irrevocably committed to the production of immunoglobulins of a single specificity, now stands as the major unresolved problem in immunology. Attempts to understand what is happening initially led to two very different hypotheses. The first postulated that diversity arises from *somatic mutations* occurring in a single germ line immunoglobulin gene. Usually the further assumption was made of hypermutability within the nucleotide sites specifying the variable regions of the light and heavy chains, particularly those directly coding for the amino acids of the "active sites." The second approach postulated *a separate gene* for every different light and heavy chain. Even with the new knowledge that most vertebrate DNA does not code for amino acid sequences (so that the size of a gene is about 30 times larger than previously guessed), the amount of genetic material needed—(1000 to 3000 light chain genes $\times$ 1000 to 3000 heavy chain genes = 10^6 to 10^7 different antibodies)—would still occupy only some 1 to 10% of the total DNA. So, *a priori,* the postulating of moderately large numbers of immunoglobulin genes does not contradict any fundamental coding concepts.

Now we realize that neither hypothesis in its all-or-none form is correct, and we suspect that when the story is finally understood, both somatic mutations and multiple germ line genes will be involved. Already enough firm facts exist to rule out the most simplified forms of each hypothesis.

TWO FORMS OF LIGHT CHAINS

Two quite distinct types of light chains exist within every human. One is called κ (kappa) and the other λ (lambda). Each has very characteristic amino acid sequences, and any given chain is unmistakably of the κ or λ type (Figure 19–30). It is thus impossible to imagine their origin by somatic mutations from a single germ line light chain gene. Instead we must postulate the existence of at least one separate germ line gene for each group. Not only is the constant region of a given chain of either κ or λ specificity, but the variable regions can also be clearly assigned. Those "variable" sequences not involved in active site formation fall into two general patterns, one of which V_κ, is always linked to the C_κ sequence while the other, V_λ, is invariably attached to the C_λ sequence.

```
   Variable      Hypervariable        Variable                         Constant

9                 1                   1                 1               1               1             1
1 2 3 4 5 6 7 8 9 0 1 2 3 4 5 6 7 8 9 0 1 2 3 4 5 6 7 8 9 0 1 2 3 4 5 6 7 8 9 0 1 2 3 4 5 6 7 8 9 0
A D Y Y C N S R D S S G K H V L F G G G T K L T V L G Q P K A A P S V T L F P P S S E E L Q A N K A
A D Y Y C S S Y V D N N N F X V F G G G T K L T V L R Q P K A A P S V T L F P P S S E E L Q A N K A
A H Y H C A A W D Y R L S A V V F G G G T Q L T V L R Q P K A A P S V T L F P P S S E E L Q A N K A
A D Y Y C Q A W D S S L N A V V F G G G T K V T V L G Q P K A A P S V T L F P P S S E E L Q A N K A
A D Y Y C Q A W D - - S M S V V F G G G T R L T V L S Q P K A A P S V T L F P P S S E E L Q A N K A
A T Y Y C Q Q Y D - - T L P R T F G Q G T K L E I K R T Y - A A P S V F I F P P S N E Q L K S G T A
A T Y Y C Q Q F D - - N L P L T F G Q G T K V D F K R T Y - A A P S V F I F P P S D E Q L K S G T A
G V Y Y C Q M R L - - E I P Y T F G Q G T K L E I R R T Y - A A P S V F I F P P S D E Q L K S G T A
G V Y Y C M Q A L - - Q T P L T F G G G T N V E I K R T Y - A A P S V F I F P P S B Z Z L K S G T A
A V Y Y C Q Q Y G - - S S P S T F G Q G T K V E L K R T Y - A A P S V F I F P P S D E Q L K S G T A
```

A = ala	E = glu	I = ile	N = asn	S = ser	Y = tyr
B = asx	F = phe	K = lys	P = pro	T = thr	Z = glx
C = cys	G = gly	L = leu	Q = gln	V = val	
D = asp	H = his	M = met	R = arg	W = trp	

Figure 19–30

Selected amino acid sequences from the V and C regions of several human λ (top) and κ (bottom) light chains. Not all chains have exactly the same number of amino acids, and the sequences are aligned for maximum homology. [Modified from H. Eisen, *Immunology*, Harper and Row, 1974.]

The various V_κ and V_λ sequences themselves can further be subdivided into distinct subpatterns. In man, at least three very distinct subgroups of V_κ sequences exist, while V_λ sequences fall into at least five different subgroups (Figure 19–31). In contrast, all the C_κ (C_λ) sequences are effectively identical. The absence of intermediate patterns of variable sequences seems to argue against the subgroups arising from somatic mutations during embryological development. So most people now believe that there exists at least one germ line gene coding for the amino acid sequences of each subgroup. More sequence data, however, must be elucidated before the matter is settled.

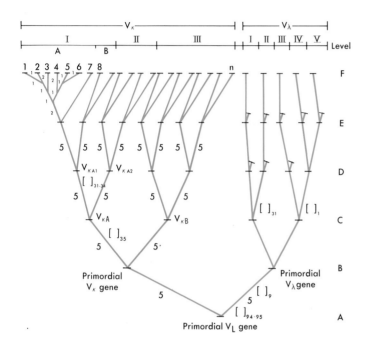

Figure 19–31
A possible genealogic tree for the variable region subgroups of human light chains. The numbers on the branches indicate the number of amino acid substitutions occurring between branch points. The genetic events responsible for the pattern may reflect mutations occurring during both vertebrate evolution and somatic differentiation. [Redrawn from L. Hood, J. H. Campbell, and J. Elgin, *Ann. Rev. Genetics,* 9, in press (1975).]

DIFFERENT GENES FOR THE VARIOUS FORMS OF HEAVY CHAINS

There is also firm evidence for the existence of a number of different genes coding for the various heavy chains. The several immunoglobulin classes (IgG, IgM, IgD, IgA, and IgE) have such characteristic heavy chains that only their coding by separate genes can be imagined. The heavy γ (IgG) chains, moreover, fall into four antigenically distinct blocks of specific amino acids (γ_1, γ_2, γ_3, and γ_4). So here again we believe that each must be coded by a separate germ line gene. Only two subclasses of the human IgA and IgM antibodies are now described, but much less work has been done with these antibody classes. With time, evidence for additional heavy chain subclasses and their respective germ line genes is likely to materialize.

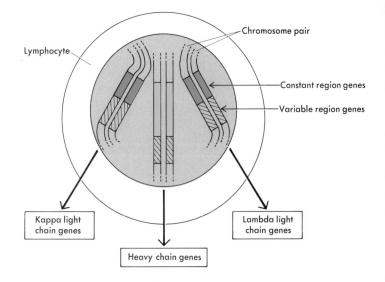

Figure 19–32
The different families of antibody genes segregate independently as if they are on different pairs of chromosomes.

ALLOTYPES

Each distinct immunoglobulin germ line gene can exist in a variety of allelic forms which respectively give rise to immunoglobulin chains with characteristic amino acid markers (*allotypes*). For example, individuals homozygous for the IgG allele Gma$^+$ have the γ_1 constant sequence . . . Asp.Glu.Leu.Thr.Lys. . . , while those carrying the alternative GMA$^-$ alleles produce γ_1 heavy chains with the sequence . . . Met.Glu.Glu.Thr.Lys. . . Extensive analysis of the inheritance of allotypic markers in the rabbit indicates that genes coding for the λ light chains are not at all linked to the genes coding for the κ chains or to the genes coding for the heavy chains (Figure 19–32). Interestingly, the genes coding for the various groups of heavy chains (e.g., γ_2, α_1, γ_1, μ_2) all lie very close to each other.

SEPARATE GERM LINE GENES FOR THE V AND C REGIONS

How the separate genes for the various subgroups could easily maintain identical sequences (allotypes) on their constant halves and let pronounced evolutionary divergence develop in specific regions of the variable halves has always seemed most tricky. For this to happen, some

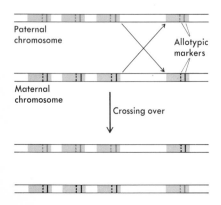

Figure 19–33
Schematic illustration of how crossing-over between a large number of tandemly arranged antibody genes should lead to a disappearance of clearcut allotype markers. Since the allotypes behave as single Mendelian factors, those genes carrying allotype markers must be present only once per haploid genome.

way would have to be found not only to restrict most spontaneous mutations to the variable half, but also to prevent crossing-over from recombining allelic markers (Figure 19–33). While it is possible to conceive of complicated *ad hoc* hypotheses that might allow tandemly linked V–C genes to maintain identical allotypes, not one has the smell of potential truth. Instead, it seems highly probable that separate germ line genes exist for the V and C regions. Moreover, the clearcut inheritance of allotypes suggests that each distinctive C sequence is coded by a single C gene. In contrast, there may exist multiple V genes, at least one for every subgroup, and maybe many, many more, depending upon whether somatic mutations are the cause of appreciable V-region diversity.

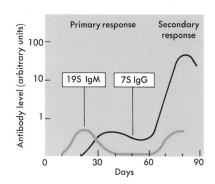
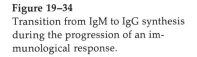

Figure 19–34
Transition from IgM to IgG synthesis during the progression of an immunological response.

CONSERVATION OF THE SPECIFICITY OF THE ACTIVE SITE DURING THE IgM → IgG TRANSITION

Strong evidence for the existence of separate genes for the V and C regions comes from study of the relationship between the IgM and IgG molecules produced during the course of an immunological response. Following a primary antigenic (or unspecific) transformation, the first antibodies that are secreted always belong to the IgM class (19S). Some cell divisions later, and usually only during the secondary response, IgG (7S) molecules begin to predominate (Figure 19–34). This switchover occurs without any change in the specificity of the active site. Not only are light chains of identical sequence found in the corresponding 7S and 19S immunoglobulins, but the V regions of their heavy μ (IgM) and γ (IgG) chains are also the same. In contrast, the corresponding heavy chain constant regions have different sequences (Figure 19–35). How a given variable sequence can be attached to two different constant regions is, as we shall discuss below, at the heart of the problem of the generation of antibody specificity.

Why a given active site should first be part of IgM and then of IgG molecules is not understood. Immunoglobulins of the highly multimeric 19S variety easily form precipitates at low concentrations, and so may be more easily taken up by macrophages when the immunological response is starting. In contrast, high antibody valency may not be needed at high antibody concentration.

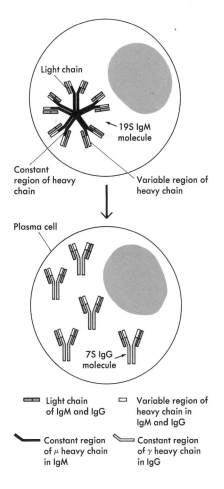

Figure 19–35
Attachment of the same V_H region to different C_H regions during an IgM → IgG transition.

623

SINGLE mRNA CHAINS CODE FOR COMPLETE IMMUNOGLOBULIN CHAINS

The fact that a given variable region can be attached to several types of constant regions, while many different variable sequences can be linked up to a given constant region, raises the obvious question of whether these joining processes might occur after synthesis of the respective V and C polypeptide chains. Conceivably, some form of polypeptide ligase joins together newly synthesized V and C chains. Now, however, several types of experiments make this proposal untenable. In the first place, kinetic experiments show that, for a given immunoglobulin chain, the N-terminal (variable) amino acids are always laid down before any C-region amino acids are linked up to each other. In the second place, *in vitro* protein synthesis experiments show that mRNA chains isolated from immunoglobulin-producing plasma cells code for either complete light or heavy chains, not for their V or C components. Those mRNA molecules coding for the 23,000 MW light chains sediment at 13S, while those coding for the 55,000 MW heavy chains sediment at 16S. The light-chain mRNA molecules have some 200 nucleotides at the 5' end and some 400 nucleotides at the 3' end which do not code for amino acids found in mature immunoglobulin chains. The functions of these sequences remain largely unknown, with some 50 nucleotides on the 5' end coding for amino acids present transiently in a light-chain precursor (Figure 19–36).

Figure 19–36
Diagrammatic representation of a mouse myeloma light-chain mRNA. The V and C lengths are based on corresponding numbers of amino acids in the V and C regions. P refers to amino acids present in a short-lived precursor to the 13S mRNA. UT = untranslated region.

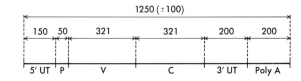

NUMBER OF GENES CODING FOR THE V AND C REGIONS

Direct measurements of the number of V and C genes using highly purified 13S (light) mRNA molecules from specific myeloma cells to hybridize cellular DNA have not yet yielded unambiguous results. In the first such experiments, about half the mRNA hybridized at the rate expected for single gene copies, leading to the speculation that it represents hybridization of RNA to C DNA. The other half hybridized some 200 times faster, hinting at the presence of several hundreds of copies of the V sequences similar to those which coded for the respective myeloma mRNA. Exactly how far V sequences can diverge before their mRNA products cease hybridizing is not clear, and so these experiments were first thought to provide a lower limit for the number of gene equivalents. As the same results were obtained when using DNA isolated from myeloma cells and when using DNA purified from liver cells, differentiation into immunoglobulin-producing cells apparently does not involve gene amplification.

Unfortunately this type of answer is not universally reported, and now there exist counterclaims (also from hybridization experiments) that most, if not all, the V region genes are present only in single copies. These reports, if true, suggest that there may be only a single copy per subgroup of the genes coding for the variable portions of immunoglobulin genes. Support for this alternative position comes from genetic analysis of the rabbit immunoglobulin genes. Allelic differences have been found that lead to amino acid differences not only in the C region but also in the V region of the heavy chain. The three alternative a_1, a_2, and a_3 alleles lead to quite different sequences along the first 30 N-terminal amino acids in the V region. If the V genes were present in hundreds of copies, the maintenance of such reproducible differences is almost impossible to imagine.

Thus, we must reserve judgment about the relative number of V and C genes until we have both more evidence about the purity of the mRNA used in the hybridization experiments and confirmation from other species of the existence of allotypic differences in the V region.

Of great potential use may be the finding of mutant myeloma cells which produce incomplete immunoglobulins containing only the constant region, since their immunoglobulin mRNA molecules are likely to be transcripts of only their respective C genes. If so, they can be used in RNA–DNA hybridization experiments to unambiguously measure the absolute C gene number.

IDIOTYPES

Immunization of an animal with an antibody produced by a genetically identical inbred strain leads to the production of antibodies directed against the active site of the immunoglobulin used as the antigen. Antibodies against specific antigen combining sites (*idiotypes*) are known as *idiotypic antibodies*. Correspondingly, those antigenic determinants in the active site that induce such antibodies are called *idiotypic markers*. Possession of anti-idiotype antibodies allows tests to be made of the capacity of various strains of the same species to produce specific idiotypes. When such experiments are done with rabbits or mice, only certain strains are found to be capable of making a specific idiotype (active site). Most likely these differences reflect differences in the immunoglobulin germ line genes, with only certain nucleotide sequences (or their somatically mutated derivatives) having the capability of coding for given idiotypes. If so, the genetic analysis of idiotypic markers eventually should provide direct information on the organization of the V genes. The breeding results obtained so far indicate that the various idiotypic markers of a given type of V chain are all closely linked to each other. Whether the very low rate of recombination observed reflects intragenic recombination between single V genes or intergenic recombination between groups of tightly linked V genes is not yet known.

COMBINATION OF V AND C GENES

When V and C genes come together is not established. It does seem likely, however, that this step is the crucial

Figure 19–37
The hypothesis that the linking of V and C regions occurs by specific chromosomal deletions. As shown above, this scheme easily allows the same V region to be attached first to a μC region and then to a γC region.

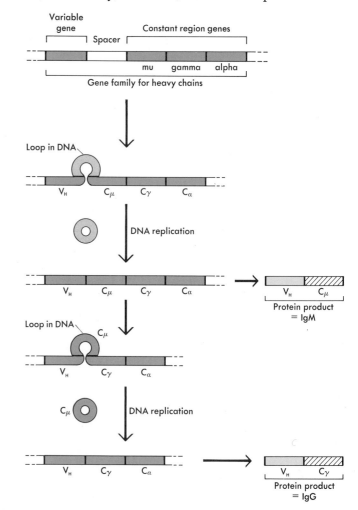

event in the commitment of a small lymphocyte toward the production of just one type of immunoglobulin. Use of allotypic markers in the rabbit indicates that V and C regions do not randomly combine with each other in diploid cells. Instead, the variable and constant chain markers of a given immunoglobulin are usually derived from the same parent, suggesting that the respective genes are closely located near each other on the same chromosome. Conceivably, some very regular deletions of genetic material bring the V and C regions together (Figure 19–37). But such speculations may make sense only when the chromosomal locations of the immunoglobulin genes are much better defined.

GENES AFFECTING THE IMMUNE RESPONSE

All animals of a given species do not give immune responses to a specific antigen. Some of these differences have recently been shown to have a genetic basis, with the discovery of a gene in the mouse that controls the response level to certain antigens. For example, homozygous possession of the allele Ir^b leads to a high response to TGAL (a synthetic polypeptide containing lysine, alanine, tyrosine, and glutamic acid residues), and a poor response to HGAL (the corresponding polypeptide in which tyrosine is replaced by histidine). In contrast, Ir^a homozygosity leads to a poor response to TGAL and a high response to HGAL. Interestingly, heterozygous Ir^b/Ir^a animals generate strong responses to both antigens. More unexpectedly, this immune response locus is located in the middle of the gene complex that codes for the major surface proteins which lead to rejection of transplanted grafts between unrelated individuals (Figure 19–38).

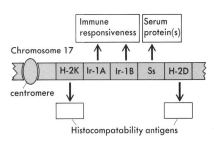

Figure 19–38
Genetic map of chromosome 9 of the mouse, showing how genes controlling the immune response are interspersed between the H-2 genes coding for the major histocompatibility antigens.

TRANSPLANTATION IMMUNITY

When skin from one individual is grafted onto a genetically distinct individual, it initially appears to be accepted and becomes infiltrated with blood vessels. Within 7 to 14 days, however, it is invaded by white blood cells and necrosis starts, leading quickly to graft rejection. That rejection is an immunological reaction is shown by comparing the speed at which a primary graft is destroyed to the rate at which a second graft is cast off. Second grafts using tissue from the same donor are rejected within 3 to 4 days, while a second graft using unrelated tissue again takes 7 to 14 days to be sloughed off (Figure 19–39). At first it was suspected that circulating antibodies mediate graft rejection; but then it was found that intact small lympho-

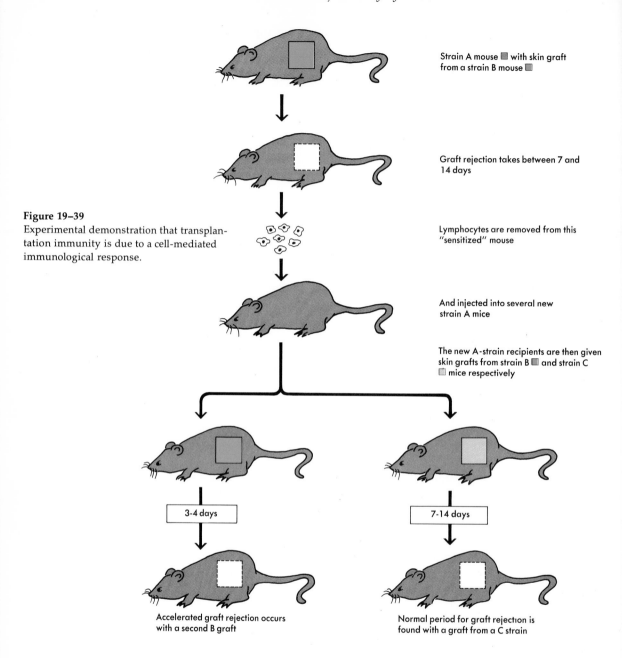

Figure 19–39
Experimental demonstration that transplantation immunity is due to a cell-mediated immunological response.

Strain A mouse ▨ with skin graft from a strain B mouse ▨

Graft rejection takes between 7 and 14 days

Lymphocytes are removed from this "sensitized" mouse

And injected into several new strain A mice

The new A-strain recipients are then given skin grafts from strain B ▨ and strain C ▨ mice respectively

3-4 days

7-14 days

Accelerated graft rejection occurs with a second B graft

Normal period for graft rejection is found with a graft from a C strain

cytes are the active agent. They can recognize the surface of grafted cells as either native or foreign and specifically destroy the cells bearing foreign antigens.

Genetic analysis of highly inbred mouse strains reveals the existence of a set of contiguous genes (the H-2 complex) which predominate in the graft rejection (*histocompatibility*) process (Figure 19–38). They determine the specificity of the major mouse histocompatibility antigens, the H-2 proteins. In humans, a similar linked group of major histocompatibility genes is called the HL-A complex. Within each H-2 (HL-A) complex, there is evidence for the coding of two distinct antigenic determinants, and so each diploid cell expresses four different specificities.

THE HL-A (H-2 PROTEIN) IS AN IMMUNOGLOBULIN-LIKE STRUCTURE

The most unexpected discovery has just been made that the major histocompatibility surface proteins (MW ~ 125,000) have a structure very like that of an immunoglobulin. They are built up from two light chains (MW ~ 12,000) and two heavy chains (MW ~ 50,000) held together by disulfide bonds. Even more suggestive, the light component is identical to β2-microglobulin, a serum protein whose amino acid sequence contains an S—S region strikingly homologous to the S—S linked C_H3 region of the IgG_1 heavy chain. Since the structure of the β2-microglobulin component appears invariant, the amino acid differences which give rise to the various HL-A (H-2) specificities must all reside within the heavy component. Direct analysis of the heavy chain is still tricky, because of the relatively low concentrations of the HL-A (H-2) proteins on the cell surface. Preliminary experiments with the H-2 antigen suggest each chain consists of three S–S bonded regions, each of which also resembles a C region of a heavy immunoglobulin chain. Given its extreme importance both theoretical and practical, the fine details of HL-A (H-2) structure are likely soon to be worked out.

This close homology between the histocompatibility antigens and the immunoglobulins suggests that, early in the evolution of vertebrates, immunoglobulins evolved by the duplication of genes coding for histocompatibilitylike surface antigens. Conceivably the normal function of the histocompatibility antigens is some form of cell–cell interaction. When their structures become worked out in three dimensions, we may see that specific cavities are formed in the region between their light and heavy chains.

IMMUNOLOGICAL TOLERANCE

Up until now, we have avoided mentioning a very important characteristic of the immunological response—the ability to recognize a molecule as *foreign*. How does an antibody-forming system know that a molecular surface is present on a foreign molecule? How is it able to avoid making antibodies against its own proteins and nucleic acids? This lack of responsiveness has nothing to do with

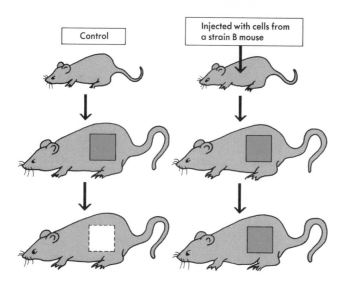

Figure 19-40
Experimental demonstration of immunological tolerance. Early exposure to foreign histocompatibility antigens enables a mouse to later accept skin grafts bearing these specific antigens.

an inherent lack of antigenicity of these molecules. For example, almost any human protein, if injected into a rabbit, will induce the formation of specific antibodies. Likewise, rabbit proteins are antigenic in humans. These facts lead to the conclusion that an immunological system learns how to recognize native molecules. Such a learning process occurs very early in life, before the development of an active immune response to foreign proteins. If a foreign antigen is injected into a newly born animal before it possesses an antibody-forming system, then, in adult life, the animal is unable to form antibodies against the early-injected foreign antigen. The foreign antigen is recognized as though it were a host protein (Figure 19-40). This lack of immunological responsiveness to all those antigens present when the antibody-forming system was being developed is called *immunological tolerance.*

It would indeed be surprising if the seemingly opposite behaviors of specific tolerance and specific antibody response were in fact completely unrelated. Perhaps very early in life there exist only a few cells genetically competent to form antibodies against a specific antigen. Immunological tolerance might result if at this time the excess presence of normal antigen resulted in immediate death of those cells (Figure 19-41).

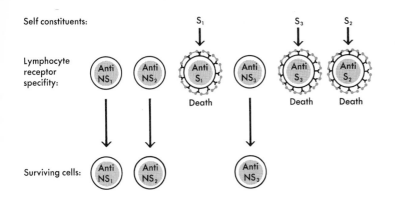

Figure 19–41
Origin of immunological tolerance to normal cellular (self) constituents
by the selective killing of lymphocytes bearing complementary immunoglobulin receptors. The cells which survive are able to react only
with foreign (nonself) antigens.

MIXED LYMPHOCYTE REACTIONS

Mixing of lymphocytes from two different individuals
(closely related species) frequently induces a small percentage (1%) of them to become transformed and commence limited proliferation. Such *mixed lymphocyte reactions* are immunological responses resulting from the presence on
certain lymphocytes of antibodies specific for the histocompatibility antigens of other individuals of the same or
closely related species. The very existence of such reactions
is not surprising by itself, considering the vast number of
different immunological specificities expressed by a given
individual's lymphocytes. Most unanticipated however, is
the magnitude of the effect. Whereas normally only one in
10^5 to 10^6 lymphocytes will strongly bind most unselected
antigens, a much larger fraction possess antibodies complementary to the histocompatibility antigens of other
members of its species.

SOMATIC GENERATION OF IMMUNE RECOGNITION

Conceivably, many of the antibody genes present in the
germ line of an individual are directed against the major
surface (histocompatibility?) antigens of its respective
species. If so, many of the immediate lymphocyte progeny
of the undifferentiated stem cells will invariably be stimulated to divide by contact with normal cells bearing the

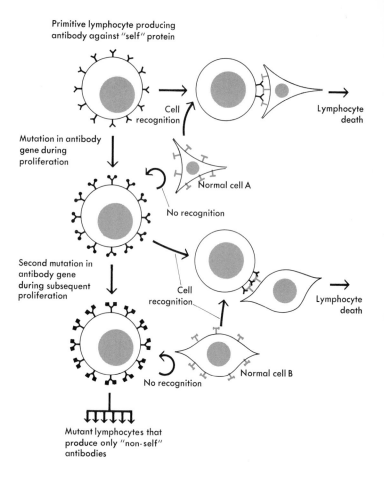

Figure 19–42

Somatic generation of immune diversity as the result of a progressive series of mutations that block the recognition of normal surface antigens.

appropriate histocompatibility antigens. Repeated such binding events might not only trigger off limited lymphocyte proliferation but also lead to killing of the resulting "memory" cells. The only lymphocytes that should survive would be those that possess *mutant antibody genes* coding for antibodies of new immunological specificity. And since a change at one of many different sites in the hyper-variable regions should reduce, if not nullify, specific binding to a surface antigen, large numbers of antibodies with new specificities should be produced (Figure 19–42). The carrying out of such continued negative selection may, in fact, be the primary role of the thymus and bursa. The vast majority of the lymphocytes in these lymphoid glands seem to divide only to die, with only a few percent ever leaving to enter the circulating system. This situation has seemed very weird until now, since constant cell division that invariably led to cellular death seemed pointless. But as soon as one postulated a need for repeated division cycles to produce mutant antibodies unable to combine with normal cell antigens, the whole picture suddenly became very clear.

A very satisfactory aspect of this theory is that it starts to explain the otherwise odd coincidence that mutations in the H-2 gene complex so directly affect the final potentiality of the immune response. Conceivably, the immune response gene(s) in some way control(s) the specificity of one or more histocompatibility proteins. When it (they) mutate(s) from one form to another, somewhat different sets of active sites will be produced by negative selection against cells bearing antibodies complementary to the respective histocompatibility genes. This hypothesis also may provide an explanation for the abnormally strong immunological response against foreign H-2 (HL-A) antigens. Those lymphocytes that bear germ line-coded antibodies complementary to "foreign" antigens will not be selected against, and in large numbers will remain essentially dormant until contact with appropriate foreign cells.

EMBRYONIC TIME COURSE OF ANTIBODY DEVELOPMENT

Several years ago, the key to the searching out of the origin of antibody diversity appeared to be the finding of *in vitro* systems for the induction of the primary and secondary responses. Now, with the discovery that the receptors which recognize antigens are themselves antibodies, this objective seems much more manageable and claims of *in vitro* responses are no longer regarded with hostile skepticism. Clearly a next step is to understand how antigen binding to surface immunoglobulin receptors triggers a proliferative response. But even when we understand this point, we shall not be any further ahead in understanding how a small lymphocyte becomes differentiated to produce only a single antibody species. All the lymphocytes currently studied have been programmed before our experiments commence.

Thus it is obvious that we must try to move earlier and earlier in embryonic development to discover when large numbers of different antibodies first appear. So far no one has fixed on the crucial stage. Tadpoles, which possess only a few million lymphocytes, have immunological responses quite similar to those of adult frogs. So one is likely to have to go back very early in the development of the blood-forming cells before fewer specificities may be seen. If, at some such early stage, far fewer antibody types are in fact seen—and they are even more preferentially directed toward surface antigens—then the somatic theory of antibody generation would obtain strong support. But if, on the contrary, most immunological specificities are present when the first antibody-synthesizing cell can be detected, then a germ-line origin of diversity will look like the correct answer.

Thus we see that the molecular biologists who have decided to concentrate on antibody synthesis as a means of understanding cell differentiation have chosen a problem that, upon close examination, seems even more complicated than embryological differentiation, seen from afar. Even so, there is no reason to believe that the really fundamental embryological problems, when properly posed, will prove any easier to solve.

SUMMARY

There is a great need to find a "simple" system in which cells become irreversibly differentiated to produce a well defined protein in response to the addition of a specific, well understood external molecule. One system, initially thought to be "simple," is the synthesis of specific antibodies as a result of the injection of specific foreign objects. Those objects that induce the synthesis of specific antibodies are called antigens. Many macromolecules, including most proteins and some carbohydrates and nucleic acids, are antigens.

Most antibodies (immunoglobulin = Ig) function after they have been secreted from the cells in which they were synthesized. Such humoral antibodies usually inactivate antigens by combining with them to form complexes that become engulfed by scavenger white blood cells. Other antibodies function while still attached to the cells in which they are made (cell-bound antibodies). As such, they are responsible for a variety of cell-mediated responses including the killing of cells bearing complementary surface antigens. Why some antigens lead to humoral responses while others cause cell-mediated responses is not known.

It was originally thought that an antigen induced the formation of a specific antibody by combining with a nascent antibody before it acquired its final three-dimensional shape. According to this hypothesis, the antibody would fold around the antigenic surface, thereby forming a region complementary in shape to the antigen's specific surface (instructive theories). Now, however, it is known that each specific antibody possesses a unique sequence of amino acids which folds in a unique three-dimensional shape. The function of the antigen is thus to set into motion a train of events that leads to the synthesis of the specific mRNA templates that code for the desired amino acid sequence (selective theories).

All antibodies are constructed from groups of four poly-peptide chains. Each group contains two identical light chains (MW ~ 22,500) and two identical heavy chains (MW ~ 53,000) which are linked together by disulfide bonds. The common form of humoral antibody, those of the IgG class, contains one such group and has MW ~ 150,000. In the IgM class, five such groups are regularly linked together, while one, two, or three groups are found in the functional IgA class molecules. The antibodies present in normal blood serum are a collection of many, many types of proteins with different amino acid sequences. Single antibody species, however, are produced by myeloma tumors, now the source of virtually all antibodies used in amino acid sequence studies. Some myelomas produce many more light chains than heavy. The excessive light chains, excreted into urine, are called the Bence-Jones proteins.

Two forms of light chain exist, κ and λ, easily distinguished by the amino acid sequences in their variable as well as their constant regions. Either a κ or a λ chain is used to form all immunoglobulins, irrespective of class. The distinctive feature that separates the various classes of immunoglobulin from each other is the nature of their heavy chains. IgM molecules contain μ chains, IgG γ chains, and IgA α chains. All the genes coding for κ chains are located on one linkage group (chromosome). Completely unlinked are the genes for the λ chains. Still another linkage group carries the heavy-chain genes.

Both the light and heavy chain have variable sequences at their amino terminal end and constant regions at their carboxyl ends. The antigen combining sites (the active sites) are formed by amino acids at the variable regions of both the light and heavy chains. Direct interaction between the antigen and antibody usually involves less than a fourth of the variable amino acids. In fact, many amino acids within the variable region are effectively invariant, reflecting the fact that these side groups are necessary to form the "basic immunoglobulin fold." In contrast, those amino acids that form the active site show very great variability from one antibody type to another (hypervariable regions).

The general immunoglobulin fold is found not only within all variable regions, but also in the constant regions. One such fold is found within each light-chain constant region, while three are linked together to form the constant region of heavy chains. The suspicion exists that today's immunoglobulins arose by an evolutionary process in which the duplication of a primitive immunoglobulin gene was later followed by divergence into specific genes for the variable and constant regions.

Antigens initiate immunological responses by combining with immunoglobulin molecules (receptors) on the surfaces of small lymphocytes. Even early in development, very large numbers of different types of small lymphocytes exist, each already committed toward the synthesis of a specific single Ig. Each small lymphocyte remains inert until combination with antigens cross-links its surface Ig molecules, somehow to lead to a cell transformation process in which the small lymphocytes enlarge and divide to become blast cells (clonal theory of selection). Two types of small lymphocytes exist, the T cells and the B cells. Transformation of T cells generates lymphoblasts, the cells responsible for cell-mediated immunity. B-cell transformation leads to the formation of the plasma cells, the sites of humoral antibody synthesis.

At the start of a B-cell response to a specific antigen, the first antibodies produced belong to the IgM class. Several weeks later IgM synthesis is replaced by IgG synthesis. Whether the same plasma cell can produce both IgG and IgM is not known. Most importantly, the specificity of the active site does not change during a given IgM-to-IgG transition. Instead, the specific V_H region initially found attached to a μC region becomes somehow translocated so that it becomes attached to a γC region. Such translocations strongly hint that in the germ line, the V and C regions are coded by separate genes which only become attached to each other when a lymphocyte becomes committed to the synthesis of a single antibody type. Now it is believed that each haploid genome may contain only one gene for each type of constant region. Originally the best guess was that a very large number of separate germ-line V genes existed. Now, however, there is growing support for the hypothesis that most immunological diversity arises by selection of somatic mutants created during the generation of small lymphocytes from their still undiscovered precursor cells.

Since antibodies are formed only against foreign proteins, an animal's immunological system must be able to recognize its own proteins. This learning process occurs early in life, before circulating antibodies exist. If a foreign protein is injected into a newborn animal, the animal in later life is unable to form antibodies against the foreign protein (immunological tolerance). Now we guess that tolerance normally results from the selective killing of lymphocytes bearing antibodies complementary to the antigens available to the immunological system at the time it comes into existence during embryological development.

REFERENCES

LANDSTEINER, K., *The Specificity of Serological Reactions*, rev. ed., Dover, New York, 1964. A paperback reprint of a scientific classic, last revised in 1943. Still a beautiful introduction to immunology.

"Antibodies," *Cold Spring Harbor Symp. Quant. Biol.,* **XXXII** (1967). Reflects a now classic meeting in the history of immunology. The concluding summary by Nils Jerne is a minor masterpiece.

BURNET, F. M., *Cellular Immunology,* Cambridge University Press, 1969. An imaginative overview of cell-mediated immunity.

ROITT, I., *Essential Immunology,* Blackwell Scientific Publications, 2nd edition, Oxford, 1974. An extremely well done introductory text.

DAVIS, B. D., R. DULBECCO, H. N. EISEN, H. S. GINSBERG, W. B. WOOD, Jr., AND M. McCARTY, *Microbiology,* 2nd ed., Harper and Row, 1973. The excellent section on immunology by Herman Eisen is also available as a separate text.

Defense and Recognition, edited by R. R. Porter, University Park Press, Baltimore-Butterworths, London (1973). An excellent collection of review articles on the nature, origin, and function of the immunoglobulins.

The Immune System: Genomes, Receptors, Signal, E. Sercarz, A. R. Williamson, and C. F. Fox (eds.), Academic Press, 1974. These quickly published proceedings of a March 1974 meeting at Squaw Valley reflect the increasing excitement that envelopes modern immunology.

EDELMAN, G. P., "The Structure and Function of Antibodies," *Scientific American,* August, 1970. The basic structure of the antibody molecule simply explained.

POLJAK, R. J., L. M. AMZEL, H. P. AVEY, B. L. CHEN, R. P. PHIZACKERLY, AND F. SAUL, "Three-Dimensional Structure of the Fab' Fragment of a Human Immunoglobulin at 2.8-Å Resolution," *Proc. Nat. Acad. Sci.,* **70,** 3305 (1973).

SEGAL, M., E. A. PADLAN, G. H. COHEN, S. RUDIKOFF, M. POTTER, AND D. R. DAVIES, "The Three-Dimensional Structure of a Phosphoryl Choline-Binding Mouse Immunoglobulin Fab and the Nature of the Antigen Binding Site," *Proc. Nat. Acad. Sci.,* **71,** 4298 (1974). This and the preceding article give the first very awaited looks at antibody structure at atomic resolution.

AMZEL, L. M., R. J. POLJAK, F. SAUL, J. M. VARGA, AND F. F. RICHARDS, "The Three-Dimensional Structure of a Combining Region–Ligand Complex of Immunoglobulin NEW at 3.5-Å Resolution," *Proc. Nat. Acad. Sci.,* **71,** 1427 (1974).

637

A precise description of the bonds linking an antigen to an antibody.

CUNNINGHAM, B. A., J. L. WANG, G. R. GUNTHER, G. N. REEKE, Jr., AND J. W. BECKER, "Molecular Analysis of the Initial Events in Mitogenesis," *Cellular Selection and Regulation in the Immune Response,* G. M. Edelman (ed.), Raven Press, New York (1974). A review paper that emphasizes lymphocyte transformations induced by lectins.

HOOD, L., J. H. CAMPBELL, AND S. C. R. ELGIN, "The Organization, Expression and Evolution of Antibodies and Other Multigene Families," *Ann. Rev. Genetics,* **9,** in press (1975). A most useful review that relates currently known facts and hypotheses about the organization of immunoglobulin genes with those of other multigenic systems.

LEVIN, A. S., H. H. FUDENBERG, J. E. HOPPER, S. K. WILSON, AND A. NISONOFF, "Immunofluorescent Evidence for Cellular Control of Synthesis of Variable Regions of Light and Heavy Chains of Immunoglobulins G and M by the Same Gene," *Proc. Nat. Acad. Sci.,* **68,** 169 (1971). An important experiment that argues for the existence of separate V and C genes.

MILSTEIN, C., G. G. BROWNLEE, E. M. CARTWRIGHT, J. M. JARVIS, AND N. J. PROUDFOOT, "Sequence Analysis of Immunoglobulin Light-Chain Messenger RNA," *Nature,* **252,** 354 (1974). Now the most complete sequence analysis of an Ig mRNA.

BARSTAD, P., V. FARNSWORTH, M. WEIGERT, M. COHN, AND L. HOOD, "Mouse Immunoglobulin Heavy Chains are Coded by Multiple Germ Line Variable Region Genes," *Proc. Nat. Acad. Sci.,* **71,** 4096 (1974). Here very detailed amino acid sequences are used to support the multiple germ line V gene hypothesis.

PREMKUMAR, E., M. SHOYAB, AND A. R. WILLIAMSON, "Germ Line Basis for Antibody Diversity: Immunoglobulin V_H- and C_H-gene Frequencies Measured by DNA–RNA Hybridization," *Proc. Nat. Acad. Sci.,* **71,** 99 (1974). Experiments that use hybridization data to argue for the existence of a very large number of different germ line V genes.

RABBITTS, T. H., AND C. MILSTEIN, "Mouse Immunoglobulin Genes: Studies on the Reiteration Frequency of Light-Chain Genes by Hybridization Procedures," *Eur. J. Biochem.,* **52,** 125 (1975).

TONEGAWA, S., C. STEINBERG, S. DUBE, AND A. BERNARDINI, "Evidence for Somatic Generation of Antibody Diversity," *Proc. Nat. Acad. Sci.*, **71**, 4027 (1974). This and the preceding paper, also using DNA–RNA hybridization data, come to the opposing conclusion that very few V genes exist and so most immunological diversity must have a somatic basis.

CAPRA, J. D., AND J. M. KEHOE, "Hypervariable Regions, Idiotypy, and the Antibody-Combining Site," in *Advances in Immunology*, F. J. Dixon and H. G. Kunkel (eds.), Academic Press, New York, **20**, 1 (1975). A very complete and most useful review.

PINK, J. R. L., AND B. A. ASKONAS, "Diversity of Antibodies to Cross-reacting Nitrophenyl Haptens in Inbred Mice," *Eur. J. Immunol.*, **4**, 426 (1974). An elegant demonstration of the large numbers of different idiotypes which can be formed by a single inbred mouse strain in response to a single hapten.

WEIGERT, M., M. POTTER, AND D. SACHS, "Genetics of the Immunoglobulin Variable Region," *Immunogenetics*, **1**, 511, Springer-Verlag, New York (1975). A much needed summary of recent genetic attempts to map idiotypic markers.

SHREFFLER, D. C., AND C. S. DAVID, "The H-2 Major Histocompatibility Complex and the I Immune Response Region: Genetic Variation, Function, and Organization," in *Advances in Immunology*, F. J. Dixon and H. G. Kunkel (eds.), Academic Press, New York, **20**, 1 (1975). An extensive review which emphasizes genetic data.

PETERSON, P. A., L. RASK, K. SEGE, L. KLARESKOG, H. ANUNDI, AND L. ÖSTBERG, "Evolutionary Relationship Between Immunoglobulins and Transplantation Antigens," *Proc. Nat. Acad. Sci.*, **72**, 1612 (1975). The most complete data yet presented about the molecular structure of the histocompatibility antigens.

JERNE, N. K., "The Somatic Generation of Immune Recognition," *Eur. J. Immunol.*, **1**, 1 (1971). A very important conceptual paper.

SPEAR, P. G., A. WANG, U. RUTISHAUSER, AND G. M. EDELMAN, "Characterization of Splenic Lymphoid Cells in Fetal and Newborn Mice," *J. Exp. Med.*, **138**, 557 (1973). A look at idiotype diversity during early stages of embryological development.

The Viral
Origins
of Cancer

The fact that we still are so lost amidst the biochemical complexity of the vertebrate cell raises the obvious question as to whether far too many scientists are studying the nature of diseased cells. Instead, perhaps most should still be homing in on normal cells. Work on, say, the scientific aspects of arteriosclerosis or mental retardation may convey a sense of social responsibility, but we must always ask whether such socially motivated commitments are the outgrowth of serious science or whether, at best, they represent the sloppy aspirations of well intentioned amateurs. In particular we must ask whether much too much attention is now given to cancer research, for it currently receives the lion's share of the money that goes toward medical research.

Too many claims of impending success in cracking cancer have created a credibility gap that will disappear only when solid, clean answers are presented to the outside world. The murky fickleness of too many past breakthroughs has created the image of cancer research as a scientific discipline whose conclusions more often responded to human hopes than to conventional standards of scientific correctness. Now, however, this situation is swiftly changing, and the quality of much current cancer research is indistinguishable from that of *E. coli* molecular biology. The main reason for this shift has been the development of systems where well defined animal viruses reproducibly convert normal cells into their cancerous equivalents. Such transformations often are due to the integration of specific viral genes into cellular chromosomes. The direct question thus can be posed: Why should a gene(s) that normally function(s) in viral multiplication also have the capacity to change a normal cell into a cancerous one? To find the answer, large numbers of scientists are now engaged in as complete a description as possible of the way tumor viruses multiply, hoping that, if the functions of their key genes are worked out, the way they make a cell malignant will also fall out. As we shall see below, we do not yet have any final answers, but the increasing elegance of much current tumor virology inspires confidence that this approach will tell us much more, not only about the cause of cancer, but also about its biochemical essence.

CANCER AS A HEREDITARY CHANGE

When a cancer cell divides, the two progeny cells are usually morphologically identical to the parental cell. The factor(s) that give(s) cancer cells their essential quality of unrestrained growth is thus regularly passed on from parent to progeny cells. These changes persist not only in tumors growing in intact animals but also in tumor cells growing in tissue culture. Hundreds of generations of growth can occur in tissue culture without appreciable reversion to a normal state. The permanence of such changes is shown not only by perpetuation of a typical morphology, but also by the ability of progeny cells to cause new tumors when injected into a tumor-free animal of genetic composition similar to the one from which the original tissue culture was obtained.

The heritability of the changes allowing unrestrained growth immediately suggests genetic changes within chromosomal DNA. Support for this notion comes from experiments in which the *Sendai* virus has been used to fuse cancer cells with normal cells. The 4N tetraploid cells which result from nuclear fusion regularly have a normal phenotype and do not form tumors when injected into genetically identical hosts. Most 4N cells, however, generally are not stable and frequently divide to produce cells with lower chromosome numbers. Many of these subtetraploid cells are cancerous, indicating that the loss of one or more cellular chromosomes leads to the reexpression of the cancerous phenotype (Figure 20–1). At first sight, this result seems to indicate that the cancerous property is a recessive genetic trait, perhaps resulting from a loss of normal genetic material. However, almost all of the

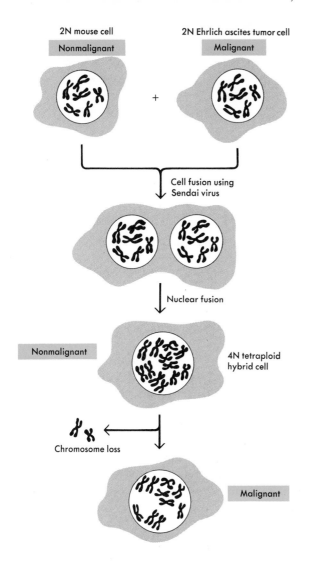

Figure 20–1
Fusing a malignant with a nonmalignant cell to determine whether the cancerous phenotype is dominant or recessive.

"normal" genetic revertants that arise from virally "transformed" cancer cells have a very large number (4N to 8N) of chromosomes. So a very high chromosome number *per se* may make a cell more normal. Further experiments clearly must be done to see whether the tetraploid cells invariably lost specific chromosomes when they became cancerous.

SOMATIC MUTATIONS AS POSSIBLE CAUSES OF CANCER

Many of the essential changes that make cells cancerous may be somatic mutations (mutations occurring in cells not destined to become sex cells). Under this scheme, a somatic mutation could cause a cancer if its occurrence upset a normal control device regulating cell division. Proponents of this hypothesis believe that, in general, several somatic mutations are necessary to cause a cancer. This idea is based on the fact that the incidence of cancer greatly increases with age. This phenomenon would be explainable if a particular cell had to accumulate several mutations, each occurring randomly in time, before becoming a full-fledged cancer cell. At present there exists no direct evidence either for or against this theory, which we might best describe as cancer due to loss of an essential gene(s) function. Even without evidence, however, it is clear that somatic mutations must occur; it would be surprising if at least some did not disrupt the normal control of cell division.

CANCER INDUCTION BY RADIATION

Indirect indication that somatic mutations are important causal factors in the induction of cancer comes from the realization that a very large number of the agents that increase the frequency of cancer are also strong mutagens. Collectively, all cancer-causing agents are called *carcinogens*. Among the most potent carcinogens are the ultraviolet and ionizing forms of radiation, agents long known to also be highly mutagenic. Exposure, for example, of skin to uv light causes skin cancer while x-rays applied to the thyroid induce thyroid cancer. None of these various forms of radiation, however, can be used to cause all the cells in an exposed population to become cancerous. This is because they also cause many other deleterious changes, many of which lead to cell death. A radiation dose large enough to make virtually all cells cancerous would also kill practically every cell. Exactly how these radiations cause cancer is very far from clear, since not only can they all cause simple mutations but the ionizing form also breaks chromosomes, and large deletions of genetic material frequently result. So it would be most surprising if they cause cancer by changes in just one specific gene. Instead, they most likely cause a variety of genetic changes, any one of which might help lead to the cancerous state.

IN VIVO CONVERSION OF CHEMICAL CARCINOGENS INTO STRONG MUTAGENS

A tight connection between many chemical carcinogens and mutagens has only lately emerged. While certain compounds that mimic ionizing radiation in their physiological consequences (e.g., the nitrogen mustards) have long been known to be both mutagens and carcinogens, many of the most powerful carcinogens long appeared devoid of any mutagenic potential. Particularly striking was the absence of any apparent mutagenicity within a variety of highly carcinogenic fused-ring hydrocarbons like benzpyrene and benzanthracene (Figure 20–2). They appeared to have no special affinity for DNA, and it was therefore postulated that their carcinogenicity somehow resulted from their binding to and inactivation of key proteins.

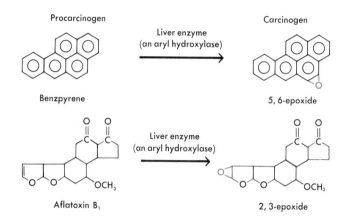

Figure 20–2
Conversion of benzpyrene (present in charcoal-broiled foods) and Aflatoxin B₁ (found in moldy peanuts and pistachio nuts) by enzymes of the liver into carcinogenic (mutagenic) epoxide derivatives.

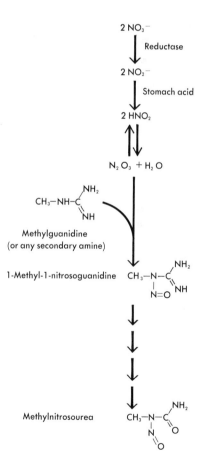

Figure 20–3
Diagram showing the *in vivo* conversion of nitrates (found in pastrami, frankfurters, bacon, etc.) into the powerfully carcinogenic (mutagenic) nitrosamines.

Over the past several years, however, overwhelming evidence has accumulated that these carcinogens, *per se*, are not mutagens. But when they enter certain cells, they become chemically transformed into derivatives which are not only highly mutagenic but also much more powerful carcinogens than their precursors. For example, nitrates by themselves cause no genetic (carcinogenic) changes, but in cells they may become converted into the very potent mutagens (carcinogens), the nitrosamines (Figure 20–3).

Much of their mutagenicity arises from their ability to chemically modify guanine residues, leading to base substitutions during DNA replication as well as to breaks in the polynucleotide backbones which can generate more extensive genetic rearrangements. In general, compounds which are transformed into strong carcinogens likewise become strong mutagens, and vice versa. There thus seems little doubt that much if not most carcinogenesis is the result of changes in DNA.

Long unexplained were strikingly different responses to the same carcinogen often shown by different individuals within a species. Now it appears that many such differences reflect genetically determined levels of the intracellular enzymes that generate the active forms of the various carcinogens. Among the most potent such enzymes are the *aryl hydroxylases,* whose normal function in the liver is the detoxification of aromatic hydrocarbons. While these enzymes act on most hydrocarbons to produce harmless metabolites, when they act on benzpyrene-like molecules, they instead produce very carcinogenic derivatives. Ingestion of significant amounts of benzpyrene thus invariably leads to the emergence of hepatomas (liver tumors). Why the human population shows such wide variation in their aryl hydroxylase levels is not known, though it is tempting to speculate on past dietary differences between different evolving human populations.

IMMUNOLOGICAL SURVEILLANCE

Since spontaneous mutations within a given gene probably occur at a frequency of 10^{-6} to 10^{-9} per cell division, and since there are some 10^9 cells per gram of tissue, the question naturally arises, Why don't tumors arise much more frequently in the larger eucaryotic organisms? Either there must be some way of specifically suppressing those particular mutations which lead to the cancerous state, or there must exist some mechanisms for specifically destroying newly emerged cancer cells. Now we believe that the latter possibility is correct, with cell-mediated immunological responses recognizing and destroying newly emerged cancer cells and their descendants (*immunological surveillance*). This defense mechanism is based on the fact that the surfaces of tumor cells almost always contain antigenic determinants not found in normal cells. Why specific tumor antigens exist is not at all obvious, unless profound changes at the cell surface are almost always obligatory features of the cancerous phenotype.

The almost obligatory presence of "tumor-specific antigens" was long a highly debatable issue. The argument was made that, if they exist, they would automatically lead to the killing of all newly arising cancer cells; and since, at first sight, tumors grow uncontrolled within their hosts, they must not have significant antigenic differences from the normal cells they have descended from. In reality, however, most tumors grow very slowly. Today, most people believe that *in vivo* tumors grow despite the acquisition of new antigenic markers. This argument further proposes that it is only the exceptional cancer cell and its descendants against which most animals cannot react. Why the host antibody response fails in these rare cases is not known, since, in many cases, the serum of animals with growing tumors contains large numbers of T-cell-derived lymphoblasts specifically directed against the respective tumor cell. One possible answer is the simultaneous existence of circulating antibodies of similar specificity. These *"blocking antibodies"* may prevent the attachment of the killer lymphoblasts by coating tumor cell surfaces. The eventual success of an immunological response against a tumor may thus depend on the relative strengths of the "B" and "T" cell responses.

In any case it is now clear that failure to mount an effective immunological response results in a great increase in the number of spontaneous tumors. Thus the evolution of vertebrates with their relatively long lives must have been dependent upon development of an immunological response against cancer. Conceivably, the use of antibodies to combat microbial infections arose as an incidental side benefit of a system primarily selected to defend animals against cancer.

USE OF NEWBORN ANIMALS (NUDE MICE) TO DEMONSTRATE THE ONCOGENIC POTENTIAL OF CANCER CELLS

Most cancer cells are thus immunologically rejected even when injected into histocompatibly identical hosts. Their *oncogenicity* (tumor-causing capacity) can generally be shown only by injecting many tens of millions of cells, so as to overwhelm the immunological potential of their host. Some cancer cells, however, have such strong tumor antigens that they never can form tumors when tested under ordinary circumstances, while the tumors formed by other cancer cells grow only to a limited size and then start regressing, eventually to disappear. Much use is thus made of tricks to temporarily suppress the immunological system of their hosts. Large doses of ionizing radiation or immunosuppressive drugs like the cortical steroids thus have been frequently applied before cells of uncertain oncogenic potential are to be tested. When this is done, orders-of-magnitude fewer cells may be required for induction of visible tumors. Another trick involves the use of newborn (neonatal) hosts, since in many species the immunological system does not become very effective until just after birth. And over the past year, increasing use has been made of "nude mice," a newly discovered mutant strain of mice which congenitally contain very little thymus tissue, and so are incapable of mounting strong T-cell responses. Even so, the injection of over a million tumor cells is generally necessary to provoke a tumor in such mice. Extreme care, however, must be taken to prevent nude mice from dying of microbial infection; and they are best employed when their cages can be kept in a semisterile environment.

The final oncogenicity of a given cancer cell is thus a reflection of both the degree to which it has lost the ability to regulate its growth and the strength of its tumor-specific antigens.

VIRUSES AS A CAUSE OF CANCER

Alternative to the somatic mutation idea is the hypothesis that many cancers are caused by viruses. In a sense, this is not really a hypothesis, since there is already convincing evidence that a number of specific cancers in animals, ranging from fish to mammals, are virus-induced. The relevant question is thus not whether viruses can cause cancer, but whether a sizable fraction of cancers are virus-induced. Until recently, there was no intellectual framework in which to consider how viruses might cause cancer. Now, largely as a result of work with bacterial viruses, we realize that, after they enter a cell, certain viruses have a choice either to multiply and kill their host cell(s), or to insert their chromosomes into host chromosomes to take up superficially inert "proviral" forms. The essential aspect of viral carcinogenesis could thus be the *introduction of new genetic material*, in contrast to somatic mutations which, we suspect, often cause a *loss* of functional genetic material.

Though the first known tumor virus, the *Rous sarcoma* virus, was discovered over sixty years ago in 1912, examples of new tumor viruses at first appeared so infrequently that it was generally believed that viruses could not be a major causal factor in cancer. But as soon (~1950–55) as newborn animals began to be used as test systems, it became much easier to show that a virus had oncogenic potential. Also initially preventing many people from accepting a viral causation is the fact that EM observation generally reveals no viruses in most tumor cells. Now, however, we realize that under many circumstances the genetic material of a virus may be inserted into a host chromosome in a "proviral" form without any production of virus particles. So the absence of detectable viruses does not provide evidence either for or against a viral origin for a given cancer cell.

At present it seems likely that not all tumor viruses act in the same way. For example, a variety of cancers are caused by RNA-containing viruses. The best known of these RNA viruses is the *Rous sarcoma* virus, which causes solid tumors in chickens (a sarcoma is a tumor of connective tissue). Certain other RNA viruses cause leukemia and sarcomas in both birds and mammals, probably including man. Acting quite differently are several groups of DNA viruses. One group is responsible for wartlike growths on the skin in many different mammals, ranging from rodents to man. Other distantly related DNA viruses, of which the best known are a mouse virus called polyoma and a monkey virus called SV40, can cause a variety of tumors when injected into newborn animals. Similarly, the *Herpes* virus, EB, appears to cause not only mononucleosis, but also a rare African childrens' tumor, the *Burkitt lymphoma*. Below we shall focus attention first on the better understood DNA tumor viruses, and then go on to consider the still very incompletely analyzed RNA tumor viruses.

THE VERY SIMPLE STRUCTURE OF AN SV40 (POLYOMA) PARTICLE

SV40 is a spherical virus of 450-Å diameter that normally multiplies in monkeys. It is among the smallest DNA-containing viruses, with a molecular weight of about 28 million. On the outside is a protein shell, which electron microscope examination shows to be built up from 72 morphologically identical subunits (Figure 20–4). Each of these subunits is built up from five or six smaller protein molecules, to give a grand total of 420 polypeptide chains in the shell. Most, if not all, of these shell polypeptides (VP 1) have a MW $\sim$ 47,000. Also found in every SV40 particle are some 40 copies of a second polypeptide (VP 2) of MW $\sim$ 34,000. Where it is located or what functions it plays are not known. Inside each shell is a single circular double-helical DNA molecule, of molecular weight of slightly more than 3 million, with a contour length of 1.6μ (Figure 20–5). Bound to this DNA are 4 of the 5 main host histones, H4, H2a, H2b, and H3 (Figure 20–6). Absent is H1, the histone thought to be involved in chromosome coiling.

Virtually identical in appearance and basic construction to SV40 is polyoma, a virus that multiplies in mice. Like SV40 it also can induce cancer when injected into newborn animals, and so is also very extensively studied.

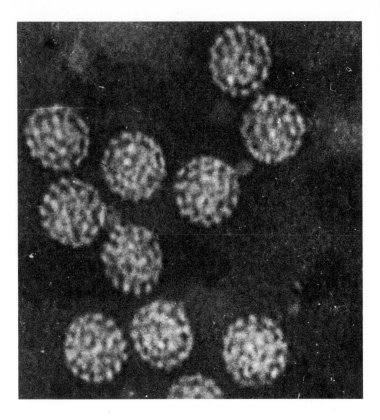

Figure 20–4
An electron micrograph [Courtesy of L. V. Crawford, Institute of Virology, Glasgow] of *Polyoma* virus particles. The viral diameter is about 500 Å. Careful observation reveals that seventy-two subunits are used to construct the external protein coat. Two of the particles (lower left) are held together by an antipolyoma antibody molecule.

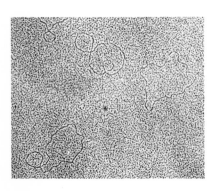

Figure 20–5
An electron micrograph of several molecules of purified *Polyoma* DNA. All clearly have a circular contour. The DNA contour length is 1.6 μ, corresponding to 3×10^6 daltons. [Courtesy of W. Stoeckenius, Rockefeller Institute.]

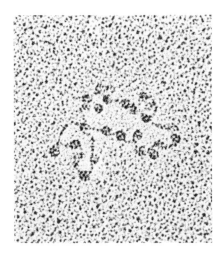

Figure 20–6
Chromatin-like structure of the SV40 DNA–histone complexes released from SV40 particles. The dense bodies resemble the "nucleosomes" of eucaryotic chromatin. [Courtesy of C. Cremisi, P. Pignatti, O. Croissant, and M. Yaniv, Institut Pasteur, Paris.]

In all important respects, SV40 and polyoma behave in a biologically identical manner; and so results gained with one virus usually hold for the other.

As only some 4800 nucleotides are present in the SV40 (polyoma) chromosome, the maximum number of amino acids that it can code for ranges near 1600. Given the plausible assumption that both the 47,000 (430AA) and 34,000 (300AA) MW proteins found in the mature virus particles are coded for by the viral DNA, approximately 45% of the total genome is used to specify structural protein. This leaves about 2600 nucleotides to code for the protein specifically involved in viral multiplication. The ~900AA which it can maximally specify is easily sufficient to form 3 to 5 average sized polypeptide chains. If so, the SV40 (polyoma) gene number would be 5 to 7, or very similar to that of the very small single-stranded DNA phages, each of whose chromosomes contains some 8 to 9 genes. Recent genetic data (see below), however, hints that the SV40 (polyoma) DNA may code for as few as 3 genes. So SV40 (polyoma) may have as simple a genetic structure as any known DNA virus (Figure 20–7).

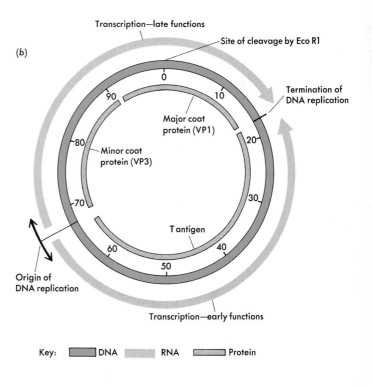

Figure 20–7
(a) A schematic map of SV40, showing the icosahedral arrangement of its 72 capsomeres. (b) The circular chromosome of SV40 (*Polyoma*) showing its three genes, the initiation points for DNA replication, and the transcription pattern of its early and late genes.

651

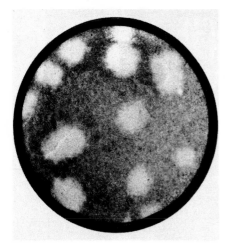

Figure 20–8
Plaques of *Polyoma* virus. *Polyoma* particles were added to embryonic mouse cells growing on a glass surface. The plaques, which become visible after twenty-five days of incubation, represent contiguous masses of dead cells which stain differently than growing cells. [From Crawford and Diamond, *Virology,* **22,** 235 (1964), with permission.]

LYTIC VS. TRANSFORMING RESPONSE

When a SV40 virus enters a susceptible monkey cell, it generally disappears without a trace. Less commonly ($\sim 10^{-2}$), it multiplies like a conventional virus, producing a large number of new virus particles. The site of final maturation within its host cell is the nucleus, which eventually becomes totally filled with a million or so progeny particles. During this process, which occupies some 24 to 48 hours at 37°C, the normal nuclear functions slowly become disrupted, and the infected cell necessarily dies (*a lytic infection*). In contrast, no progeny are made when an SV40 particle enters a mouse or hamster cell. Instead, a very small percentage ($\sim 10^{-5}$) of such infected cells become *transformed* into morphologically distinguishable cancer cells.

Both lytic and transforming responses can be observed outside living animals, in cell culture. In these experiments, cultured cells (often derived from embryonic animals) are allowed to grow on a glass surface bathed in nutrient solution. A number of virus particles are then added. Some of these particles multiply in the cells to produce progeny particles which then adsorb to nearby cells, producing a circular region of dead cells similar to the plaques formed by bacterial viruses (Figure 20–8). If, on the other hand, the virus transforms one of the cultured cells, the transformed cell then begins to multiply in the disorganized and easily identifiable fashion of a cancer cell (Figure 20–9). Particularly noticeable in the case of the transformed cells is their disordered growth pattern, which results in the formation of groups of cells piled irregularly on top of each other.

Figure 20–9
Clones formed by transformed (left) and normal (right) mouse embryo cells. The transformed cells pile on top of each other, thereby forming a thicker (darker) mass than do normal cells. [Courtesy of M. Stoker, Institute of Virology, Glasgow, Scotland.]

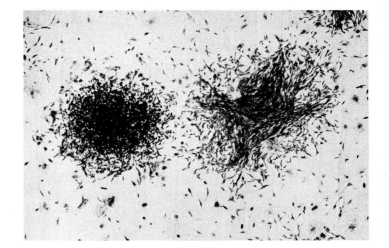

PERMISSIVE VS. NONPERMISSIVE CELLS

Cells in which a specific virus multiplies are called *permissive* cells, while cells in which viral growth does not occur are known as *nonpermissive* cells. A permissive cell line usually is derived from an animal in which a given virus normally reproduces. In contrast, nonpermissive cell lines usually originate from animals incapable of multiplying a given virus. Thus, depending upon the specific virus, a cell can be either permissive or nonpermissive. For example, the very well-known mouse strain 3T3 is permissive for polyoma, a mouse virus, but nonpermissive for SV40, a monkey virus. In contrast, when SV40 infects a 3T3 cell, it never multiplies but occasionally induces cell transformation. And when polyoma infects the nonpermissive hamster cell line BHK (baby hamster kidney), it can transform but not multiply, and so BHK cells are frequently used in polyoma transformation experiments.

PHYSICAL MAPPING OF SV40 DNA

DNA molecules now can be routinely cleaved into specific small fragments by *restriction endonucleases* that produce double-stranded cuts in the midst of very specific sequences. One such enzyme, the *E. coli* R_1 restriction enzyme, recognizes the sequence

GAATTC

CTTAAG

which occurs just once in SV40 DNA, and so converts the circular SV40 DNA molecule into a linear rod with unique ends (Figure 20–10). Other such enzymes recognize sequences occurring more frequently in SV40 DNA and generate a correspondingly larger number of defined fragments. For example, the combined restriction enzymes of *Hemophilus influenzae* break at 11 sites, while those from *Hemophilus parainfluenzae* make only three cuts. These various fragments can be ordered by analysis of the composition of larger fragments produced by incomplete digestion. Use of this order (Figure 20–11), together with the current rapid progress in determining the nucleotide sequences of the individual fragments, should allow the complete sequencing of all of the $\sim 4{,}800$ SV40 nucleotides within the next several years.

The R_1-created linear SV40 rods can be further characterized by the technique of *partial denaturation mapping*. This method relies on the fact that DNA molecules rich in A–T base pairs denature more rapidly than DNA rich in the stronger G–C base pairs. Thus, when SV40 DNA is

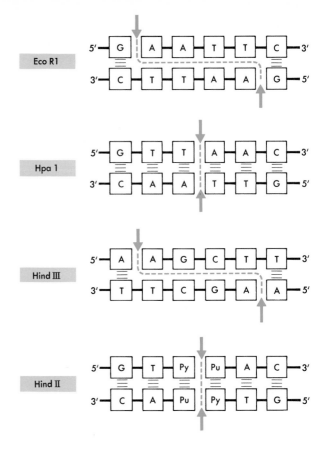

Figure 20–10

A diagram to show the basic sequences recognized by some restriction endonucleases. The colored arrows indicate where the strands are cut. The enzymes from *Hemophilus influenzae* dII and from *Hemophilus parainfluenzae* cut both strands in the same place. Those enzymes from *E. coli* R1 and from *H. influenzae* dIII cut at different places and leave cut strands with "sticky ends." The base sequences are palindromic about the center.

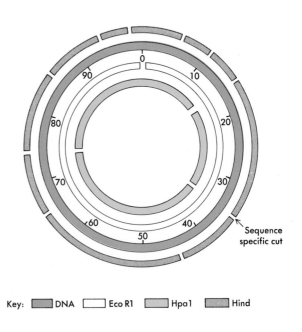

Figure 20–11

Sites of cutting of the SV40 chromosome by several widely used restriction enzymes. For orientation of the cuts, compare with the site of *E. coli* R1 cleavage in the genetic map shown in Figure 20–7.

partially denatured by exposure to alkali, AT-rich regions tend to separate while the GC-rich sections remain intact. In the electron microscope, such molecules display single-stranded bubbles connected by double-stranded linear segments (Figure 20–12). These denaturation patterns are very reproducible, and easily distinguish one DNA from another.

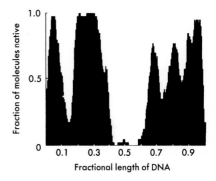

Figure 20–12
(a) Electron microscope view of SV40 DNA that has been converted to its linear form by treatment with endonuclease R1 and partially denatured at pH 10.9. [Courtesy of H. Delius.] (b) A histogram of the native regions, derived from maps of the linear SV40 DNA shown in (a). [Reproduced with permission from Mulder and Delius, 1972.]

INFECTIVITY OF SV40 DNA

Addition of highly purified DNA to the appropriate permissive (nonpermissive) cell also leads to the lytic (transforming) response. The frequencies with which these respective events occur are usually several orders of magnitude lower than when intact virus particles are used to infect, most likely because free DNA molecules are much less capable of adsorbing to their host cells. When they do initiate infection (transformation), the final result, however, is indistinguishable from that started by mature virus particles, showing still again that all genetic specificity resides within DNA.

EARLY STAGE IN THE LIFE CYCLE IS DOMINATED BY THE SYNTHESIS OF THE T ANTIGEN

The life cycle of SV40, like that of phages, can be subdivided into *early* and *late stages*. The early stage, which lasts some 8 to 10 hours, prepares for the synthesis of the SV40 DNA and structural proteins which occur in the late stage. At first it was suspected that a number of viral-specific enzymes were synthesized in the early phase, but now there is a growing conviction that perhaps only one SV40-specified protein is made early. This protein, called the

"*T antigen*" (Figure 20–13) moves into the nucleus soon after its synthesis in the cytoplasm. As yet we are still very, very uncertain about its physical form since, when it is carefully isolated, it moves as part of a large molecule (MW $> 500,000$), but when less care is taken it appears to have a MW $\sim 75,000$.

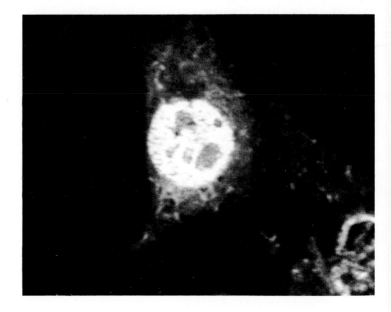

Figure 20–13
Demonstration, by the fluorescent antibody technique, of the presence of the SV40-specific nuclear antigen within non-permissive hamster cells transformed by SV40. [Courtesy of V. Defendi, Wistar Institute, Philadelphia.]

The early mRNA species coding for the T antigen sediments at 19S and has a MW $\sim 800,000$. It is synthesized counterclockwise (when the SV40 chromosome is oriented as in Figure 20–7) off the "e" (early) strand. Its total coding capacity is at best some 750AA or approximately the number found in the lower-weight form of the T antigen. It thus follows that the larger form of the T antigen represents either aggregation of many identical T antigen polypeptides or the attachment of the "T antigen" polypeptide to a much larger host molecule.

GENETIC EVIDENCE FOR THREE SV40 (POLYOMA) GENES

Over the past 5 years a very large number of different temperature sensitive mutants have been isolated in SV40 (polyoma) after treatment with nitrous acid. They appear to fall in only three complementation groups, two of which correspond to the late genes coding for the two SV40 structural proteins VP 1 and VP 2. The third is the early gene which codes for the T antigen. Mutants in it not only block accumulation of T antigen, but also SV40 DNA replication and the synthesis of the SV40 structural proteins. Too little crossing-over occurs to permit ordering of these genes by purely genetic means. However, they now can be precisely located on the SV40 DNA by experiments in which small DNA "restriction fragments" are used in DNA transformation experiments to complement temperature sensitive members of the various complementation groups (Figure 20–14).

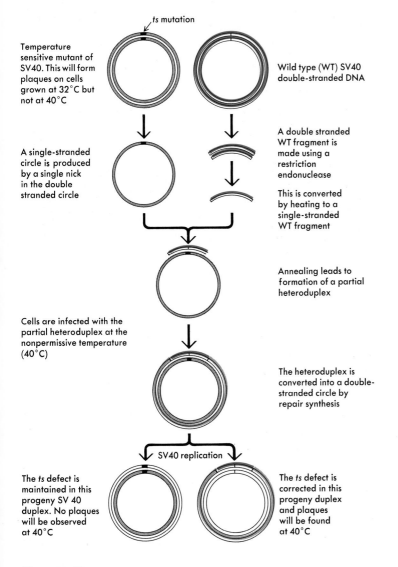

ts mutation

Temperature sensitive mutant of SV40. This will form plaques on cells grown at 32°C but not at 40°C

Wild type (WT) SV40 double-stranded DNA

A single-stranded circle is produced by a single nick in the double stranded circle

A double stranded WT fragment is made using a restriction endonuclease

This is converted by heating to a single-stranded WT fragment

Annealing leads to formation of a partial heteroduplex

Cells are infected with the partial heteroduplex at the nonpermissive temperature (40°C)

The heteroduplex is converted into a double-stranded circle by repair synthesis

SV40 replication

The *ts* defect is maintained in this progeny SV 40 duplex. No plaques will be observed at 40°C

The *ts* defect is corrected in this progeny duplex and plaques will be found at 40°C

Figure 20–14
Location of the site of an SV40 temperature sensitive mutation through transformation by partial heteroduplexes created from a mutant circular single strand and a specific wild type single-stranded DNA fragment made by using a restriction endonuclease.

INDUCTION OF HOST ENZYMES INVOLVED IN DNA SYNTHESIS

How the "T antigen" is able to direct so many late processes seems now to be the prime obstacle in the understanding of SV40 replication. One potential clue comes from the observation that a number of host enzymes involved in DNA synthesis can strikingly increase in amounts after viral infection. They do so when SV40 infects G_1-blocked cells whose DNA synthesis has stopped. Growth in such inhibited cells occurs despite the fact that, before infection, these cells lack many of the enzymes (e.g., CDP reductase, dTMP synthetase, and DNA polymerase) used to make the various DNA nucleotide precursors and to link them together. Successful SV40 infection must thus somehow unlock the cellular control device which normally shuts off the synthesis of the "DNA enzymes" when their presence is no longer wanted. Most likely, the T antigen is involved in the induction process, since the various enzymes begin increasing in amount just after T antigen has appeared in the nucleus (Figure 20–15).

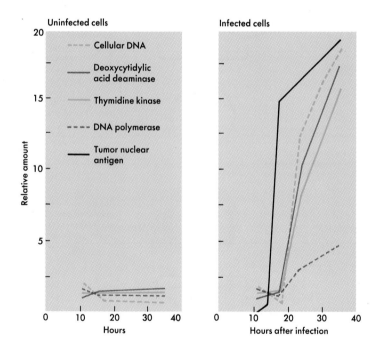

Figure 20–15
Induction of enzymes involved with DNA synthesis following infection of permissive monkey cells by SV40 virus. Before these cellular-coded enzymes appear, synthesis of the viral-specific protein, the "T" nuclear antigen, has begun. [Redrawn from Dulbecco, *Scientific American*, April, 1967.]

Induction of the "DNA enzymes" starts off not only SV40 DNA synthesis, but also large-scale synthesis of host DNA. The specific signals which cause a circular SV40 chromosome to begin its replication also act upon the many host chromosomes. This observation, surprising at first, makes sense when the genetic origin of the "DNA enzymes" is considered. They must be coded for by host chromosomes since the SV40 chromosome is not nearly large enough to have this task. Coding for all these enzymes would require a size much more like that of λ DNA (10 times larger than SV40).

No induction occurs if RNA synthesis is blocked, and so T antigen must either directly or indirectly affect the specificity of transcription. Very recent evidence showing that it tightly binds to DNA at or near the starting points for DNA and RNA synthesis argues for a direct effect. Conceivably it acts like the CAP factor (see page 391) by altering the state of DNA so that it becomes more easily transcribed (replicated). Or it may in fact be a σ-like specificity factor that directly combines with RNA polymerase and determines which genes are to be transcribed. The very fast sedimentation rate of native T antigen may be the result of its binding to the form of eucaryotic RNA polymerase that makes mRNA (initiates DNA synthesis).

SV40 DNA REPLICATION STARTS AT A UNIQUE SITE

EM examination of replicating SV40 reveals that it retains a covalently closed circular form throughout replication. Replication starts at a single point (see Figures 9–28 and 20–7), and the two replicating forks move in opposite directions until they meet. Interestingly, the starting point is close to, if not identical to, the promoters that control the synthesis of the early and the late SV40 mRNA (see below). This raises the question of whether, in fact, the nucleotide sequence signalling "start DNA replication" is recognized by an RNA polymerase.

TURNING ON OF LATE SV40 RNA POLYMERASE

The RNA species which codes for the SV40 structural protein begins to accumulate at the same time as DNA replication starts. It is coded for by the ℓ (late) DNA strand, and so is synthesized in the clockwise direction, as opposed to the counterclockwise direction of early mRNA synthesis. A given DNA segment codes only for early or late RNA; and most, if not all, of the DNA appears to directly code for amino acid sequences. This may mean that the apparent 5' ends of both the early and the late SV40 DNA are coded for by closely adjacent nucleotides (Figure 20–7).

The matter may not be straightforward, however, since late mRNA isolated from nuclei is much larger than that found in the cytoplasm attached to ribosomes. The size of polysomal SV40 mRNA may thus be partially a reflection of processing events occurring in the nucleus which somehow degrade all those RNA sequences that do not code for viral-specific proteins. So neither the 5' nor 3' ends of the mRNA isolated from polyosomes need reflect the true starting or stopping site of transcription.

Nothing is known so far about why late SV40 mRNA accumulates only in the late phase. At first we thought it must signify the early phase synthesis of a new σ-like specificity factor (T antigen?) for transcription; but now we must also consider the possibility that late mRNA is synthesized at an early time but quickly broken down before it can combine with ribosomes. Work to settle this question by using purified eucaryotic RNA polymerase II has not been successful, but this may have been due to use of purified SV40 DNA. Now we suspect that SV40 DNA never exists free within a nucleus, being always quickly converted into chromatin by binding to cellular histones. Hopefully, *in vitro* study of the transcription of SV40 chromatin will at last tell us how late SV40 mRNA synthesis is controlled.

ABORTIVE INFECTIONS PRECEDE TRANSFORMATION

Many of the SV40 particles infecting nonpermissive 3T3 cells go through the early stage of reproduction. Early SV40 mRNA is made, as is the SV40 T antigen and some SV40 DNA. No viral structural proteins, however, are synthesized even though much late SV40 mRNA is detected, especially in the nucleus. For some as yet unknown reason, the nonpermissive cells do not permit late mRNA to function.

ONE PARTICLE CAN TRANSFORM A CELL

When SV40 particles are added to nonpermissive cells, the number of cells transformed is directly proportional to the amount of input virus (Figure 20–16). If a certain number of particles transforms one in a thousand cells, then ten times that number transforms one in a hundred cells. This relationship tells us that one particle by itself is completely sufficient to do the job. Furthermore, provided enough particles are added, virtually all the exposed cells can be transformed, indicating that all the cells growing in the culture are capable of being transformed. The probability of a single SV40 particle causing transformation, however, is very, very low. Some tens of thousands of infecting particles ordinarily must be present to give the average cell a 50-percent probability of conversion to the cancer state.

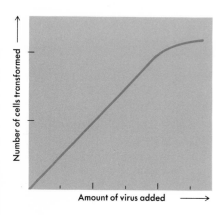

Figure 20–16
The linear relation between the amount of SV40 added to nonpermissive cells and the number of resulting transformed cells. The plateau reached at high viral doses reflects the inability of a fraction (usually between 50 percent and 99 percent) of nonpermissive cells to be transformed. The significance of such a refractory population is unclear.

ABSENCE OF INFECTIOUS SV40 PARTICLES FROM TRANSFORMED CELLS

No infectious SV40 particles are present in transformed cells. It is, of course, not surprising that the nuclei of transformed cells are not filled with viral particles. This would probably result in cell death. The fact is, however, that no particles *at all* can be detected. Here there is an obvious analogy to lysogenic bacterial viruses (see Chapter 7). Lysogenic phages also have two possible fates. They may multiply lytically or they may become part of the host chromosome by crossing-over.

TRANSFORMATION INVOLVES INTEGRATION OF SV40 DNA INTO HOST CHROMOSOMES

DNA–RNA hybridization studies were first used to reveal the presence of SV40 DNA sequences in the DNA isolated from transformed cells. In contrast, when DNA from normal cells was tested, no SV40 DNA sequences were detected. This integrated SV40 DNA is present within the nucleus, not within a cytoplasmic (mitrochondrion?) organelle. Moreover, within the nucleus, it must be covalently bound to chromosomal DNA, since careful extraction techniques indicate that the DNA sequences that hybridize with SV40 mRNA are part of exceedingly long DNA molecules (MW > 10⁹).

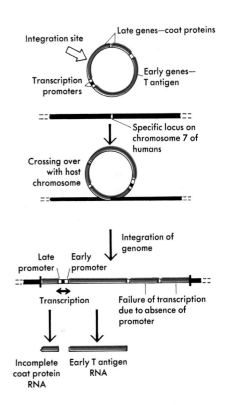

Figure 20–17
Schematic illustration of how SV40 may integrate into a specific host chromosome. Integration within the late region has the important consequence that intact late SV40 mRNA cannot be synthesized. In contrast, intact early mRNA molecules can be made in such transformed cells.

Each transformed cell contains some 1 to 20 SV40 genomes, the exact number of which may reflect the number of SV40 particles used to transform the respective nonpermissive cell. When only one (several?) SV40 molecule(s) is integrated into a human cell, it frequently becomes inserted into chromosome 7. The enzyme(s) involved in the insertion process is (are) not known, though it seems simplest to imagine a crossing-over process like that sketched in Figure 20–17. The fact that integration occurs so preferentially into one chromosome argues that crossing-over probably involves very specific nucleotide sequences on both the SV40 and human chromosomes.

LIBERATION OF INFECTIOUS PARTICLES FOLLOWING FUSION OF A TRANSFORMED NONPERMISSIVE CELL WITH A NONTRANSFORMED PERMISSIVE CELL

For many years, efforts have been made to induce polyoma-transformed cells to yield infectious virus particles. But despite application of the several agents that induce the λ prophage to detach from the *E. coli* chromosome, all such attempts have been unsuccessful. The cell-fusion technique, however, works well with SV40. When an SV40-

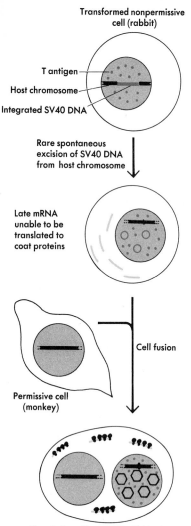

Transformed nonpermissive
cell (rabbit)

T antigen

Host chromosome

Integrated SV40 DNA

Rare spontaneous
excision of SV40 DNA
from host chromosome

Late mRNA
unable to be
translated to
coat proteins

Cell fusion

Permissive cell
(monkey)

Translation of late SV40 mRNA using
translation factors from permissive cell
yields structural coat proteins and the
subsequent assembly of mature SV40
particles in the nucleus

Figure 20–18
Rescue of infectious SV40 particles from a
transformed nonpermissive rabbit cell following fusion with a permissive monkey
cell. In a small percentage of the nonpermissive cells, the SV40 genome is recombined out of the host chromosome and
temporarily exists as a free SV40 molecule
(episome). It is only in these cells that
SV40 is found after fusion with permissive
cells.

transformed 3T3 cell is fused with a nontransformed cell,
permissive for SV40, within 24 hours many of the fused
cells yield SV40 progeny (Figure 20–18). Something
present in the permissive cell allows SV40 to carry out a
complete multiplication cycle. As yet, this finding has not
been duplicated with polyoma-transformed cells, but most
likely the reason for this is trivial, not fundamental. With
polyoma, as well as SV40, we believe that all the viral
genome, not just an incomplete portion, is present within
the transformed cells.

DO PERMISSIVE CELLS PROVIDE FACTORS NEEDED TO TRANSLATE LATE mRNA?

The fact that fusion with a permissive cell induces the
transformed nuclei of nonpermissive cells to yield SV40 indicates that, unlike the λ prophage, SV40 genomes are
under positive, not negative, control. In the case of λ, the
prophage genome is kept inactive by the presence of its
repressor, which specifically prevents transcription of
early, as well as late, λ mRNA. But if the failure to transcribe the late SV40 genomes were due to a repressor,
fusion with a permissive cell should not lead necessarily to
induction. The repressor should continue to bind to the
promoter for the late SV40 genes. Instead, we now suspect
that the permissive cell may contain a molecule necessary
for the translation of late SV40 mRNA. Development of *in
vitro* systems where addition of SV40 DNA to cell-free extracts leads to synthesis of the SV40 structural proteins
may be necessary to test this hypothesis.

VIRAL-SPECIFIC mRNA IN TRANSFORMED CELLS

Only the early gene coding for T antigen functions well in
transformed cells. Late genes function either not at all or at
very low rates, and so are not involved in making normal
cells into cancer cells. The active agent must thus be the T
antigen which, as earlier mentioned, has the capacity to
turn on a group of genes involved in DNA replication.
This particular set of functionally related genes may represent, however, only a minor fraction of the genes brought
into action by the T antigen and so need not be the genes
whose functioning, *per se*, makes a cell cancerous. SV40-
transformed cells, for example, secrete much more "cell
factor," the proteolytic enzyme that rapidly converts fi-
brinogen into fibrin, and whose active secretion (see page
583) seems to be characteristic of almost all cells that
form solid tumors. So, conceivably, secretion of cell factor

662

alone can make a cell cancerous. The alternative possibility, however, must also be considered that many of the genes turned on by the T antigen collectively act to create the cancerous phenotype.

TUMOR-SPECIFIC SURFACE ANTIGENS

Not routinely explained by the simple hypothesis that all early SV40 and polyoma mRNA codes for the T antigen is the presence on the surface of transformed cells of tumor-specific antigens. These antigens, which appear to be specific for a given virus and not for the cell they transform, are detected by the failure of SV40- (polyoma)-transformed cells to form tumors when they are injected into adult animals of identical genetic background. For example, BHK cells transformed by SV40 have a different surface antigen than that found in polyoma-transformed BHK cells. So, many people have postulated the existence of a second early gene to code for these tumor-specific transplantation antigens (TSTA). But in the absence of any genetic evidence for a second early gene and, conceivably, any leftover nucleotides in the early region to code for a second protein, the alternative possibility exists that the T antigens of polyoma and SV40 turn on somewhat different host gene sets, which in turn lead to the presence of different molecules at the cell surface.

In any case, the antigenicity of virus-induced tumors must be a necessary feature for ensuring the survival in nature of the host of the virus and hence of the virus itself. If the tumors provoked did not contain specific antigens, then nearly all viral infections might result in host death due to cancer. This argument in no way, however, tells us why the tumor antigen need be viral-specific since, *a priori*, a generalized cancer antigen would seem to do the job.

ADENOVIRUSES AS AN ALTERNATIVE SYSTEM TO PINPOINT AN ONCOGENE

Equally promising as a model system to understand how an integrated viral gene can cause cancer are the adenoviruses. These are large animal viruses, the human varieties of which normally multiply to produce common cold-like symptoms. Each adenovirus particle has a diameter of ~ 800 Å, a MW $\sim 175 \times 10^6$, and an icosahedral-shaped (20 faces) protein shell formed by the regular aggregation of several types of specific protein subunits (Figure 20–19). Inside each shell is a double-stranded

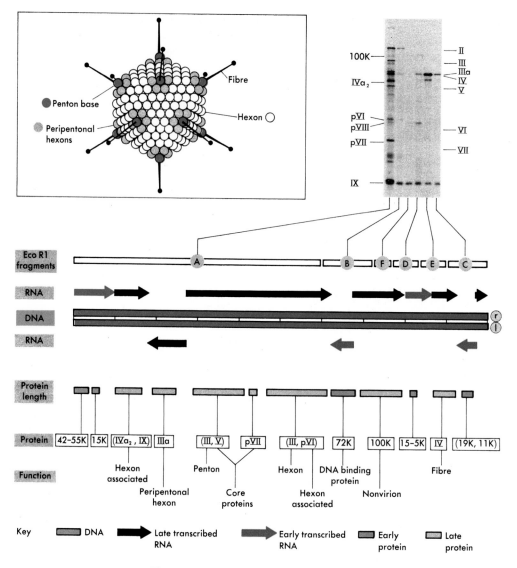

Figure 20–19

Schematic diagram of an adenovirus particle together with a representation of its linear DNA molecule, showing the fragments created by *E. coli* R1 digestion. Also shown are locations of the several early and late adeno mRNA species, as well as the proteins thought to be coded by these mRNA species. In the upper right is shown the SDS polyacrylamide gel pattern of the proteins made of mRNA molecules transcribed from adeno DNA fragments that have been made by *E. coli* R1 cuts.

DNA molecule of MW ~ 25×10^6 bound to two types of "core" proteins. Altogether some nine (ten?) different proteins are located within each virus particle, and each probably plays some vital functional role.

Infection of susceptible human cells by human adenoviruses leads to a lytic infection which culminates in the appearance of $\sim 10^5$ new viral particles. These progeny particles accumulate in the nucleus, which becomes functionally destroyed as a consequence of viral multiplication. The take-over of host cell synthetic machinery by the adenovirus genome is much more complete than that following successful SV40 infection, where the host's chromosomes continue to function normally until very late in the infectious process. In contrast, after adenovirus infection, both host cell DNA and RNA synthesis quickly slow down, and within fifteen hours most newly synthesized proteins are coded by the viral DNA.

THE ADENOVIRUS GENOME CODES FOR SOME TWENTY DIFFERENT PROTEINS

Some twenty different viral-specific proteins (genes) have already been identified within adenovirus-infected cells. While some are relatively small (MW $\sim$ 10,000), others are quite large (MW $\sim$ 100,000); and their combined MW's almost add up to the total coding capacity of the 25×10^6-MW adenovirus DNA molecule. So almost all the adenovirus genome is used to code for specific amino acids, with very little DNA left over for possible regulatory functions. The relative order of the many adeno genes can be determined by genetic methods, since infection of a single cell by several genetically distinct particles leads to the appearance of recombinant virus particles. Recently a variety of genetic and biochemical tricks have been used to assign specific locations along the DNA molecule to certain genes (Figure 20–19); and over the next several years the absolute location of every adenovirus gene should be worked out.

EARLY VS. LATE GENES

During the first several hours of adeno infection, only a few adeno genes (the early genes) function, with the vast majority of genes (the late genes) coming into action only after the early genes have played some vital and yet-to-be-worked out role. Early mRNA does not disappear as the late mRNA begins to accumulate, but continues to be present throughout the viral life cycle. Unlike with phage T7, which has a DNA molecule of similar size, the early adenovirus mRNA's are not specified by contiguous DNA regions. Two different early mRNA molecules are coded by the "h" strand and two by the "ℓ" strand (Figure 20–19). Late mRNA also is coded by both strands, though the vast majority is made on the ℓ strand. How the switchover from exclusively early mRNA to a mixture of early and late

665

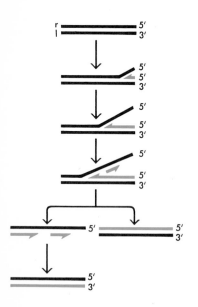

Figure 20-20
A schematic diagram showing the temporary creation of 5'-ended single-stranded DNA tails during the initial stage of adenovirus DNA replication. Here replication is shown starting only from the left end, though most likely it also begins at the right end.

adenovirus mRNA is accomplished is not known. Conceivably, processing of a much larger precursor is involved. Although discrete mature mRNA species are seen in the cytoplasm attached to ribosomes, the newly synthesized mRNA chains found in the nucleus are much larger and heterogeneous, with some appearing to be complete transcripts of either the ℓ or $\hbar$ strand.

INVERTED SEQUENCES AT THE ENDS OF ADENO DNA MOLECULES

The way the linear adenovirus DNA molecules are replicated is still very mysterious. DNA synthesis is usually asymmetric with only one strand serving as a template, thereby generating single-stranded tails coming off double-stranded parental molecules. Later, the resulting single strands must somehow become converted to double helices (Figure 20–20).

Conceivably replication can start at either terminal, since the nucleotide sequences at the two ends are identical except for their opposite (inverted) polarities. Why the two terminal sequences (~ 100 base pairs) should be identical remains to be worked out. Most likely they have some role during DNA replication, since no adenovirus mRNA's complementary to the inverted sequences exist.

While highly purified adeno DNA always has a linear form, the DNA molecules released from gently disrupted virus particles are initially circular (Figure 20-21). A structural protein(s) present in the virus particle must hold the two ends of the linear adeno DNA together, since linear DNA molecules are generated when the circular complexes are treated with proteolytic enzymes. Circular forms are also seen in infected cells, leading to the conjecture that circularity, *per se*, may play some important role in DNA replication.

TRANSFORMED CELLS NEVER CONTAIN COMPLETE ADENOVIRUS GENOMES

Infection by adenoviruses of cells in which they cannot multiply (nonpermissive) occasionally creates transformed

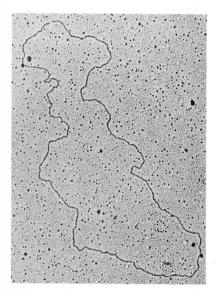

◀ **Figure 20–21**
Electron micrograph of the circular form of the adenovirus 2 chromosome, seen after gentle disruption of infectious particles. Treatment with proteolytic enzymes converts the circular form into the linear form of purified adenovirus DNA. [Courtesy of Ellen Daniell, Cold Spring Harbor Laboratory.]

cells capable of forming tumors. As with SV40, the transforming process is the result of the insertion of adenovirus genes into host chromosomes. How this process occurs is not known, since only incomplete viral genomes are ever seen in transformed cells. Specific blocks of genes always are missing, most likely because their continued functioning while on a host chromosome would immediately kill the host cell (for example, by shutting off all host cell DNA and RNA synthesis). While many transformed cells contain up to half the viral genome, some contain as little as 4% of the total genome. In these cases, the adenovirus genetic material is exclusively derived from the extreme left end of the genetic map, and always encompasses an early gene read off the ℓ strand (Figure 20–22).

Large amounts of mRNA transcribed off this early gene(s) are seen in transformed cells, but as yet its corresponding protein product(s) has not been molecularly well characterized. It can, however, be easily detected immunologically, since all adeno-transformed cells, as well as those lytically infected, contain a viral-specific ("T") antigen that is found in the cytoplasm as well as the nucleus. Whether it binds to DNA like the SV40 "T" antigen is not yet known. Because the adeno "T" is relatively stable, it soon should be possible to thoroughly purify it and hopefully begin to understand how the expression of its function can change a normal cell into a malignant one.

ADENOVIRUS-TRANSFORMED CELLS ARE EASILY DISTINGUISHABLE FROM THEIR SV40-TRANSFORMED COUNTERPARTS

The morphology of hamster cells transformed by adenoviruses is reproducibly distinct from hamster cells transformed by SV40 (polyoma). Most adeno-transformed cells are highly rounded throughout the cell cycle, closely resembling the spherical cells found in the middle of mitosis (Figure 20–23). Almost no intracellular muscle-like fibers exist and, possibly as a consequence, adeno-transformed cells cannot attach to and later flatten down on solid surfaces. While SV40-transformed cells are less flat than their normal equivalents, they seldom resemble mitotic cells; and many, more organized muscle fibers are always present throughout the cell cycle. It is thus probable that the cancerous phenotype can be reached by more than one biochemical pathway, and that the adeno T antigen functions in a different way than the SV40 (polyoma) T antigen.

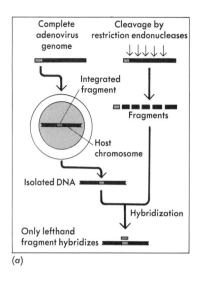

(a)

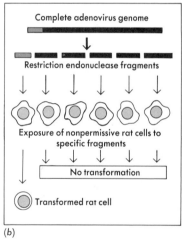

(b)

Figure 20–22
Experiments demonstrating that only the extreme left end of the adenovirus DNA molecule is involved in cell transformation. (a) Here we see that the adenovirus-specific DNA present in transformed cells forms DNA–DNA hybrids only with DNA fragments from the extreme left end of the genetic map. (b) Shows that only the DNA fragment at the left end is capable of transforming nonpermissive cells.

Figure 20–23
(a) A scanning electron micrograph of adenovirus-transformed hamster cells. The spherical cells do not separate from each other after mitosis, and they form clumps in which the individual cell boundaries cannot be seen. (b) Electron micrograph of a thin section made through a clump of transformed cells. A cell in mitosis is present at the upper part of the clump. [Reproduced with permission from R. Goldman *et al., Cold Spring Harbor Symp.,* **XXXIX,** 601 (1974).]

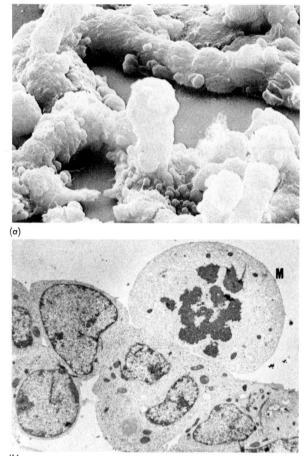

(a)

(b)

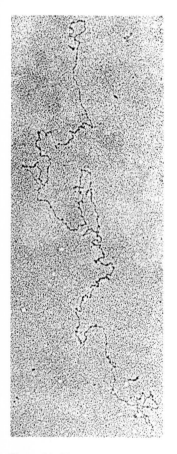

Figure 20–24
Electron micrograph of a linear single-stranded *Herpes simplex* DNA chain (MW ~45 × 10⁶) found after denaturation of the double-helical form (MW ~ 90 × 10⁶). [Reproduced with permission from P. Sheldrick and N. Berthelot, *Cold Spring Harbor Symp.,* **XXXIX,** 607 (1974).]

HERPES VIRUSES AS ONCOGENIC AGENTS

Though solid evidence now links a number of *Herpes* viruses with specific forms of cancers, only in the past several years has serious study begun on the way they act at the molecular level. Complicating such studies has been the unavailability, until recently, of cell culture systems where *in vitro* transformation can be reproducibly studied. Also inhibiting research has been the realization that the great structural complexity of *Herpes* viruses means that the molecular approach may be much, much harder to pull off than with SV40-like or adeno-like viruses.

All *Herpes* viruses contain a single linear double-helical DNA molecule of MW almost 10⁸ (Figure 20–24). This places *Herpes* viruses among the genetically most complex viruses, with sufficient nucleotides to code for some 100 average sized proteins. Approximately 20% of the genome is used to code for the some 25 different proteins used to form its central core, its surrounding capsid, and the outer lipid-containing membrane (Figure 15–1). Most likely much of the remaining genome codes for large numbers of specific enzymes needed for the replication of the viral DNA and protein. For example, most Herpes viruses contain a gene that codes for a viral-specific thymidine kinase. Genetic analysis of the various Herpes viruses, however, has just begun; and some years may pass before it becomes possible to assign many genes to specific sites along the DNA genome. Likewise, analysis of their mRNA products has only begun; and while there are early–late distinctions, it may be a long time before the exact sizes and functions of all the many different Herpes virus mRNA molecules can be ascertained.

Clearly the vast complexity of individual Herpes genomes means that the molecular approach is only likely to succeed if attention is focused on only selected viruses and not spread out on the increasingly large number of Herpes viruses implicated as the causative agents of specific animal cancers. Now it appears probable that most future experiments at the molecular level will focus either on the common *Herpes simplex* viruses or on *EB*, the human virus that causes mononucleosis and which also may be the causative agent of the Burkitt tumor found in children living in certain tropical regions.

CELL TRANSFORMATION BY INACTIVATED *HERPES* VIRUSES

Herpes simplex viruses (HSV) normally carry out a typical lytic cycle after they infect permissive cells. In this process, the DNA of the infecting viruses moves into the nucleus, where it is first transcribed into RNA and later replicated to form large numbers of progeny DNA molecules (Figure 20–25). When, however, the infecting DNA is damaged, for example by uv light or by shearing into smaller fragments, the viral multiplication process is blocked and most infected cells continue to grow normally. A small fraction, however, become transformed into cells capable of forming tumors (Figure 20–26). Here again, transformation results from integration of viral genetic material into host chromosomal DNA. Preliminary experiments suggest that most transformed cells contain only a small fraction of the viral genome; but as yet its location on the Herpes chromosome is not known. Recent experiments, however, show that Herpes DNA fragments made by "restriction" nucleases can also transform permissive cells; and so we should soon

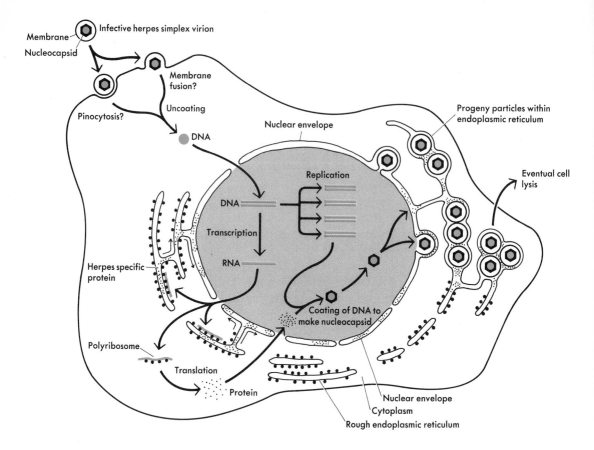

Figure 20–25
Schematic diagram to show the multiplication of *Herpes simplex* virus in a permissive cell.

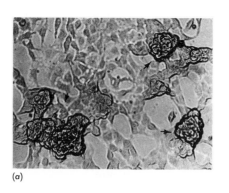

(a)

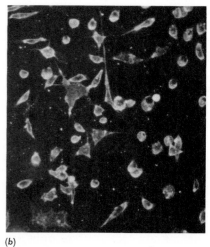

(b)

Figure 20–26
(a) Foci of rat embryo cells transformed by sheared fragments of HSV DNA. (b) Fluorescent antibody demonstration of the presence of HSV-specific antigen(s) in the HSV-transformed rat cells. [Reproduced with courtesy of Wilkie *et al.*, *Cold Spring Harbor Symp.*, **XXXIX,** 657 (1974).]

know exactly how much and what part of the Herpes genome is necessary for transformation. Hopefully, as with SV40 and the adenoviruses, only one or possibly several viral genes may be responsible for producing the cancerous behavior.

Whether complete *Herpes simplex* genomes are ever integrated into host DNA as proviruses is not yet established. Once a human has a Herpes infection (e.g., a cold sore), the viral genome seems to persist indefinitely in a latent state that can be quickly activated by external stimuli, like excessive sunlight. But whether the latent state involves Herpes DNA recombined into a host chromosome in a λ prophage-like fashion is still not known despite several decades of active speculation. Conceivably a Herpes-specific repressor keeps most of the integrated Herpes genomes in a nonfunctional state. If this is the case, then cells containing a complete *Herpes simplex* genome would have a normal phenotype, with a cancerous state resulting only from the integration of incomplete genomes not subject to repressor control.

EB VIRUS AND ITS RELATIONSHIP TO THE BURKITT LYMPHOMA AND MONONUCLEOSIS

EB is a ubiquitous Herpes virus that infects most humans in early life without producing any serious disease. It grows best in lymphoid cells and, under still-not-understood conditions, is the cause of infectious mononucleosis, a self-limiting disease characterized by excessive proliferation of lymphoid tissue. Most likely, it is also involved in the origin of the Burkitt tumor, a cancer of lymphoid cells that occurs predominantly in humid tropical regions. While normal lymphoid cells will not grow in continuous culture, permanent cell lines can easily be derived from the lymphoid cells characteristic of mononucleosis or of the Burkitt tumor. Their unique ability to grow in culture is the result of their possession, in a proviral form, of one to many copies of complete EB genomes. Whether such genomes exist in the nucleus as free DNA molecules or as proviruses integrated into host chromosomal DNA is still controversial. Occasionally the EB proviruses become activated to multiply, and most lymphoid cell cultures contain a small percentage of dying cells harboring mature EB virus particles. The factors that control the percentage of cells in which EB is multiplying are still very unclear, though, in many strains, growth in arginine-free medium or in the presence of halogenated pyrimidines (e.g., 5-BUdR) greatly increases the percentage of virus-containing cells.

The EB virus isolated from Burkitt cells appears identical to that which causes mononucleosis; but why one type of virus leads to two such different diseases is unclear. One possibility is that the Burkitt tumor occurs only in humans simultaneously infected with the malarial parasite and with EB. Malaria can dampen the immunological response, and so may inhibit the "T" cell response which normally destroys the proliferating cells that characterize mononucleosis.

Currently inhibiting molecular work on EB has been the failure to find a host cell in which EB infection immediately leads to virus multiplication. The only known response is the transforming one. When EB is added to a population of dormant small lymphocytes, a tiny percentage enlarges, begins synthesizing DNA, and commences continuous cycles of cell division. So no doubt exists that the presence of the EB genome gives perpetual life to transformed lymphocytes. But since it is not yet possible to have a population of cells, all of which are producing mature EB, molecular analysis of EB multiplication is beset with many experimental hurdles. One important fact, however, has emerged. All cells harboring EB genomes contain a nuclear antigen that is specifically bound to chromatin. Conceivably it affects the generalized transcription pattern and so might be directly involved in the proliferating capacity of cells bearing EB genomes.

RNA TUMOR VIRUSES

While the DNA tumor viruses encompass a large variety of structurally diverse entities, those RNA viruses which have the capacity to cause cancer are morphologically very similar, and probably are descended from a single ancestral virus. All are approximately spherical in shape, with diameters ~ 1000 Å, and have as their basic morphological design a lipid-containing membranous envelope surrounding an RNA-containing core (Figure 20–27). The outer membrane contains the usual phospholipid components of cellular plasma membranes as well as large numbers of two different viral-specific glycoproteins, aggregates of which usually project out as knobs from the membrane surface. The inner core consists of a central RNA-containing nucleoid surrounded by an icosahedral-shaped protein shell built up from protein capsomeres. Each of the ~ 390 capsomeres is itself formed by the regular aggregation of several viral-specific polypeptide chains. The nucleoid is normally tightly coiled within the shell, and its filamentous form is seen only after gentle disruption of an RNA tumor virus. Conceivably, only one of the

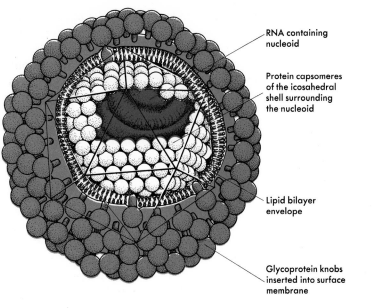

(a)

RNA containing nucleoid

Protein capsomeres of the icosahedral shell surrounding the nucleoid

Lipid bilayer envelope

Glycoprotein knobs inserted into surface membrane

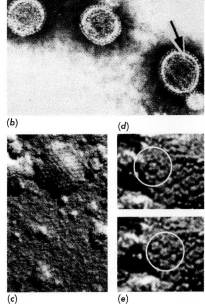

virally specified proteins is found in the nucleoid. In contrast, some three to four different types of protein are found in the shell (Table 20–1).

While RNA tumor viruses have been found to cause cancer in a large variety of different vertebrates, those which have been best studied normally multiply in the chicken (e.g., *Rous sarcoma* virus) or the mouse (e.g., *Rauscher leukemia* virus and the mouse mammary tumor virus). On the whole, results obtained from one virus are interchangeable with those from all other RNA tumor viruses. Below, we shall largely refer to the *Rous sarcoma virus* (RSV) and its relatives, because up until now they have been the most intensively studied.

Table 20-1 Proteins of Avian RNA Tumor Viruses

Name	Molecular weight	Approximate number per virion	Location
gp70	70,000	600	Outer envelope
gp35	35,000	1500	Outer envelope
p27	27,000	4000	Core
p19	19,000	3600	Core
p15	15,000	5500	Core(?)
p12	12,000	6000	Nucleoid
p10	10,000	< 1500	Core(?)

Figure 20–27
(a) Schematic view of a generalized RNA tumor virus. In some viruses (e.g., the mouse mammary-tumor virus), the individual glycoprotein knobs (spikes) are very distinct, while in others they closely interpenetrate and are not easily seen as discrete bodies. (b) An electron micrograph of mouse mammary-tumor virus particles, showing the glycoprotein knobs outlined by the phosphotungstic acid stain. [Reproduced with permission from Sarker *et al., Cancer Res.,* **55**, 740 (1975).] (c) An electron micrograph of disrupted Friend mouse leukemia virus particles showing the icosahedral-shaped shell constructed by the regular packing of large protein subunits (capsomeres). (d) A five-neighbored penton (circled) found on the corner of a shell. (e) A six-neighbored hexon. [(c), (d), and (e) reproduced with permission from M. V. Nermut *et al., Virology,* **49**, 345 (1972).]

GENERALIZED LIFE CYCLE OF RNA TUMOR VIRUSES

All RNA tumor viruses have a life cycle which starts with the adsorption of an infectious particle to the surface of a susceptible host cell. It then penetrates the outer cell membrane, usually by a pinocytotic process which brings the still intact particle into the cytoplasm (Figure 20–28). Here the RNA genome, after separation from its encompassing membrane, becomes transcribed into a complementary DNA strand using the viral-specific enzyme *reverse transcriptase*. In turn, the single-stranded DNA chain becomes enzymatically converted to a circular double-helical DNA

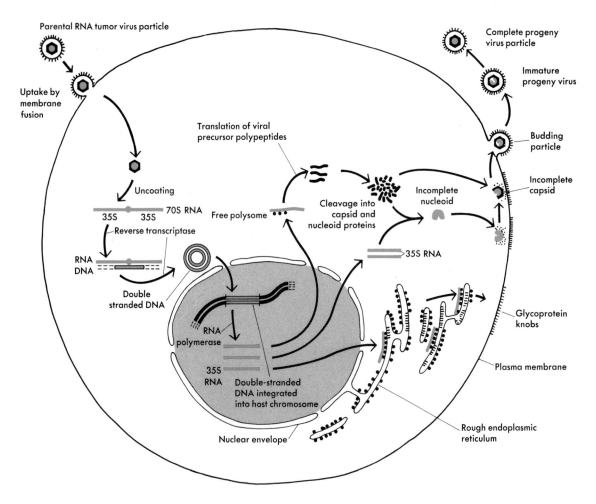

Figure 20–28
Schematic diagram of the major steps in the replication of an RNA tumor virus.

provirus. The provirus then integrates into host chromosomal DNA, most likely by a recombination process similar to that which places the λ DNA into the *E. coli* chromosome.

After integration, proviral DNA is transcribed into RNA chains, most of which initially serve as templates for the synthesis of viral-specific proteins. As infection progresses, more and more of the newly made RNA chains combine with newly synthesized nucleoid and capsomeric proteins to form "cores" which move to the cell surface. There the cores become enveloped by sections of the outer cell membrane into which the viral-specific "membrane glycoproteins" have been inserted. When their outer envelopes become complete, the newly made RNA tumor viruses detach from the cell surface as mature virus particles.

Most importantly, none of the steps in the multiplication of an RNA tumor virus necessarily interferes with normal cellular processes; and the infected cells do not obligatorily die, as is the case when the various DNA tumor viruses multiply. Each RNA tumor virus-infected cell thus can release thousands of progeny particles from its surface as it goes through a division cycle.

ISOLATION OF MUTANTS THAT CAN MULTIPLY BUT NOT TRANSFORM

The integration of only a single provirus is sufficient to transform a susceptible normal cell into its cancerous equivalent. The transition occurs relatively quickly; and, for example, within 24 hours of an RSV infection, a culture of normal chicken fibroblasts can become synchronously converted into rounded transformed cells (Figure 20–29).

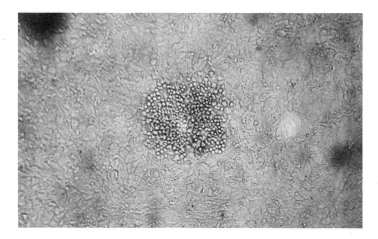

Figure 20–29
Photograph of a group (focus) of chicken cells (magnification 30X) transformed by RSV infection. The spherical transformed cells are easily distinguished from the background of normal cells. [Photograph by H. Rubin, Department of Molecular Biology, University of California, Berkeley.]

RNA tumor viruses thus provide an ideal model system for studying the chemical events which underlie the cancerous conversion. Most useful to this analysis has been the isolation of temperature-sensitive RNA tumor virus mutants which are unable to transform at high temperature ($\sim 39°C$) but will transform at lower temperature (e.g., 32°C). Such mutants undergo normal viral multiplication at the high temperature, with loss of only the transforming ability. They are being used increasingly to study transformation, since cells transformed by such a mutant lose their cancerous appearance within several hours of transfer from 32° to 39°C. Conversely, if the temperature is changed from 39°C back to 32°C, hitherto normally acting cells acquire a cancerous behavior. Such temperature shifts have no effect on the multiplication of the tumor virus, which continues at the same rate irrespective of whether its respective cell behaves normally or cancerously.

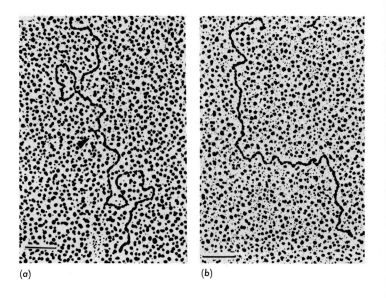

Figure 20–30
Electron micrographs of Feline Leukemia Virus RNA. (a) A 70S molecule in which two 35S chains are held together in the region marked by the arrow. (b) A 35S chain formed by the heating of the 70S form. [Reproduced with permission from Kung *et al., Cold Spring Harbor Symp.,* **XXXIX,** 827 (1974).]

(a)　　　　　　　　　(b)

THE UNRESOLVED PARADOX OF THE 70S GENOME

The RNA genome isolated from a gently disrupted RNA tumor virus particle sediments at 70S, a value which initially suggested a MW slightly greater than 10^7. Direct measurements of its MW, however, using the electron microscope to measure its contour length, now give MW's $\sim 6 \times 10^6$ (Figure 20–30). 70S RNA must thus be more tightly compacted than most RNA molecules and sediment proportionally faster. Heating of a 70S RNA molecule (Figure 20–30) to temperatures which break hydrogen bonds but not covalent polynucleotide backbone bonds yields two 35S RNA molecules, each of which has a molecular weight of about 3×10^6 (Figure 20–30).

The question thus arises whether all the 35S sub-units have identical sequences, or whether 2 chains with different sequences are present within each particle. Now it appears that all 35S chains have the same base sequences and that for reasons totally mysterious, each mature virus particle contains two chains only.

Realization that the fundamental genome size is only $\sim 3 \times 10^6 (\sim 9 \times 10^3$ nucleotides, or a coding capacity of ~ 3000 amino acids) means that the working out of the basic genetic organization of RNA tumor viruses should not be a Herculean task. Summing up of the molecular weights of the eight known polypeptides (the seven structural polypeptides + the reverse transcriptase) yields $\sim 300,000$ daltons (2700AA), or already more than 80% of the ~ 3000AA that can be coded by the viral genome. So only some 1 to 3 more virally specified proteins may have to be described before all the viral genes are known. Furthering this objective should be the ease with which large numbers of genetic mutants can be obtained. Crossing-over occurs following mixed infection, most likely at the DNA proviral stage, and a genetic map should soon be available.

tRNA ATTACHED TO GENETIC RNA

Large numbers of transfer RNA (tRNA) molecules are found within each mature RSV particle. Many different tRNA species are so enclosed, and while some may be contaminants accidentally caught by the budding process, over half such molecules are firmly bound to the 70S RNA complex and are only released by hydrogen bond-breaking agents, which break down the 70S complex into its component 35S chains. So, some of the four to eight tRNA molecules bound to each 35S chain help glue the two chains together. As we shall discuss below, another possible function of the bound tRNA is to serve as primers on which reverse transcriptase can initiate DNA synthesis.

FORMATION OF COMPLEMENTARY DNA CHAINS BY REVERSE TRANSCRIPTASE

Each RSV particle contains one to several copies of the enzyme *reverse transcriptase*. This virally coded enzyme (MW $\sim 110,000$) initially functions to make DNA complements of the infecting RNA genomes and later to convert these single-stranded chains into double-helical molecules. Discovery of this enzyme has been the decisive step in understanding how a virus with an RNA genome could provide stable genetic information to dividing cells. Before its existence was known, it often seemed necessary to postulate very large numbers of free RNA genomes within each transformed cell. If the number were small, normal cells which lacked the tumor virus genome would be continously segregated off. This argument, however, never was particularly convincing since inhibitors of DNA synthesis invariably block RNA tumor virus synthesis, an unexplained fact unless DNA synthesis was an obligatory step in tumor formation. The alternative hypothesis was thus proposed that the genetic information in the infecting RNA genome was transferred to a DNA chain during the transformation process. Initially, however, this idea seemed most heretical, and only with the discovery of reverse transcriptase did the existence of a DNA proviral stage become the generally accepted hypothesis.

Exactly how reverse transcriptase catalyzes the formation of the initial DNA complement is still up in the air. Like other DNA polymerases it cannot initiate new DNA chains but can only add deoxynucleotides to short RNA primers. Most unexpectedly, the primers seem to be some of the tRNA molecules bound to the 70S RNA. Most, if not all, DNA chains commence growth by adding onto the 3' ends of tryptophanyl tRNA. How many tRNA primers are used to make the complement of a 35S RNA chain is not known, although the fact that 1 to 2 tryptophanyl tRNA molecules are bound to each 35S component hints that only one primer may be used.

When the RNA primer is removed from the 5' ends has not yet been established, nor do we know how this is enzymatically accomplished. The only clue comes from the unexpected finding that the 110,000-MW polypeptide that carries the reverse transcriptase activity also has a second active site that catalyzes the breakdown of RNA in RNA–DNA hybrid molecules. But how such a specificity could allow the removal of RNA primers is far from clear.

CIRCULAR DOUBLE-HELICAL PROVIRUSES

While still within the cytoplasm, the resulting single-stranded DNA complements become converted into circular double helices. The way this happens is only partially known. They have MW's of 6×10^6, and so their immediate precursors are DNA complements of the 35S RNA chains, not those of the entire 70S complex. Reverse transcriptase is again the active agent, but what the primers are is not known. Nor do we understand how the double helix becomes circularized, since no evidence exists that the single-stranded 35S RNA genomes ever assume circular configurations.

After their formation, the circular proviruses move into the nucleus. There they integrate into host chromosomes during all phases of the cell cycle. So this process does not seem to involve synthesis of daughter chromosomal DNA. Instead it may require only the presence of enzymes promoting genetic recombination. Whether one or more such enzymes are RSV-specific remains to be established, but the great efficiency of the process hints that it (they) may be virally coded.

DNA TRANSFORMATION EXPERIMENTS THAT PROVE THE DNA PROVIRAL HYPOTHESIS

Mere existence of an RNA-to-DNA enzyme by itself does not prove that its DNA products go on to become integrated into chromosome DNA. Clean proof that this does happen comes from experiments in which DNA isolated from the chromosomes of RSV-transformed cells is added to normal chicken fibroblasts. Following DNA absorption, some of the treated cells not only become transformed, but also begin to produce progeny RSV particles. The genetic information necessary to make a complete RSV particle must have been present as a provirus integrated into chromosomal DNA. In contrast, when DNA obtained from normal cells is added to other normal cells, transformation never results. DNA molecules of the size that can encompass complete viral genomes appear to be necessary not only to transmit the ability to produce active progeny virus but also to convey the cancerous phenotype. This latter requirement is puzzling, since it is not clear why most of the RSV genome should be needed to effect transformation.

TRANSCRIPTION OF PROVIRAL DNA

After integration, proviral RSV is transcribed by eucaryotic RNA polymerase II, the enzyme responsible for cellular mRNA synthesis. Compared to other genetic DNA it is preferentially transcribed, with ~1% of total cellular mRNA being RSV-specific. So it is relatively easy to study these transcripts, the majority of which sediment at 35S and most likely are complete transcripts of the RSV provirus. Other transcripts sediment at 20S, and are perhaps partial transcripts; up until now, however, no one has shown whether the 35S and 20S forms share common nucleotide sequences.

Initially most transcripts bind to ribosomes and direct RSV-specific protein synthesis. Still very unclear is how the internal RSV proteins are released into the cytoplasm while the two RSV membrane proteins are exclusively inserted into the membrane of the endoplasmic reticulum. The simplest explanation would be the existence of two different transcripts, one of which attaches to membrane-bound ribosomes, where it serves as the template for the RSV proteins that are found in the outer membranous envelope. All the internal protein would be made by the second transcript, which should bind exclusively to free ribosomes. No evidence, however, exists for such a dichotomy; and preliminary research seems to show 20S and 35S RSV-specific transcripts attached to both membrane-bound and free ribsomes. But such experiments are much affected by contaminating ribonucleases, and later experiments may give different answers.

INTERNAL STRUCTURAL PROTEINS ARE DERIVED FROM A COMMON POLYPEPTIDE PRECURSOR

The five RSV internal proteins are not synthesized as separate proteins, but first as part of a large polypeptide precursor that quickly is proteolytically cleaved into proteins of the sizes found in the core of mature virus particles. No firm evidence exists for a still larger precursor whose cleavage yields the surface glycoproteins as well as the internal proteins. Each RSV mRNA template may have at least two independent initiating (ribosome-binding) and termination sites: one for the synthesis of the precursor of the internal polypeptides and one or more for the surface glycoproteins. RSV RNA would then more closely resemble bacterial mRNA than eucaryotic mRNA. Obviously, more experiments need to be done before the transcription and translation of RNA tumor virus DNA is molecularly understood.

CONVERSION OF 35S RNA INTO 70S RNA IN BUDDING VIRUS PARTICLES

After they become inserted into cell membranes, the viral glycoproteins begin to aggregate. Normal cellular membrane proteins become excluded from such aggregates, leading to the formation of extensive membrane patches where all the protein components are of viral specificity (Figure 20–28). These patches have strong affinity for the RNA-containing internal cores, leading to the rapid envelopment of newly assembled cores by sections of viral-specific membrane. Mature virus particles thus always form as buds off cellular membranes. Usually most buds come off the exterior plasma membrane, but others pinch off into the internal cavity of the endoplasmic reticulum.

When budding starts, the genetic RNA is still present as separate 35S chains; but as the mature form comes into existence, the 35S chains come together to form the 70S complex.

TRANSFORMATION IN THE ABSENCE OF VIRAL MULTIPLICATION

For several years, changes in the cell surface brought about by the budding off of progeny virus particles were considered a possible way that RNA tumor viruses induce cancer. But now many cases have been found where RNA viruses can transform but not multiply. Often they involve infection of foreign cells. For example, RSV particles are never budded off from RSV-transformed rat cells, despite the integration of complete RSV genomes. Transformation, but not multiplication, also follows infection of chicken cells at 39° by certain classes of temperature sensitive RSV mutants unable to multiply at that temperature. With some of these mutants, little if any viral structural protein is made, making most unlikely the hypothesis that the insertion of RSV-specific protein into cell membranes is the cause of RSV-induced cancers.

MUTANTS THAT MULTIPLY NORMALLY BUT CANNOT TRANSFORM

During RSV multiplication, mutants arise that have completely lost the capacity to transform, but that multiply normally. Many of these variants are deletion mutants, and now it appears that these deletions always involve a specific segment of the wild type genome. This transforming segment or *oncogene* (*cancer gene*) amounts to some 15% of the wild type genome, and so has the capability to code for 450 AA. Here again we have a very different picture than that found with the DNA tumor viruses, where the viral genes necessary for transformation appear to play necessary roles in viral multiplication. The study of RNA tumor virus multiplication, *per se*, may tell us nothing about how these RNA viruses cause cancer. To learn this, we must eventually focus attention on the specific genetic material present in transforming viruses and absent in nontransforming viruses.

One way to zoom directly in on the oncogene is to use reverse transcriptase to make DNA probes (Figure 20–31) that will specifically detect the oncogenic segment. Such probes can then be used to see whether unexpressed oncogene-like DNA sequences might also be present in normal cells. Most importantly, recent experiments suggest that oncogene-like DNA sequences appear to be present in

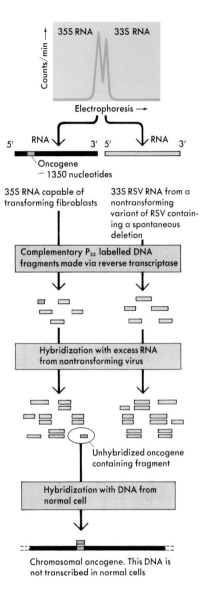

Figure 20–31
Use of reverse transcriptase to make DNA probes specific for oncogenes.

all cells, normal or transformed. The oncogenes of normal cells, however, are not transcribed, in contrast to abnormal cells where they are most active.

It thus becomes increasingly likely that the oncogenicity of many, if not most, RNA tumor viruses is the result of their accidental acquisition of one or more "normal" genes of their host cell. An oncogene may thus be a host gene whose functioning is no longer subject to an effective control mechanism when it becomes inserted (transduced) into a viral genome.

Now we only have hints as to the biochemical role of these potentially oncogenic normal genes. One important clue may be the recent observation that transformation by RSV induces synthesis of hemoglobin mRNA by chicken fibroblasts. In contrast, nontransforming variants do not stimulate hemoglobin mRNA synthesis. Interestingly, only the mRNA's characteristic of early life (fetal hemoglobin mRNA) are increased in amount. No stimulation of adult hemoglobin mRNA synthesis occurs. This suggests that the oncogenic passenger in RSV may be a host chicken gene involved in turning on large sets of the genes that usually function only in embryonic development.

RNA TUMOR VIRUS-LIKE GENOMES AS NORMAL CELLULAR CONSTITUENTS

All vertebrate cells seem to contain one to several latent RNA tumor virus-like genomes integrated into their chromosomes. These genomes normally are transcribed at very low rates, and only rarely are virus particles budded off from the respective cell surfaces. Normally, these progeny particles are unable to multiply in cells of the same species in which they were produced. In contrast, they often multiply very well in cells from other species. For example, a latent cat virus normally cannot multiply in cat cells but grows very well in human cells. The general failure of endogenous viruses to multiply in their own cells may be due to the presence of repressor-like molecules, which specifically inhibit endogenous genomes but which are unable to block the expression of foreign genomes to which they are not normally exposed.

Expression of certain endogenous proviral genomes is greatly increased by growth of their respective cellular hosts in the presence of halogenated pyrimidines. Somehow the incorporation of BUdR or IUdR into cellular (proviral?) DNA nullifies the normal control of proviral transcription, leading to the production of large numbers of mature virus particles. Other endogenous proviruses become activated when their host cells are exposed to specific steroid hormones. For example, estrogen treatment greatly increases the number of RNA tumor virus-like particles present within uterine cells. Hormone-stimulated placental tissue likewise contains many RNA tumor virus-like particles.

Unlike the RSV or the *Rauscher leukemia* virus, the endogenous RNA tumor-like viruses are not true tumor viruses. While they multiply very well in foreign cells, the average particle does not cause cell transformation and lacks any oncogenes. Only when very, very large numbers are added to susceptible animals do cancers, usually leukemias, result. So, most likely it is only the very rare variant, that has accidentally incorporated cellular genes, which can induce the leukemias, etc. The question thus arises whether most normal endogenous proviruses are passengers without a function, or whether they play a necessary role at some stage of normal cellular existence. The fact that virtually all known vertebrate cells contain them argues that they cannot be dispensed with and that we should actively search out their normal role.

SELECTIVE EXPRESSION OF ENDOGENOUS GENOMES DURING EMBRYOLOGICAL DEVELOPMENT

In fact there already exists strong evidence that endogenous viral genomes code for a specific surface antigen that appears during the development of the mouse. A major surface protein of the mouse thymocyte (G_{1X}) appears identical to the gp70 glycoprotein coded by one of its endogenous RNA viral genomes. It is first found in liver hematopoietic cells during the fourteenth day of development, and by the eighteenth day its expression is similar to that of the adult. Then, besides being characteristic of certain adult lymphoid tissue, it is also found in the epididymis (Figure 20–32) and on the surfaces of spermatozoa. So we now guess that the expression of this endogenous RNA viral genome is tightly regulated during differentiation and development. The mystery, of course, remains why these surface components should be coded by a genome with the capacity for independent self-replication.

SEARCHING FOR HUMAN TUMOR VIRUSES

Since so many different cancers of birds and rodents have viral etiologies, it has seemed common sense to suppose that many human cancers also are caused by viruses. Therefore much effort has gone, over the past decade, into linking up major human cancers with specific viruses. On the whole, these experiments have led nowhere, with a

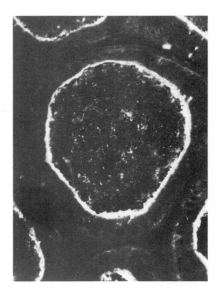

Figure 20–32
Expression of an endogenous RNA virus glycoprotein in the columnar epididymal cells of a New Zealand mouse as demonstrated by immunofluorescent microscopy. The circular rings of fluorescence outline boundaries of sections cut through the convoluted epididymis. [Photo courtesy of R. A. Lemer, B. C. Del Villano, and C. W. Wilson, Scripps Clinic, LaJolla, California.]

viral origin for most human cancer still an open affair. At first it seemed obvious to look for mature viruses within tumor cells, but we now know enough about DNA tumor viruses to realize that their multiplication always kills their host cells. So the absence of DNA-containing particles by itself says nothing about a possible viral etiology. That must be looked for by more indirect means, for example, by searching for viral-specific tumor antigens or for viral-specific DNA or RNA.

Though the tumorigenic RNA tumor viruses multiply without killing their host cells, EM searches for human RNA tumor viruses are complicated by the realization that human cells also contain nontumorigenic endogenous RNA viruses that are morphologically identical to those that cause cancer. So the sighting of large numbers of RNA tumor virus-like particles within a cancer cell may mean only that the cancerous condition induces the expression of an endogenous viral genome. And conversely, if no RNA tumor viruses are seen, this does not rule out an RNA tumor virus origin, since we now know of many model systems where such RNA viruses transform but do not multiply when they infect foreign cells. Thus much human cancer might be due to infection by RNA viruses which normally multiply in other species without producing cancer. The only moral now to draw is that the more we learn about model tumor virus systems, the more intelligently we can search for viral causes of human cancer.

STUDY OF CANCER AT THE MOLECULAR LEVEL

It is all too clear that a chasm of uncertain size still exists between the molecular study of tumor viruses and the more applied experiments that attempt to understand the biochemical essence(s) of the cancer cell(s). On the one side, there are many, all conceivably superficial, reasons for believing that the cancer-causing genes (*oncogenes*) of the various tumor viruses set off cancer in quite different ways. While all oncogenes may somehow activate repressed stretches of cellular chromatin, there is no reason to believe that the same inactive host genes are set into action by all tumor viruses. On the other hand, most cancer cells which have the ability to grow into solid tumors (e.g., carcinomas and sarcomas) seem to have the common property of secreting abnormally large amounts of specific proteolytic enzymes. Regardless of whether transformation is due to a DNA or an RNA tumor virus, the same specific proteases are released from the respective host cells. Moreover, experiments with temperature sensitive viral mutants show that protease release is dependent upon the continued functioning of the tumor virus oncogenes, and so some direct chain of events must link oncogenic function and protease release. But we have at present no way to know whether this chain involves only a few biochemical steps, or whether we will have to understand eucaryotic cell biochemistry in great detail before the pathways to cancer are clearly defined.

So here we have to face up to the same message that we reached earlier when thinking about embryology and immunology. No matter either how cleverly we think, or how great our capacity to spend vast sums of money for crash programs, the inherent complexity of eucaryotic cells still exceeds our intuitive powers. So all too frequently we must be satisfied with modest objectives. This sobering reflection, however, should not obscure the more relevant fact that the study of tumor viruses is now an almost exact science; and we have every reason to anticipate continued rapid, if not sensational, progress over the next one to two decades.

SUMMARY

A cancer cell is a cell that has lost the ability to control its growth and division. It divides both when and where it should not, and tumors are the result of its disorganized growth within an animal host. There exist two main hypotheses for the origin of cancer cells; both are essentially genetic. One states that they might be due to the accumulation of somatic mutations, with the most extreme cancerous phenotypes being the result of a progressive series of mutations. Alternatively, most cancers could result from the insertion of new genetic material as a consequence of the infection of normal cells by tumor (cancer) viruses. Now we believe that neither explanation will explain all cancers, and it seems likely that some forms of cancer largely result from somatic mutations, while others are a consequence of infection by tumor viruses.

Many, if not most, of the somatic mutations that lead to cancer are the results of carcinogenic (cancer-promoting) agents present in the environment. Exposure to ultraviolet rays, for example, creates cancer cells through genetic changes in cellular DNA. Carcinogenic hydrocarbons, like benzpyrene, become powerful mutagens after transformation by liver enzymes, the aryl hydroxylases. Likewise, nitrates, by themselves nonmutagenic, after conversion into nitrites can react with a variety of secondary amines to form the highly mutagenic nitrosamines. The fact that these carcinogens are all general mutagens makes it likely that mutations in a large number of different genes can lead to the cancerous phenotype.

Once a cancer cell arises in an animal, it does not necessarily form a tumor. Most are destroyed by immunological reactions mediated by cell-bound antibodies (immunological surveillance). These immunological responses are directed against tumor-specific antigens present on the surface of the cancer cells. Apparently the change to the cancerous state almost invariably results in a drastic reorganization of the outer cell surface that creates new surface configurations that are recognized as "foreign" by the host immunological system. Why the rare cancer cell occasionally escapes surveillance and grows into a tumor is not at all clear. In any case, the oncogenicity of a given tumor cell is best tested by its injection into an animal with impaired immunological responses.

Many types of viruses can cause cancer. Those that contain DNA are a very heterogeneous collection ranging from the very small **Papova** *viruses to the very large* **Herpes** *and* **Pox** *viruses. The most simple DNA viruses that induce cancer belong to the* **Papova** *group and include SV40, a monkey virus, and* **Polyoma,** *a mouse virus. Both SV40 and* **Polyoma** *have MW's of about 25 million and contain single circular DNA molecules of MW* $\sim 3 \times 10^6$ *which may code for as few as three genes. Each have life cycles that are divided into early and late phases. During the early phase, an early mRNA molecule is made which codes for the "T(tumor) antigen," a protein that localizes in the nucleus where it preferentially binds to SV40 (Polyoma) DNA at the site of initiation of DNA synthesis. How the T antigen functions is not known, but the best current guess is that it is involved in the initiation of viral DNA synthesis. It may also function in promoting the synthesis of the late SV40 (Polyoma) mRNA, which codes for two structural viral proteins, VP 1 and VP 2.*

When SV40 (Polyoma) infects a susceptible (permissive) cell, it multiplies lytically. In contrast, when it infects a nonpermissive cell, abortive infections occur in which only early mRNA and T antigen (and viral DNA?) are made. A very small fraction of abortive infections lead to transformation of the nonpermissive cell to a cancer cell. Transformation is marked by the insertion of intact viral genomes (proviruses) in specific host chromosomes, most likely by a λ-like crossing-over event. While inserted into a host chromosome, T antigen but not VP 1 and VP 2, is synthesized. The specific gene product that leads to the cancerous phenotype thus is the "T antigen." Most likely the T antigen acts by turning on the transcription of many gene sets whose functions are normally blocked in the adult animal. Fusion of an SV40-transformed nonpermissive cell with a normal permissive cell induces the transformed nucleus to produce progeny SV40 particles. The essence of cells nonpermissive for SV40 may be their inability to translate late SV40 mRNA.

Adenoviruses also induce cancer through insertion of viral genes into host chromosomes. These much larger DNA viruses have linear DNA molecules of MW ∼ 25 × 10⁶. Their life cycles also are divided into early and late phases, with some four species of early and five (or six) species of late mRNA being made in infected permissive cells. Transformation results from the insertion of only a fraction of the adeno genome into the chromosome of a nonpermissive cell. Only the extreme left end (∼7 percent) of the adeno chromosome need be inserted, suggesting that only one to two early genes function to produce the transformed phenotype.

Transformation by the **Herpes** *viruses is much less well understood. Only a fraction of their DNA genomes is required, but where it is located is not yet known. Possession of the intact* **Herpes** *chromosome need not kill a host cell, since* **Herpes**-*transformed cell lines occasionally have the capacity to initiate a viral multiplication cycle that produces large numbers of progeny virus and eventually kills the respective host cells.*

All the RNA tumor viruses have essentially the same structure: a central RNA-containing core surrounded by an outer lipid membrane into which viral-specific glycoproteins are inserted. In mature virus particles, the RNA is present as 70S molecules of MW ∼ 60 × 10⁶. Each 70S molecule is composed of two identical 35S chains of some 9000 nucleotides held together by secondary bonds. Penetration of a virus particle into the cytoplasm of a susceptible permissive cell is followed by uncoating of the 70S RNA genome. The 70S RNA then serves as a template for the synthesis of a complementary DNA strand, using the enzyme reverse transcriptase, several copies of which are present in every viral particle. In turn, the DNA serves as a template for the synthesis of its complement, becomes circular, and is inserted by crossing over as a linear rod into a host chromosome. The integrated viral-specific DNA provirus subsequently becomes transcribed into 35S viral-specific RNA chains that serve as templates for the various viral-specific proteins. Virus progeny then form by a budding process from the cell membrane.

Viral mutants exist which multiply normally but which have lost the capacity to transform. These mutants often are the result of deletions of the gene (oncogene) which codes for the protein that creates the cancerous phenotype. The oncogenes of RNA tumor viruses appear to be the result of genetic accidents which insert (transduce) genes governing key cellular control processes into the genomes of the endogenous RNA viruses normally found in all vertebrate cells. Normally, these endogenous viral genomes do not function to make progeny particles. But under certain specific conditions (e.g., hormonal stimuli), they become transcribed and virus particles bud off the plasma membrane. Usually these particles cannot multiply in the cells from which they are released. In contrast, they frequently multiply in the cells of foreign species. The real cellular role of these endogenous genomes is unclear, though some selectively function in certain differentiated cells, coding for specific surface glycoproteins.

REFERENCES

BURNET, F. M., "Cancer: Biological Approach. I. Processes of Control." *Brit. Med. J.*, **1**, 779 (1957). A superb appraisal of the complexity of the cancer field, with emphasis on the consequences of somatic mutation.

SUSS, R., F. KINGEL, AND J. D. SCRIBNER, *Cancer: Experiments and Concepts*, Springer-Verlag, 1973. A broad and quite good introduction to cancer research.

KLEIN, G., U. BREGULA, F. WIENER, AND H. HARRIS, "The Analysis of Malignancy by Cell Fusion," *J. Cell Sci.* **8**, 659 (1971). Experiments that show that the malignant character of tumor cells is suppressed by fusion with normal cells.

BURNET, F. M., *Immunological Surveillance*, Pergamon Press, 1970. A semipopular introduction to the concept that cancer can be controlled by cell-mediated immunological responses.

GROSS, L., *Oncogenic Viruses*, 2nd ed., Pergamon Press, 1970. Slightly outdated but still fascinating for its very complete description of early tumor virus research.

TOOZE, J. *The Molecular Biology of Tumor Viruses*, Cold Spring Harbor Laboratory, 1973. An introduction to modern tumor virus research, as seen by molecular biologists.

FENNER, F., B. R. McAUSLAN, C. D. MIMS, J. SAMBROOK, AND D. O. WHITE, *The Biology of Animal Viruses*, Academic Press, 1974. An advanced text that covers virtually all aspects of animal virus multiplication. Several chapters specifically describe the tumor viruses.

RNA Viruses and Host Genome in Oncogenesis, ed. P. Emmelot and P. Bentvelzen, North-Holland, Amsterdam, 1972. Papers from a meeting devoted to RNA tumor viruses.

TEMIN, H. M., "RNA-Directed DNA Synthesis," *Scientific American*, 1972. The experiments that led to and then proved the author's now famous proposal for information transfer from RNA to DNA.

Selected Papers in Tumor Virology, edited by J. Tooze and J. Sambrook, Cold Spring Harbor Laboratory, 1974. A very large collection of original articles that emphasize the molecular approach.

The Herpes *Viruses,* edited by Albert Kaplan, Academic Press, 1973. A useful compilation of review articles.

Tumor Viruses, Cold Spring Harbor Symposium on Quantitative Biology XXXIX, 1974. One hundred thirty-seven different articles giving the definitive June 1974 view of tumor virus research.

Cancer, F. F. Becker (ed.), Plenum Press, 1975. A comprehensive review series in four volumes, with much information on cancer viruses and mutagenesis.

AMES, B. N., H. O. KOMMEN, AND E. YAMSAKI, "Hair Dyes are Mutagenic: Identification of a Variety of Mutagenic Ingredients," *Proc. Nat. Acad. Sci.,* **72,** 2423 (1975).

FRAENKEL-CONRAT, H., AND R. R. WAGNER, *Comprehensive Virology,* Plenum Press, 1974. A thorough, multi-volume review containing much information on tumor viruses.

Glossary

A muscle protein (MW ~ 95,000) found in the Z line of muscle fibers where it attaches to actin filaments. — **α-Actinin**

An expression of the relative amount of adenine-thymine pairs to guanine-cytosine pairs in a molecule of DNA. — **A + T/G + C Ratio**

A high-energy ester of acetic acid (active acetate) which is an important metabolite for both the Krebs Cycle and fatty acid biosynthesis. — **Acetyl CoA**

Amino acids having a net negative charge at neutral pH. — **Acidic Amino Acids**

Globular protein molecules (MW ~ 70,000) which polymerize to form very long, thin filaments. — **Actin**

Antibiotic that blocks elongation of RNA chains. — **Actinomycin D**

(See *Amino Acyl Synthetase*.) — **Activating Enzyme**

The energy that must be supplied to a system to allow a chemical reaction to proceed. — **Activation Energy**

Region of protein directly involved in interaction with other molecules. — **Active Site**

Small RNA molecules (tRNA) that bind amino acids to their proper positions on an mRNA template during protein synthesis. Each is specific for both an amino acid and a template codon. — **Adaptor Molecules**

A high-energy phosphate ester which serves as the principal energy-storage compound of the cell. — **Adenosine Triphosphate (ATP)**

Adenovirus

Animal viruses of ~80 nm diameter, which contain linear duplex DNA within an icosahedral protein shell. Many serotypes are pathogenic, causing common cold-like symptoms. During infection the virus particles appear intranuclearly.

Adenylcyclase

Enzyme that catalyzes production of cyclic AMP from ATP.

Affinity Chromatography

A technique of molecular separation in which molecules are attached to an insoluble (e.g., sepharose) matrix. Only those molecules which show affinity to the bound molecule (e.g., an antibody for its antigen) are retained. These trapped molecules can be subsequently eluted.

Allele

One of two or more alternative forms of a gene.

Allergy

Increased sensitivity to an antigen brought about by previous exposure.

Allosteric Effectors

Those small molecules which reversibly bind to allosteric proteins at a site distinct from the active site, so as to cause a change in protein shape.

Allosteric Proteins

Proteins whose biological properties are changed by the binding of specific small molecules (allosteric effectors) at sites other than the active site.

Allotypes

Term used by immunologists in place of alleles.

Amino Acids

The building blocks of proteins. There are twenty common amino acids, each present as the L-stereoisomer. All amino acids have the same basic structure, but they differ in their side groups (R):

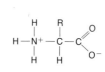

Amino Acid Sequence

The linear order of the amino acids in a peptide or protein.

Amino Acid Side Group(s)

(See *Amino Acids.*)

Amino Acyl Adenylate (AA ~ AMP)

In protein synthesis, an activated compound that is an intermediate in the formation of a covalent bond between an amino acid and its tRNA adaptor.

Amino Acyl Synthetase

Any one of at least twenty different enzymes that catalyze (1) the reaction of a specific amino acid with ATP to form amino acyl-AMP (activated amino acids) and pyrophosphate and (2) the transfer of the activated amino acid to tRNA forming amino acyl-tRNA and free AMP.

Amino Group

$-NH_2$, a chemical group, characteristically basic because of the addition of a proton to form $-NH_3^+$.

The end of a polypeptide chain that has a free α-amino group. **Amino Terminal**

Cellular movement that involves cytoplasmic streaming into cellular extensions called *pseudopodia*. **Amoeboid Movement**

Chromosome number that is not an exact multiple of the haploid number. **Aneuploidy**

A unit of length convenient for describing atomic dimensions; equal to 10^{-8} cm. **Angstrom (Å)**

(See *Immunoglobulins*.) **Antibodies**

The three-base group on a tRNA molecule that recognizes and pairs with a three-base codon of mRNA, leading to the correct positioning of an amino acid into a growing protein. **Anticodon**

Any object that, upon injection into a vertebrate, is capable of stimulating the production of neutralizing antibodies. **Antigen**

Chemical structure (small compared to macromolecule) recognized by the active site of an antibody. Determines specificity of antibody-antigen interaction. **Antigenic Determinant**

Protein that prevents the recognition of initiation sites by the specificity factor of RNA polymerase. An anti-specificity factor is synthesized during phage T4 infection. **Anti-Specificity Factor**

Protein that prevents normal termination of RNA synthesis, perhaps by interfering with action of ρ factor. Bacteriophage λ "N" gene product may be an antitermination factor. **Antitermination Factor**

Amino acids whose side chains include a derivative of a phenyl group. The aromatic amino acids found in protein are phenylalanine, tyrosine, and tryptophan. **Aromatic Amino Acids**

A region of an operon within which most RNA polymerase molecules stop elongation. Only with the receipt of a specific molecular signal (antitermination factor) will transcription proceed normally. **Attenuator Region**

When a photographic emulsion is placed in contact with radioactive material (e.g., thin sections of a cell), the radiation exposes the film, revealing details of the location and geometry of the radioactive components. **Autoradiographs**

Detection of radioactive label in cytological preparations and macromolecules by exposure of photographic film. **Autoradiography**

Rigid, linear cellular projections composed largely of microtubules. **Axopodia**

693

B Lymphocytes   Small lymphocytes which synthesize and secrete humoral antibodies upon antigenic stimulation.

Backbone   The atoms in a polymer that are common to all its molecules (e.g., the sugars and phosphates in RNA).

Bacterial Viruses   Viruses that multiply in bacteria.

Bacteriophages (Phages)   (See *Bacterial Viruses.*)

Balbiani Rings   Particularly large chromosomal puffs.

Basal Body   A microtubule-containing organelle located at the base of cilia, and thought to be involved in the organization of ciliary microtubules.

Base Analogs   Purines and pyrimidines that differ slightly in structure from the normal nitrogenous bases. Some analogs (e.g., 5-bromouracil) may be incorporated into nucleic acids in place of the normal constituents.

Base-Pairing Rules   The requirement that adenine must always form a base pair with thymine (or uracil) and guanine with cytosine, in a nucleic acid double helix.

Basic Amino Acids   Amino acids having a net positive charge at neutral pH.

Bence–Jones Protein   Light chains of a single antibody species produced by myeloma cells. Commonly detected in urine of human multiple myeloma patients.

β-Galactosidase   An enzyme catalyzing the hydrolysis of lactose into glucose and galactose; in *E. coli,* the classic example of an inducible enzyme.

Biological Clocks   Mechanisms that allow expression of certain biological structures (genes) at periodic intervals. May involve specificity and anti-specificity factors.

Blocked (Capped) 5′ Ends   The 5′ ends of most eucaryotic mRNA's are post-transcriptionally modified by the addition of GTP in a 5′–5′ condensation.

Breakage and Reunion   The classical model of crossing-over by physical breakage and crossways reunion of completed chromatids during meiosis.

5-Bromouracil (5-BUdR)   A mutagenic base analog of thymine in which the 5-CH_3 group is replaced by bromine.

Bursa   Any pouch or sac in an animal. The Bursa of Fabricius is the chief lymphoid organ of birds, where B lymphocytes become immunologically competent.

A radioactive carbon isotope emitting a weak β particle (electron). Its half-life is 5700 years.	^{14}C
A class of RNA viruses defined on the basis of their appearance in the electron microscope. "C" type particles have a centrally placed, spherical RNA-containing nucleoid and seem to be responsible for many sarcomas and leukemias.	"C" type particles
A measure of energy, defined as the amount of energy necessary to raise 1 cc of water $1C°$.	Calorie
The name given to a group of diseases that are characterized by uncontrolled cellular growth.	Cancer
A highly regular, shell-like structure, composed of aggregated protein subunits, that encloses the nucleic acid component of viruses.	Capsid
Protein subunits which aggregate to form the capsid.	Capsomeres

A chemical group, characteristically acidic, as a result of the dissociation of the hydroxyl H to form

Carboxyl Group

$$-\overset{\overset{\displaystyle O}{\|}}{C}-O^-$$

The end of a polypeptide chain that has a free α-carboxyl group.	Carboxyl Terminal
An agent that induces cancer.	Carcinogen
Decreased synthesis of certain enzymes in bacteria grown on glucose or other very good catabolite source. Caused by low levels of cyclic AMP in such cells.	Catabolite Repression
A dimeric protein of MW $\sim$ 44,000. It is the positive control element for all the glucose-sensitive operons.	Catabolite Activator Protein (CAP)
Compounds that are breakdown products of food molecules.	Catabolites
A substance that can increase the rate of a chemical reaction without being consumed (e.g., enzymes catalyze biological reactions).	Catalyst
The fundamental unit of life; the smallest body capable of independent reproduction. Cells are always surrounded by a membrane.	Cell
The *in vitro* growth of cells isolated from multicellular organisms.	Cell Culture
The timed sequence of events occurring in a cell in the period between mitotic divisions.	Cell Cycle

Cell Differentiation The process whereby descendants of a common parental cell achieve and maintain specialization of structure and function.

Cell Factor Protease of MW $\sim$ 40,000 which is selectively secreted in large amounts by tumor cells. Its major action in blood appears to be the proteolytic activation of plasminogen to plasmin.

Cell-Free Extract A fluid containing most of the soluble molecules of a cell, made by breaking open cells and getting rid of remaining whole cells.

Cell Fusion Formation of a single hybrid cell with nuclei and cytoplasm from different cells. Often induced by treatment of a mixed cell culture with killed *Sendai* virus.

Cell Line Those secondary cells that survive the crisis period by somehow acquiring the ability to multiply indefinitely.

Cell-Mediated Immunity Immune responses directly mediated by T lymphocytes rather than by circulating (humoral) antibody molecules.

Cellular Affinity Tendency of cells to adhere specifically to cells of the same type, but not of different types. This property is lost in cancer cells.

Central Dogma The basic relationship between DNA, RNA, and protein: DNA serves as a template for both its own duplication and the synthesis of RNA; and RNA, in turn, is the template in protein synthesis.

Centrioles Paired, cylindrical structures composed of nine sets of microtubule triplets that lie just outside the nuclei of animal cells. May be involved in the formation of the spindle.

Centromere A heterochromatic region of the eucaryotic chromosome which is the chromosomal site of attachment of the kinetochore. It divides just before replicated chromosome separation, and so holds together the paired chromatids.

Ceramide A lipid. The fatty acid amide of sphingosine, a long-chain unsaturated amino alcohol.

Cerebrosides Lipid molecules which contain a polar carbohydrate group attached to the amino alcohol sphingosine. They are abundant in the myelin sheath of nerve cells.

Chemical Carcinogen Any chemical substance capable of causing cancer.

Chemiosmotic Theory A hypothesis which accounts for the coupling of electron transfer and ATP formation by postulating the formation of H^+ gradients across the mitochondrial membrane.

The two daughter strands of a duplicated chromosome that are still joined by a single centromere. **Chromatids**

The nucleoprotein fibers of which eucaryotic chromosomes are composed. **Chromatin**

A concentrated chromatin "bead" on a eucaryotic chromosome. May be a region of gene redundancy. Results from local coiling of a continuous thread. **Chromomere**

Spindle microtubules which originate at the kinetochores. **Chromosomal Tubules**

Threadlike structures into which the hereditary material of cells and viruses is associated. **Chromosomes**

Locomotory organelles located on the cell surface, composed of a precise arrangement of microtubules within an outpocketing of the cell. **Cilia**

Mitotic divisions of fertilized egg until the stage when regions of egg shift relative to one another. **Cleavage Divisions**

A group of cells all descended from a single common ancestor. **Clone**

The external structural protein(s) of a virus. **Coat Protein(s)**

A sequence of three adjacent nucleotides that code for an amino acid (or chain termination). **Codon**

Small molecules which associate with proteins to form active enzymes (e.g., NAD^+). **Coenzymes**

A poisonous alkaloid found in the meadow saffron plant. It binds tubulin on a molar basis and thereby causes breakdown of microtubules. **Colchicine**

A group of contiguous cells, usually derived from a single ancestor, growing on a solid surface. **Colony**

A series of sequentially acting proteins present in the blood serum which, when activated, lyse foreign cells. **Complement**

Polynucleotide sequences that are related by the base-pairing rules. **Complementary Base Sequences**

Two structures, each of which defines the other; for instance, the two strands of a DNA helix. **Complementary Structures**

The introduction of two mutant chromosomes (or sections of chromosomes) into the same cell for the purpose of seeing whether their respective mutations occurred in the same gene. **Complementation Test**

A mitogenic lectin derived from the pokeweed. **Concanavalin A**

Concatemer Structure formed by the aggregation of unit-size components.

Conditional Lethal Mutations A class of mutants whose viability is dependent on growth conditions (e.g., temperature-sensitive lethals).

Constitutive Enzymes Enzymes that are synthesized in fixed amounts, irrespective of the growth conditions.

Contact Inhibition The cessation of cell movement (division?) that is often observed when freely growing cells from a multicellular organism come into physical contact with each other.

Coordinated Enzyme Synthesis Enzymes whose rates of production are observed to vary together. (For example, in *E. coli* cells growing in the absence of β-galactosides, the addition of lactose to the medium causes the coordinated induction of β-galactosidase and β-galactoside permease.)

Copolymer A polymeric molecule containing more than one kind of monomer unit.

Copy Choice A now discarded model for crossing-over, which doesn't involve breakage of the recombining chromosomes. This mechanism has a new DNA molecule being formed on a paternal chromosome switch to the maternal one and continue to grow on that template. Similarly, the maternally growing strand switches to the paternal chromosome.

Corepressors Metabolites which, by their combination with repressors, specifically inhibit the formation of the enzyme(s) involved in their metabolism.

Coupled Reaction A thermodynamically unfavorable reaction, which by association with a thermodynamically favorable reaction is driven in the direction of product formation.

Covalent Bonds Strong chemical bonds formed by the sharing of electrons between atoms.

Crisis Period After a number of divisions, depending on the specific cell being cultured, most secondary cells die, with only the exceptional cell surviving.

Crossing-Over The process of exchange of genetic material between homologous chromosomes.

CsCl Centrifugation (See *Equilibrium Centrifugation.*)

Cyclic AMP Adenosine monophosphate with phosphate group bonded internally (phosphodiester bond between 3' and 5' carbon atoms) to form cyclic molecule. Active in regulation of gene expression in bacterial and eucaryotic cells.

Guanosine monophosphate with phosphate group internally bonded (phosphodiester bond between 3' and 5' carbon atoms) to form a cyclic molecule.

Cyclic GMP

An inhibitor of protein biosynthesis.

Cycloheximide

A unit of weight equal to the weight of a single hydrogen atom.

Dalton

Two or more codons that code for the same amino acid.

Degenerate Codons

Loss of a section of the genetic material from a chromosome. The size of the deleted material can vary from a single nucleotide to sections containing a number of genes.

Deletions

The loss of the native configuration of a macromolecule resulting, for instance, from heat treatment, extreme pH changes, chemical treatment, or other denaturing agents. It is usually accompanied by loss of biological activity.

Denaturation

A method of molecular or organelle separation which relies upon differential sedimentation through a preformed density gradient upon centrifugation.

Density Gradient Centrifugation

The condensation product of a purine or pyrimidine with the five-carbon sugar, 2-deoxyribose.

Deoxynucleoside

A compound which consists of a purine or pyrimidine base bonded to the sugar, 2-deoxyribose, which in turn is bound to a phosphate group.

Deoxyribonucleotide

Structure resulting from association of two identical subunits.

Dimer

The chromosome state in which each type of chromosome except the sex chromosomes is always represented twice (2N).

Diploid State

Covalent bond between two sulfur atoms in different amino acids of a protein. Important in determining secondary and tertiary structure.

Disulfide Bond

A polymer of deoxyribonucleotides. The genetic material of all cells.

DNA (Deoxyribonucleic Acid)

The first enzyme found to catalyze the formation of the 3'–5' phosphodiester bonds of DNA. It also possesses 3' to 5' single-strand proofreading and 5' to 3' double-strand exonuclease activities, for use in DNA repair, most likely its chief biological function.

DNA Polymerase I

DNA Polymerase III Enzyme which catalyzes the formation of 3'–5' phospho-diester bonds at a very rapid rate. It also possesses a 3' to 5' exonuclease activity for "proofreading." Its chief role seems to be DNA replication.

DNA–RNA Hybrid A double helix that consists of one chain of DNA hydrogen bonded to a chain of RNA by means of complementary base pairs.

Dominant An allele which exerts its phenotypic effect when present either in homozygous or heterozygous form.

Early vs. Late Genes Genes transcribed early and late after infection of bacteria by bacteriophage. May require different specificity factors for recognition of promoters.

Early vs. Late Proteins During viral infection, viral-specific proteins are synthesized at characteristic times after infection, in groups that can be classed as "early" and "late." Often under positive control of bacterial and viral specificity factors.

Electron Microscopy A technique for visualizing material that uses beams of electrons instead of light rays and that permits greater magnification than is possible with an optical microscope. Resolutions of ~ 10 Å are attainable with biological materials.

Electronegative Atom An atom with a tendency to gain electrons.

Elongation Factor G (EF-G) Also called translocase, this protein brings about the movement of the peptidyl tRNA from the "A" site to the "P" site of the ribosome.

Elongation Factor T (EF-T) The protein responsible for the positioning of the AA $\sim$ tRNA complexes in the "A" site of the ribosome.

Embden–Meyerhof Pathway (See *Glycolysis.*)

Endogenous Virus Virus which exists in a proviral, partially inactive form within a host cell.

End Product A chemical compound that is the final product of a sequence of metabolic reactions.

Endonuclease An enzyme that makes internal cuts in DNA backbone chains.

Enzymes Protein molecules capable of catalyzing chemical reactions.

Episome A genetic element that can exist either free or as part of the normal cellular chromosome. Examples of episomes are the sex(F^+) factor and lysogenic phage DNA.

That tissue which acts as a covering or lining for any organ or organism. **Epithelium**

A technique used to separate molecules by their density: Most often used to separate nucleic acid fragments which, when ultracentrifuged in a heavy salt solution characterized by a density gradient, move to the region of the centrifuge cell in which their density equals that of the solution. **Equilibrium Centrifugation**

Nucleated cell in bone marrow that differentiates into a red blood cell. **Erythroblast**

Hemoglobin-rich, anucleate red blood cell involved in oxygen transport. **Erythrocyte**

Cultured cells of single origin capable of stable growth for many generations. **Established Cell Line**

Hormone produced by ovary. **Estrogen**

Organism with cells that have nuclear membranes, membrane-bound organelles, 80S ribosomes, and characteristic biochemistry. **Eucaryote**

Active chromatin which normally is not visible throughout interphase. Euchromatic regions of chromosomes are actively transcribed into RNA. **Euchromatin**

Also called reverse pinocytosis. Exocytosis involves the merging of internal membranes with the external plasma membrane. This process counterbalances the transfer of external membranes to a cell's interior by pinocytosis and phagocytosis. **Exocytosis**

An enzyme that digests DNA from the ends of strands. **Exonuclease**

Inhibition of the enzymatic activity of the first enzyme in a metabolic pathway by the end product of that pathway. **Feedback (End-Product) Inhibition**

Anaerobic metabolism of glucose. **Fermentation**

An iron storage protein primarily found in the liver and spleen which contains up to 20 percent of its weight in the form of iron. **Ferritin**

An episome that determines the sex of a bacterium. The presence of this factor in the cell makes it a male. (Female cells are called F^-.) **Fertility Factor (F^+)**

Fusion of gametes of opposite sexes to produce diploid zygote. **Fertilization**

Fibroblasts	Differentiated cells which grow very well in culture. They have the spindle shape and growth rate of connective tissue cells.
Fine Structure Mapping	Genetic analysis which reveals the location of mutations on the chromosome with high resolution.
Fluorescent Antibody Technique	Detection of specific antigen in cells by staining with a specific antibody conjugated with a fluorescent dye.
Formamide	A small organic molecule used in double-helical DNA denaturation. Formamide combines with the free NH_2 groups of adenine and prevents the formation of A–T base pairs. Its structure:

$$H-\overset{\overset{\textstyle O}{\|}}{C}-NH_2$$

Frameshift Mutations	Insertions or deletions in the hereditary DNA molecule which shift the normal reading frame for translation, often leading to nonfunctional protein products.
Free Energy	Energy that has the ability to do work.
Freeze-Fracture	Frozen samples are fractured with a knife edge, and then complementary surfaces are cast in metal. The casting can then be viewed in the electron microscope.
Fusidic Acid	Antibiotic that inhibits the working of elongation factor G.
Gametes	Haploid cells (ova and sperm) which unite at fertilization to produce a diploid zygote.
Gangliosides	Glycolipids that contain sialic acid. They are most often found on the outer surface of the cell membrane, and are extremely prevalent in nerve cells.
Gene	A stretch along a chromosome that codes for a functional product. (Either RNA or its translation product, a polypeptide.)
Generation Time	The time necessary for growing cells to double their mass under specified conditions.
Gene Redundancy	Presence in a cell of many copies of a single gene. Multiple copies may be inherited or result from selective gene duplication during development.
Genetic Information	The information contained in a sequence of nucleotide bases in a DNA (or RNA) molecule.
Genetic Map	The arrangement of mutable sites on a chromosome as deduced from genetic recombination experiments.

Haploid set of chromosomes, with their associated genes. **Genome**

The genetic constitution of an organism (to be distinguished from its physical appearance or phenotype). **Genotype**

Cells which differentiate early in embryogenesis and which later are exclusively able to undergo meiosis to form sex cells (ova and sperm). **Germ Line**

Steroid hormones made by the adrenal cortex that have, among their other effects, modifications of carbohydrate metabolism. An example of a glucocorticoid is cortisone. **Glucocorticoids**

Those operons whose functioning is blocked by the presence of glucose. Glucose indirectly lowers the cyclic AMP level, thereby blocking a necessary positive control signal. **Glucose-Sensitive Operons**

A polymer of indefinite length formed by 1–4 glycosidic linkages between adjacent glucose residues. **Glycogen**

Lipids that contain polar, hydrophilic carbohydrate head groups. **Glycolipids**

The process of glucose catabolism. **Glycolysis**

A polypeptide to which sugar residues are attached. **Glycoprotein**

A complex series of flattened, parallel membranes which appear to function in granule formation and molecular processing. Here, lysosomes are formed and secretory products are packaged into vacuoles. **Golgi Apparatus**

Leukocytes which possess distinct cytoplasmic granules. Includes: eosinophils, basophils, and neutrophils. **Granulocytes**

Covalently bonded groups of atoms that behave as a unit in chemical reactions. **Group (Functional)**

Reactions (excluding oxidations or reductions) in which molecules exchange functional groups. **Group-Transfer Reactions**

The change in the number of cells in a growing culture as a function of time. **Growth Curve**

A specific substance that must be present in the growth medium to permit a cell to multiply. **Growth Factor**

A radioactive isotope of hydrogen, a weak β emitter, with a half-life of 12.5 years. **^{3}H (Tritium)**

Regions of double helix formed by the pairing of two contiguous complementary stretches of bases on the same single DNA or RNA strand. **Hairpin Loops**

Haploid State The chromosome state in which each chromosome is present only once.

Haptens Small molecules, nonantigenic by themselves, that are capable of stimulating specific antibody synthesis when chemically coupled to a larger molecule.

Heavy Chains Generally refers to the heavy polypeptide components (MW $\sim$ 55,000) of the immunoglobulin.

Heavy Isotopes Forms of atoms containing greater than the common number of neutrons and thus more dense than the commonly observed isotope (e.g., ^{15}N, ^{13}C).

HeLa Cells An established line of human cervical carcinoma (cancer) cells. Has been used for many years in study of biochemistry and growth of cultured human cells.

Helix A spiral structure with a repeating pattern described by two simultaneous operations—rotation and translation. It is the natural conformation of many regular biological polymers.

Hemoglobin Protein carrier of oxygen found in red blood cells. Composed of two pairs of identical polypeptide chains and an iron-containing heme group.

Hepatoma A specific form of liver cancer.

Hereditary Disease A pathological condition whose cause is a gene mutation and that can therefore be transferred from one generation to the next.

Herpes An enveloped animal virus of $\sim$200 nm diameter, which possesses a linear duplex DNA molecule within an icosahedral capsid. *Herpes*-type viruses have been implicated as the causative agents of cold sores, shingles, and mononucleosis.

Heterochromatin Highly compacted chromatin which is not transcribed into RNA. Heterochromatic regions of chromosomes remain tightly compacted throughout interphase.

Heteroduplex Double-stranded DNA molecule in which the two strands do not have completely complementary base sequences. Can arise from mutation, recombination, or by annealing DNA single strands *in vitro*.

Heterozygous Gene Pair The presence of different alleles, for a given gene, on the homologous chromosomes of a diploid organism.

Hfr (High Frequency of Recombination) Strains of *E. coli* that show unusually high frequencies of recombination. In these cells the F factor is integrated into the bacterial chromosome, where it is thought to play some part in the transfer of the chromosome from Hfr to F⁻ cells. (See also *Fertility Factor*.)

A bond that yields a large (at least 5 kcal/mole) decrease in free energy upon hydrolysis.

High-Energy Bond

Specific cell proteins involved in transplantation immunity. Foreign histocompatibility antigens, introduced, for example, by a skin graft from a genetically distinct individual, provoke a cell-mediated immune response.

Histocompatibility Antigens

Proteins rich in basic acids (e.g., lysine) found in chromosomes of all eucaryotic cells except fish sperm, where the DNA is specifically complexed with another group of basic proteins, the protamines.

Histones

The active (complete) form of an enzyme.

Holoenzyme

Chromosomes that pair during meiosis, have the same morphology, and contain genes governing the same characteristics.

Homologous Chromosomes

The presence of identical alleles, for a given gene, on the homologous chromosomes of a diploid organism.

Homozygous Gene Pair

Chemical substances (often small polypeptides) synthesized in one organ of body that stimulate functional activity in cells of other tissues and organs. Many act by stimulating adenylcyclase in cell membrane to produce cyclic AMP.

Hormones

A cell whose metabolism is used for growth and reproduction of a virus.

Host Cell

Sites in genes at which mutations occur with unusually high frequency.

"Hot Spots"

Immunity mediated by freely circulating immunoglobulins.

Humoral Immunity

The reannealing of single-stranded nucleic acid chains. The formation of double-stranded regions indicates complementarity of sequence.

Hybridization of Nucleic Acid

Amino acid side chains consisting of carbon and hydrogen only.

Hydrocarbon Side Groups

A weak attractive force between one electronegative atom and a hydrogen atom that is covalently linked to a second electronegative atom.

Hydrogen Bond

The breaking of a molecule into two or more smaller molecules by the addition of a water molecule:

Hydrolysis

$$H_2O + A\!\!-\!\!B \rightarrow H\!\!-\!\!A + HO\!\!-\!\!B$$

Hydrophilic	Pertaining to molecules or groups that readily associate with H_2O.
Hydrophobic	Literally, water hater. Used to describe molecules or certain functional groups in molecules that are, at best, only poorly soluble in water.
Hydrophobic Bonding	The association of nonpolar groups with each other in aqueous solution, arising because of the tendency of water molecules to exclude nonpolar molecules.
Hypervariable Sites	Regions of the variable section of an antibody chain which show great variability among antibodies of differing specificity. These sites fold together to form the active site where antigen is bound.
Icosahedron	A regular polyhedron composed of 20 equilateral triangular faces.
Idiotype	The binding specificity of an immunoglobulin for a specific antigen.
Idiotypic Antibodies	Antibodies directed against specific antigen binding sites (idiotypes).
Idiotypic Markers	Those antigenic determinants in the antigen binding site of an immunoglobulin which stimulate idiotypic antibody formation.
Immunoglobulins	Y-shaped protein molecules which bind to and neutralize antigen. They are composed of units of four polypeptide chains (2 heavy and 2 light) linked together by disulfide bonds. Each of the chains has a constant and a variable region. They can be divided into five classes, IgG, IgM, IgA, IgD, and IgE, based on their heavy-chain component.
Immunological Surveillance	An immunological defense mechanism which involves cell-mediated recognition and destruction of newly emerged cancer cells and their descendants.
Immunological Tolerance	Absence of immune response to antigens, resulting from recognition of "self" or induced by very large antigen dose.
Immunology	The study of antibodies and their interactions with antigen.
Immunosuppressive Drug	Drug that blocks normal response of antibody-producing cells to antigen.
Inducers	Molecules that cause the production of larger amounts of the enzymes involved in their uptake and metabolism, compared to the amounts found in cells growing in the absence of an inducer.

706

Enzymes whose rate of production can be increased by the presence of inducers in the cell.	**Inducible Enzymes**
Purified viral nucleic acid that can infect a host cell and cause the production of progeny viral particles.	**Infectious Viral Nucleic Acid**
Three proteins (IF1, IF2, IF3) required for the initiation of protein synthesis.	**Initiation Factors**
A mutation in which one or more new bases are added between preexisting bases on a nucleic acid chain.	**Insertion**
Restoration of a lost function by a second mutation that is located in a different gene than the primary mutation.	**Intergenic Suppression**
The chemical reactions in a cell that transform food molecules into molecules needed for the structure and growth of the cell.	**Intermediary Metabolism**
Restoration of a lost function by a second mutation that is located within the same gene as the primary mutation.	**Intragenic Suppression**
Pertaining to experiments done in a cell-free system. Currently, the term is sometimes modified to include the growth of cells from multicellular organisms under cell culture conditions.	**In Vitro (Latin: in glass)**
The incorporation of amino acids into polypeptide chains in a cell-free system.	**In Vitro Protein Synthesis**
Pertaining to experiments done in a system such that the organism remains intact, either at the level of the cell (for bacteria) or at the level of the whole organism (for animals).	**In Vivo (Latin: in life)**
Body which attaches laterally to the chromosomal centromere and is the site of chromosomal tubule attachment.	**Kinetochore**
A radioactive atom, introduced into a molecule to facilitate observation of its metabolic transformations.	**Label (Radioactive)**
Giant diplotene chromosome found in oocyte nucleus, with loops projecting in pairs from most chromomeres. Loops are sites of active gene expression.	**Lampbrush Chromosome**
A protein coded by a mutant gene that shows some residual activity.	**Leaky Protein**
Cell-agglutinating proteins. Most so far studied have been isolated from plant seeds.	**Lectins**
Form of cancer characterized by extensive proliferation of nonfunctional immature white blood cells (leukocytes).	**Leukemia**

707

Light Chains	Generally refers to the light polypeptide components (MW ~ 23,000) of the immunoglobulin.
Linked Genes	Genes that are located on the same chromosome and that therefore tend to segregate together.
Lipids	A large, varied class of water-insoluble organic molecules. Includes steroids, fatty acids, prostaglandins, terpenes, and waxes.
Lipid Bilayer	An early model for the structure of cell membranes based upon the hydrophobic interactions between phospholipids. The polar head groups face outwardly to the solvent, with the hydrophobic tails clustered in the interior.
Lymphatic Tissues	Those tissues, including the lymph nodes and vessels, the thymus, the spleen, and the Bursa of Fabricius (in birds), which produce and contain the lymphocytes.
Lymphoblasts	The cytoplasm-rich cells differentiated from antigenically stimulated T lymphocytes.
Lymphokines	Factors released by antigenically stimulated T lymphocytes which attract phagocytic monocytes.
Lymphoma	Cancer of lymphatic tissue.
Lysis	The bursting of a cell by the destruction of its cell membrane.
Lysogenic Bacterium	A bacterium that contains a prophage.
Lysogenic Viruses	Viruses that can become prophages.
Lysosomes	Internal cellular granules which contain a large variety of hydrolytic enzymes. These granules fuse with ingested food vacuoles and break down the contents.
Lysozymes	Enzymes that degrade the polysaccharides found in the cell walls of certain bacteria.
Lytic Infection	Viral infection leading to lysis of cell and "burst" of progeny virus.
Lytic Viruses	Viruses whose multiplication leads to lysis of the host cell.
Macromolecules	Molecules with molecular weights ranging from a few thousand to hundreds of millions.
Macrophage	Large, phagocytic white blood cell.
Map Units	A number proportional to the frequency of recombination between two genes. One map unit corresponds to a recombination frequency of 1 percent.

Process whereby germ line diploid cells undergo division to form haploid sex cells. **Meiosis**

Small lymphocytes which originate as a result of a primary response, and which probably form the basis for the secondary response. **Memory Cells**

An invagination of the bacterial cell membrane. Associated with DNA replication. **Mesosome**

RNA that serves as a template for protein synthesis. **Messenger RNA (mRNA)**

A set of consecutive intercellular enzymatic reactions that converts one molecule to another. **Metabolic Pathway**

The sum total of the various chemical reactions occurring in a living cell. **Metabolism**

Long, 60Å diameter intracellular fibers that contain polymerized actin, and which are thought to function in maintenance of cell structure and movement. **Microfilaments**

A unit of length convenient for describing cellular dimensions; it is equal to 10^{-3} cm or 10^5 Å. **Micron (μ)**

Narrow ($\sim 0.2\ \mu$) cytoplasmic projections which can reach $20\ \mu$ in length. They extend and retract rapidly from the cell surface, perhaps playing a sensory role. **Microspikes**

Hollow, cylindrical tubules of 250Å diameter formed by helical aggregation of tubulin molecules. **Microtubules**

Long, narrow cytoplasmic projections which increase cellular surface area and form the brush borders of certain (e.g., intestinal) cells. **Microvilli**

A mutation that changes a codon coding for one amino acid to a codon corresponding to another amino acid. **Missense Mutation**

Organelle found in the cytoplasm of all aerobic eucaryotic cells which is the center of ATP generation through oxidative phosphorylation. **Mitochondrion**

Substances which provoke cell division (mitosis). **Mitogens**

Process whereby chromosomes duplicate and segregate, accompanied by cell division. **Mitosis**

Crossing-over between homologous chromosomes during mitosis, which leads to the segregation of heterozygous alleles. **Mitotic Recombination**

The sum of the atomic weights of the constituent atoms in a molecule. **Molecular Weight**

The largest leukocytes (macrophages) found in the blood; they are phagocytic, amoeboid cells. **Monocytes**

Monolayer   A layer of cells that is uniformly one cell thick.

Monomer   The basic subunit from which, by repetition of a single reaction, polymers are made. For example, amino acids (monomers) condense to yield polypeptides or proteins (polymers).

Muscle Proteins   Those proteins present within muscle fiber which are involved in the contractile process. Includes actin, myosin, tropomyosin, and α-actinin.

Mutable Sites   Sites along the chromosome at which mutations can occur. Genetic experiments tell us that each mutable site can exist in several alternative forms.

Mutagens   Physical or chemical agents, such as radiation, heat, or alkylating or deaminating agents, which raise the frequency of mutation greatly above the spontaneous background level.

Mutation   An inheritable change in a chromosome.

Myeloma   Cancer of antibody-producing cells characterized by proliferation of a single clone of plasma cells producing a pure immunoglobulin.

Myoblasts   Precursor cells which aggregate to form the multinucleated striated muscle cell.

Myosin   Protein molecules, each composed of two coiled subunits (MW $\sim$ 220,000), which can aggregate to form a thick filament which is globular at each end.

Negative Control   Prevention of biological activity by presence of a specific molecule; a prominent example is inhibition of mRNA initiation by binding of specific repressor to specific sites along a DNA molecule.

Nitrogenous Base   An aromatic N-containing molecule having basic properties (tendency to acquire an H atom). Important nitrogenous bases in cells are the purines and pyrimidines.

Nitrous Acid (HNO_2)   A very powerful mutagen that replaces the amino groups of DNA bases with keto groups.

Nonbasic Chromosomal Proteins   The acidic nonhistone proteins associated with chromosomes (e.g., DNA Polymerase).

Nonmuscle Cells   Cells that are not specifically differentiated to perform a contractile function.

Nonsense Mutation   A mutation that converts a codon that specifies some amino acid into one that does not specify any amino acid (a nonsense codon). Nonsense codons have the function of terminating the polypeptide chain.

Enzymes which cleave the phosphodiester bonds of nucleic acid chains.	**Nucleases**
A nucleotide polymer. (See also *DNA* and *RNA*.)	**Nucleic Acid**
The inner core of an RNA tumor virus particle, consisting of RNA surrounded by an icosahedral protein shell.	**Nucleoid**
Round, granular structure found in nucleus of eucaryotic cells, usually associated with specific chromosomal site. Involved in rRNA synthesis and ribosome formation.	**Nucleolus**
A newly discovered strain of mice which congenitally contain very little thymus tissue, and so are incapable of mounting strong T cell responses.	**Nude Mice**
Spherical, ~ 100Å diameter masses seen along partially dissociated chromatin.	**Nu Particles (Nucleosomes)**
The gene of cancer-causing viruses responsible for inducing the transformed phenotype.	**Oncogene**
Unfertilized egg cell.	**Oocyte**
A chromosomal region capable of interacting directly (or indirectly?) with a specific repressor, thereby controlling the functioning of an adjacent operon.	**Operator**
A genetic unit consisting of adjacent genes that function coordinately under the joint control of an operator and a repressor.	**Operon**
Membrane-bound structure found in eucaryotic cell containing enzymes for specialized function. Some organelles, including mitochondria and chloroplasts, have DNA and can replicate autonomously.	**Organelle**
Chemical reactions which involve electron transfer from a reductant to an oxidant. The reductant is said to be *oxidized* and the oxidant *reduced*, as a result of the transfer.	**Oxidation–Reduction Reactions**
Coupled electron transfer and ATP formation which occurs on the mitochondrial membrane.	**Oxidative Phosphorylation**
A radioactive isotope of phosphorus that emits strong β particles and has a half-life of 14.3 days.	**^{32}P**
The sideways attachment of two homologous chromosomes prior to crossing-over.	**Pairing**
A stretch of DNA in which identical (or almost identical) base sequences run in opposite directions.	**Palindrome**

Partial Denaturation	The partial unwinding of the double helix. Those regions that remain intact last are probably GC-rich, since G–C base pairs, held together by three hydrogen bonds, are more stable than A–T base pairs (two hydrogen bonds).
Peptide Bond	A covalent bond between two amino acids in which the α-amino group of one amino acid is bonded to the α-carboxyl group of the other with the elimination of H_2O.
Permissive vs. Nonpermissive Cells	Permissive cells support lytic infection by a specific virus. Nonpermissive cells do not.
Peroxisomes	Intracellular organelles which contain a fine granular matrix and often crystal-like cores. They have been found to contain four enzymes involved in hydrogen peroxide metabolism, including the H_2O_2-degradative enzyme catalase. They may be important in purine degradation, photorespiration, and the metabolic pathway known as the glyoxylate cycle.
Phage	(See *Bacterial Viruses.*)
Phage Cross	Multiple infection of a single bacterium by bacteriophages that differ at one or more genetic sites. This leads to the production of recombinant progeny phage, which carry genes derived from both parental phage types.
Phagocytosis	A process for food gathering employed by many cells. This process involves the surrounding of cellular-sized objects by pseudopod-like projections.
Phase-Contrast Microscope	An instrument that translates differences in the phase of transmitted or reflected light into gradations of contrast with the background.
Phenotype	The observable properties of an organism; produced by the genotype in cooperation with the environment.
Phosphodiester	Any molecule that contains the linkage

$$R-O-\overset{\overset{\displaystyle O}{\|}}{\underset{\underset{\displaystyle O^-}{|}}{P}}-O-R',$$

where R and R′ are carbon-containing groups (e.g., nucleosides), O is oxygen, and P is phosphorus.

Phospholipids	Lipids that contain charged, hydrophilic phosphate head groups. These are a primary component of cell membranes.
Pinocytosis	A process for capturing macromolecular-sized food objects. Its mechanism is thought to be basically similar to that employed in phagocytosis.

Number of base pairs per turn of the double helix.

Pitch

Round clear areas in a confluent cell sheet that result from the killing or lysis of contiguous cells by several cycles of virus growth.

Plaques

Highly proliferative cells which are intermediates in the differentiation of B lymphocytes to plasma cells.

Plasmablasts

The cells differentiated from antigenically stimulated B lymphocytes. They are characterized by much rough endoplasmic reticulum and secretion of massive amounts of antibody.

Plasma Cells

Physical barrier that surrounds the cell surface and encloses the cytoplasm. The membrane is semipermeable and largely composed of lipid and protein.

Plasma Membrane

Cytoplasmic, autonomously replicating chromosomal elements found in bacteria.

Plasmids

An enzyme formed by the proteolysis of plasminogen which acts to hydrolyze the blood-coagulating protein fibrin.

Plasmin

An enzymatically inactive protein found in blood that is activated by cleavage of a single arg–val bond to form plasmin.

Plasminogen

Quantitative effect of polar mutation in one gene on expression of later genes of operon. Function of the distance between the nonsense condon and the next polypeptide chain-initiation signal.

Polarity Gradient

Mutation in one gene that reduces expression of genes further from promoter in the same operon. Nonsense mutations frequently are polar.

Polar Mutation

Spindle microtubules which originate at the centriolar regions just outside the nuclear membrane.

Polar Tubules

The 3′ ends of most eucaryotic mRNA's contain relatively long stretches of poly A which is enzymatically added after transcription.

Poly A of Eucaryotic mRNA

A method of molecular separation which relies upon the differential migration of molecules through a polyacrylamide matrix upon application of an electrical potential.

Polyacrylamide Gel Electrophoresis

A regular, covalently bonded arrangement of basic subunits (monomers) that is produced by repetitive application of one or a few chemical reactions.

Polymer

(See *DNA, RNA Polymerase.*)

Polymerase

Polynucleotide	A linear sequence of nucleotides in which the 3′ position of the sugar of one nucleotide is linked through a phosphate group to the 5′ position on the sugar of the adjacent nucleotide.
Polynucleotide Ligase	Enzyme that covalently links DNA backbone chains.
Polynucleotide Phosphorylase	A bacterial enzyme which catalyzes the polymerization of ribonucleoside diphosphates to yield free phosphate and RNA. Its physiological function remains unknown.
Polypeptide	A polymer of amino acids linked together by peptide bonds.
Polyribosome	Complex of a messenger-RNA molecule and ribosomes (number depending on size of mRNA), actively engaged in polypeptide synthesis.
Polytene Chromosome	Giant chromosome composed of many fibrils (up to 2000) arising from successive rounds of chromatid duplication. Pairing of many identical chromomeres gives rise to characteristic banding pattern.
Positive Control	Control by regulatory protein—the presence of which, in correct conformation, is required for gene expression.
Primary Cells	Those cells obtained from multicellular organisms and seeded onto culture plates.
Primary Protein Structure	The number of polypeptide chains in a protein, the sequence of amino acids within them, and the location of inter- and intrachain disulfide bridges.
Primary Response	The rather weak response of the immunological system upon its first exposure to a given antigen, characterized by first IgM and then IgG synthesis.
Primer	A structure which serves as a growing point for polymerization.
Procaryote	Simple unicellular organism, such as bacterium or blue-green alga, with no nuclear membrane, no membrane-bound organelles, and characteristic ribosomes and biochemistry.
Promoter	Region on DNA at which RNA polymerase binds and initiates transcription.
Proofreading	A process by which an incorrectly selected nucleotide or amino-acyl tRNA can be removed and replaced by the correct molecule.
Prophage	The proviral stage of a lysogenic phage. (See *Provirus.*)
Prosthetic Groups	Coenzymes that are very strongly bound to their enzymes, e.g., heme.

A class of proteins rich in the basic amino acid arginine. They are found complexed to the DNA of the sperm of many invertebrates and fish.	**Protamines**
The state of a virus in which it is integrated into a host cell chromosome and is thus transmitted from one cell generation to another.	**Provirus**
Open loops, like those of lambrush chromosomes, found in polytene chromosomes. The larger and more diffuse their appearance, the higher the rate of transcription.	**Puffs**
A radioactively labeled compound is added to living cells or a cell extract (pulse) and, a short time later (e.g., five seconds), an excess of unlabeled compound is added to dilute out the "hot" compound. Samples are then taken at periods after the pulse to follow the course of the label as the compound is metabolized (chase).	**Pulse-Chase Experiment**
Antibiotic that inhibits polypeptide synthesis by competing with amino-acyl tRNA's for ribosomal binding site "A."	**Puromycin**
An isotope with an unstable nucleus that stabilizes itself by emitting ionizing radiation.	**Radioactive Isotope**
The incorrect placement of an amino acid residue in a polypeptide chain during protein synthesis.	**Reading Mistake**
An allele which exerts its phenotypic effect only when present in homozygous form, being otherwise masked by the dominant allele.	**Recessive**
The appearance in the offspring of traits that were not found together in either of the parents.	**Recombination**
Genes whose primary function is to control the rate of synthesis of the products of other genes.	**Regulatory Genes**
Specific proteins involved in the reading of genetic stop signals for protein synthesis.	**Release Factors**
The return of a protein or nucleic acid from a denatured state to its "native" configuration.	**Renaturation**
Enzymatic excision and replacement of regions of damaged DNA. Repair of thymine dimers by uv irradiation is best understood example.	**Repair Synthesis**
Y-shaped region of chromosome that is a growing point in DNA replication.	**Replicating Fork**
The structure of a nucleic acid at the time of its replication—the term most frequently used to refer to double-	**Replicating Forms (RF)**

helical intermediates in the replication of single-stranded DNA and RNA viruses.

Repressible Enzymes — Enzymes whose rates of production are decreased when the intracellular concentration of certain metabolites increases.

Repressor — The product of a regulatory gene, now thought to be a protein and to be capable of combining both with an inducer (or corepressor) and with an operator (or its mRNA product).

Resistance Transfer Factors — Among the most common plasmids, these confer resistance to antibiotics.

Restriction Enzymes — Components of the restriction-modification cellular defense system against foreign nucleic acids. These enzymes cut unmodified (e.g., methylated) double-stranded DNA at specific sequences which exhibit twofold symmetry about a point.

Reticulocyte — Immature red blood cell active in hemoglobin synthesis.

Reverse (Back) Mutation — A heritable change in a mutant gene that restores the original nucleotide sequence.

Reverse Transcriptase — An enzyme coded by certain RNA viruses which is able to make complementary single-stranded DNA chains from RNA templates and then to convert these DNA chains to double-helical form.

ρ Factor — Protein involved in correct termination of synthesis of RNA molecules.

Ribonuclease — An enzyme that can cleave the phosphodiester bonds of RNA.

Ribonucleotide — A compound that consists of a purine or pyrimidine base bonded to ribose, which in turn is esterified with a phosphate group.

Ribosomal Proteins — A group of proteins that bind to rRNA by noncovalent bonds to give the ribosome its three-dimensional structure.

Ribosomal RNA (rRNA) — The nucleic acid component of ribosomes, making up two-thirds of the mass of the ribosome in *E. coli*, and about one-half the mass of mammalian ribosomes. Ribosomal RNA accounts for approximately 80 percent of the RNA content of the bacterial cell.

Ribosomes — Small cellular particles (~ 200 Å in diameter) made up of rRNA and protein. Ribosomes are the site of protein synthesis.

Small, disease-causing bacteria-like objects which are obligate intracellular parasites. They contain both DNA and RNA, as well as a protein synthesizing machinery. Q fever, typhus, and Rocky Mountain spotted fever are caused by Rickettsiae.	**Rickettsiae**
An inhibitor of bacterial RNA polymerase.	**Rifampicin**
A polymer of ribonucleotides.	**RNA (Ribonucleic Acid)**
An enzyme that catalyzes the formation of RNA from ribonucleoside triphosphates, using DNA as a template.	**RNA Polymerase**
A mechanism for the replication of circular DNA. This model begins with the specific cut of one parental strand of the double helix, which is thus converted to a chain bearing distinct 5′ and 3′ ends. The 3′ end serves as a primer for DNA replication upon the uncut, circular strand (the template) while the 5′ end is rolled off as an increasingly long free tail.	**Rolling Circle**
Extensive inner membranous sacs (endoplasmic reticulum) which have bound ribosomes. Secretory proteins appear to be synthesized on these membrane-bound ribosomes.	**Rough Endoplasmic Reticulum**
Extensive, lamellar projections which are involved in cell attachment to solid surfaces.	**Ruffled Edges (Lamellipodia)**
A radioactive isotope of sulfur, a β emitter with a half-life of 87 days; very useful in studying protein systems, since it can be incorporated into proteins via the sulfur-containing amino acids.	^{35}**S**
Cancer of connective tissue.	**Sarcoma**
The ~2.5 μ repeating unit within striated muscle fibers which contains a set of interacting actin and myosin filaments.	**Sarcomere**
Eucaryotic DNA that bands at a different density than that of most cellular DNA upon equilibrium centrifugation. In many cases it consists of highly repetitive DNA; in other cases it arises from organelles.	**Satellite DNA**
Electron microscope technique which permits three-dimensional structures rather than thin sections to be viewed.	**Scanning Electron Microscopy**
Those cells arising from proliferation of primary cells. They generally are able to divide only a finite number of times; then most die.	**Secondary Cells**

717

Secondary Protein Structure   The helical or extended structure of a polypeptide chain (e.g., α helix, β sheet).

Secondary Response   The vigorous response of the immunological system upon exposure to an antigen previously encountered. Largely characterized by IgG synthesis.

Serum Protein   Protein found in serum (cell-free) component of blood. Includes immunoglobulins, albumin, clotting factors, and enzymes.

Sex-Linked   Those genes present on the sex chromosomes (chromosomes which determine sex).

Sialic Acid   Derivatives of neuraminic acid in which the free amino group is acetylated.

σ Factor   Subunit of RNA polymerase that recognizes specific sites on DNA for initiation of RNA synthesis.

Sliding Filament Model   A mechanism which explains muscle contraction on the basis of the making and breaking of cross bridges between adjacent thick (myosin) and thin (actin) filaments.

Small Lymphocytes   White blood cells, found in large numbers in higher vertebrates, which are characterized by a seemingly inert nucleus surrounded by only traces of an endoplasmic reticulum-poor cytoplasm. They fall into two groups, B and T cells, which are indistinguishable under the light microscope.

Smooth Endoplasmic Reticulum   Extensive inner membranous sacs (endoplasmic reticulum) which are free of ribosomes. The smooth ER appears to be a major site for attachment of sugar residues to nascent proteins to form glycoproteins.

Smooth Muscle   Involuntary, nonstriated muscle cells which are found in the lining of internal organs (e.g., gizzard) and blood vessels.

Somatic Mutation   A mutation occurring in any cell that is not destined to become a germ cell.

Specificity Factors   Proteins which reversibly associate with the core component of RNA Polymerase and determine which promoters the polymerase will bind to.

Spindle   Cellular structure, largely composed of microtubules, involved in eucaryotic chromosomal segregation.

Spleen   Lymphoid organ in the upper left abdomen which functions as a storage site for red blood cells and, due to the large number of resident macrophages, as a blood filter.

Spontaneous Mutations   Mutations for which there is no "observable" cause.

The formation from vegetative cells of bacteria and other organisms of dry, metabolically inactive cells with thick surface coats (spores), which can resist extreme environmental conditions. **Sporulation**

A cell from which other cells stem or arise by differentiation. **Stem Cell**

Molecules that have the same structural formula but different spatial arrangement of dissimilar groups bonded to a common atom. Many of the physical and chemical properties of stereoisomers are the same, but there are differences in their crystal structures, in the direction in which they rotate polarized light, and, very importantly for biological systems, in their ability to be used in an enzyme-catalyzed reaction. **Stereoisomers**

Pertaining to the arrangement in space of the atoms in molecules. **Steric (Stereochemical)**

Compounds which are derivatives of a tetracyclic structure composed of a cyclopentane ring fused to a substituted phenanthrene nucleus. Many steroids have intense biological activity; e.g., the sex hormones and the adrenal cortex hormones (corticoids). **Steroids**

Complementary single-stranded tails projecting from otherwise double-helical nucleic acid molecules which are terminally redundant. **"Sticky" Ends**

An antibiotic ($C_{21}H_{39}N_7O_{12}$), isolated from *Streptomyces griseus* (a soil bacterium), which binds specifically to bacterial 30S ribosomal subunits, thereby blocking protein biosynthesis. **Streptomycin**

Bundles of microfilaments aligned parallel just under the plasma membrane. These fibers contain myosin, tropomyosin, actin, and α-actinin, and seem to be involved in cell motility. **Stress Fibers**

Muscle cells which possess myofibrils that are regularly arranged so as to create visible striations. **Striated Muscle**

A molecule whose chemical conversion is catalyzed by an enzyme. **Substrate**

Twisted forms taken by covalently closed, circular double-stranded DNA molecules when purification has removed the protein components of the chromosome, thereby slightly changing the pitch of the double helix. **Supercoils**

A gene that can reverse the phenotypic effect of a variety of mutations in other genes. **Suppressor Gene**

Suppressor Mutation	A mutation that totally or partially restores a function lost by a primary mutation and is located at a genetic site different from the primary mutation.
Svedberg	The unit of sedimentation (S). S is proportional to the rate of sedimentation of a molecule in a given centrifugal field and is thus related to the molecular weight and shape of the molecule.
Sympathetic Neurons	Cells of the autonomic nervous system which function to depress secretion, and to decrease muscle tone and blood vessel contraction.
Synapsis	Zipper-like pairing of homologous chromosomes in the first meiotic prophase.
Synthetic Polyribonucleotides	RNA made *in vitro* without a nucleic acid template, either by enzymatic or by chemical synthesis.
"T" Antigen	Antigen found in nuclei of cells infected or transformed by certain tumor viruses (e.g., polyoma and SV40). May be an early viral-specific protein.
T Lymphocytes	Small lymphocytes which possess the surface θ antigen and direct cell-mediated immune responses.
Tautomeric Shifts	Reversible changes in the localization of a proton in a molecule that alter the chemical properties of the molecule.
Temperature-Sensitive Mutation	Mutation-yielding protein that is functional at low (high) temperature, but that is inactivated by temperature elevation (lowering).
Template	The macromolecular mold for the synthesis of another macromolecule.
Template RNA	(See *Messenger RNA*.)
Tertiary Structure (of a Protein)	The three-dimensional folding of the polypeptide chain(s) that characterize(s) a protein in its native state.
Tetramer	Structure resulting from association of four subunits.
Three-Factor Crosses	Mating experiments involving three distinguishable genetic markers (e.g., $a^+b^+c^+ \times abc$).
Thymus	A lymphoid gland in the upper chest, most prominent in young animals, in which T cells become immunologically competent.
Transcription	A process involving base pairing, whereby the genetic information contained in DNA is used to order a complementary sequence of bases in an RNA chain.

The transfer of bacterial genes from one bacterium to another by a bacteriophage particle.
Transduction

Any of at least twenty structurally similar species of RNA, all of which have a MW $\sim$ 25,000. Each species of tRNA molecule is able to combine covalently with a specific amino acid and to hydrogen bond with at least one mRNA nucleotide triplet. Also called *adaptor RNA*.
Transfer RNA (tRNA)

Enzymes that catalyze the exchange of functional groups.
Transferases

The genetic modification induced by the incorporation into a cell of DNA purified from cells or viruses.
Transformation

The process whereby the genetic information present in an mRNA molecule directs the order of the specific amino acids during protein synthesis.
Translation

Regulation of gene expression by controlling the rate at which a specific mRNA molecule is translated.
Translational Control

(See *Elongation Factor G.*)
Translocase

A muscle protein which associates with actin to form long thin fibers and which plays a role in the regulation of muscle contraction.
Tropomyosin

A proteolytic enzyme (MW $\sim$ 23,800), secreted by the pancreas, that cleaves peptide chains where the basic amino acids arginine and lysine appear.
Trypsin

Globular protein subunits (MW's 55,000 and 57,000) whose regular helical packing forms the hollow, cylindrical microtubules.
Tubulin

Mass formed by the uncontrolled proliferation of cancerous cells.
Tumor

A virus that induces the formation of a tumor.
Tumor Virus

The number of molecules of a substrate transformed per minute by a single enzyme molecule, when the enzyme is working at its maximum rate.
Turnover Number (of an Enzyme)

A genetic recombination experiment involving two markers (e.g., $a^+b^+ \times ab$).
Two-Factor Cross

A high-speed centrifuge that can attain speeds up to 60,000 rpm and centrifugal fields up to 500,000 times gravity and thus is capable of rapidly sedimenting macromolecules.
Ultracentrifuge

Polypeptides that bind to and thus stabilize single-stranded DNA. In doing so, they tend to unwind the double helix.
Unwinding Protein

Urea	An organic compound used as a protein denaturing agent. Its formula is $CO(NH_2)_2$.
uv Radiation	Electromagnetic radiation with wavelength shorter than that of visible light (3900–2000 Å). Causes DNA base-pair mutations and chromosome breaks.
van der Waals Force	A weak attractive force, acting over only very short distances, resulting from attraction of induced dipoles.
Viral-Specific Enzyme	An enzyme produced in the host cell, after viral infection, from viral genetic information.
Viroids	Pathogenic agents thought to consist only of very short RNA molecules.
Viruses	Infectious disease-causing agents, smaller than bacteria, which always require intact host cells for replication and which contain either DNA or RNA as their genetic component.
Weak Interactions (also called Secondary Bonds)	The forces between atoms that are less strong than the forces involved in a covalent bond (includes ionic bonds, hydrogen bonds, and van der Waals forces.)
Wild-Type Gene	The form of a gene (allele) commonly found in nature.
Wobble	Ability of third base in tRNA anticodon (5′ end) to hydrogen bond with any of two or three bases at 3′ end of codon. Thus, a single tRNA species can recognize several different codons.
X-Ray Crystallography	The use of diffraction patterns produced by x-ray scattering from crystals to determine the 3-D structure of molecules.
Zygote	The result of the union of the male and female sex cells. The zygote therefore has a diploid number of chromosomes.

Index